BIOPHYSICAL CHEMISTRY

BIOPHYSICAL CHEMISTRY

PART

II

TECHNIQUES FOR THE STUDY OF BIOLOGICAL STRUCTURE AND FUNCTION

Charles R. Cantor
COLUMBIA UNIVERSITY

Paul R. Schimmel
MASSACHUSETTS INSTITUTE OF TECHNOLOGY

W. H. FREEMAN AND COMPANY
San Francisco

Cover drawing after R. D. B. Fraser and T. P. McRae, in *Physical Principles and Techniques of Protein Chemistry*, part A, ed. S. J. Leach (New York: Academic Press, 1969).

Sponsoring Editor: Arthur C. Bartlett
Project Editor: Pearl C. Vapnek
Manuscript Editor: Lawrence W. McCombs
Designer: Robert Ishi
Production Coordinator: Linda Jupiter
Illustration Coordinator: Cheryl Nufer
Artists: Irving Geis and Eric Hieber
Compositor: Syntax International
Printer and Binder: R. R. Donnelley & Sons Company

Library of Congress Cataloging in Publication Data

Cantor, Charles R 1942–
Techniques for the study of biological structure and function.

(Their Biophysical chemistry; pt. 2)
Includes bibliographies and indexes.
1. Molecular biology—Technique. 2. Biological chemistry—Technique. I. Schimmel, Paul Reinhard, 1940– joint author. II. Title.
QH345.C36 pt. 2 [QH506] 574.1′9283s [574.1′9283]
ISBN 0-7167-1189-3 79-24854
ISBN 0-7167-1190-7 pbk.

Printed in the United States of America

9 8 7 6 5 4 3 2 1

To Louis and Ida Dianne Cantor
and Alfred and Doris Schimmel

Contents in detail of Part II

Contents in brief of Parts I and III

PART III THE BEHAVIOR OF BIOLOGICAL MACROMOLECULES

Preface

Biophysical Chemistry is concerned with biological macromolecules and complexes or arrays of macromolecules. The work deals with the conformation, shape, structure, conformational changes, dynamics, and interactions of such systems. Our goal is to convey the major principles and concepts that are at the heart of the field. These principles and concepts are drawn from physics, chemistry, and biology.

We have aimed at creating a multilevel textbook in three separately bound parts. The material covers a broad range of sophistication so that the text can be used in both undergraduate and graduate courses. It also should be of value to general scientific readers who simply wish to become familiar with the field, as well as to experienced research scientists in the biophysical area. For example, perhaps half of the material requires only the background provided by a one-semester undergraduate course in physical chemistry. A somewhat smaller fraction necessitates the use of concepts and mathematical techniques generally associated with a more sophisticated background, such as elementary statistical thermodynamics and quantum mechanics.

Biophysical Chemistry is organized into three parts. The first part deals with the structure of biological macromolecules and the forces that determine this structure. Chapter 1 introduces the fundamental questions of interest to biophysical chemists, Chapters 2–4 summarize the known structures of proteins, nucleic acids, and other biopolymers, and Chapters 5–6 treat noncovalent forces and conformational analysis.

Part II summarizes some of the techniques used in studying biological structure and function. The emphasis is on a detailed discussion of a few techniques rather than an attempt to describe every known technique. Chapters 7–9 cover spectroscopic methods, Chapters 10–12 deal with hydrodynamic methods, and Chapters 13–14 discuss x-ray and other scattering and diffraction techniques.

Part III demonstrates how techniques and principles are used in concert to gain an understanding of the behavior and properties of biological macromolecules. The focus is on the thermodynamics and kinetics of conformational changes and ligand interactions. New techniques are introduced as needed, and a few selected case histories or systems are discussed in considerable detail. The topics include ligand interactions (Chapters 15–17), the special theories and techniques used to study molecules that are statistical chains rather than definite folded conformations (Chapters 18–19), protein conformational changes (Chapters 20–21), nucleic acid conformational changes (Chapters 22–24), and membranes (Chapter 25).

We have made every effort to keep the chapters as independent as possible, so that the reader has a wide choice of both the material to be covered and the order in which it is to be treated. Extensive cross-references to various chapters are included to help the reader find necessary background material if the parts are not read in sequence. Where possible, examples are taken repeatedly from a small number of systems, so that the reader can have the experience of contrasting information gained about the same protein or nucleic acid from a variety of different approaches.

Within each chapter, we have attempted to maintain a uniform level of rigor or sophistication. Short digressions from this level are segregated into boxes; longer digressions are indicated by a bullet (•) preceding the section or subsection heading. Readers with a less rigorous background in physics, mathematics, and physical chemistry should find helpful the many boxes that review elementary material and make the text fairly self-contained; Appendix A provides a basic review of principles of matrix algebra. Other boxes and special subsections are aimed at advanced readers; in many cases, these discussions attempt to illuminate points that we ourselves found confusing.

In different sections, the level of mathematical sophistication varies quite significantly. We have tried to use the simplest mathematical formulation that permits a clear presentation of each subject. For example, hydrodynamic properties are treated in one dimension only. The form of a number of the fundamental equations is extracted by dimensional analysis rather than through lengthy (and not particularly instructive) solutions of hydrodynamic boundary-value problems. On the other hand, x-ray and other scattering phenomena are treated by Fourier transforms, and many problems in statistical mechanics are treated with matrix methods. These advanced mathematical techniques are used in only a few chapters, and numerous boxes are provided to assist the reader with no previous exposure to such methods. The remaining sections and chapters are self-contained and can be understood completely without this advanced mathematical formalism.

Some techniques and systems are not covered in any fair detail. This represents a biased choice by the authors, not necessarily of which techniques we feel are important, but simply of which are instructive for the beginning student in this field.

Each chapter concludes with a summary of the major ideas covered. In addition, each chapter is heavily illustrated, including some special drawings by Irving Geis. Certainly, much can be learned simply by reading the chapter summaries and by studying the illustrations. Also, we believe the illustrations convey some of the excitement of the field.

Problems are provided at the end of each chapter. These vary in difficulty from relatively simple to a few where the full answer is not known, at least to the authors. Answers to problems are provided in Appendix B.

Detailed literature citations are not included, except to acknowledge the source of published material reproduced or adapted here. However, a list of critical references for each chapter is included. In virtually all cases, these articles will provide an immediate entrée to the original papers needed for more detailed study.

The problem of notation and abbreviations in this field is a difficult one. In drawing together material from so many different types of research, we have had to adapt the notation to achieve consistency and to avoid confusion among similar symbols. Wherever possible, we have followed the recommendations of the American Chemical Society, but inevitably we have had to develop some conventions of our own. A glossary of some of the more frequently used symbols is provided.

At MIT some of this material has been used in an undergraduate course in biophysical chemistry. The course was designed to meet the needs of students wishing a second course in physical chemistry, but developed in a biochemical framework. The idea was to construct a course that covered much of the same material with the same rigor as a parallel, more traditional course. The only preparation required was a one-semester course in undergraduate physical chemistry, which at MIT is largely concerned with chemical thermodynamics.

Over the years graduate courses in biophysical chemistry at MIT and at Columbia have made use of much of the material presented here. In addition, a special-topics course in protein structure has used some of the material. Because a broad range of subjects is covered, its usefulness as a text will hopefully meet a variety of individual teaching tastes and preferences, as well as enable instructors to vary content as needs develop and change.

It is obvious that a work of this complexity cannot represent solely the efforts of its two authors. As we sought to master and explain the wide range of topics represented in biophysical chemistry, we learned why so few books have been written in this field in the past two decades. We owe a great debt to many who helped us in ways ranging from sharing their understanding to providing original research data.

We give special thanks to Irving Geis, for his effort on a number of complex illustrations and for his helpful advice on numerous other drawings; to Wilma Olson, for reading a major portion of the entire manuscript; to Robert Alberty and Gordon Hammes, for their influence, through teaching and discussions, on the material on biochemical equilibria and kinetics; to Richard Dickerson, for providing material and advice that were essential for the preparation of Chapter 13; to Paul Flory, for inspiring our treatment of conformational energies and configurational statistics of macromolecules; to Howard Schachman, whose course at Berkeley inspired parts of several chapters; to R. Wayne Oler, for bringing the authors together for this undertaking, and to Bruce Armbruster, for sealing the commitment; to the helpful people at W. H. Freeman and Company, including Ruth Allen, Arthur Bartlett, Robert Ishi, Larry McCombs, and Pearl Vapnek; to Kim Engel, Karen Haynes, Marie Ludwig, Joanne Meshna, Peggy Nelson, Cathy Putland, and Judy Schimmel, for typing and related work associated with the manuscript; and to Cassandra Smith and to Judy, Kathy, and Kirsten Schimmel, for their patience with the intrusion this work has made on the authors' lives.

Many people read and commented on specific chapters, provided figures, notes and materials, and spent much time with us in helpful discussions. We gratefully thank these people: Robert Alberty, Arthur Arnone, Struther Arnott, P. W. Atkins, Robert Baldwin, Larry Berliner, Bruce Berne, Richard Bersohn, Sherman Beychok, Victor Bloomfield, David Brandt, John Brandts, John Chambers, Sunney Chan, Patricia Cole, Robert Crichton, Francis Crick, Donald Crothers, Norman Davidson, Richard Dickerson, David Eisenberg, Robert Fairclough, Gerry Fasman, George Flynn, David Freifelder, Ronald Gamble, Robert Gennis, Murray Goodman, Jonathan Greer, O. Hayes Griffith, Gordon Hammes, John Hearst, Ellen Henderson, James Hildebrandt, Wray Huestis, Sung Hou Kim, Aaron Klug, Nelson Leonard, H. J. Li, Stephen Lippard, Richard Lord, Brian Matthews, Harden McConnell, Peter Moore, Garth Nicolson, Leonard Peller, Richard Perham, Michael Raftery, Alexander Rich, Frederick Richards, David Richardson, Wolfram Saenger, Howard Schachman, Harold Scheraga, Benno Schoenborn, Verne Schumaker, Nadrian Seeman, Robert Shulman, Mavis Shure, Louise Slade, Cassandra Smith, Hank Sobell, Thomas Steitz, Robert Stroud, Lubert Stryer, Serge Timasheff, Ignacio Tinoco, Jr., Richard Vandlen, Jerome Vinograd, Peter von Hippel, Christopher Walsh, James Wang, Gregorio Weber, Peter Wellauer, Barbara Wells, Robert Wells, William Winter, Harold Wyckoff, Jeffries Wyman, and Bruno Zimm.

November 1979

Charles R. Cantor
Paul R. Schimmel

Glossary of symbols

This glossary includes some of the symbols used extensively throughout the text. In many cases, the same or very similar symbols are used in certain contexts with other meanings; the meaning of a symbol is explained in the text where it is introduced.

Symbol	Meaning
A	Absorbance.
A_{ij}	Amplitude of kinetic decay.
Å	Angstroms.
a	Hyperfine splitting constant. Long semiaxis of ellipse. Persistence length.
$\mathbf{a}$	Unit cell basis vector.
$\mathbf{a}^*$	Reciprocal cell basis vector.
a_{ij}	Parameters composed of rate constants.
a_s	Exponent relating sedimentation to chain length.
a_η	Exponent relating viscosity to chain length.
b	Short semiaxis of ellipse.
$\mathbf{b}$	Unit cell basis vector.
$\mathbf{b}^*$	Reciprocal cell basis vector.
C	Molar concentration.
C_n	Rotational symmetry group element. Characteristic ratio.
C_∞	Limiting characteristic ratio.
ΔC_p^0	Standard constant pressure heat capacity change per mole.
c	Velocity of light in vacuum. Ratio of k_R/k_T. Weight concentration.
c_p	Plateau weight concentration.
$\hat{c}_i$	Weight concentration of ith species or component.
$\mathbf{c}$	Unit cell basis vector.
$\mathbf{c}^*$	Reciprocal cell basis vector.
D	Debye.
D	Translational diffusion constant.
D_n	Dihedral symmetry group element.
D_{rot}	Rotational diffusion constant.
$D_{20,w}$	D extrapolated to 20° C, water.
E_a	Activation energy.
E_d	Interaction energy between two dipoles.
E_{kl}	Nonbonded pair interaction potential.
E_{tor}	Torsional potential energy.
$E(\Phi_i, \Psi_i)$, E_i	Total rotational potential for residue i.
$\mathbf{E}$	Electric field.
e	Exponential function. Unit of charge on electron.
F	Frictional coefficient ratio.
$F(\mathbf{S})$	Structure factor.
$F_H(\mathbf{S})$	Structure factor, heavy-atom contribution.

Symbol	Meaning
$F_{Tot}(\mathbf{S})$	Structure factor for an array.
$F_m(\mathbf{S})$	Molecular structure factor.
$\mathbf{F}$	Force.
$\mathcal{F}$	The Faraday.
f	Translational frictional coefficient.
f_{app}	Apparent fractional denaturation.
f_D	Fraction in denatured state.
f_N	Fraction in native state.
f_{min}	Translational friction coefficient of anhydrous sphere.
f_{rot}	Rotational friction coefficient for sphere.
f_{sph}	Translational friction coefficient for sphere.
f_a, f_b	Rotational friction coefficient around a, b axis of ellipse.
G	Gibbs free energy.
ΔG^0	Standard Gibbs free energy change per mole.
$\Delta \bar{G}^0$	Intrinsic standard free energy change (with statistical component removed).
$\Delta G_{I,ij}$	Free energy of interaction between two ligands.
ΔG_r	ΔG per residue.
ΔG_{Tot}	Total free energy change per mole.
ΔG_{el}	Change in electrostatic free energy.
ΔG_T	Total free energy of formation of configuration.
$\Delta\Delta G_T$	Difference in ΔG_T between two configurations.
$\Delta \bar{G}_{gr}$	Average helix growth free energy change per residue pair.
g	g value for free electron, 2.00232.
g_x, etc.	Component of g-factor tensor.
H	Enthalpy.
H_{xy}	Magnetic field in xy plane.
ΔH	Enthalpy change per mole.
ΔH^0	Standard enthalpy change per mole.
ΔH_r	ΔH per residue.
ΔH_D	Enthalpy change for conversion from fully native to fully denatured state.
ΔH_{app}	Apparent enthalpy change for conversion from fully native to fully denatured state.
$\mathbf{H}$	Magnetic field.
$\mathbf{H}_{res}$	Magnetic field at which resonance occurs.
$\underset{\sim}{\mathbf{H}}$	Hamiltonian operator.
$\Delta\mathbf{H}_{loc}$	Magnetic field generated by local environment.
h	Planck's constant.
$\hbar$	$h/2\pi$
I	Intensity of radiation. Nuclear spin quantum number. Ionic strength.
$I(\mathbf{S})$	Scattering intensity relative to a single electron at the origin.
i	$\sqrt{-1}$
$\hat{\mathbf{i}}$	Cartesian unit vector.
J	NMR coupling constant.
$\mathbf{J}_2$	Solute flux.
$\underset{\sim}{\hat{\mathbf{j}}}$	Cartesian unit vector.
K_D	True equilibrium constant for conversion from fully native to fully denatured state.
K_p	Michaelis constant for product.
K_S	Michaelis constant for substrate.
K_η	Coefficient relating viscosity to chain length.
K_s	Coefficient relating sedimentation to chain length.
K_{app}	Apparent equilibrium constant for conversion from fully native to fully denatured state.
K_i	Macroscopic equilibrium constant. Equilibrium constant for forming ith configuration. Equilibrium constant for transition from native state to intermediate state i.
$\tilde{K}_i$	Apparent dissociation constant, one-ligand system.
$\tilde{K}_{ij}$	Apparent dissociation constant, two-ligand system.
k	Boltzmann's constant. Microscopic equilibrium dissociation constant.
k_R	Microscopic dissociation constant for R state.
k_T	Microscopic dissociation constant for T state.
k_i	Microscopic equilibrium constant.
$\hat{\mathbf{k}}$	Cartesian unit vector.

Symbol	Meaning
L_c	Contour length.
L, L'	Equilibrium constant for R_0 T_0.
$\mathbf{L}$	Angular momentum.
l	Length of one polymer bond.
l_e	Length of statistical segment.
M	Molecular weight.
$\bar{M}_n$	Number-average molecular weight.
$\bar{M}_w$	Weight-average molecular weight.
$\bar{M}_i$	Molecular weight of ith macromolecular species.
M_{ij}	Species with i bound L_1 and j bound L_2.
$M^{(j)}$	Set of all species with j bound L_2.
$\mathbf{M}$	Magnetization.
$\mathbf{M}_{xy}$	Magnetization in xy plane.
$\underset{\sim}{\mathbf{M}}$	Statistical weight matrix.
m	Colligative molality. Mass of object.
m_e	Mass of electron.
m_i	Molality of ith species.
m_s	Quantum number of electron spin along z axis.
m_I	Quantum number of nuclear spin along z axis.
m'	Total molality.
$\underset{\sim}{\mathbf{m}}$	Magnetic dipole operator.
N_0	Avogadro's number.
N_C	Number of carbons in amphiphile R chain.
N'_C	Number of carbons in amphiphile that are imbedded in hydrocarbon core of micelle.
N_e	Number of statistical segments.
N_{Ch}	Number of chains in micelle.
N_{hg}	Number of head groups in micelle.
n	Refractive index. Number of sites. Number of bonds in polymer.
n_i	Number of moles of component i. Number of sites of certain type.
n_w	Weight-average degree of polymerization.
P	Pitch of helix. Pressure. Patterson function.
P_0	Solvent vapor pressure.
P_v	Solvent vapor pressure in presence of solute.

Symbol	Meaning
P_r	Axial ratio.
pK_a	$-\log_{10} K_a$
pO_2	Partial pressure of oxygen.
$(pO_2)_{1/2}$	Partial pressure of oxygen at half saturation.
$\underset{\sim}{\mathbf{p}}$	Momentum operator.
Q	Configurational partition function.
R	Gas constant.
R_G	Radius of gyration.
$\langle R_G^2 \rangle_0$	Unperturbed mean square radius of gyration.
$\bar{R}$	Fraction of molecules in R state.
$\underset{\sim}{\mathbf{R}}$	Nuclear position operator.
$\underset{\sim}{\mathbf{R}}(\alpha,\beta)$	Coordinate transformation matrix.
r	Distance of separation.
r_D	Donnan ratio.
r_e	Radius of equivalent sphere.
$\langle r^2 \rangle_0$	Unperturbed mean square end-to-end distance.
$\mathbf{r}$	Polymer end-to-end vector.
$\underset{\sim}{\mathbf{r}}$	Electron position operator.
S	Svedberg (unit of sedimentation coefficient).
S_A	Partial molal entropy.
S'_A	Unitary part of S_A.
ΔS_r	ΔS per residue.
ΔS^0	Standard entropy change.
ΔS_u^0	Unitary standard entropy change.
$\mathbf{S}$	Scattering vector.
s	Sedimentation coefficient. Statistical weight. Equilibrium constant for helix growth. Equilibrium constant for base-pair formation.
$s_{20,w}$	Sedimentation coefficient corrected to 20° C, water
$\hat{\mathbf{s}}$	Unit vector along scattered radiation.
$\hat{\mathbf{s}}_0$	Unit vector along incident radiation.
T	Temperature (in degrees Kelvin usually).
T_m	Melting temperature.
T_1	Longitudinal relaxation time.
T_2	Transverse relaxation time.
$\underset{\sim}{\mathbf{T}}_i$	Transformation matrix.
t	Time.

Symbol	Meaning
U^0_{mic}	Attractive part of μ^0_{mic}.
u	Component of M_{xy} in phase with H_{xy}. Electrophoretic mobility.
V	Volume.
V_h	Hydrated volume.
$\bar{V}_i$	Partial specific volume of component i.
V_p	Maximum reaction velocity in reverse direction.
V_s	Maximum reaction velocity in forward direction.
v	Speed (also called velocity). Component of M_{xy} out of phase with H_{xy}.
v_i	Initial reaction velocity.
$\langle v_2 \rangle$	Effective average solute velocity.
$\bar{v}$	Partial molar volume.
$\bar{v}_s$	Partial molar volume of pure solvent.
$\mathbf{v}$	Velocity.
$W(r)$	Radial distribution function of end-to-end distance.
$W(x,y,z)$	End-to-end distance distribution function.
W^0_{mic}	Repulsive part of μ^0_{mic}.
$(\bar{X}_i)$	Equilibrium concentration.
$\Delta(X_i)$	Difference between temporal and equilibrium concentration.
x_b	Bottom of cell.
x_m	Meniscus position.
y	General physical property.
y_D	Physical property of denatured state.
y_N	Physical property of native state.
$\bar{y}$	Fractional saturation of site.
$\bar{y}_F$	Fractional saturation with ligand F.
z	Charge on macromolecule or ion in units of e.
z_i	Ionic valence of ith ion.
α	Degree of association. Dimensionless binding parameter like $(F)/k_R$.
α_H	Hill constant.
β	Dimensionless binding parameter.
β_e	Bohr magneton.
β_n	Nuclear magneton.
β_s	Mandelkern–Flory–Scheraga parameter.
β'	Scheraga–Mandelkern parameter.

Symbol	Meaning
Γ	Parameter affecting relaxation amplitudes.
γ	Magnetogyric ratio. $(A)/K_{AR}$ binding parameter. Velocity gradient dv_x/dz.
λ_1, λ_2	Parameters composed of rate constants.
δ	Chemical shift parameter. Phase shift.
$\delta(x)$	Dirac delta function of argument x.
δ_1	Hydration (in grams per gram).
δ_{ij}	Kronecker delta.
ε	Dielectric constant. Molar decadic or residue extinction coefficient.
$\Delta\varepsilon$	Circular dichroism $(\varepsilon_L - \varepsilon_R)$.
η	Solution viscosity.
η_0	Solvent viscosity.
η_{rel}	Relative viscosity.
η_{sp}	Specific viscosity.
$[\eta]$	Intrinsic viscosity.
Θ_i	Fractional saturation of ith site.
θ	Scattering angle. Fractional helicity.
$[\theta]$	Molar ellipticity.
$\underset{\sim}{\Lambda}$	Matrix of λ_i's.
λ	Eigenvalue. Wavelength. Kinetic decay time.
λ_j	jth kinetic decay time of jth eigenvalue.
μ_i	Chemical potential per mole.
μ_i^0	Standard chemical potential per mole.
$\hat{\mu}_i$	Chemical potential per gram.
$\hat{\mu}_i^0$	Standard chemical potential per gram.
μ^0_{mic}	Standard chemical potential of amphiphile in micelle.
μ^0_w	Standard chemical potential of amphiphile in aqueous phase.
$\boldsymbol{\mu}_m$	Magnetic moment.
$\underset{\sim}{\boldsymbol{\mu}}$	Electric dipole moment operator.
ν	Frequency. Simha factor in viscosity. Moles of ligand bound per mole of macromolecule.
ν_N	Saturation density for lattice with N units.
π	Osmotic pressure.
ρ	Mass density (in grams per cm^3).
$\rho(\mathbf{r})$	Electron density.
σ	Nucleation constant.

Symbol	Meaning
σ_h	Superhelix density.
τ	Number of supercoils.
τ_F	Fluorescence decay time.
τ_a, τ_b	Rotational relaxation time for a-, b-axis orientation.
τ_c	Rotational correlation time.
τ_r	Rotational relaxation time of sphere.
τ, τ_j	Reaction relaxation times.
Φ	Electrical potential. Voltage difference.
Φ_c	Universal constant for random coils $2 \cdot 1 \times 10^{23}$.
ϕ	N–C′ torsional angle. Phase of complex number.
ϕ_{1a}, ϕ_{20}, etc	Monomer wave functions.
ϕ_F	Fluorescence quantum yield.
ϕ_p	Practical osmotic coefficient.
ϕ', ϕ''	Nucleic acid backbone torsional angles.
$[\phi]$	Molar rotation per residue.
χ	Mole fraction of all solute species.
χ_i	Mole fraction of ith component.
χ_A	Mole fraction of Ath component.
χ_{gc}	Mole fraction G + C.
χ	Glycosidic bond torsional angle.
ψ	C′–C torsional angle.
ψ', ψ''	Nucleic acid backbone torsional angles.
Ω_{jk}	Number of ways of putting k helical units into j separated sequences.
Ω_k	$(n - k + 1)$ number of ways of placing k helical units in one sequence within chain of n residues.
$\Omega_{n,i}$	Number of ways of assorting i items (ligands) in n boxes (sites).
ω	Circular frequency or angular velocity.
ω_0	Larmor frequency.
ω', ω''	Nucleic acid backbone torsional angles.
$\Delta\omega_{1/2}$	Line width.
$\boldsymbol{\omega}$	Angular velocity.
imag	Imaginary part of.
$\langle\rangle$	Average.
$\langle\vert\rangle$	Overlap integral.
$\langle\Vert\rangle$	Expectation value integral.
*	Superscript, complex conjugate, as in F^*
$\vert\vert$	Amplitude of complex number or length of vector, as in $\vert F\vert$.
∇	Vector differential.
( )	Molar concentration, as in (A).
†	Superscript, transpose of matrix, as in $A^\dagger$.
⌒	Superscript, convolution product, as in $\widehat{AB}$.

General Rules

K	Macroscopic equilibrium constant.
k	Microscopic equilibrium constant or rate constant.
C	Molar concentration.
c	Weight concentration.
$\underset{\sim}{\mathbf{M}}$	All matrices and operators.
$\hat{\imath}$	All unit vectors.
R_G	Radius of gyration.
χ	Mole fraction.
Φ	Voltage or electrical potential.

PART II

TECHNIQUES FOR THE STUDY OF BIOLOGICAL STRUCTURE AND FUNCTION

The problem of determining the structure of a macromolecule is awesome, and no known technique is completely adequate. The long-range goal, in many cases, is to specify the location in space of each of the component atoms of a protein or nucleic acid. In a molecular weight of 30,000 daltons (d), there are roughly 4,000 atoms. For each of these N atoms, three Cartesian coordinates would have to be specified. Because we care nothing about the location or orientation of the whole molecule, six coordinates are irrelevant. The $3N - 6$ parameters that remain must be measured to define the structure. Also needed is the chemical identity of the element occupying each position, requiring N more parameters.

A typical 30,000 d protein or nucleic acid would require about 16,000 pieces of data to complete a structure determination. If one agrees to ignore hydrogen atoms, this number will be halved. Even so, this is an overpowering demand. No single technique currently available can provide so much detailed information about a single system. In fact, some of the most popular techniques provide only one or two useful parameters for a macromolecular system.

Fortunately, in practice, the situation is rarely as bleak as just described. It is typical to know, at the outset, the primary structure of a macromolecule of interest.

For each atom, then, one must determine a bond length, a bond angle, and (for all atoms except hydrogen) an internal rotation angle; this requires about 10,000 parameters for a 4,000-atom structure. With many biological polymers, accurate atomic structures (i.e., bond lengths and bond angles) of each residue are known. Only the spatial location of the residues remains to be specified; this task requires about 2,000 parameters. Thus, the task of the physical chemist is brought within a reasonable (if not always practical) range.

Further simplifications are possible with proteins and nucleic acids. Two angles are sufficient to describe the relative locations of two adjacent peptide residues in a polypeptide chain. Probably an average of two more parameters can define the orientation of each side chain. For a protein of 30,000 d, there are 300 residues, so about 1,200 parameters must be fixed. A nucleic acid of the same molecular weight would require only 800 parameters.

If any real symmetry exists, such as identical subunits or extensive helices, the problem becomes easier. Still, even in the most favorable cases, only a single technique (x-ray diffraction of single crystals) can provide sufficient information to define the entire three-dimensional structure of a biopolymer. However, x-ray work is laborious and has a number of serious limitations. Rarely is the quality of the data so good that precise atomic positions are obtained.

Many interesting questions can be explored and even answered without a knowledge of the complete three-dimensional structure. The following chapters provide many examples. In order to proceed, one has to simplify the problem. Two general approaches can be used. The first is to consider the system at a level much less precise than atomic resolution. Simplest are techniques that seek information about rough size and shape. In many of these, all molecular detail is put aside. The macromolecule is modeled as an ellipsoid of revolution, a coil with uniform stiffness or lack of it, a rod, or other simple shapes. Most hydrodynamic and certain scattering techniques fall into this category.

At the other extreme are techniques that look at only one small part of a macromolecule. These can sometimes attain the precision of x-ray diffraction (or even better). What is sacrificed is any overview of the entire structure. Usually, these types of techniques make use of probe molecules, whose properties allow them to be singled out from the rest of the system. Sometimes they are an intrinsic part of the macromolecule;

in other cases they are added extrinsically by the investigator. In principle, by numerous repetitions of probes, most of a structure could be mapped out. In practice, it is usually feasible to concentrate on only a few selected regions. These are chosen either to be of special biological interest, or else for expediency (that is, a probe naturally exists or can easily be introduced). Examples of successful and widely used probe techniques are fluorescence, EPR, some aspects of NMR, and chemical modification.

In between these two extremes are techniques that can selectively examine certain general aspects of a structure while ignoring others. These provide an average picture that can offer a lead to powerful constraints on the possible three-dimensional structure. Very often such information is accessible for regions of regularly ordered secondary structure. Experimental techniques such as CD/ORD, UV, IR, Raman spectroscopy, and tritium exchange afford a reasonable overview of the amounts of helix in a protein or nucleic acid. Sometimes certain details about the nature, size, and composition of the helices are accessible. None of these techniques can permit the definition of an entire 3-D structure, but they often yield a convenient and accurate approximate picture for the expenditure of relatively little effort.

Virtually all of the techniques in the biophysical chemist's arsenal gain in strength when used in concert. In many cases, parallel studies from a variety of approaches on a single macromolecule have led to a wealth of detailed information. In no case yet has a sufficient body of data ever been accumulated to match the total static structural picture that x-ray techniques can provide. However, many insights have been obtained that would be available from pure x-ray studies only with a great deal of luck and extrapolation. Table II-1 compares some of the basic characteristics of all the techniques discussed. It should be readily apparent that a detailed discussion of the principles and practice of each of these is beyond the scope of a single textbook. We have chosen to concentrate on only about half of them. Even within this limited set, the coverage is far from comprehensive. The principal focus of this book is molecules, not methods. Therefore, we stress experimental results and how they are interpreted, rather than methods. Practical aspects of techniques will be introduced in detail only when they are absolutely necessary in order to appreciate the meaning of the measured quantities. A large body of texts and reference works exists for each of the techniques treated. It is a virtual necessity that the reader who wishes to use a particular technique consult some of these.

Table II-1
Techniques for structural analysis of biopolymers

Technique	Information available												Experimental constraints					
	Complete 3° structure	Detailed subunit arrangement	Specific site structure, bonding	Proximity between specific sites	Orientation of 2°-structure units	Type and extent of 2° structure	Molecular weight	Shape, topology	Flexibility	Ionization of individual residues	Amounts & sites of small-molecule binding	Side-chain exposure, environment	Purity of sample needed	Size of sample needed	Ease of environmental variation	Effort involved in experiment	Need for 1°-structure knowledge	Possibilities for kinetic measurements
X-ray diffraction	3	3	2	2	3	3	3	3	–	1	2	2	–	–	–	–	–	–
Electron diffraction	3	3	2	2	3	3	3	3	–	1	2	2	–	–	–	–	–	–
Electron microscopy	1	2	–	1	–	–	1	3	1	–	–	–	3	3	–	2	3	–
Autoradiography	1	–	–	1	–	–	1	2	–	–	–	–	3	3	–	1	3	–
Neutron scattering	–	2	–	–	–	–	1	2	–	–	–	–	1	–	1	–	2	–
X-ray scattering	–	1	–	–	–	–	1	2	1	–	–	–	1	–	1	–	2	–
EXAFS	–	–	3	1	–	–	–	–	–	–	2	–	–	–	1	2	–	–
Rayleigh scattering	–	1	–	–	–	–	2	1	1	–	–	–	1	2	2	1	2	3
Inelastic light scattering	–	1	–	–	–	–	2	1	1	–	–	–	1	1	2	2	2	–
Sedimentation velocity	–	–	–	–	–	–	2	1	1	–	–	–	1	3	2	2	2	–
Sedimentation equilibrium	–	–	–	–	–	–	3	–	–	1	–	–	1	3	2	2	2	–
Diffusion	–	–	–	–	–	–	2	1	–	–	–	–	1	2	–	–	2	–
Viscosity	–	–	–	–	–	–	2	2	2	–	–	–	1	3	3	3	2	1
Gel filtration	–	–	–	–	–	–	2	1	–	–	1	–	3	3	3	3	2	1
Absorption: UV/vis	–	–	–	–	–	1	–	–	–	1	1	1	1	3	3	3	1	3
Linear dichroism, birefringence	–	–	–	–	2	1	1	2	2	–	–	–	1	1	1	1	1	1

Technique	Information available												Experimental constraints					
	Complete 3° structure	Detailed subunit arrangement	Specific site structure, bonding	Proximity between specific sites	Orientation of 2°-structure units	Type and extent of 2° structure	Molecular weight	Shape, topology	Flexibility	Ionization of individual residues	Amounts & sites of small-molecule binding	Side-chain exposure, environment	Purity of sample needed	Size of sample needed	Ease of environmental variation	Effort involved in experiment	Need for 1°-structure knowledge	Possibilities for kinetic measurements
CD/ORD	–	–	1	–	–	2	–	–	–	1	2	1	1	3	3	3	1	1
Absorption: IR	–	–	–	–	–	1	–	–	–	–	–	1	–	1	1	2	1	1
Dichroism: IR	–	–	–	–	2	1	–	–	–	–	–	–	–	1	1	1	1	–
MCD/MORD	–	–	2	–	–	–	–	–	–	1	–	–	1	3	3	3	1	1
Fluorescence	–	–	–	2	–	–	–	–	–	1	1	1	3	3	3	3	1	3
Polarized fluorescence	–	1	–	–	1	–	1	2	2	–	–	–	3	3	3	2	1	2
Phosphorescence	–	–	1	1	–	–	–	–	–	–	–	–	3	3	–	2	1	–
Raman scattering	–	–	1	–	–	1	–	–	–	–	–	1	1	1	2	2	1	1
Resonance Raman	–	–	2	–	–	–	–	–	–	–	–	–	3	3	3	2	1	1
EPR	–	–	2	2	1	–	–	2	2	–	2	1	3	3	2	2	1	1
NMR–^{1}H	1	–	3	2	–	–	–	–	1	3	3	1	–	1	1	1	1	1
NMR–^{13}C	–	–	2	1	–	–	–	–	2	2	2	–	–	–	2	–	2	–
^{3}H–Exchange	–	–	1	–	–	2	–	–	1	–	–	1	1	2	2	1	1	1
Potentiometric titration	–	–	1	–	–	–	–	–	2	1	2	–	1	–	1	–	1	1
Electrophoresis	–	–	–	–	–	–	1	1	–	1	–	–	3	3	1	3	2	–
Photoelectron spectroscopy	–	–	2	–	–	–	–	–	–	1	1	–	2	1	–	2	1	–
Mössbauer	–	–	2	–	–	–	–	–	–	1	1	–	2	1	–	2	1	–

NOTE: Under "Information available," larger numbers indicate more powerful techniques, and a dash indicates not usually applicable; under "Experimental constraints," larger numbers indicate easier measurements or interpretations, and a dash indicates difficult experimental hurdles.

7
Absorption spectroscopy

7-1 BASIC PRINCIPLES

The range of techniques

Conceptually, a typical spectroscopic experiment is extremely simple. Electromagnetic radiation at a certain nominal wavelength λ (or frequency $\nu = c/\lambda$) is allowed to impinge on the sample. Then some properties of the radiation that emerges from the sample are measured. One of the simplest properties is the fraction of the incident radiation absorbed or dissipated by the sample. (Typical techniques are optical absorption spectroscopy, some modes of NMR spectrometry, and various elastic scattering techniques.) Instead, one can examine the radiation emitted by the samples at wavelengths other than that used for excitation. (Fluorescence, phosphorescence, Raman scattering, and inelastic light scattering are examples.) Not only the emergent intensity, but also the distribution of emergent frequencies, are sources of information. In more complex techniques, not just intensity is detected, but also the kind and degree of polarization of the radiation emitted by a sample. (ORD, CD, and fluorescence polarization fit into this category.)

The range of wavelengths used in spectroscopy of biological molecules is impressive. It is summarized in Table 7-1, along with a list of the corresponding energies of the photons involved. These range from millions of kcal mole^{-1} (more than sufficient to break the strongest covalent chemical bonds if the energy could be localized) to less than 10^{-3} kcal mole^{-1} (far smaller than typical thermal energies—RT is about 0.6 kcal mole^{-1} at room temperature).

Table 7-1
Biologically useful spectroscopic regions

Typical wavelength (cm)	Approximate energy (kcal $mole^{-1}$)	Spectroscopic region	Techniques and applications
10^{-11}	3×10^{8}	γ-Ray	Mössbauer
10^{-8}	3×10^{5}	X-ray	X-ray diffraction, scattering
10^{-5}	3×10^{2}	Vacuum UV	Electronic spectra
3×10^{-5}	10^{2}	Near UV	Electronic spectra
	Carbon–carbon bond energy		
6×10^{-5}	5×10^{3}	Visible	Electronic spectra
10^{-3}	3×10^{0}	IR	Vibrational spectra
	RT at ambient temperature		
10^{-2}	3×10^{-1}	Far IR	Vibrational spectra
10^{-1}	3×10^{-2}	Microwave	Rotational spectra
10^{0}	3×10^{-3}	Microwave	Electron paramagnetic resonance
10	3×10^{-4}	Radio frequency	Nuclear magnetic resonance

In practice, these diverse regions of the spectrum offer an almost endless series of challenges to the experimental spectroscopist. When examined in detail, the various techniques seem quite different. This difference results from quite distinct aspects of molecular structure detected in various kinds of spectroscopy, and from large variations in the experimental approaches that must be used to gather data. However, all techniques share a number of common features, as we shall see.

Qualitative description of spectroscopy

There is no simple way to explain the interaction of light with matter. Light is a rapidly oscillating electromagnetic field. Molecules contain distributions of charges and spins that have electrical and magnetic properties. These distributions are altered when a molecule is exposed to light. In a typical spectroscopic experiment, light is sent through a sample, either continuously or in a pulse. What one must deal with is the *rate* at which the molecule responds to this perturbation. One must explain why only certain wavelengths cause changes in the state of the molecule. One must calculate how the presence of the molecule alters the radiation that emerges from the sample.

Calculation of properties of molecules by quantum mechanics

The principles of quantum mechanics best explain the energy states a molecule is allowed to occupy, and the mechanisms by which a molecule can change from one state to another. Some readers of this book will already have had an introduction to quantum mechanics, the theory that currently represents our understanding of the properties of molecules. For them, the following summary should serve as a review and as an introduction to the nomenclature used later in the chapter. For the reader who has had no prior exposure to quantum mechanics, this summary will at least introduce some of the terms and basic concepts. Such a reader should not be too concerned about the mathematical formalism because, although it is the language of quantum mechanics, it will be used sparingly in this book. However, it is important to grasp several key concepts.

1. The state of a system (e.g., an atom, a molecule, a crystal) is described by a wavefunction.
2. An observable quantity (e.g., the energy, the dipole moment, the location in space) is governed by a mathematical device known as an operator.
3. The result of measurement on a state (e.g., what is the energy?) can be computed by taking the average value of the operator on that state. This result is called an expectation value.
4. A transition between two states of a system can be induced by a perturbation, which is measured by an operator. Its effectiveness in producing the transition is governed by the extent to which it can deform the initial state to make it resemble the final state.
5. The ability of light to induce transitions in molecules can be calculated according to its ability to induce dipole moments that oscillate with the light.
6. The preferred directions for inducing dipole moments are fixed with respect to the geometry of the molecule.

Description of a molecule by wavefunctions

The state of a molecule or system is described by a wavefunction, Ψ. In general, Ψ is a function of the positions and spin of all electrons and nuclei and the presence of any external fields. It is a complex number and is time dependent. The wavefunction Ψ is not a directly measurable quantity. It is related to the probability of finding the system in a particular position, spin, and so on, and it is often referred to as a probability amplitude.

The probability of finding the system at a particular position or spin can be computed by squaring the amplitude Ψ for that particular position or spin. Because Ψ is a complex number, it must be multiplied by its complex conjugate Ψ^* to give a real probability:[§]

$$P = \Psi^*\Psi \tag{7-1}$$

[§] Because ψ is a complex number, it can be written as $a + bi$, where a and b are both real, and $i = \sqrt{-1}$. The complex conjugate ψ^* is $a - bi$, and the probability $\psi^*\psi$ is $a^2 + b^2$.

Because the system must be in some position and spin, P is normalized. If it is integrated over all space and spin, symbolized by the variable τ, then

$$\int P\,d\tau = \int \Psi^*\Psi\,d\tau = \langle\Psi|\Psi\rangle = 1 \tag{7-2}$$

Here we have introduced the Dirac notation, in which angle brackets are used to symbolize integration, and the complex conjugate of the function on the left is implicit.

Consider a system that has two possible states described by the wavefunctions Ψ_a and Ψ_b. If the system is actually in state a, the probability of finding it in that state, $\langle\Psi_a|\Psi_a\rangle$, is one, by Equation 7-2. Similarly, if the system is in state b, then $\langle\Psi_b|\Psi_b\rangle = 1$. These statements are called normalization, and the wavefunctions that obey them are called normalized wavefunctions.

A measure of the similarity of the two states is given by the integral symbolized by $\langle\Psi_a|\Psi_b\rangle$, which is sometimes called an overlap integral. This integral is a pure number that measures the projection of state b onto state a. If a and b are considered as vectors, than $\langle\Psi_a|\Psi_b\rangle$ is analogous to the dot product $\mathbf{a}\cdot\mathbf{b}$. The maximal value occurs when a is identical to b; then $\langle\Psi_a|\Psi_b\rangle = 1$. The smallest possible value for $\langle\Psi_a|\Psi_b\rangle$ is zero; when this occurs, the two wavefunctions are said to be orthogonal. (In the vector analogy, two perpendicular vectors have a dot product $\mathbf{a}\cdot\mathbf{b} = 0$.)

Suppose there are only two states, a and b, and one is not sure whether a system is in state a or in state b. Then the system can be described as a linear combination of two wavefunctions:

$$\Psi = C_a\Psi_a + C_b\Psi_b \tag{7-3}$$

Because the system must be in one state or the other, $\langle\Psi|\Psi\rangle = 1$. The coefficients C_a and C_b are related to the probabilities that the system will be found in state a or in state b, but the relationship is not simple.[§] In the special case applicable to all the systems we are interested in, where $\langle\Psi_a|\Psi_a\rangle = 1$ and $\langle\Psi_a|\Psi_b\rangle = 0$, the probability that the system will be in state a is

$$P_a = C_a^*C_a = |C_a|^2 \tag{7-4}$$

Often it is convenient to try to describe aspects of a system separately. For example, one may want to treat the spin (s) of an electron separately from its spatial location (r). Then one attempts to write the wavefunction as a product of two functions; $\Psi(r, s) = \Psi(r)\sigma(s)$ is such a factorization of an electron wavefunction. If spin

[§] In general, the probability amplitude that the system is in state a is $\Psi_a\langle\Psi_a|\Psi\rangle$, where the integral $\langle\Psi_a|\Psi\rangle$ is a number that describes the overlap between Ψ and Ψ_a. The probability that the system is in state a can be computed by analogy with Equation 7-2 as

$$\langle\langle\Psi_a|\Psi\rangle\Psi_a|\Psi_a\langle\Psi_a|\Psi\rangle\rangle = \langle\Psi_a|\Psi_a\rangle\,|C_a\langle\Psi_a|\Psi_a\rangle + C_b\langle\Psi_a|\Psi_b\rangle|^2$$

after inserting Equation 7-3 for Ψ. The C_b term is an interference term that arises from the wavelike properties of quantum mechanical states. [See A. Messiah, *Quantum Mechanics*, vol. 1 (New York: Wiley, 1962), pp. 296–298.]

and position are truly independent, such a factorization into a product of two wavefunctions is exact. However, in many cases this kind of description is only approximate. Thus, for molecules, one usually attempts to separate (1) the parts of the wavefunction dealing with electron position for fixed nuclei from (2) those that depend on nuclear motions:

$$\Psi = \Psi_e(r, R)\Phi_N(R) \tag{7-5}$$

This separation, in which r refers to electrons and R to nuclei, is called the Born–Oppenheimer approximation. It is the basis of most analysis of molecular spectra. The basic idea is that electrons move so much faster than nuclei that the electronic wavefunction Ψ_e can be evaluated for a *fixed* nuclear configuration; the nuclear wavefunction Φ_N can be evaluated as R varies for a *time-average* electron configuration.

In a similar way, the state of two electrons can be described as the product of two one-electron wavefunctions

$$\Psi = \Psi_1(r_1)\Psi_2(r_2) \tag{7-6}$$

providing that the electrons maintain distinct identities and don't interact.

Operators and values of observable quantities

Properties of the system can be calculated by examining the effect of various mathematical operations on the wavefunction. For example, $\underset{\sim}{\mathbf{I}}$ is the identity operator. Applied to a wavefunction, it just returns the same function: $\underset{\sim}{\mathbf{I}}\Psi = \Psi$. Another common operator is d/dx, the differential operator; sin, $\sqrt{}$, and many other mathematical forms also are operators. They are not physical observables, nor is the result of their application to a wavefunction directly observable.

Many of the operators of interest in quantum mechanics satisfy eigenvalue equations:

$$\underset{\sim}{\mathbf{Q}}\Psi = \Lambda\Psi \tag{7-7}$$

Here Λ is a pure number: the eigenvalue. Any state Ψ that satisfies this equation is called an eigenstate (or, equivalently, an eigenfunction) of the operator, with eigenvalue Λ. The significance of Λ can best be seen by computing the expectation value of an operator. For operators that correspond to physically observable quantities, the expectation value is defined by analogy to Equation 7-2:

$$\langle O \rangle = \int \Psi^* \underset{\sim}{\mathbf{Q}} \Psi \, d\tau = \langle \Psi | \underset{\sim}{\mathbf{Q}} | \Psi \rangle \tag{7-8}$$

Substituting Equation 7-7, and using the result of Equation 7-2, we obtain

$$\langle O \rangle = \langle \Psi | \Lambda | \Psi \rangle = \Lambda \tag{7-9}$$

So, if the system is in a state Ψ that satisfies Equation 7-7, the value Λ is what will actually be measured for the observable corresponding to $\mathbf{O}$.

For example, the momentum operator $\mathbf{p}$ is $-i\hbar\nabla$ (or, in one dimension, $-i\hbar\, d/dx$), where i is $\sqrt{-1}$, and $\hbar$ is Planck's constant divided by 2π. The momentum of a system can be calculated as $\langle \Psi | \mathbf{p} | \Psi \rangle$. Other examples of operators are the position operator, $\mathbf{r}$, and the dipole moment operator, $e\mathbf{r}$ (where e is the charge on the electron).

The Schrödinger equation

The time dependence of the behavior of the system is given by an equation that is deceptively simple in appearance:

$$i\hbar\, d\Psi/dt = \mathbf{H}\Psi \tag{7-10}$$

In this time-dependent Schrödinger equation, $\mathbf{H}$ is the Hamiltonian operator of the system. It is defined such that

$$\langle \Psi | \mathbf{H} | \Psi \rangle = E \tag{7-11}$$

where E is the total energy of the system. In general, $\mathbf{H}$ can be written as $\mathbf{T} + \mathbf{V}$, where $\mathbf{T}$ is the kinetic energy operator, and $\mathbf{V}$ is the potential energy operator. The form of $\mathbf{T}$ can be inferred from classical mechanics, where kinetic energy is just

$$\sum_i \tfrac{1}{2} m_i \mathbf{v}_i^2 = \sum_i \mathbf{p}_i^2/2m_i$$

where the sum is taken over the masses (m_i) and velocities ($\mathbf{v}_i$) or momenta ($\mathbf{p}_i$) of all particles in the system. Thus, $\mathbf{T} = \sum_i \mathbf{p}_i^2/2m$, where $\mathbf{p}_i^2 = -\hbar^2\nabla^2$. The form of $\mathbf{V}$ depends on the system. It contains terms due to interactions among electrons and nuclei of molecules of the system, as well as interactions between these particles and any external fields.

Consider the special case in which $\mathbf{H}$ itself is not time-dependent. Certain states of the system will satisfy the eigenvalue equation

$$\mathbf{H}\Psi = E\Psi \tag{7-12}$$

This is called the time-independent Schrödinger equation. States that satisfy this equation have certain interesting properties. Their energy (by Equation 7-7) is E, and it is constant with time. The wavefunctions Ψ are still time-dependent. Substituting Equation 7-12 into Equation 7-10, we obtain

$$i\hbar\, d\Psi/dt = E\Psi \tag{7-13}$$

This simple first-order differential equation integrates easily to yield

$$\Psi(t) = \Psi(0)e^{-iEt/\hbar} \tag{7-14}$$

where $\Psi(0)$ is the wavefunction at zero time. The corresponding probability is time-independent:

$$P = \Psi^*(t)\Psi(t) = |\Psi(0)|^2 e^{-iEt/\hbar}e^{+iEt/\hbar} = |\Psi(0)|^2$$

Thus, any eigenstate Ψ that satisfies Equation 7-12 is a *stationary state* of the system. Its observable properties do not change with time.

Another important property of states that satisfy Equation 7-12 is orthonormality. If Ψ_1 and Ψ_2 are two eigenstates with corresponding energies E_1 and E_2, it can be shown in complete generality that

$$\langle\Psi_1|\Psi_2\rangle = \langle\Psi_2|\Psi_1\rangle = 0 \tag{7-15}$$

as long as $E_1 \neq E_2$.

Consider a system that may be in either state 1 or state 2. Its wavefunction from Equation 7-3 is $\Psi = C_1\Psi_1 + C_2\Psi_2$. Suppose a measurement is made on the system to see if it is in state 1. Using Equation 7-4, the probability of finding the system in state 1 is

$$|\langle\Psi_1|\Psi\rangle|^2 = |C_1|^2 \tag{7-16}$$

if Ψ_1 is normalized. Note that this result is independent of any properties of state 2. Thus the states that satisfy Equation 7-12 correspond to pure observable states of the system with well-defined energy and other properties.[§]

A final property of the states that satisfy Equation 7-12 is that they form a complete set. That is, any state of the system, however prepared, can be described as a

[§] However, a state that is an eigenfunction of an operator such as $\underset{\sim}{\mathbf{H}}$ is not necessarily an eigenfunction of some other operator. This question is tested by examining whether or not the two operators commute. For example, the Hamiltonian operator $\underset{\sim}{\mathbf{H}}$ and the angular momentum operator $\underset{\sim}{\mathbf{L}}$ can be applied in either order without changing the value of the result $\underset{\sim}{\mathbf{H}}\underset{\sim}{\mathbf{L}}\Psi = \underset{\sim}{\mathbf{L}}\underset{\sim}{\mathbf{H}}\Psi$. This means that wavefunctions Ψ can be found that are simultaneously eigenstates of both $\underset{\sim}{\mathbf{H}}$ and $\underset{\sim}{\mathbf{L}}$. Thus a state Ψ can simultaneously have a well-defined angular momentum and a well-defined energy. In contrast, the position and linear momentum operators do not commute: $\underset{\sim}{\mathbf{r}}\underset{\sim}{\mathbf{p}}\Psi \neq \underset{\sim}{\mathbf{p}}\underset{\sim}{\mathbf{r}}\Psi$. Thus one cannot measure simultaneously a well-defined position and a well-defined momentum for a molecule. This observation is the basis of the Heisenberg uncertainty principle.

linear combination of the eigenfunctions:

$$\Psi = \sum_i C_i \Psi_i \tag{7-17}$$

In principle, the sum must be carried out over all of the states Ψ_i. This may be an infinite set. In practice, in many cases, most coefficients are small. Suppose the system is initially in state Ψ_a, and a small perturbation is applied. The result will be given by Equation 7-17, in which C_a is almost one, and only a few other C_i values are finite. This observation is the basis of perturbation theory, which is used to describe responses of molecules to weak electromagnetic fields or weak interactions with other molecules.

Suppose the perturbation is a potential $\underset{\sim}{\mathbf{V}}$. One can say that the potential causes mixing of other states into the original state Ψ_a to form the final state Ψ. The mixing can be examined by computing the expectation value of the potential between the original state and the perturbed state:

$$\langle \Psi | \underset{\sim}{\mathbf{V}} | \Psi_a \rangle = \sum_i C_i \langle \Psi_i | \underset{\sim}{\mathbf{V}} | \Psi_a \rangle \tag{7-18}$$

In the absence of $\underset{\sim}{\mathbf{V}}$, all the terms in Equation 7-18 would be zero except $\langle \Psi_a | \Psi_a \rangle$ because the states Ψ_i are orthogonal. In the presence of $\underset{\sim}{\mathbf{V}}$, other terms become nonzero. Thus the potential is capable of causing transitions between state Ψ_a and other states. Application of the potential operator to state Ψ_a results in the possibility of converting this state to other states. The amplitude of this probability is given by terms such as $\langle \Psi_i | \underset{\sim}{\mathbf{V}} | \Psi_a \rangle$.

Interaction of light with molecules

To explain spectroscopic measurements, we must calculate the effect of light on a molecule. For simplicity, we shall restrict attention only to the electric field of the light, although more rigorous treatments include magnetic effects as well. Light is a transverse wave that oscillates periodically in time and space.[§] A typical chromophore (a molecular moiety that interacts with light) is small (say 10 Å) compared to the wavelength of light (say 3,000 Å). Thus, one can ignore the spatial variation of the electric field of the light within the molecule. The electric field felt by a molecule in the presence of light can be written as

$$\mathbf{E}(t) = \mathbf{E}_0 e^{i\omega t} \tag{7-19}$$

where $\mathbf{E}_0$ is the maximum amplitude; it is a vector that describes the polarization

[§] More details about the characteristics of electromagnetic radiation are given later in this chapter, and also in Chapter 13.

direction of the light. The circular frequency ω is equal to $2\pi\nu = 2\pi c/\lambda$, where ν is the frequency in cycles sec^{-1}.

Suppose our system is originally in state Ψ_a, an eigenstate of the time-independent Hamiltonian $\underset{\sim}{\mathbf{H}}$, with energy E_a. Light perturbs the system such that, if it is examined after exposure to light for a certain time, some molecules are no longer in state Ψ_a, but are in other states, as earlier described for the general case. What we need to compute is the *rate* at which light causes transitions between Ψ_a and other states. To keep things simple, we shall consider a molecule with only two states, Ψ_a and Ψ_b. Because light is a time-dependent interaction, the time-dependent Schrödinger equation must be solved. The Hamiltonian can be written as

$$\underset{\sim}{\mathbf{H}}' = \underset{\sim}{\mathbf{H}} + \underset{\sim}{\mathbf{V}}(t) \tag{7-20}$$

where the effect of the light appears entirely in $\underset{\sim}{\mathbf{V}}(t)$. In the absence of light, Equation 7-14 gives the time behavior of the two eigenstates. The wavefunction in the presence of light must be a linear combination of these two states, but the coefficients C_a and C_b are now time-dependent:

$$\Psi(t) = C_a(t)\Psi_a e^{-iE_a t/\hbar} + C_b(t)\Psi_b e^{-iE_b t/\hbar} \tag{7-21}$$

Inserting this expression into the time-dependent Schrödinger equation (Eqn. 7-10), we obtain (after cancellation of a few terms)

$$i\hbar(\Psi_a e^{-iE_a t/\hbar}\, dC_a/dt + \Psi_b e^{-iE_b t/\hbar}\, dC_b/dt) = \underset{\sim}{\mathbf{V}}(t)[\Psi_a e^{-iE_a t/\hbar} C_a(t) + \Psi_b e^{-iE_b t/\hbar} C_b(t)] \tag{7-22}$$

We need to generate equations to evaluate $C_a(t)$ and $C_b(t)$. This can be done by multiplying by $\Psi_a^* e^{+iE_a t/\hbar}$ or $\Psi_b^* e^{+iE_b t/\hbar}$. The result is

$$i\hbar\, dC_a/dt = \langle\Psi_a|\underset{\sim}{\mathbf{V}}|\Psi_a\rangle C_a + \langle\Psi_a|\underset{\sim}{\mathbf{V}}|\Psi_b\rangle C_b e^{-i(E_b - E_a)t/\hbar} \tag{7-23a}$$

$$i\hbar\, dC_b/dt = \langle\Psi_b|\underset{\sim}{\mathbf{V}}|\Psi_a\rangle C_a e^{-i(E_a - E_b)t/\hbar} + \langle\Psi_b|\underset{\sim}{\mathbf{V}}|\Psi_b\rangle C_b \tag{7-23b}$$

The integrals symbolized by the angle brackets are taken over spatial coordinates only.

Equation 7-23a,b represents two equations in the two unknowns $C_a(t)$ and $C_b(t)$. To proceed further, it is necessary either to make some approximations using initial conditions on C_a and C_b, or else to insert an explicit form for the perturbing potential $\underset{\sim}{\mathbf{V}}(t)$. Most textbooks on quantum mechanics use the former approach. Here we illustrate the latter approach.

A molecule is perturbed by light because its distribution of electric charge is altered by the presence of the oscillating electric field **E**. To describe the charge distribution of a molecule, we could consider each individual charge, but it is easier to expand the charge distribution in a multipole series (as we did in Chapter 5 to describe intermolecular interactions). For electrically neutral molecules, the leading term in this expansion is the electric dipole. In quantum mechanics, this is described by the operator $\underset{\sim}{\mu} = \sum_i e_i \underset{\sim}{r}_i$, where the sum is taken over each electronic charge (e_i) at position $\underset{\sim}{r}_i$. We shall restrict attention to the electronic part of the wavefunction, as described in Equation 7-5. The positions of nuclei are assumed to be fixed and thus can be ignored.

The interaction energy between a molecule and light is given simply by $\underset{\sim}{V}(t) = \underset{\sim}{\mu} \cdot \mathbf{E}_0 e^{i\omega t}$. It is easy to show that the terms $\langle \Psi_a | \underset{\sim}{V} | \Psi_a \rangle$ and $\langle \Psi_b | \underset{\sim}{V} | \Psi_b \rangle$ in Equation 7-23 are zero for this choice of $\underset{\sim}{V}(t)$. This is because the integrals in the angle brackets are evaluated over *all* space. The dipole operator $\underset{\sim}{\mu}$ changes sign as $x \to -x$, $y \to -y$, and $z \to -z$. But the spatial part of $\Psi_a^* \Psi_a$ and $\Psi_b^* \Psi_b$ must be identical at x and $-x$, at y and $-y$, and at z and $-z$ because it is the square of a function. Hence the integrals vanish. The two equations (7-23a,b) now become

$$i\hbar \, dC_a/dt = C_b \langle \Psi_a | \underset{\sim}{\mu} | \Psi_b \rangle \cdot \mathbf{E}_0 e^{-i(E_b/\hbar - E_a/\hbar - \omega)t} \tag{7-24a}$$

$$i\hbar \, dC_b/dt = C_a \langle \Psi_b | \underset{\sim}{\mu} | \Psi_a \rangle \cdot \mathbf{E}_0 e^{-i(E_a/\hbar - E_b/\hbar - \omega)t} \tag{7-24b}$$

where $\mathbf{E}_0$ has been removed from the integrals because it is constant over the dimensions of the molecules. (Note that $\langle \Psi_a | \underset{\sim}{\mu} | \Psi_b \rangle = \langle \Psi_b | \underset{\sim}{\mu} | \Psi_a \rangle$.)

These two equations (7-24a,b) can now be solved simultaneously.[§] We choose to compute the probability P_b that the system is in state b at time t. From Equation 7-4, we have $P_b = |C_b(t)|^2$. The result, for small $\mathbf{E}_0$ and arbitrarily starting with $|C_b(0)|^2 = 0$, *is*

$$|C_b(t)|^2 = \frac{|\langle \Psi_b | \underset{\sim}{\mu} | \Psi_a \rangle \cdot \mathbf{E}_0|^2}{\hbar^2} \frac{t^2 \sin^2[(E_b/\hbar - E_a/\hbar - \omega)t/2]}{2[(E_b/\hbar - E_a/\hbar - \omega)t/2]^2} \tag{7-25}$$

Thus $|C_b(t)|^2$ is large only when the energy denominator $(E_b/\hbar - E_a/\hbar - \omega)$ is small. Because $\hbar\omega$ is the energy of the light, transitions from a to b will be induced only when the light energy is approximately equal to the energy separation between the two states $(E_b - E_a)$:

$$h\nu = \hbar\omega = E_b - E_a \tag{7-26}$$

That is, spectral bands (light absorption or light-induced transition) will occur only at certain narrow wavelength (or frequency) intervals.

We need to know the rate at which molecules in state a are transformed to state b by the presence of light. This is just the rate of change of $|C_b(t)|^2$ in response to illumination with radiation centered about frequency ν. If the frequency distribution is narrow, the result for polarized light and oriented molecules can be shown to be

$$\frac{dP_b}{dt} = \frac{d}{dt} \int d\nu |C_b(t)|^2 = \frac{1}{2\hbar^2} |\langle \Psi_b | \underset{\sim}{\mu} | \Psi_a \rangle \cdot \mathbf{E}_0|^2 \tag{7-27}$$

[§] In practice, this solution is fairly tricky; see Eyring et al. (1944).

A similar equation holds for the rate at which light stimulates $b \rightarrow a$ transitions.

To compute absorption intensities, we must know the rate at which energy is taken up from the incident light beam. We can write the transition rate dP_b/dt as a product of two terms

$$dP_b/dt = B_{ab}I(\nu) \tag{7-28}$$

where B_{ab} is the transition rate per unit energy density of the radiation, and $I(\nu)$ is the energy density incident on the sample at frequency ν. It is well known from the study of electricity and magnetism that the energy density of light is $I(\nu) = |\mathbf{E}_0|^2/4\pi$. Equation 7-27 was derived for oriented molecules illuminated by polarized light. For solutions, one must average over all orientations. The average of $|\langle \Psi_b|\underset{\sim}{\boldsymbol{\mu}}|\Psi_a\rangle \cdot \mathbf{E}_0|^2$ is just $(1/3)|\langle \Psi_b|\underset{\sim}{\boldsymbol{\mu}}|\Psi_a\rangle|^2|\mathbf{E}_0|^2$ because it requires just an average over $\cos^2\theta$, where θ is the angle between $\underset{\sim}{\boldsymbol{\mu}}$ and $\mathbf{E}_0$. Using this result, we can evaluate B_{ab} from Equations 7-27 and 7-28:

$$\boxed{B_{ab} = (2/3)(\pi/\hbar^2)|\langle \Psi_b|\underset{\sim}{\boldsymbol{\mu}}|\Psi_a\rangle|^2} \tag{7-29}$$

The rate at which energy is removed from the light will depend on the number of $a \rightarrow b$ absorption transitions stimulated by the light, on the number of $b \rightarrow a$ emission transitions, and on the energy per transition ($E_b - E_a = h\nu$). Using Equation 7-28, this rate is

$$-dI(\nu)/dt = h\nu(N_aB_{ab} - N_bB_{ba})I(\nu) \tag{7-30}$$

where N_a and N_b are the number of molecules per cm^3 in states a and b, respectively. Thus, light absorption (or other optical properties) depends on concentration through the factors N_a and N_b.

The quantities B_{ab} and B_{ba} are called Einstein coefficients for stimulated absorption and emission, respectively. For simple cases, such as the two-state system described here, $B_{ab} = B_{ba}$.

Transition dipoles

The molecular parameter that determines the magnitude of the transition probability B_{ab} (and thus the intensity of light absorption) is proportional to $\langle \Psi_b|\underset{\sim}{\boldsymbol{\mu}}|\Psi_a\rangle$. This integral describes the ability of light to distort a molecule in state a so as to produce elements that resemble state b. Classically, one treats the interaction of light with matter as the induction of dipoles by the light electric field:

$$\boldsymbol{\mu}_{\text{ind}} = \underset{\sim}{\boldsymbol{\alpha}} \cdot \mathbf{E} \tag{7-31}$$

where $\underset{\sim}{\alpha}$ is the polarizability, and $\boldsymbol{\mu}_{\text{ind}}$ is the induced dipolar moment. Because **E** is a function that fluctuates with time, $\boldsymbol{\mu}_{\text{ind}}$ fluctuates also. Thus, by analogy, $\langle\Psi_b|\underset{\sim}{\boldsymbol{\mu}}|\Psi_a\rangle$ is the dipole induced by light on a quantum mechanical system.

The term $\langle\Psi_b|\underset{\sim}{\boldsymbol{\mu}}|\Psi_a\rangle$ is called a transition dipole moment. It also is called a dipole matrix element (Box 7-1) and abbreviated as $\boldsymbol{\mu}_{ba}$. These quantities can be calculated for molecules by performing the integral $\int\Psi_b^*\underset{\sim}{\boldsymbol{\mu}}\Psi_a\,d\tau$. The transition dipole for each pair of states is in a fixed orientation relative to the structure of the molecule. It can be drawn as a vector in the coordinate system defined by the location of the nuclei of the atoms. Because $\boldsymbol{\mu}_{ba}$ is like a vector, it has three Cartesian components, μ_x, μ_y, and μ_z.

Box 7-1 WAVEFUNCTIONS AS VECTORS, AND OPERATORS AS MATRICES

A system in the state Ψ can be expressed as a linear combination of wavefunctions, $\Psi = \sum_i C_i\Psi_i$, where each Ψ_i is an eigenfunction of a particular operator. If we choose to work with a complete set of eigenfunctions, the set Ψ_i is a basis (a coordinate system) in which all states can be described. The state Ψ is then like a vector with components $(C_1, C_2, \ldots, C_i, \ldots)$. The result of applying the operator $\underset{\sim}{\mathbf{O}}$ to a state can be calculated by examining individually its result on each of the basic eigenfunctions. For example, $\underset{\sim}{\mathbf{O}}\Psi = \sum_i C_i\underset{\sim}{\mathbf{O}}\Psi_i$. All of the expectation values of the operator for various states can be collected together into a matrix. Each component is an integral such as $O_{ij} = \langle\Psi_i|\underset{\sim}{\mathbf{O}}|\Psi_j\rangle$. The overall matrix is

$$\begin{pmatrix} O_{11} & O_{12} & O_{13} & \cdots \\ O_{21} & O_{22} & O_{23} & \cdots \\ O_{31} & O_{32} & O_{33} & \cdots \\ \vdots & \vdots & \vdots & \end{pmatrix}$$

In the special case where the states Ψ_i are eigenfunctions of the operator with eigenvalues Λ_i, we know that $\langle\Psi_i|\underset{\sim}{\mathbf{O}}|\Psi_j\rangle = \Lambda_i$ if $i = j$, and $\langle\Psi_i|\underset{\sim}{\mathbf{O}}|\Psi_j\rangle = 0$ if $i \neq j$. Therefore, the matrix is diagonal:

$$\begin{pmatrix} \Lambda_1 & 0 & 0 & \cdots \\ 0 & \Lambda_2 & 0 & \cdots \\ 0 & 0 & \Lambda_3 & \cdots \\ \vdots & \vdots & \vdots & \end{pmatrix}$$

This formalism is so convenient and useful that it is common practice to call the expectation values $\langle\Psi_i|\underset{\sim}{\mathbf{O}}|\Psi_j\rangle$ *matrix elements* of the operator $\underset{\sim}{\mathbf{O}}$, and to symbolize them by O_{ij}. For example, using this formalism, Equation 7-27 can be written as

$$dP_b/dt = (1/2\hbar^2)|\boldsymbol{\mu}_{ba} \cdot \mathbf{E}_0|^2$$

where $\boldsymbol{\mu}_{ba}$ is now understood as a matrix element of the dipole moment operator.

Parameters available from spectral measurements

The wavelength or frequency that shows maximal absorption is a measure of the energy separation of two states (Eqn. 7-26). The shape of the spectral band of an isolated molecule is predicted (Eqn. 7-25) to be very narrow. In reality, factors we have not yet considered result in broader bands. These factors include environmental effects, state heterogeneity, and molecular motion.

The intensity of absorption (or emission) associated with a transition is described by Equations 7-27, 7-28, and 7-29. From intensity measurements, we can evaluate $|\langle\Psi_b|\underset{\sim}{\boldsymbol{\mu}}|\Psi_a\rangle|^2$. This value, in turn, provides information about the electronic distribution within the molecule. The intensity depends on the relative orientation of the molecule with the light-polarization direction: $\langle\Psi_b|\underset{\sim}{\boldsymbol{\mu}}|\Psi_a\rangle\cdot\mathbf{E}_0$. Thus, for oriented molecules, some geometrical information about the system also is available.

The light-induced dipole represented by $\langle\Psi_b|\underset{\sim}{\boldsymbol{\mu}}|\Psi_a\rangle$ oscillates with the electric field of the applied radiation. Thus, although $\langle\Psi_b|\underset{\sim}{\boldsymbol{\mu}}|\Psi_a\rangle$ is like a vector, it does not have a preferred direction, only a preferred orientation. Therefore, $\langle\Psi_b|\underset{\sim}{\boldsymbol{\mu}}|\Psi_a\rangle$ sometimes is represented by a double-headed arrow (↔) rather than the single-headed arrow usually used for a vector. When two dipoles are simultaneously excited by light, however, their relative phase is important. Depending on the properties of the system, they may be excited in phase or out of phase, or with some arbitrary phase shift. For example,

	In phase	Out of phase by 180°
Time $= t$	↑↑	↑↓
Time $= t + 1/2\nu$	↓↓	↓↑

This effect is important when we consider the optical properties of molecules with more than one chromophore. Dipoles induced by light in one chromophore can interact with those induced in neighboring chromophores. Depending on the phases and orientations, these interactions can be attractive (as in the out-of-phase example just shown) or repulsive.

7-2 ABSORPTION SPECTROSCOPY OF ELECTRONIC STATES

The measurement most frequently performed on biopolymers is the absorption of visible or ultraviolet light. This technique is used for purposes ranging from simple concentration determinations to resolution of complex structural questions. In this section, we consider first some of the basic features of these measurements, and then particular aspects of absorption relevant to the properties of large molecules.

Energy states of molecules

Figure 7-1 shows a section through the potential energy surfaces of the two lowest electronic states of a typical simple molecule. Superimposed on each of these states

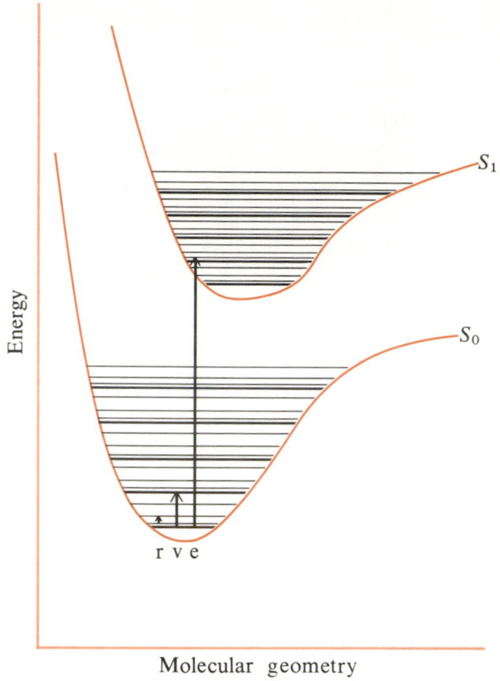

Figure 7-1

Energy levels of a small molecule. Selected rotational sublevels of the vibrational levels of each of two electronic states are shown. Transitions corresponding to electronic (e), vibrational (v), and rotational (r) spectra are indicated.

is a series of vibrational levels that, in turn, are subdivided into a myriad of rotational levels. The energy spacing between the lowest rotation–vibration states of the two electronic states S_0 and S_1 typically is 80 kcal mole^{-1}. This energy is much greater than the thermal energies at room temperature. Therefore one knows from statistical mechanics that, for all practical purposes, in the absence of radiation that can excite a transition, all molecules in a solution are in the lowest electronic state, S_0. The energy spacing between vibrational levels is of the order of 10 kcal mole^{-1}. This energy also is larger than thermal energies so, at least approximately, we can consider only the lowest vibrational level of S_0 to be appreciably populated. However, rotational energy spacings are only 1 kcal mole^{-1} or less; therefore many rotational levels are populated.

When light of the correct frequency is absorbed, the molecule can be excited to one of many rotation–vibration levels of the electronic state S_1. Thus, in the absence of other effects, one should observe a spectrum that is composed of a huge number of closely spaced sharp spectral bands. Their individual intensities depend on the magnitude of the transition dipole $\langle S_1, \text{vib}, \text{rot}|\underset{\sim}{\boldsymbol{\mu}}|S_0\rangle$, where S_1, vib, rot is the wavefunction of the particular rotation–vibration state of S_1 occupied, and S_0 is the ground-state wavefunction. In practice, each of these spectral bands is so broad that one observes only a relatively smooth spectral envelope (Fig. 7-2). The causes of band broadening include environmental heterogeneity, Doppler shifts, and other effects.

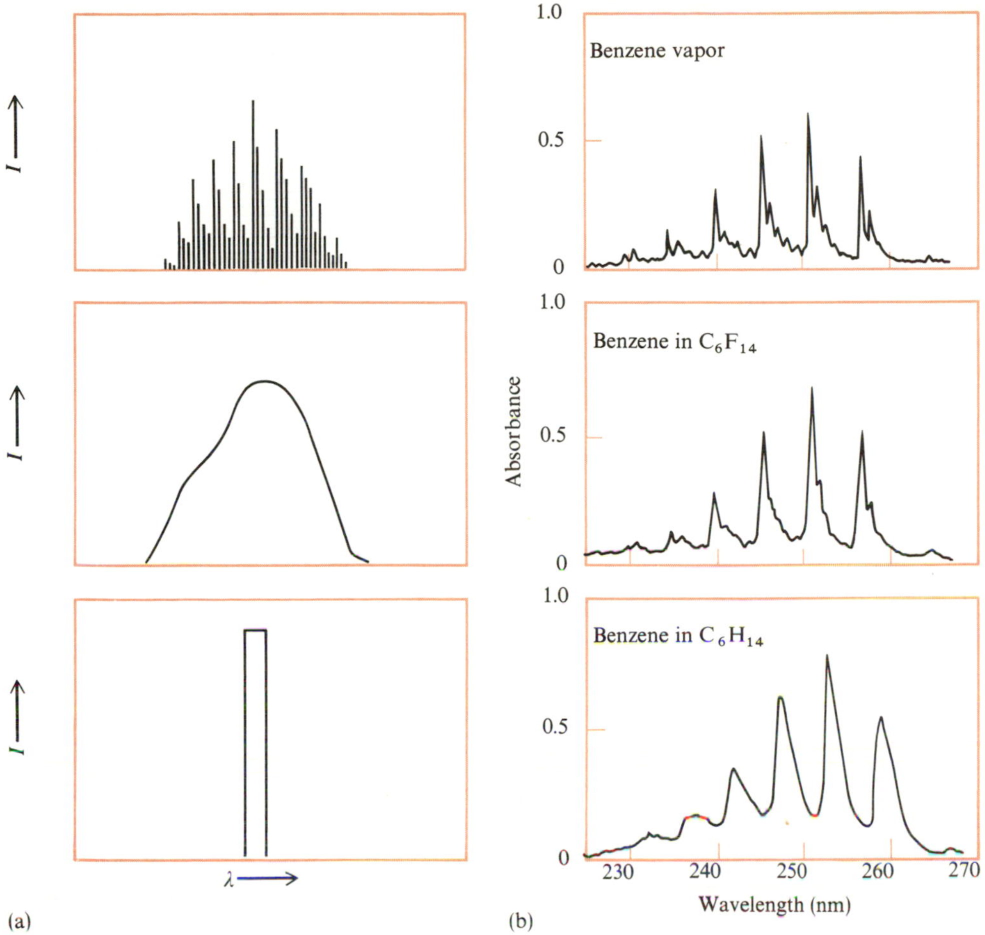

Figure 7-2

Electronic absorption spectra of small molecules. **(a)** Spectra of a typical small molecule. Shown, from top to bottom, are the gas-phase spectrum, the solution spectrum, and the theoretical spectrum usually dealt with in calculations. **(b)** Absorption spectra of benzene, showing solvent-induced broadening. [After J. B. Birks, *Photophysics of Aromatic Molecules* (New York: Wiley, 1970), p. 117.]

The result of all of this is that a potentially complex and very informative spectrum is reduced to a very simple one. Although much information is lost, there is a compensating advantage. In trying to gain some theoretical understanding of the spectrum, we can ignore all of the rotation–vibration states. The spectrum is modeled as a single electronic band (Fig. 7-2a), and usually no rigorous attempt is made to deal with its shape. In certain special cases (such as in dilute gases or at very low temperatures in rigid matrices), sharp spectral bands can be resolved for large molecules (Birks, 1970).

The extinction coefficient

Figure 7-3a is a simple schematic diagram of a light-absorption measurement. The incident light has intensity I_0 at wavelength λ. It impinges on a sample (usually a solution of absorbing molecules at a concentration of C moles liter^{-1}) for a path length of l cm. The light that is not absorbed by the sample emerges with intensity I.

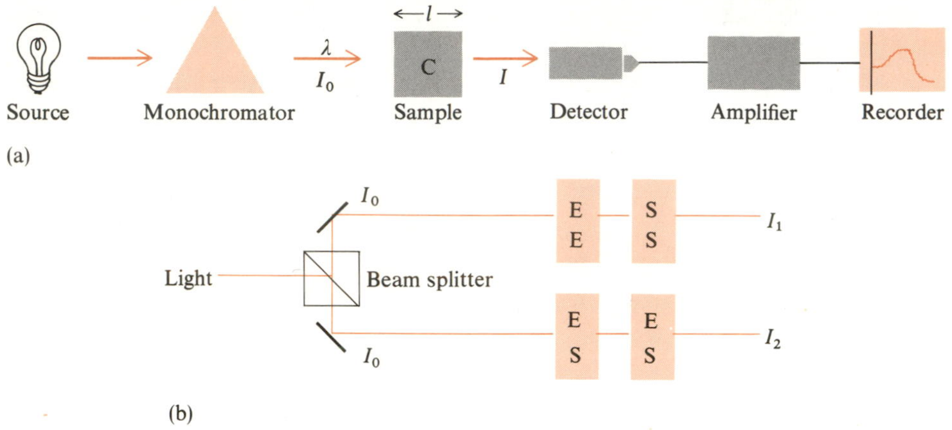

Figure 7-3

Schematic diagrams of spectroscopic experiments. **(a)** For measuring light absorption. **(b)** For difference spectrometry.

Consider a sample of molecules in a layer perpendicular to the direction of light propagation, and sufficiently thin (dl) so that the light intensity within this layer is essentially constant. Then the fraction of light absorbed ($-dI/I$) should be simply proportional to the number of absorbing molecules. The resulting equation is

$$-dI/I = C\varepsilon'\,dl \qquad (7\text{-}32)$$

where ε' is a proportionality constant called the molar extinction coefficient. It is independent of concentration for a set of noninteracting molecules, and it contains the wavelength or frequency dependence of the absorption spectrum. If we integrate this equation over the entire sample (integrating the left-hand side from initial intensity I_0 to final intensity I, and the right-hand side from zero to l), we obtain

$$\ln\,(I_0/I) = C\varepsilon' l \qquad (7\text{-}33)$$

Converting to log base 10, we have the common form of the Beer–Lambert law:

$$A(\lambda) \equiv \log\,(I_0/I) = C\varepsilon(\lambda)l \qquad (7\text{-}34)$$

where $\varepsilon = \varepsilon'/2.303$. The quantity A is called the absorbance or (sometimes) the optical density.

When a spectrum is measured, the quantity of real interest is ε. A common practice is to tabulate the wavelength of maximal extinction (λ_{max}) and the extinction at this wavelength (ε_{max}). The value of ε_{max} for typical single chromophores varies over a wide range from as little as $1\ \text{M}^{-1}\ \text{cm}^{-1}$ to more than $10^5\ \text{M}^{-1}\ \text{cm}^{-1}$ (see Box 7-2).

The most accurate measurements of A usually are obtained in the range of 0.1 to 2. Smaller values mean that only a tiny fraction of the incident light is absorbed; larger values mean that only a small fraction of the incident light reaches the detector. For a chromophore with ε_{max} of $10^4\ \text{M}^{-1}\ \text{cm}^{-1}$ (such as a nucleic acid base), Equation 7-34 means that concentrations of 10^{-5} to 10^{-4} molar should be used for samples with a 1 cm path length.

With macromolecules, it frequently is inconvenient to use *molar* extinction, either because it is a very large number or because the molecular weight is not known. To circumvent these problems, a standard practice is to employ average *extinction per residue*. For example, for a polynucleotide, one would calculate the concentration as the moles of total phosphate per liter. Beware: in the scientific literature, the distinction between molar and residue extinction is often implicit. In most cases, these quantities differ by a factor of at least 10^2, so there is not too much chance of real confusion.

What is of interest is the spectrum of the macromolecule and not that of the solvent. Therefore it is the usual practice to measure the *difference* in absorbance between a macromolecule solution and a solvent blank. This is most convenient if a double-beam spectrophotometer is used that automatically records this difference. In the sample beam, the intensity I_s that emerges is (from Eqn. 7-34)

$$\log I_s = \log I_0 - A_s \tag{7-35}$$

Similarly, in the reference beam, the transmitted intensity I_R is

$$\log I_R = \log I_0 - A_R \tag{7-36}$$

Therefore, an instrument that detects $\log I_R/I_s$ will measure $A_s - A_R$.

A variation on this theme is difference spectroscopy. Suppose one would like to measure precisely a small spectral change induced in a system—say, when a substrate is added to an enzyme. If the enzyme–substrate mixture is put into one beam of a double-beam instrument, and separate solutions of enzyme (E) and substrate (S) are placed in a reference beam, the spectrum recorded will be just the differences induced by interaction of two species to form a complex (ES). Figure 7-3b is a schematic of such an experiment using four sample cells, each with a path length l. One measures

$$\log(I_2/I_1) = A_1 - A_2 = 2[(\varepsilon_E C_E + \varepsilon_S C_S) - (\varepsilon_E C'_E + \varepsilon_S C'_S + \varepsilon_{ES} C'_{ES})]l$$

where the primes refer to concentrations of species in the samples containing both E and S.

Box 7-2 EXTINCTION COEFFICIENTS AND CROSS SECTIONS

Here we shall derive the relationship between molecular size and the extinction coefficient. Consider a slab of solution with an area A (in cm^2) and a thickness dl. If the molar concentration of solute is C, the number of solute molecules per cm^3 is $CN_0/1{,}000$, and the number of solute molecules in the slab is $CAN_0dl/1{,}000$. If each solute molecule has a radius r, the fraction of the cross-sectional area of the slab occupied by solute molecules is

$$f_{max} = (\pi r^2 CAN_0 dl/1{,}000)/Adl = \pi r^2 CN_0/1{,}000$$

If each molecule absorbed all the light that impinged on it, this would be the most efficient possible absorption process. The fraction of the incident light absorbed would be f_{max}, as shown in the figure. Thus there would be a relationship between light absorption and molecular size.

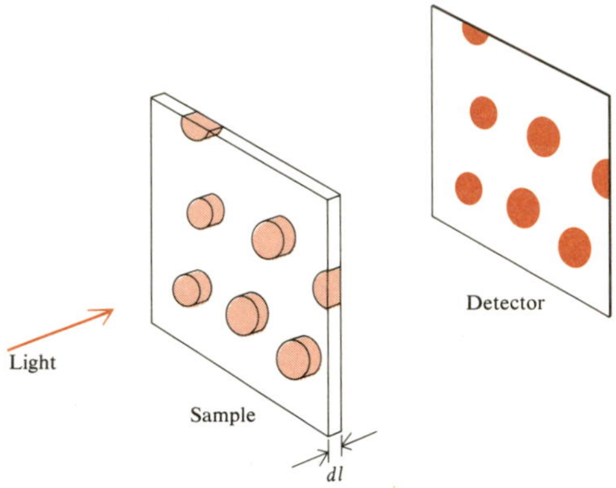

If molecules are less than perfect absorbers, let P be the probability that light impinging on a molecule is absorbed. Then the fraction of incident light absorbed will be $f_{max}P$. The contribution of each individual molecule is

$$\sigma = P\pi r^2$$

This factor σ is called the cross section of the molecule. Using σ, we can write the fraction of light absorbed as $dI/I = (\sigma CN_0/1{,}000)dl$. Comparing the above equations with Equations 7-32, 7-33, and 7-34, it is easy to show that

$$\varepsilon = (\sigma CN_0/1{,}000)/2.303C = \sigma N_0/2{,}303 = \pi r^2 PN_0/2{,}303$$

A typical aromatic ring is a little over 1 Å in radius. Thus we can estimate that the highest possible extinction coefficient for such a compound ($P = 1$) will be about 10^5 M^{-1} cm^{-1}. A more typical value of ε for a strong absorber is 10^4 M^{-1} cm^{-1}, which means that about 10% of the light passing through the chromophore is absorbed. (This discussion adapted from notes provided by Robert Gennis.)

The extinction coefficient in calculating molecular properties

The molar extinction coefficient is defined empirically by Equation 7-32. To relate it to molecular properties, we start from Equation 7-30. In all common samples, the number of excited molecules (N_b) is negligible. Therefore, for a one-molar solution of absorbing molecules, the rate of energy uptake per cm^3 of sample is

$$-dI(\nu)/dt = (h\nu N_0 B_{ab}/1{,}000)I(\nu) \tag{7-37}$$

where ν is the frequency used to excite the sample, and N_0 is Avogadro's number. Because light is traveling at a velocity c, the loss in intensity in a distance dl is

$$dI(\nu) = (1/c)[dI(\nu)/dt]\,dl = (h\nu N_0 B_{ab}/1{,}000c)I(\nu)\,dl \tag{7-38}$$

This allows B_{ab} to be related to ε' by comparing Equations 7-32 and 7-38 for a one-molar concentration. At first glance, it might appear that $B_{ab} = 1{,}000\varepsilon' c/N_0 h\nu$. However, because observed spectra occur over a band of frequencies (or wavelengths), and because B_{ab} was derived from two states separated by only a single frequency (ν), it is necessary to integrate ε' over an entire absorption band to compute B_{ab} (see Birks, 1970, pp. 48–50):

$$B_{ab} = (1{,}000c/N_0 h)\int(\varepsilon'/\nu)\,d\nu \tag{7-39}$$

The integral is performed over frequency, which means that spectra taken as a function of λ must be converted to frequency spectra prior to integration.

The molecular quantity of interest that can be derived from B_{ab} is $|\langle\Psi_b|\underset{\sim}{\mu}|\Psi_a\rangle|^2$. Using Equations 7-29 and 7-39, and converting ε' to ε, we obtain

$$D_{ab} \equiv |\langle\Psi_b|\underset{\sim}{\mu}|\Psi_a\rangle|^2 = 9.180 \times 10^{-3}\int(\varepsilon/\nu)\,d\nu \quad (\text{debye})^2 \tag{7-40}$$

once all the fundamental constants are evaluated. D_{ab} is called the dipole strength; clearly, it can be determined by integrating the area under an absorption band unless the band of interest is poorly resolved from others nearby.

The transition dipole length $|\underset{\sim}{\mu}_{ba}|$ usually is expressed in debyes (10^{-18} esu cm cgs) and, for an intense electronic absorption, it typically has a magnitude of several debyes. Note that from D_{ab} it is possible to compute the length of the transition dipole $|\underset{\sim}{\mu}_{ba}| = |\langle\Psi_b|\underset{\sim}{\mu}|\Psi_a\rangle|$. It is not possible to compute the direction.

Another useful measure is the oscillator strength, f_{ab}, which compares the intensity of absorption to that expected from a three-dimensional harmonic oscillator. This can be shown to be

$$f_{ab} = (8\pi^2 mc/3h\nu)D_{ab} = 4.315 \times 10^{-9}\int\varepsilon(\nu)\,d\nu \quad (\text{dimensionless}) \tag{7-41}$$

where m is the mass of the electron. For a strongly allowed transition, f_{ab} may be in the range of 0.1 to 1. The quantities D_{ab} and f_{ab} are very important in understanding some of the special optical effects observed in polymers.

Linear dichroism from oriented samples

In solution, molecular orientations are randomized. Even if polarized light is employed, still what is measured is

$$|\langle\Psi_b|\underset{\sim}{\boldsymbol{\mu}}|\Psi_a\rangle\cdot\mathbf{E}_0|^2=(1/3)(\mu_x^2+\mu_y^2+\mu_z^2)\mathbf{E}_0^2$$

which is the spatial average of the transition dipole moment. Oriented samples, where absorption depends on $\boldsymbol{\mu}\cdot\mathbf{E}_0$ are required if any information is to be obtained on the direction of $\boldsymbol{\mu}_{ba}$ within a molecule. There are several ways to orient molecules. The macromolecule or small molecule can be observed in a crystal or fiber. Long asymmetric molecules can be oriented by flow or by electric fields (Chapter 12). Molecules can be oriented by stroking a viscous solution with a brush as it dries. Except for use of crystals, none of these techniques produces precise, fully oriented molecular distributions.

Consider a system of ellipsoidal molecules (Fig. 7-4). Flow orientation might produce a pattern like this. The molecules are quite well oriented along their long axis z. Thus the component of $\boldsymbol{\mu}_{ba}$ along this direction (μ_z) can be measured. However, x and y axes are randomly placed. This means that one still cannot measure μ_x and

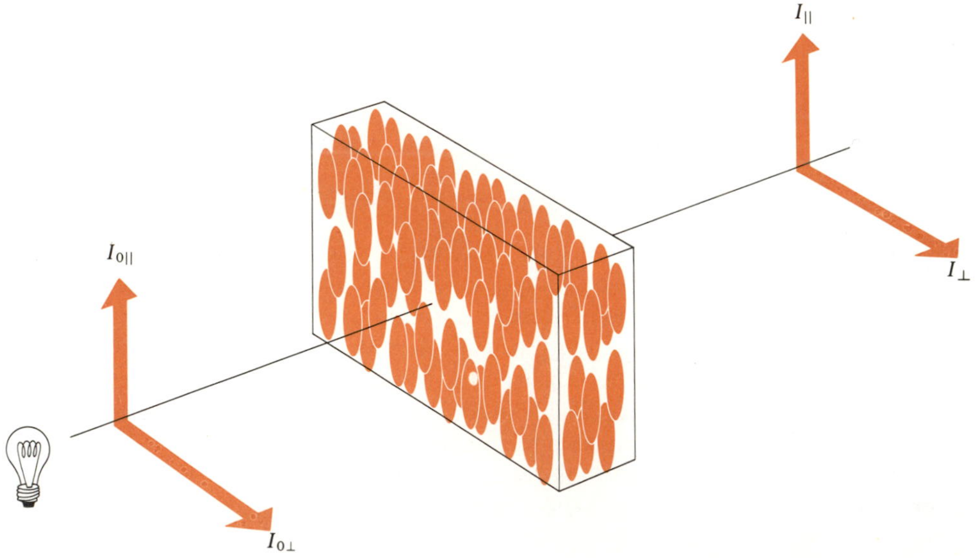

Figure 7-4

Schematic diagram of a linear dichroism experiment. Molecules in the sample all are oriented with their long axes in the direction of $I_{\|}$.

μ_y uniquely. What usually is done is to measure two extinction coefficients ($\varepsilon_{||}$ and $\varepsilon_{\perp}$) for light polarized parallel and perpendicular to the z axis. These values allow one to calculate $D_{||} = \mu_z^2$ and $D_{\perp} = (1/2)(\mu_x^2 + \mu_y^2)$ from the integrated spectra, using Equation 7-40.

The concentrations of oriented molecules may not be known accurately. In such cases, one can measure only the absorbance at each wavelength, not the extinction coefficient. Thus it often is convenient to express the results as a dichroic ratio:

$$d = (A_{||} - A_{\perp})/(A_{||} + A_{\perp}) \tag{7-42}$$

where $A_{||}$ is the absorbance of light polarized parallel to the z axis, and $A_{\perp}$ is the absorbance of light polarized perpendicular to the z axis.

There are three basic uses of dichroism measurements. The simplest is to determine whether a spectral band corresponds to a single electronic transition. That is, are only two electronic states (a and b) involved? If so, the dichroic ratio should be the same regardless of the wavelength used. Suppose instead that the absorption wavelengths of two transitions $a \rightarrow b$ and $a \rightarrow c$ overlap partially. It is highly likely that the transition dipoles $\boldsymbol{\mu}_{ba}$ and $\boldsymbol{\mu}_{ca}$ are not parallel. In this case, the dichroic ratio will change in value as λ is varied, and it may even change sign. (An example is shown in Fig. 7-5.)

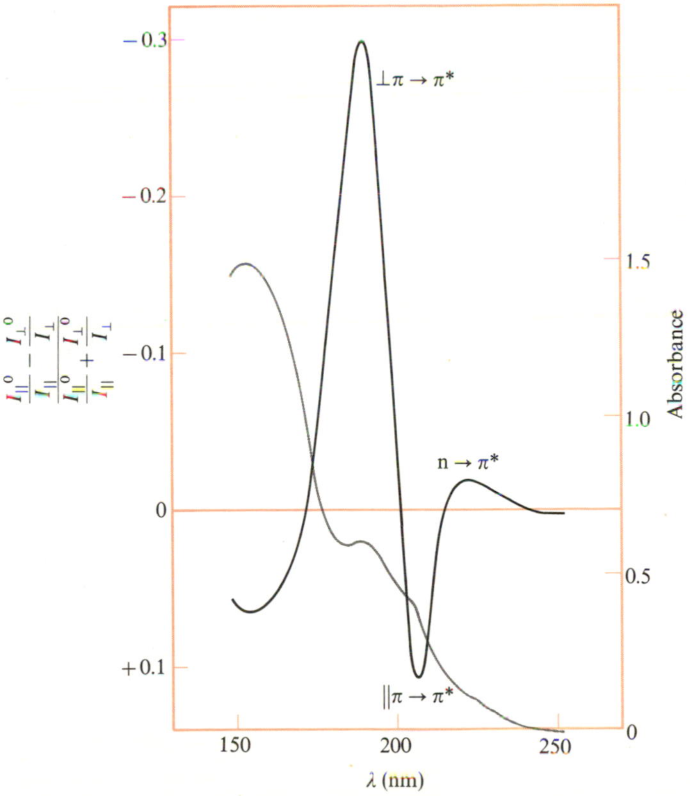

Figure 7-5

Linear dichroism (black curve) *and absorbance* (gray curve) of a poly-L-glutamic acid film. The polypeptide has an α-helical structure. Bands assigned to the n → π^* and to the parallel and perpendicular polarized $\pi \rightarrow \pi^*$ transitions are indicated on the dichroism spectrum. Note that this spectrum is plotted in terms of raw light intensities rather than absorbance. [After J. Brahms, J. Pilet, H. Damang, and V. Chandrasekharan, *Proc. Natl. Acad. Sci. USA* 60:1130 (1968).]

If the transition dipole direction of a chromophore is known, dichroism measurements can be used to learn important aspects of macromolecular structure. For example, the transition dipoles of most intense transitions of planar aromatic molecules lie in the plane of the conjugated rings. In this case, the orientation of these rings within the structure can be assessed. However these findings are only as reliable as the knowledge about the direction and extent of orientation of the sample.

When the structure of a sample is known precisely, as in the case of a thin slice of a crystal, linear dichroism measurements can be used to measure the transition dipole direction. One must be cautious in interpreting such measurements though, because intermolecular interactions between closely packed molecules can change the apparent dipole orientation. This problem can be circumvented, when necessary, by using a doped crystal containing a small amount of the strongly absorbing molecule of interest, cocrystallized in a matrix of a nonabsorbing molecule of similar shape.

Spectral properties of a simple molecule: formaldehyde

We want to consider briefly light absorption by the residues that make up proteins and nucleic acids. Some of these are quite complex and, before treating them, it is worthwhile to examine in more detail the properties of a smaller molecule. The concepts derived here will be generally applicable.

What do the electronic states of a simple molecule look like? In general, to predict this, one must solve the time-independent Schrödinger equation (Eqn. 7-12) for a many-particle system. This is still impractical to sufficient accuracy for systems with more than several heavy atoms. However, a wide variety of approximate calculations have led to some useful generalizations. For example, most of the electrons are fairly localized near one nucleus, and they occupy orbitals not terribly different from those in an isolated atom. Some are more delocalized into orbitals that encompass more than one nucleus. These electrons are frequently the ones involved in spectral transitions at low energies.

The formaldehyde molecule (Fig. 7-6) will be used to illustrate the results of approximate quantum mechanical calculations. First we shall describe the bonding in the ground electronic state of formaldehyde. Then we shall characterize a few of

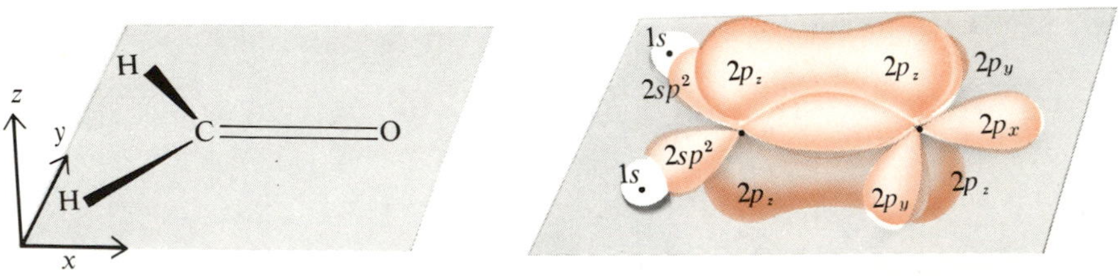

Figure 7-6

The formaldehyde molecule and a schematic diagram of its bonding.

the lowest-lying excited states. Next, we shall estimate the directions and magnitudes of the transition dipoles between the ground states and these excited states. From these results, we shall be able to predict the relative intensities of some of the absorption bands of formaldehyde and, also, the expected linear dichroism.

The four atoms of formaldehyde contain a total of 16 electrons. First, consider the four atoms separately. Each hydrogen has one electron in a $1s$ atomic orbital. This orbital is spherically symmetrical about the nucleus. The six electrons of carbon can be described as $1s^2 2s^2 2p_y 2p_x$, and the eight electrons of oxygen can be described as $1s^2 2s^2 2p_y^2 2p_x 2p_z$. The $2p$ orbitals are cylindrically symmetrical about the nucleus. The subscripts x, y, and z denote the directions of the symmetry axes. The assignment of two oxygen electrons to the $2p_y$ orbital and only one electron to each of the other $2p$ orbitals is arbitrary. Note that each $2p$ orbital has a node (a region with no electron density).

Next, consider the changes in these atomic orbitals when the atoms bond together to make formaldehyde. In order to account for the 120° bond angles in formaldehyde, we must describe the carbon as $1s^2 (2sp^2)^3 2p_z$. Here the $2s$ and two $2p$ atomic orbitals have been combined to make three hybrid $2sp^2$ orbitals that point toward the vertices of an equilateral triangle. The $1s^2$ orbitals of carbon and the $1s^2$ and $2s^2$ orbitals of oxygen are too tightly bound to have much role in bonding. The two C—H bonds are formed by carbon hybrid sp^2 orbitals overlapped with hydrogen $1s$. The third sp^2 orbital forms a single bond with the oxygen $2p_x$ orbital. This is called a σ bond because the electron distribution has no node. What we have left are one $2p_z$ orbital on carbon and the $2p_z$ and $2p_y$ oxygen orbitals. A π molecular orbital can be formed by overlap of the singly occupied $2p_z$ orbitals on oxygen and carbon. This π orbital is occupied by two electrons, forming a π bond (a bond with a single node). The remaining two oxygen electrons then must lie in a $2p_y$ atomic orbital. Therefore, the ground state of H_2CO has a double bond between carbon and oxygen. The highest-energy occupied orbitals are the π bonding orbital and the nonbonding (n) $2p_y$ orbital located on oxygen.

In describing the ground state of H_2CO, we formed a molecular orbital by taking a linear combination of carbon and oxygen $2p_z$ orbitals. The wavefunction for this orbital is $\Psi_\pi = 2p_z$ (carbon) $+ 2p_z$ (oxygen). A second linear combination also is possible: $\Psi_{\pi^*} = 2p_z$ (carbon) $- 2p_z$ (oxygen). This π^* orbital will have higher energy than the π orbital, and it will have an extra node. The energy-level scheme we have predicted is the following:

$$
\begin{array}{ccccc}
\pi^* \ \underline{\downarrow\ \ } & & \pi^* \ \underline{\ \ \ } & & \pi^* \ \underline{\downarrow\ \ } \\[4pt]
n \ \underline{\uparrow\downarrow} & \xleftarrow{\ \pi\text{–}\pi^*\ } & n \ \underline{\uparrow\downarrow} & \xrightarrow{\ n\text{–}\pi^*\ } & n \ \underline{\uparrow\ \ } \\[4pt]
\pi \ \underline{\uparrow\ \ } & & \pi \ \underline{\uparrow\downarrow} & & \pi \ \underline{\uparrow\downarrow} \\[6pt]
\text{excited state} & & \text{ground state} & & \text{excited state}
\end{array}
$$

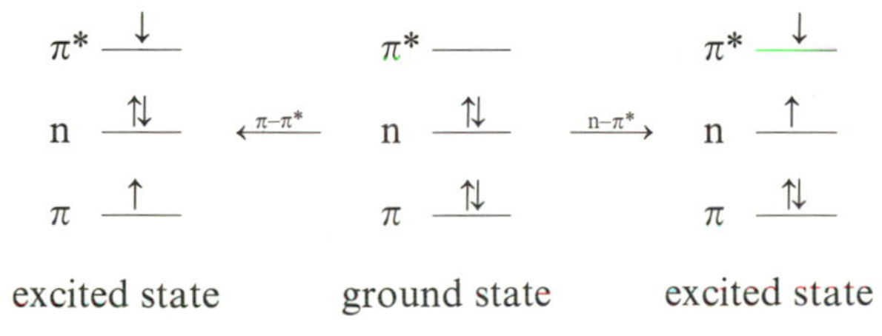

The two lowest-energy electronic transitions one could expect to see result from the promotion of an electron from the n orbital to the π^* orbital (called an $n \rightarrow \pi^*$ transition) and the promotion of an electron from the π orbital to the π^* orbital (called a $\pi \rightarrow \pi^*$ transition).

To estimate the absorption intensities associated with these transitions, we must evaluate the transition dipole moment $\langle i|\boldsymbol{\mu}|f\rangle$, where i is the initial state, and f is the final state. Because all of the electrons but one remain unchanged, $\langle i|\boldsymbol{\mu}|f\rangle$ need be evaluated only for the two orbitals directly involved in the transition. We shall adopt the simple notation in which the wavefunctions for n, π, and π^* orbitals are symbolized by the names of the orbitals. Then the integrals that must be evaluated are $\langle n|\boldsymbol{\mu}|\pi^*\rangle$ and $\langle \pi|\boldsymbol{\mu}|\pi^*\rangle$. It is easiest to evaluate these expressions by the use of symmetry arguments.

The integral represented by the angle brackets is taken over all space. It cannot depend on the choice of coordinate system. If the integrand is odd (changes sign) when reflected through any of the three planes defined by the Cartesian axes, the integral will vanish. Figure 7-7 shows the shapes of the highest occupied and lowest empty orbitals of formaldehyde. First, we use symmetry arguments to examine the properties of these wavefunctions. Consider the integral $\langle \pi|\pi^*\rangle$. As shown in Figure 7-7, π is even when reflected through the yz plane $[\pi(x) = \pi(-x)]$, whereas π^* is odd $[\pi^*(x) = -\pi^*(-x)]$. Thus their product $(\pi\pi^*)$ is odd. Any contribution to the integral of $\pi\pi^*(x)$ is exactly canceled by the contribution at $\pi\pi^*(-x)$, which is equal to $-\pi\pi^*(x)$. Thus, $\langle \pi|\pi^*\rangle = 0$; the two wavefunctions are orthogonal. You also can easily show that $\langle \pi|n\rangle = \langle \pi^*|n\rangle = 0$.

To evaluate $\langle i|\boldsymbol{\mu}|f\rangle$, it is convenient to represent it as the sum of three components: $\boldsymbol{\mu} = \hat{\mathbf{i}}\boldsymbol{\mu}_x + \hat{\mathbf{j}}\boldsymbol{\mu}_y + \hat{\mathbf{k}}\boldsymbol{\mu}_z$ and to consider the three Cartesian components, ($\boldsymbol{\mu}_x$, $\boldsymbol{\mu}_y$, and $\boldsymbol{\mu}_z$) separately. Their symmetry is shown in Figure 7-7. For a $\pi \rightarrow \pi^*$ transition, we evaluate the components of $\langle \pi|\boldsymbol{\mu}|\pi^*\rangle$:

1. $\langle \pi|\boldsymbol{\mu}_x|\pi^*\rangle$ is not zero—the product of the symmetries of π, π^*, and μ_x is even, whether we reflect through the xy, yz, or xz planes;
2. $\langle \pi|\boldsymbol{\mu}_y|\pi^*\rangle$ is zero, because the net symmetry is odd when reflected through either the xz or the yz plane;
3. $\langle \pi|\boldsymbol{\mu}_z|\pi^*\rangle$ also is zero, because it has odd symmetry when reflected through either the xy or yz plane.

The result is that $\langle \pi|\boldsymbol{\mu}|\pi^*\rangle = \langle \pi|\boldsymbol{\mu}_x|\pi^*\rangle$. The $\pi \rightarrow \pi^*$ transition is called an allowed transition because its transition dipole is nonzero. Intense absorption can occur, but only when the electric field vector of the light is parallel to the x axis of the molecule. This is described by saying that the transition is polarized along the C=O bond.

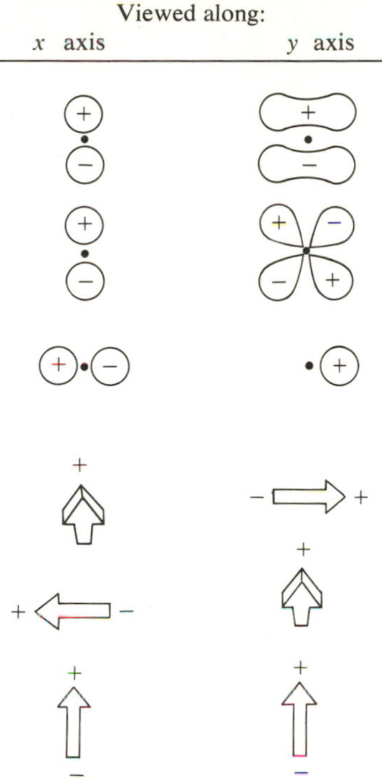

Function:	Viewed along:		Symmetry reflected through:		
	x axis	y axis	xy	xz	yz
Molecular orbitals					
π			Odd	Even	Even
π^*			Odd	Even	Odd
n			Even	Odd	None
Dipole operators					
$\underset{\sim}{\mu}_x$			Even	Even	Odd
$\underset{\sim}{\mu}_y$			Even	Odd	Even
$\underset{\sim}{\mu}_z$			Odd	Even	Even

Figure 7-7

The symmetry of molecular orbitals and dipole operators of formaldehyde.

For an n → π^* transition, we must evaluate the components of $\langle n|\underset{\sim}{\mu}|\pi^*\rangle$:

1. $\langle n|\underset{\sim}{\mu}_x|\pi^*\rangle$ is zero, because it has odd symmetry when reflected in either the xy or the xz plane;

2. $\langle n|\underset{\sim}{\mu}_y|\pi^*\rangle$ is zero, because it is odd in the xy plane;

3. $\langle n|\underset{\sim}{\mu}_z|\pi^*\rangle$ is zero, because it is odd in the xz plane.

Therefore, $\langle n|\underset{\sim}{\mu}|\pi^*\rangle = 0$, and the n → π^* transition is called symmetry-forbidden. However, this does not mean that it is unobservable. The treatment just given is oversimplified because it ignores vibrational motion and uses approximate wavefunctions. In practice, an n → π^* transition can lead to an absorption band, but its intensity will be very weak, typically less than 1% of the intensity of a typical $\pi \rightarrow \pi^*$ transition. Figure 7-8 illustrates such an experimental result for a simple carbonyl compound.

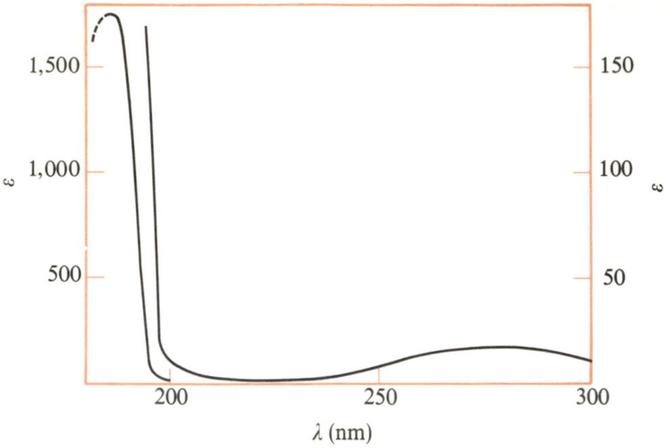

Figure 7-8

An absorption spectrum of acetone (CH_3—CO—CH_3) in *n*-hexane. There is an n → π* transition at 280 nm and a π → π* transition at 180 nm. Note that an expanded scale is needed to show the n → π* transition. The spectrum of formaldehyde (H—CO—H) should be very similar to that of acetone. Unfortunately, it is difficult to obtain a formaldehyde spectrum in solution. If hydroxylic solvents are used, hydrates or hemiacetals result. If less polar solvents are used, formaldehyde trimerizes to a six-membered ring structure with alternating O and CH_2. [Data from B. Wells, unpublished.]

7-3 SPECTROSCOPIC ANALYSIS OF BIOPOLYMERS

Restricted wavelength range of biopolymers

Any molecule absorbs light in some wavelength range. However, for any selected wavelength, certain types of chemical groups usually dominate the observed spectrum. These groups are called chromophores. Typical chromophores found in proteins and nucleic acids absorb light only at wavelengths less than 300 nm. These groups, such as peptides or nucleic acid bases, are much more complex than formaldehyde. A detailed discussion of their electronic structures is beyond the scope of a book at this level. Thus, all we will give is a brief summary of some of their particular properties and the ways these either simplify or complicate optical studies on proteins and nucleic acids.

One of the major constraints in biopolymer spectroscopy is the need to work in a solvent. For such large molecules, gas-phase spectroscopy clearly is out of the question. Spectra of solids are complicated by dichroic effects, and also by the difficulty of correlating the properties of a molecule in a solid film to those in a solution, where biological assays usually are carried out. For the overwhelming majority of proteins and nucleic acids, the most useful and biologically significant solvent is water, buffered at a pH near 7.0 and containing sufficient electrolyte (~0.15 M NaCl, for example) to mimic conditions in vivo. The use of water as a solvent immediately restricts absorption spectral measurements to wavelengths longer than 170 nm. Below this, the absorbance of even micron-thin films of water is too great to permit any increment due to macromolecules to be measured with accuracy.[§] Because water is strongly polar, electronic absorption bands in water tend to be broader than in most other solvents. The energies of individual molecules differ from one another because of strong interactions with solvent molecules at various orientations and distances. This minimizes the chance of observing vibronic substructures of electronic bonds. A further constraint is the narrow temperature range over which water is a liquid. For proteins or intact nucleic acids, this is not much of a problem—but, for many model compounds, the inability to make measurements below 0°C or above 100°C is sometimes quite frustrating.

Peptide-group domination of far UV absorption for proteins

Protein chromophores can be conveniently divided into three classes: the peptide bond itself; amino acid side chains, and any prosthetic groups. The properties of the isolated peptide chromophore can be studied conveniently in model compounds such as formamide or *N*-methylacetamide.

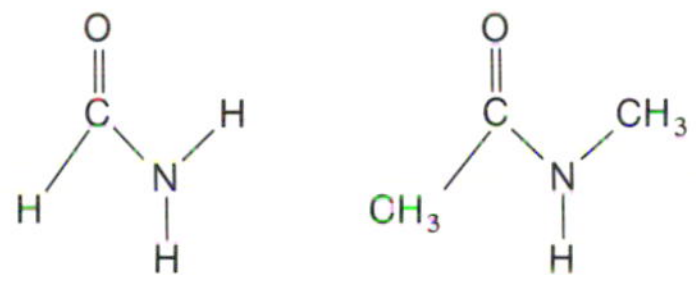

Some aspects of the electronic structure of the peptide group were discussed in Chapter 5. The π electrons of the peptide group are delocalized to some extent over three atoms: the peptide nitrogen, carbon, and oxygen. The lowest-energy electronic transition observable for a peptide is an $n \rightarrow \pi^*$ transition. As in formaldehyde, the n electron is essentially localized on the oxygen atom, and the transition is symmetry-forbidden.

The peptide $n \rightarrow \pi^*$ absorption band is typically observed at 210–220 nm, with very weak intensity ($\varepsilon_{max} \approx 100$). For example, the absorption spectrum of poly-L-lysine in various structural forms is shown in Figure 7-9. The $n \rightarrow \pi^*$ transition appears in the α-helical polymer as a small shoulder near 220 nm on the tail of a much stronger absorption band centered at 190 nm. This intense band, the main observable peptide absorbance at easily accessible wavelengths, is a $\pi \rightarrow \pi^*$ transition ($\varepsilon_{max} \approx 7000$). In formaldehyde, this lowest $\pi \rightarrow \pi^*$ transition is polarized along the C–O axis. In a peptide, because the nitrogen atom participates in the π and π^* orbitals, the $\pi \rightarrow \pi^*$ transition dipole does not lie along any particular bond. In myristamide, it lies in

[§] To penetrate below 200 nm, care must be taken to choose salts and buffers that are transparent at these wavelengths. Perchlorates and fluorides usually are most satisfactory.

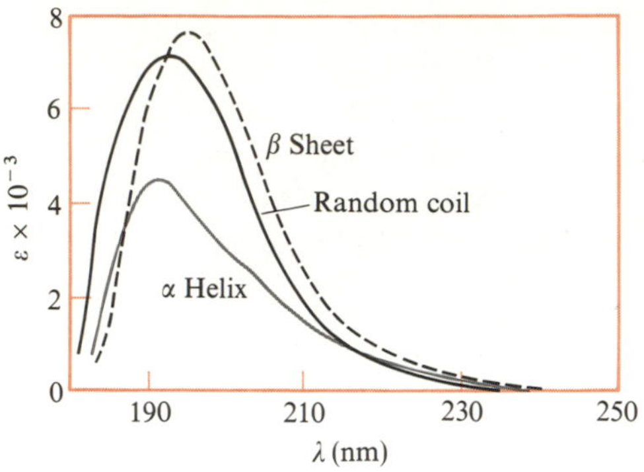

Figure 7-9
Ultraviolet absorption spectrum of poly-L-lysine in aqueous solution: random coil, pH 6.0, 25°C; α helix, pH 10.8, 25°C; β sheet pH 10.8, 52°C. [After K. Rosenheck and P. Doty, *Proc. Natl. Acad. Sci. USA* 47:1775 (1961).]

the plane of the peptide, near a line between the oxygen and nitrogen:

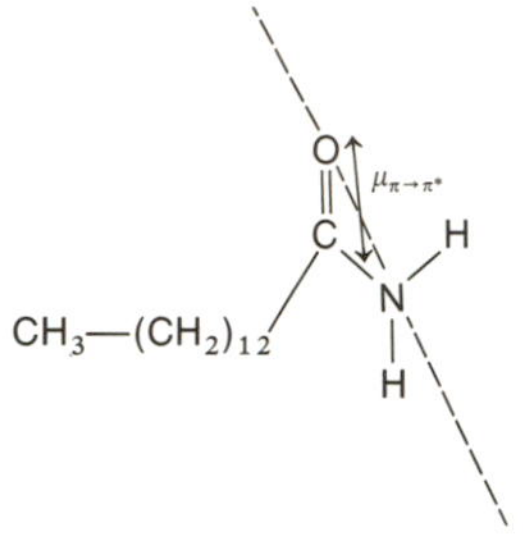

At still higher energies, a third peptide transition can be observed at 175 nm. This is tentatively assigned to an n → σ^* transition.

Aromatic amino acid domination of near UV absorption for proteins

A number of amino acid side chains—including Asp, Glu, Asn, Gln, Arg, and His—have electronic transitions in the same spectral region where strong peptide absorption occurs. It is nearly impossible to detect these in an actual protein or polypeptide, because they are not as strong as the peptide $\pi \rightarrow \pi^*$ band and because the number of these side chains is usually less than the number of peptide residues. Thus, the most useful side-chain optical properties are those that occur at wavelengths longer than 230 nm, where the peptide absorption is reduced to negligible values. Between 230 nm and 300 nm, in the near ultraviolet, one must consider the effects of the aromatic amino acids—Phe, Tyr, and Trp—and two other side chains, histidine and disulfides (cystine).

The absorption spectra of the three aromatic amino acids at neutral pH are shown in Figure 7-10. A log scale is used because the intensities are very different. By far the most intense is tryptophan. However, this amino acid is often not present

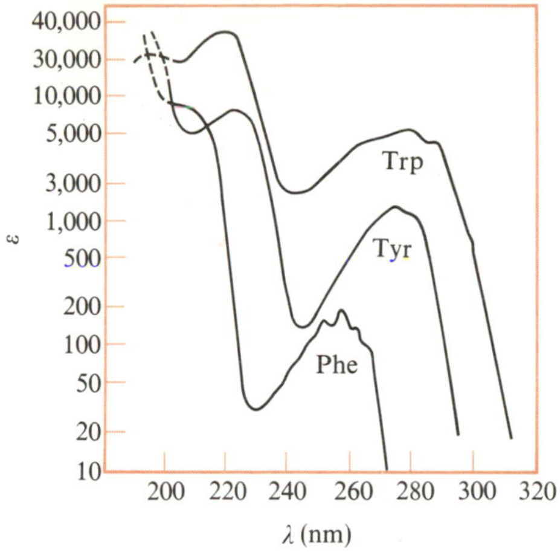

Figure 7-10

Absorption spectra of the three aromatic amino acids. A log scale has been used in order to display all three conveniently on one graph. [After D. B. Wetlaufer, *Adv. Protein Chem.* 17:303 (1962).]

in large amounts in common proteins. Therefore, the tyrosine absorbance also makes a very significant contribution. The maximum extinction of phenylalanine is much weaker, only about 200 M^{-1} cm^{-1} at 250 nm. As a result, it is nearly impossible to observe any optical contribution from phenylalanine in a protein if the other aromatic amino acids are present. The phenylalanine absorbance at 250 nm derives from a symmetry-forbidden $\pi \rightarrow \pi^*$ transition, analogous to the 256 nm band of benzene ($\varepsilon_{max} \approx 400$). In tyrosine, the local symmetry is much lower, and a much stronger transition ($\varepsilon_{max} \approx 1400$) is observed at 274 nm, in analogy to the 271 nm absorption of phenol ($\varepsilon_{max} \approx 2000$). The absorption spectrum of the indole side chain of tryptophan is more complex, and even the narrow wavelength region 240 nm to 290 nm consists of three or more electronic transitions. (See Weinryb and Steiner, 1971, for a more detailed discussion.)

Variations in pH have little effect on the absorption spectrum of an isolated peptide chromophore. In contrast, major effects are seen with tyrosine and tryptophan, because the sites of protonation directly affect the conjugated electronic system of the chromophore. Most dramatic is the spectral shift of tyrosine when the OH proton is removed ($pK_a \approx 10.9$) (See Fig. 7-11). This spectral shift can be measured very sensitively simply by monitoring the absorbance at 295 nm. Difference spectra measured at this wavelength can be used to follow the titration of tyrosines in a protein with considerable precision. The effects of pH on tryptophan absorbance are smaller and have not been exploited as much. Also, the pK_a values of tryptophan lie outside the pH range in which most proteins can be handled safely.

The cysteine side chain does not have an easily measurable absorption band in proteins because the largest-wavelength strong transition ($\lambda_{max} \approx 230$ nm) is submerged in the peptide region. Disulfides (cystine) have longer-wavelength transitions, with λ_{max} values around 250 and 270. These are too weak ($\varepsilon \approx 300$) to be of much use in optical absorption studies. However, they do contribute significantly to protein optical activity.

(a)

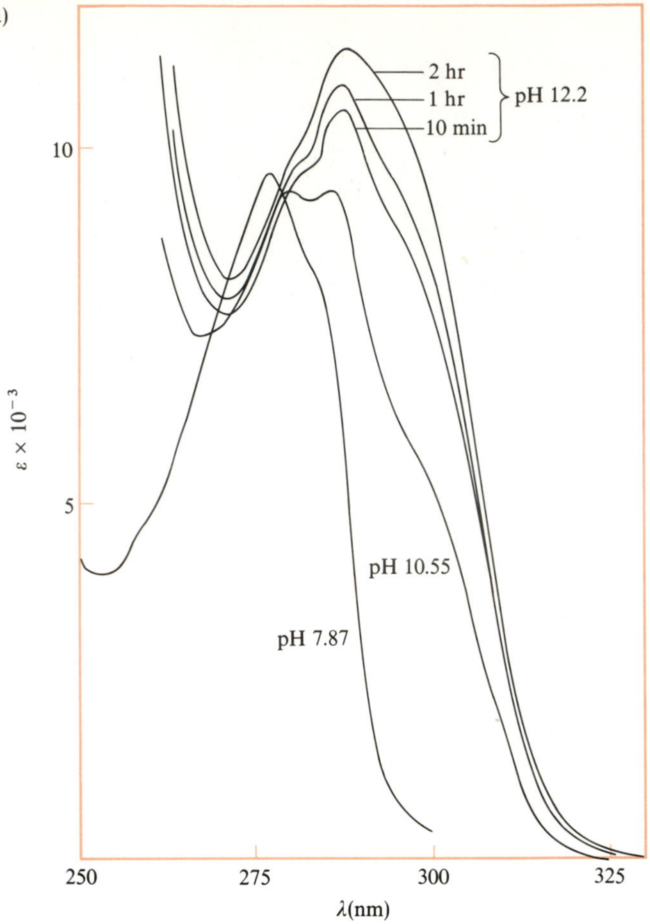
2 hr
1 hr
10 min
pH 12.2
10
5
ε × 10⁻³
pH 10.55
pH 7.87
250
275
300
325
λ(nm)

(b)

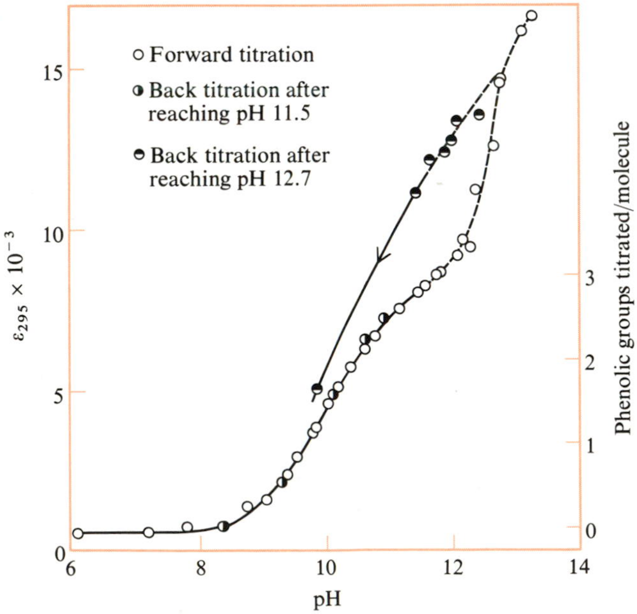
Forward titration
Back titration after reaching pH 11.5
Back titration after reaching pH 12.7
15
10
5
0
$\varepsilon_{295} \times 10^{-3}$
Phenolic groups titrated/molecule
3
2
1
0
6
8
10
12
14
pH

Figure 7-11

Spectrophotometric titration of bovine pancreatic RNase. **(a)** The ultraviolet spectrum as a function of pH and (at pH 12.2) as a function of time. There is no time dependence at the lower pH values. **(b)** Extinction coefficient at 295 nm as a function of increasing pH. Where solid lines are drawn, results are independent of time. RNase has six tyrosines. These results show that roughly three of the tyrosines titrate reversibly with a pK_a near 10.2, a normal value. The other three do not titrate until higher pH values, and this titration is accompanied by denaturation of the protein that causes the hysteresis seen when back titration is attempted. [After C. Tanford, J. D. Hauenstein, and D. G. Rands, *J. Am. Chem. Soc.* 77:6409 (1955).]

Because some of these side chains also make contributions to the absorption in the peptide region, their presence is not always an asset in spectroscopic studies of proteins. If a protein has only a few strongly absorbing side chains, an unequivocal assignment and analysis of the peptide contribution usually is feasible. With large numbers of aromatic side chains, it is still to difficult in most cases to resolve the region near 200 nm into separate contributions from the side chains and the peptides.

Effects of prosthetic groups

A protein that contains prosthetic groups creates the same sort of problem that arose from the presence of aromatic amino acids. For example, hemes, flavins, pyridoxal phosphate, and some metal–protein complexes have intense absorption bands in the near-UV or visible range. These bands usually are sensitive to the local environment and can be used with great success to monitor, for example, the state of oxidation or oxygenation. For many physical studies of enzyme action, these conveniently located chromophores are almost a necessity.

However, prosthetic groups have a built-in limitation. Almost all chromophores that absorb in the near UV or visible range have equally intense transitions between 200 and 300 nm. If these chromophores are present in a protein in sufficient quantity to be useful for direct study, they can severely interfere with the study of the optical properties of the polypeptide moiety. One solution is to prepare an apoprotein by removal of the prosthetic group. However, then the investigator must show that this conversion has not seriously altered the structure of the polypeptide chain. It is rare that direct optical measurements can suffice for this. If the sum of the spectra of the isolated prosthetic group and the apoprotein were precisely equal to the spectrum of the native holoprotein, one could be somewhat confident. However, it is virtually always true that such additivity is lacking. Even in the absence of conformational changes, electronic interactions perturb the spectra of both the prosthetic group and the protein.

To be useful, a prosthetic group must have a high enough molar extinction coefficient to be detectable at typical protein concentrations. It is usually desirable to work with protein concentrations of less than the order of 2.5 mg ml^{-1} to avoid the formation of intermolecular aggregates. A prosthetic group with ε_{max} of 10^3 to

10^4 provides more than ample experimental sensitivity over a wide range of concentrations. Table 7-2 is a brief summary of the spectral properties of some proteins that contain typical prosthetic groups.

Table 7-2
Spectroscopic properties of proteins containing prosthetic groups

		Longest-wavelength absorption band		Second-longest absorption band	
Protein	Prosthetic group	λ_{max} (nm)	ε_{max} ($\times 10^{-4}$)	λ_{max} (nm)	ε_{max} ($\times 10^{-4}$)
Amino acid oxidase, rat kidney	FMN	455	1.27	358	1.07
Azurin, *P. fluorescens*	Cu^{II}	781	0.32	625	0.35
Ceruloplasmin, human	8 Coppers (3 distinct classes)	794	2.2	610	1.13
Cytochrome *c*, reduced, human	Fe^{II}-heme	550	2.77	—	—
Ferredoxin, *Scenedesmus*	(2 Fe^{III}, 2 sulfide) cluster	421	0.98	330	1.33
Flavodoxin, *C. pasteurianium*	FMN	443	0.91	372	0.79
Monoamine oxidase, bovine kidney	Flavins plus Cu	455	4.7	—	—
Pyruvic dehydrogenase, *E. coli*	FAD	460	1.27	438	1.46
Rhodopsin, bovine	Retinal-Lys	498	4.2	350	1.1
Reubredoxin, *M. aerogenes*	(Fe^{III}, 4 Cys) tetrahedron	570	0.35	490	0.76
Threonine deaminase, *E. coli*	4 Pyridoxal phosphates	415	2.6	—	—
Xanthine oxidase	Fe, Mo	550	2.2	—	—

Estimates of protein concentration from UV absorbance

An average protein with no prosthetic groups has a λ_{max} in the near UV, at about 280 nm. Tryptophan and tyrosine have extinction coefficients at this wavelength of about 5,700 and 1,300 $\text{M}^{-1}\ \text{cm}^{-1}$, respectively. As a rough approximation, one can assume that the extinctions of these amino acids do not change drastically when they are incorporated into a protein. Because no other residues make an appreciable contribution to the absorbance at 280 nm, the extinction coefficient of a protein can be estimated if the number of tyrosines (n_{Tyr}) and tryptophans (n_{Trp}) are known. The absorbance of a 1 cm path length in a 1 mg ml^{-1} solution is given by $(5{,}700\ n_{Trp} + 1{,}300\ n_{Tyr})/M$, where M is the molecular weight. When 14 proteins were chosen at random and absorbances were computed, the result was 1.1 ± 0.5. Thus, to within a factor of two, protein weight concentrations can be determined by measuring the absorbance at 280 nm and assuming that a 1 mg ml^{-1} solution of any protein will have an absorbance of 1.0. The method is nondestructive, but it is about an order of magnitude less sensitive than the standard colorimetric Lowry test. Greater sensitivity in straight absorption measurements can be obtained by working at 230 nm. The molar extinction of each peptide residue in this range is about 300 (an average value including side-chain contributions and spillover from the peptide $\pi \rightarrow \pi^*$ transition), so a 1 mg ml^{-1} protein solution will have $A_{230} = 3$. However, many common solvents also absorb at this short wavelength, severely complicating concentration measurements.

Nucleic acid absorption dominated by bases

The strong near-UV absorption of all nucleic acids resides almost exclusively in the purine and pyrimidine bases. The sugar phosphate backbone of RNA or DNA has an insignificant contribution to the ultraviolet absorption spectrum at wavelengths greater than 200 nm. In spite of this, the free bases are imperfect models for the chromophores in nucleic acids. The absence of the glycosidic C–N bond introduces a significant perturbation of the electronic states. For example, the proton on N-9 on a free purine can tautomerize to the N-7 position. N^9-methyl purines and N^1-methyl pyrimidines are more appropriate reference compounds. These compounds have optical absorption spectra very similar to those of nucleosides.

When the spectra of the four nucleosides in water at pH 7 are examined, they seem deceptively simple at first (Fig. 7-12). However, the electronic states of purines and pyrimidines are much more complex than those of the chromophoric groups of proteins. The bases have very low symmetry and many nonbonded electrons. By making careful correlations with spectral bands in a series of increasingly simple compounds, it is possible to show that several different $\pi \rightarrow \pi^*$ and $n \rightarrow \pi^*$ transitions are expected to occur for each base in the region of the spectrum between 200 and 300 nm. The apparently simple Gaussian bands of A and U near 260 nm are really composites of more than one electronic transition. G and C show clear evidence of at least two bands, although more are probably hidden.

The transition dipoles associated with various electronic transitions of the nucleic acid bases lie in the planes of the aromatic rings. Directions measured or calculated for some of these are shown in Figure 7-13. It is evident that the near-UV absorption of each base must be represented by several different transition dipoles. This severely complicates attempts to analyze the effect of conformation on the spectra of nucleic acids.

The spectra of all four nucleosides are sensitive to pH. Protonation of C and G results in large absorbance shifts to longer wavelengths (red shifts). Deprotonation of U or T at alkaline pH also results in a large red shift in the absorption maximum. Protonation of A is accompanied by a much smaller, but still detectable, spectral change. These effects are all highly useful monitors of the extent of ionization of nucleic acid constituents. However, in polymers these spectral effects are complicated

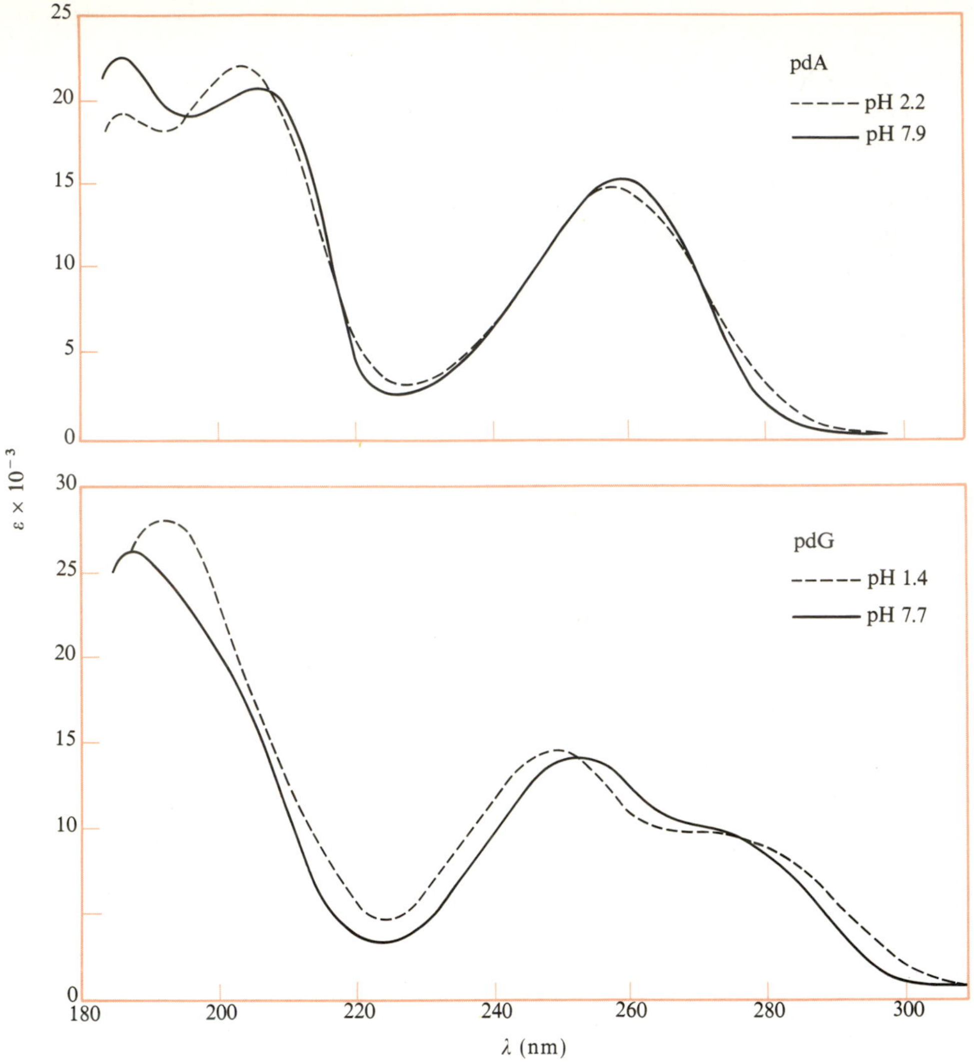

Figure 7-12

Absorption spectra of four deoxynucleotides as a function of pH. Spectra of the corresponding ribo compounds are extremely similar, except that uridine (U) has a near-UV maximum at 260 nm rather than the 268 nm maximum of dT. [After D. Voet, W. B. Gratzer, R. A. Cox, and P. Doty, *Biopolymers* 1:193 (1963).]

Figure 7-13 (facing page, bottom)

Calculated and observed electric transition dipole moments of *n*-alkyl nucleic acid bases. [After V. Bloomfield, D. Crothers, and I. Tinoco, Jr., *Physical Chemistry of Nucleic Acids* (New York: Harper & Row, 1975), p. 50.]

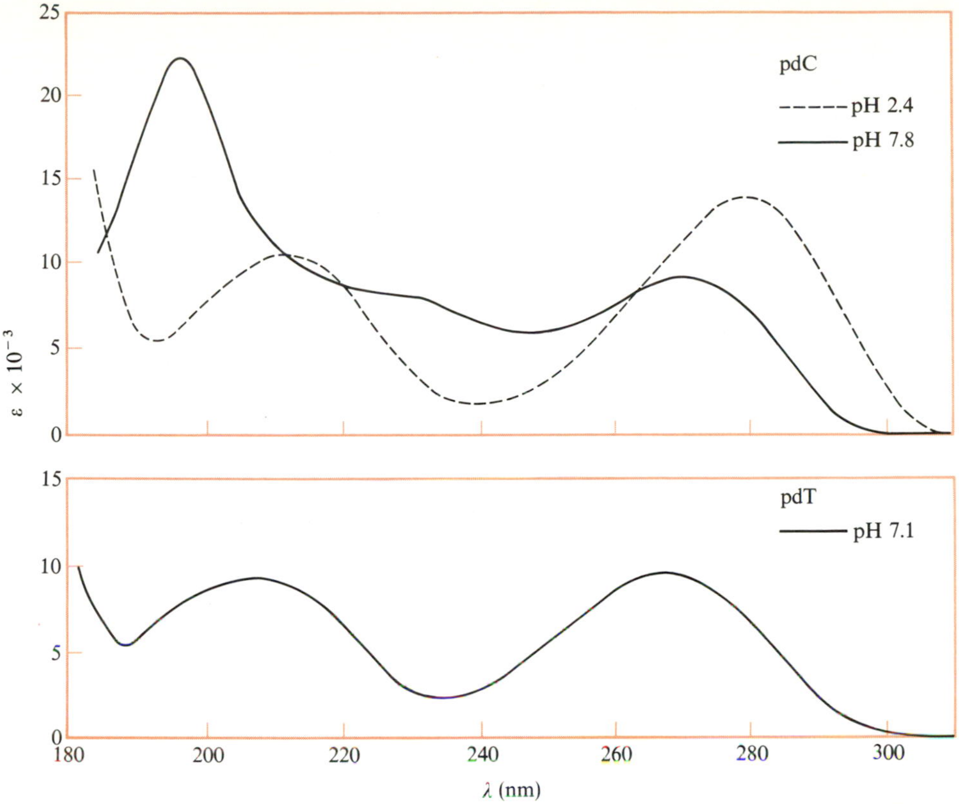
25
20
15
10
5
0
ε × 10⁻³
pdC
pH 2.4
pH 7.8
15
10
5
0
pdT
pH 7.1
180
200
220
240
260
280
300
λ (nm)

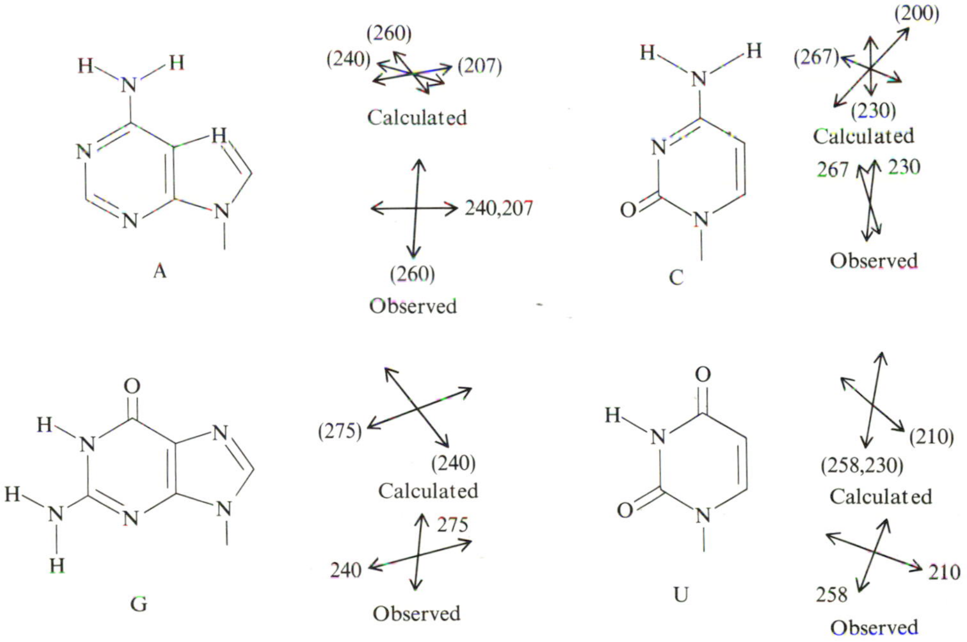
A
(260)
(240)
(207)
Calculated
240,207
(260)
Observed
C
(200)
(267)
(230)
Calculated
267
230
Observed
G
(275)
(240)
Calculated
275
240
Observed
U
(210)
(258,230)
Calculated
210
258
Observed

by the strong electronic interactions between nucleic acid bases. This topic will be discussed later in this chapter. The presence of phosphates has virtually no detectable influence on the molar extinction of nucleic acid constituents. For example, the λ_{max} and ε_{max} of ATP, ADP, AMP, adenosine, and 3′- and 2′-adenylic acid are, within experimental error, the same.

The energies of the longest-wavelength strong electronic transitions of the five common bases are nearly identical. This is a serious impediment to detailed analysis of the electronic spectrum of a DNA or RNA. In a typical nucleic acid, the properties of the isolated chromophores merge into a deceptively smooth single band with λ_{max} at about 260 nm (Fig. 7-14). The properties of the individual bases are sufficiently similar that, even in the absence of any strong base–base interactions, it is difficult to resolve the 260 nm band of a nucleic acid into separate weighted contributions from the four bases.

The average molar extinction coefficient of a nucleoside at 260 nm is about 1.0×10^4. Hence, the ultraviolet spectrum of a nucleic acid can be measured quite accurately at concentrations as low as 3 μg ml^{-1}. This considerable sensitivity has lead to widespread use of UV absorption for many analytical procedures. A few of the unusual bases found in tRNA have long-wavelength transitions shifted considerably to the red of the normal nucleic acid bases. These include the Y base (λ_{max} =

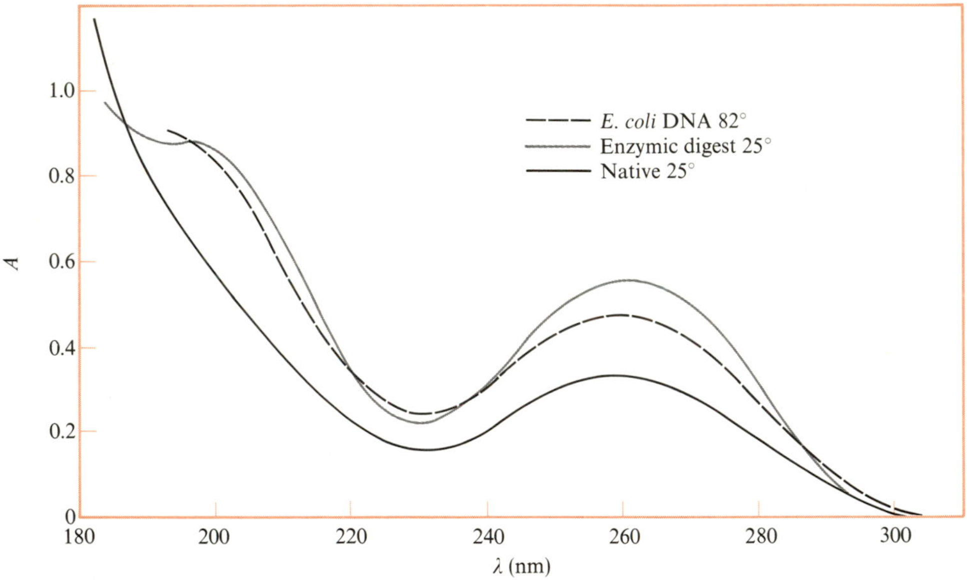

Figure 7-14

Spectrum of DNA as a function of temperature. The enzymic digest should result in a solution of only mononucleotides and possibly a few short oligomers. Note that the high-temperature spectrum is fairly close to the digested spectrum. [After D. Voet, W. B. Gratzer, R. A. Cox, and P. Doty, *Biopolymers* 1:193 (1963).]

325 nm) and 4-thiouridine ($\lambda_{max} = 340$ nm). Because the normal bases have effectively no absorbance at these wavelengths, these unusual bases can be monitored independently of the rest of the molecule by examining the spectrum at wavelengths above 300 nm. Nature has been less than completely generous, however, because the extinction coefficients of these bases are not high, and their frequency of occurrence (when they are present at all) is usually 1 per 80 bases in tRNA. Thus, much higher sample concentrations must be used to make accurate measurements above 300 nm.

Interactions between proteins and nucleic acids are of great importance in biology. Naturally, it is of considerable interest to study the physical aspects of these interactions, and absorption spectroscopy could, in principle, be of great use. However, the properties discussed above illustrate the difficulty of simultaneously studying both protein and nucleic acid components of a mixture. Consider the interaction of a 25,000 mol wt tRNA (molar extinction about 560,000 at 260 nm) with a 100,000 mol wt aminoacyl tRNA synthetase (molar extinction about 100,000 at 280 nm). Even though the synthetase has four times the mass of the tRNA, the absorbance of the latter pretty well dominates the near-UV spectrum. Whereas spectral ratios such as A_{280}/A_{260} can yield an estimate of the amounts of protein and nucleic acid present, a more detailed analysis of the spectrum to yield properties of the separate components is very hard because the protein bands are so badly overlapped by stronger nucleic acid transitions.

7-4 EFFECTS OF CONFORMATION ON ABSORPTION

How do the individual chromophores of a protein or nucleic acid know they are part of a polymer? How are they affected by the three-dimensional arrangement of other residues of the polymer around them? It is easiest to consider a polymer with a regular structure, such as a helix, where all residues have similar environments. Let's compare the measured absorbance of double-helical DNA with the spectrum expected from a mixture of monomer components. The latter spectrum could be calculated as $Cl \sum_i \chi_i \varepsilon_i$, a weighted sum of the monomer spectra, where χ_i is the mole fraction of residue of type i in the polymer; C is the total concentration of monomer residues; and ε_i is the known molar extinction of pure monomer of type i. However, it is easier to enzymatically degrade a DNA to a mixture of monomers. The spectrum of such a mixture is compared with native double helical DNA B in Figure 7-14.

Whereas the two spectra are qualitatively similar, it is apparent that there are substantial quantitative differences. These differences could have arisen from the conformation of the polymer or just from the presence of the covalent linkages between the monomers. To distinguish, one can disrupt the double helical structure by heating to 90°C. The result (also shown in Fig. 7-14) is now very close to the monomer spectra. In principle, one should measure the spectrum of the mixture of monomers at 90°C for a fair comparison. In practice, however, the absorption of the individual separate monomers is virtually independent of temperature. Generally the conformation of the polymers mainly determines deviations of the polymer absorption spectrum from that of a simple monomer. With polypeptides, one must be careful to define the monomer unit as a peptide because single amino acids lack the peptide chromophore.

How does the polymer structure alter the optical properties of its individual chromophores? One can imagine several possibilities. The actual geometry of a chromophore could be perturbed by direct strong interaction with other groups. These effects are not unknown, but they are rare and we shall ignore them.[§] All of the other possibilities fall into the class of weaker interactions. Chromophores can be perturbed by the general local environment or by electronic interaction with nearby chromophores. A number of different kinds of interactions can be observed. Let us first consider general environmental effects. These can include the local pH, dielectric constant, rigidity of the medium, and the presence of nearby groups capable of specific chemical interaction (such as proton or charge transfer, or metal-ion binding). Only the role of solvent polarity is treated here.

Spectrum sensitivity to local environment

The spectra of many chromophores are sensitive to solvent. This sensitivity can be manifested as changes in intensity, band shape, or wavelength of the absorption. We discuss only the effect on wavelength here. When a single chromophore (such as tryptophan) is dissolved in aqueous solution, the surrounding medium is (by and large) water. This would also be true for a tryptophan located at the surface of a protein. However, because tryptophan is rather hydrophobic, one can anticipate that the side chain usually will be found buried in an apolar region. There, its environment will most closely resemble a hydrocarbon solution rather than a water solution.

Changes in solvent from such polar to nonpolar media have large effects on the energy differences between electronic states, and thus they change absorption wavelengths. In general, the effect of a solvent will be to lower the average energy of each energy level. Solvent–solute interactions will occur in favorable orientations for attractive forces more frequently than in unfavorable orientations, in the same way that low-energy conformations of individual molecules predominate over high-energy conformations. The more polar the solvent, the stronger the interactions will be. To produce net shifts in absorption frequencies, however, the *relative* energies of two states must be changed.

A quantitative analysis of the effect of solvent on the absorption spectrum is an extremely difficult task. When the chromophore is in its ground state, solvent molecules will assume low-energy configurations. The excitation process occurs in about 10^{-15} sec, which is much faster than any molecular motions. Thus, to calculate the observed absorption wavelength, one must know the energy of the *excited* chromophore in the presence of the same solvent configuration as that associated with the

[§] An example is the determination of preferred dihedral angles in disulfides of cystine by the tertiary structure.

ground state. Two aspects of the solvent must be considered for this calculation: the high-frequency dielectric constant (governed by the polarizability), and the permanent dipole moment. These factors differ because polarizability effects occur rapidly enough to respond to the changed electronic distribution in the excited molecule, whereas permanent dipole effects are much slower.

When solvent polarizability effect dominate, it is safe to conclude that the electronic state of the chromophore with the highest dipole moment will be preferentially lowered by a polar solvent. Because this state almost always is the excited state, we obtain the result soon to be shown for $\pi \rightarrow \pi^*$ transitions that polar solvents lead to red shifts.

When solvent–solute dipole moment interactions are significant, the prediction of solvent-induced spectral shifts becomes very difficult. Solvent molecules cannot reorient during the absorption process, and a solute–solvent orientation that is attractive in the ground state could be more attractive, neutral, or even repulsive in the excited state. (For a more detailed discussion, see J. Birks, 1970, p. 109, and R. Marcus, *J. Chem. Phys.* 43(1965):1261.)

For a simple illustration of solvent effects, we consider the chromophore mesityl oxide: $(CH_3)_2C{=}C{-}CO{-}CH_3$. Three energy levels ($\pi$, π^*, and n) contribute to the near-UV spectrum of this molecule. The two lowest transitions are a $\pi \rightarrow \pi^*$ and an $n \rightarrow \pi^*$. We shall consider only solvent polarizability effects on these transitions. Because a π orbital is bonding, an electron in this orbital is localized mainly between the nuclei involved in the bond. It should tend to be less affected by solvent than is the much more expanded π^* orbital. Changing from a nonpolar to a polar solvent should decrease the energy separation between π and π^* levels and cause a red shift in the absorption spectrum (Table 7-3; Fig. 7-15). An electron in an n orbital interacts even

Table 7-3
Near-UV spectrum of mesityl oxide

	$n \rightarrow \pi^*$		$\pi \rightarrow \pi^*$	
Solvent	λ_{max} (nm)	ε_{max} ($\times 10^{-3}$)	λ_{max} (nm)	ε_{max} ($\times 10^{-1}$)
Hexane	229.5	12.6	327	9.8
Ethanol	237	12.6	315	7.8
Water	244.5	10.0	305	6.0

SOURCE: Data from M. Orchin and H. H. Jaffe, *Theory and Applications of Ultraviolet Spectroscopy* (New York: Wiley, 1960).

more strongly with solvent molecules than does one in a π^* orbital. For example, occupied n orbitals can serve as hydrogen-bond acceptors. Thus, one can predict that, in a more polar solvent, the energy difference between n and π^* will increase, resulting in a shift to shorter wavelength (a blue shift).

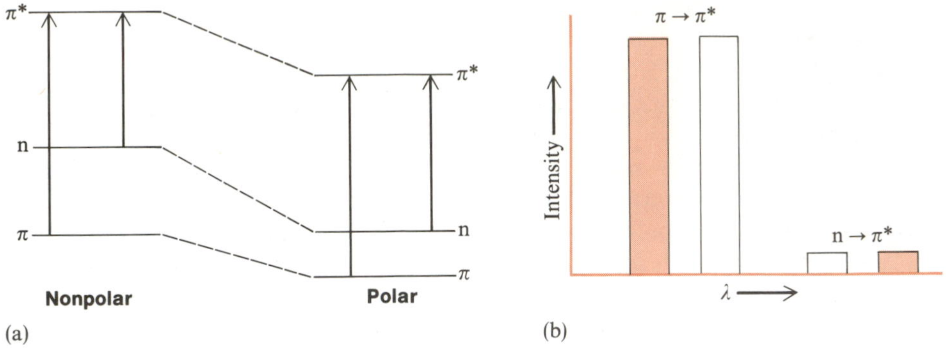

Figure 7-15

Typical effects of solvent polarity on transition energies for $\pi \rightarrow \pi^*$ and $n \rightarrow \pi^*$ transitions. **(a)** Energy levels. **(b)** Corresponding schematic spectra for polar (*unshaded bars*) and nonpolar (*shaded bars*) solvents.

At first inspection, one might predict from the example just given that the near-UV spectra of proteins should be blue-shifted with respect to the sums of the spectra of aqueous solutions of individual amino acids. The predominant transitions in the 250 to 300 nm region are $\pi \rightarrow \pi^*$. At least some of the aromatic residues usually are found buried in nonpolar regions of the protein. One might expect, by a careful analysis of the detailed shape and position of a protein absorption band, to be able to analyze the numbers of "buried" and "exposed" aromatic residues. In practice, things are much more complicated. For example, the spectrum of bovine serum albumin is red-shifted by 3 nm when compared with a mixture of amino acids at acid pH, but this shift is difficult to interpret. The free amino acids are not such good models for residues in a protein because they are highly charged. The environment inside a protein is not a relatively homogeneous isotropic nonpolar milieu resembling a cyclohexane solution. Dipolar effects must be considered in addition to polarizability. The aromatic side chains frequently are involved in specific interactions, with their polar parts still making hydrogen bonds with atoms on the peptide backbone or on other polar groups. Thus, the precise shape of the aromatic region of the ultraviolet spectrum is different for each protein. It is quite difficult to analyze the environmental effects to account for these specific structural details.

A somewhat less direct approach is needed for the spectral analysis. The method of solvent-perturbation spectroscopy has been quite successful. For this technique, one records spectra (or difference spectra) in a gradually changing series of solvents. A less polar solvent (such as ethanol) is added in small increments. If no conformational change is induced, one expects smooth variation in the spectral properties, as exposed residues feel the change in environment. By looking only at the changes in spectra, one can concentrate on exterior residues. These changes will show similar patterns of shifts in different proteins because the external environment is not sensitive to detailed protein structure. Standard curves have been developed for spectral changes between solvents for tryptophan residues and tyrosine residues. By using these curves,

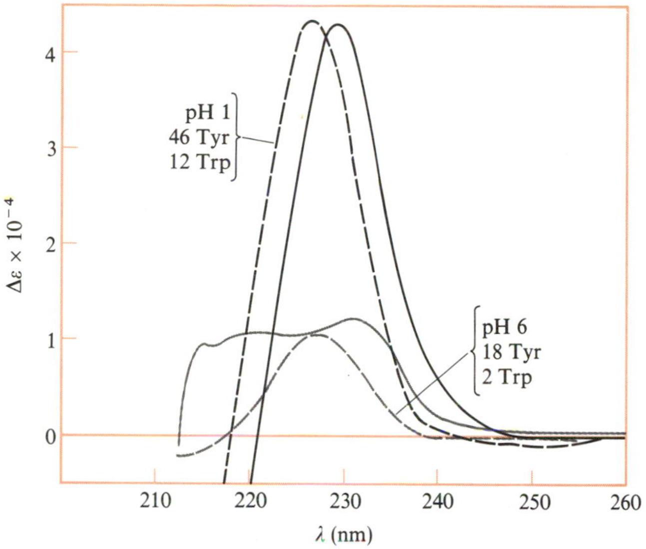

Figure 7-16

Solvent-perturbation difference spectrum, produced by treating rabbit-muscle aldolase with 20% ethylene glycol. The observed spectra (*solid curves*) are compared with calculated difference spectra (*dashed curves*) generated using data obtained on individual amino acids. Clearly, the protein has many fewer exposed aromatic groups at pH 6 than it does at pH 1. [After J. W. Donovan, *J. Biol. Chem.* 244:1961 (1969).]

it is possible to analyze the composition of exposed aromatic residues in a given protein. From the total amino acid composition, one can then learn about the number of buried residues.

Figure 7-16 shows an example of solvent-perturbation spectra for aldolase. At pH 6, there are 2 Trp and 18 Tyr side chains exposed. If a conformational change occurs as a result of changes in solvent, a fairly abrupt spectral change should result. At still higher concentrations of nonpolar solvent, one can again analyze the exposed residues. However, it must be realized that now one is no longer looking at the native protein structure. Figure 7-16 indicates that aldolase has many more aromatic residues exposed at pH 1 than at pH 6. Apparently, at pH 1 the protein is substantially unfolded.

The general sensitivity of aromatic chromophores to solvent changes is tremendously useful. Even in those cases where the spectra are too complex to analyze, the spectral changes can always be used as a convenient monitor of conformational changes. Many examples are given in subsequent chapters.

In nucleic acids, it is much more difficult to use solvent-perturbation methods. The strong base–base interactions dominate the spectra. In addition, the structures alter all too readily when the solvent composition is changed. Some progress has been made with a very delicate spectral perturbation—substituting D_2O for H_2O. However, the spectral changes induced are minute and difficult to analyze.

Interactions between different chromophores

Let us now ignore environmental effects and concentrate instead on what happens when specific chromophores get close enough together to interact electronically. This will occur with nucleic acid bases in a polynucleotide, and with the peptide chromophores of a protein or polypeptide. It is a rarer phenomenon with aromatic side chains, because there usually are few of them in a typical protein. For simplicity, we first consider a very simple polymer: a dimer containing two identical chromophores. The following analysis can easily be generalized to more complex cases. Suppose the monomeric chromophore has only two electronic states (0 and a) and a single absorption band from the transition from state 0 to a. We can ignore the shape of the absorption and just describe it as a line at a frequency $\nu_{0a} = (E_a - E_0)/h$. The ground-state wavefunction is ϕ_0; the excited state is ϕ_a. These are normalized and orthogonal: $\langle\phi_0|\phi_0\rangle = \langle\phi_a|\phi_a\rangle = 1$; and $\langle\phi_a|\phi_0\rangle = 0$. The absorption intensity of the monomer will be given by the dipole strength $|\langle\phi_0|\underset{\sim}{\mu}|\phi_a\rangle|^2$.

If one examines a dimer containing two of these monomer chromophores, one sees in general two absorption bands, one displaced a frequency $\Delta\nu$ above the monomer band, and the other displaced the same frequency below. The intensities of the two bands are not necessarily equal (Fig. 7-17). Our goal here is to explain this observation.

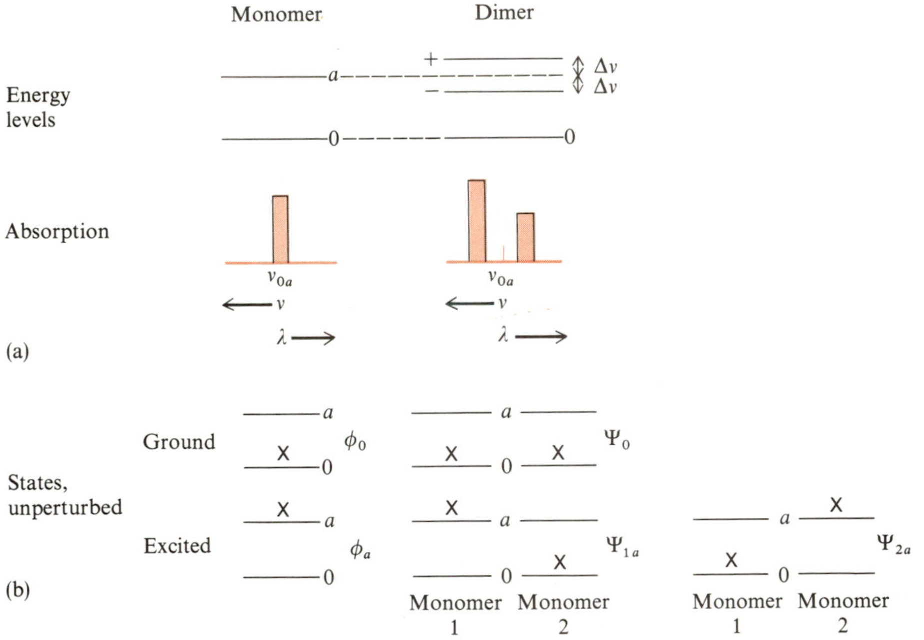

Figure 7-17

Energy levels, states, and absorption spectra of a simple monomer and a dimer of two identical chromophores. **(a)** The excited state of the monomer is split into two states by electronic interactions. The absorption intensities of transitions to the two states usually are not the same. **(b)** States and wavefunctions expected for the monomer and for the dimer in which no monomer–monomer interaction occurs.

A dimer of noninteracting monomers

Let us first describe the states of a dimer if there is no interaction between the two monomers. In the ground state of the dimer, both monomers are in their ground states. The wavefunction is written as a product of the ground-state wavefunctions of the two separate monomers, ϕ_{10} and ϕ_{20} (see Eqn. 7-6). Thus, the dimer ground-state wavefunction is $\Psi_0 = \phi_{10}\phi_{20}$. The first excited state of the dimer will contain one monomer in state a and one in state 0 (Fig. 7-17). There are two possible wavefunctions: $\Psi_{1a} = \phi_{1a}\phi_{20}$, which corresponds to monomer 1 excited; and $\Psi_{2a} = \phi_{10}\phi_{2a}$, which corresponds to monomer 2 excited. Note that these two states have orthogonal wavefunctions: $\langle\phi_{1a}\phi_{20}|\phi_{10}\phi_{2a}\rangle = 0$. The second excited state of the dimer is $\phi_{1a}\phi_{2a}$. This state lies at twice the energy above the ground state as that for the states Ψ_{1a} and Ψ_{2a}. The absorption transition from the ground state to $\phi_{1a}\phi_{2a}$ will occur at such a short wavelength that this excited state need not be considered further.

The energy of each state of the dimer is computed by evaluating the expectation value of the dimer Hamiltonian. This is $\underset{\sim}{\mathbf{H}} = \underset{\sim}{\mathbf{H}}_1 + \underset{\sim}{\mathbf{H}}_2$, where $\underset{\sim}{\mathbf{H}}_1$ and $\underset{\sim}{\mathbf{H}}_2$ are the Hamiltonian operators of monomers 1 and 2, respectively. We have the following relationships:

$$\underset{\sim}{\mathbf{H}}_1\phi_{10} = E_0\phi_{10}, \qquad \underset{\sim}{\mathbf{H}}_1\phi_{1a} = E_a\phi_{1a} \tag{7-43a}$$

$$\underset{\sim}{\mathbf{H}}_2\phi_{20} = E_0\phi_{20}, \qquad \underset{\sim}{\mathbf{H}}_2\phi_{2a} = E_a\phi_{2a} \tag{7-43b}$$

Therefore, for the dimer, one has the following energies.

Ground state Ψ_0:
$$\begin{aligned}\text{Energy} &= \langle\phi_{10}\phi_{20}|\underset{\sim}{\mathbf{H}}_1 + \underset{\sim}{\mathbf{H}}_2|\phi_{10}\phi_{20}\rangle \\ &= \langle\phi_{20}|\phi_{20}\rangle\langle\phi_{10}|\underset{\sim}{\mathbf{H}}_1|\phi_{10}\rangle + \langle\phi_{10}|\phi_{10}\rangle\langle\phi_{20}|\underset{\sim}{\mathbf{H}}_2|\phi_{20}\rangle \\ &= E_0 + E_0 = 2E_0\end{aligned} \tag{7-44a}$$

Excited state Ψ_{1a}:
$$\begin{aligned}\text{Energy} &= \langle\phi_{1a}\phi_{20}|\underset{\sim}{\mathbf{H}}_1 + \underset{\sim}{\mathbf{H}}_2|\phi_{1a}\phi_{20}\rangle \\ &= E_a + E_0\end{aligned} \tag{7-44b}$$

Excited state Ψ_{2a}:
$$\begin{aligned}\text{Energy} &= \langle\phi_{10}\phi_{2a}|\underset{\sim}{\mathbf{H}}_1 + \underset{\sim}{\mathbf{H}}_2|\phi_{10}\phi_{2a}\rangle \\ &= E_0 + E_a\end{aligned} \tag{7-44c}$$

The two excited states have the same energy. The difference in energy between the ground state and either excited state is just $E_a - E_0$. Therefore, the absorption frequency of the dimer is exactly the same as that of the monomer. However, the dimer absorption is twice as intense because it contains two monomers. We can calculate absorption intensity from the dipole strength, using the dipole operator of the dimer: $\underset{\sim}{\boldsymbol{\mu}} = \underset{\sim}{\boldsymbol{\mu}}_1 + \underset{\sim}{\boldsymbol{\mu}}_2$. For example,

$$|\langle\Psi_0|\underset{\sim}{\boldsymbol{\mu}}|\Psi_{1a}\rangle|^2 = |\langle\phi_{10}\phi_{20}|\underset{\sim}{\boldsymbol{\mu}}_1 + \underset{\sim}{\boldsymbol{\mu}}_2|\phi_{1a}\phi_{20}\rangle|^2 = |\langle\phi_{10}|\underset{\sim}{\boldsymbol{\mu}}_1|\phi_{1a}\rangle|^2 = D_{0a} \tag{7-45}$$

Thus, state Ψ_{1a} contributes an absorption intensity that is exactly equal to the intensity of an isolated monomer. State Ψ_{2a} contributes an identical intensity.

Note that, for the dimer, $\underset{\sim}{\mathbf{H}}$ and $\underset{\sim}{\boldsymbol{\mu}}$ are simply the sums of individual monomer operators. Therefore, the integrals represented by Equations 7-43, 7-44, and 7-45 always can be factored into the sums and products of individual monomer integrals. The two wavefunctions Ψ_{1a} and Ψ_{2a} describe dimers in which excitation energy is localized on a single monomer. It also is possible to describe the dimer excited state as $C_1\Psi_{1a} + C_2\Psi_{2a}$, in which case we admit some uncertainty in our knowledge as to which monomer actually is excited. This state also is a solution to the Schrödinger equation of the dimer:

$$(\underset{\sim}{\mathbf{H}}_1 + \underset{\sim}{\mathbf{H}}_2)(C_1\Psi_{1a} + C_2\Psi_{2a}) = (E_a + E_0)(C_1\Psi_{1a} + C_2\Psi_{2a}) \tag{7-46}$$

Therefore, we really cannot say which monomer is excited if our only information comes from observing a spectrum at frequency $\nu_{0a} = (E_a - E_0)/h$.

A dimer of interacting monomers

Now suppose the monomers of a dimer are close enough in space to interact electronically. We consider the case of interactions that are weak enough so that the individual monomers retain their identities. That is, electrons still are localized either on one monomer or on the other, but each monomer is perturbed by the electric field of the other. Because the monomers are uncharged, the leading term in the electronic interaction between them will be a dipole–dipole interaction, just as described earlier for the ground-state interaction between two residues in a polymer (Chapter 5). The Hamiltonian of the dimer becomes

$$\underset{\sim}{\mathbf{H}} = \underset{\sim}{\mathbf{H}}_1 + \underset{\sim}{\mathbf{H}}_2 + \underset{\sim}{\mathbf{V}} \tag{7-47}$$

where $\underset{\sim}{\mathbf{H}}_1$ and $\underset{\sim}{\mathbf{H}}_2$ are just the monomer Hamiltonians used earlier for noninteracting moments, and $\underset{\sim}{\mathbf{V}}$ is the operator describing the dipole–dipole interaction:

$$\underset{\sim}{\mathbf{V}} = (\underset{\sim}{\boldsymbol{\mu}}_1 \cdot \underset{\sim}{\boldsymbol{\mu}}_2)\underset{\sim}{\mathbf{R}}_{12}^{-3} - 3(\underset{\sim}{\boldsymbol{\mu}}_1 \cdot \underset{\sim}{\mathbf{R}}_{12})(\underset{\sim}{\mathbf{R}}_{12} \cdot \underset{\sim}{\boldsymbol{\mu}}_2)\underset{\sim}{\mathbf{R}}_{12}^{-5} \tag{7-48}$$

where $\underset{\sim}{\boldsymbol{\mu}}_1$ and $\underset{\sim}{\boldsymbol{\mu}}_2$ are the dipole operators of monomers 1 and 2, respectively, and $\underset{\sim}{\mathbf{R}}_{12}$ is the distance between the two monomers. In principle, the distance should be measured between the centers of electron density of the two monomers; in practice, the geometric centers often are a close enough approximation.

To compute the electronic states of the dimer, one must solve the Schrödinger equation using the Hamiltonian given by Equation 7-47. This is a difficult task. Instead, we will make an approximation and say that the interaction $\underset{\sim}{\mathbf{V}}$ is so weak that the ground-state wavefunction is still Ψ_0, and that wavefunctions of the types Ψ_{1a} and Ψ_{2a} are still good approximations to the excited states of the dimer. We demand only that the eigenfunctions used for the excited state be stationary. This means that the Hamiltonian must not be able to produce transitions among the excited states. For the case of two *noninteracting* monomers, note that

$$\begin{aligned}\langle \Psi_{1a}|\underset{\sim}{\mathbf{H}}|\Psi_{2a}\rangle &= \langle \phi_{1a}\phi_{20}|\underset{\sim}{\mathbf{H}}_1 + \underset{\sim}{\mathbf{H}}_2|\phi_{10}\phi_{2a}\rangle \\ &= (E_0 + E_a)(\langle \phi_{1a}|\phi_{10}\rangle \cdot \langle \phi_{20}|\phi_{2a}\rangle) = 0\end{aligned} \tag{7-49}$$

Thus, these states are stationary. However, for *interacting* monomers,

$$\langle \Psi_{1a}|\underset{\sim}{\mathbf{H}}|\Psi_{2a}\rangle = \langle \phi_{1a}\phi_{20}|\underset{\sim}{\mathbf{H}}_1 + \underset{\sim}{\mathbf{H}}_2 + \underset{\sim}{\mathbf{V}}|\phi_{10}\phi_{2a}\rangle = V_{12} \tag{7-50}$$

where V_{12} is given (from Eqn. 7-48) by

$$\begin{aligned}V_{12} &= \langle \phi_{1a}\phi_{20}|\underset{\sim}{\mathbf{V}}|\phi_{10}\phi_{2a}\rangle \\ &= (\langle \phi_{1a}|\underset{\sim}{\boldsymbol{\mu}}_1|\phi_{10}\rangle \cdot \langle \phi_{20}|\underset{\sim}{\boldsymbol{\mu}}_2|\phi_{2a}\rangle)R_{12}^{-3} \\ &\quad - (3\langle \phi_{1a}|\underset{\sim}{\boldsymbol{\mu}}_1|\phi_{20}\rangle \cdot \mathbf{R}_{12})(\mathbf{R}_{12} \cdot \langle \phi_{20}|\underset{\sim}{\boldsymbol{\mu}}_2|\phi_{2a}\rangle)R_{12}^{-5}\end{aligned} \tag{7-51}$$

The two states Ψ_{1a} and Ψ_{2a} are not stationary in the presence of the perturbation $\underset{\sim}{\mathbf{V}}$. Transitions back and forth occur rapidly. Physically, these transitions are caused by interactions between the transition dipoles of the two monomers (Eqn. 7-51). Fluctuating electric fields in one monomer are felt by the other, causing a transfer of excitation. Note that the interaction between the two monomers is proportional to R_{12}^{-3}, so it dies off rapidly with increasing separation.

Because we still want to use the wavefunctions Ψ_{1a} and Ψ_{2a}, we look for linear combinations of these two states ($C_1\Psi_{1a} + C_2\Psi_{2a}$ and $C_1'\Psi_{1a} + C_2'\Psi_{2a}$) that satisfy the following three criteria.

Normalization: $\langle C_1\Psi_{1a} + C_2\Psi_{2a}|C_1\Psi_{1a} + C_2\Psi_{2a}\rangle = 1$

Orthogonality: $\langle C_1\Psi_{1a} + C_2\Psi_{2a}|C_1'\Psi_{1a} + C_2'\Psi_{2a}\rangle = 0$

Stationary: $\langle C_1\Psi_{1a} + C_2\Psi_{2a}|\underset{\sim}{\mathbf{H}}|C_1'\Psi_{1a} + C_2'\Psi_{2a}\rangle = 0$

It is easy to demonstrate that only two wavefunctions satisfy these conditions:

$$\begin{aligned}\Psi_{A^+} &= (1/\sqrt{2})(\Psi_{1a} + \Psi_{2a}) \\ \Psi_{A^-} &= (1/\sqrt{2})(\Psi_{1a} - \Psi_{2a})\end{aligned} \tag{7-52}$$

These states are the best approximations to the excited-state wavefunctions of the dimer that we can construct using the nonperturbed-monomer wavefunctions $\phi_{10}\phi_{2a}$ and $\phi_{1a}\phi_{20}$.

The energies of the states of the dimer now can be found by evaluating the expectation values of the Hamiltonian $\underset{\sim}{\mathbf{H}} = \underset{\sim}{\mathbf{H}}_1 + \underset{\sim}{\mathbf{H}}_2 + \underset{\sim}{\mathbf{V}}$. Using results shown previously and evaluating integrals term by term, the reader should be able to show that

$$\text{Ground state:} \quad \langle \Psi_0 | \underset{\sim}{\mathbf{H}} | \Psi_0 \rangle = \langle \phi_{1a}\phi_{2a} | \underset{\sim}{\mathbf{H}}_1 + \underset{\sim}{\mathbf{H}}_2 + \underset{\sim}{\mathbf{V}} | \phi_{10}\phi_{20} \rangle = 2E_0 \quad \text{(7-53a)}$$

(Integrals of the type $\langle \phi_{10} | \underset{\sim}{\boldsymbol{\mu}}_1 | \phi_{10} \rangle$ that contribute to $\underset{\sim}{\mathbf{V}}$ are all zero, because $\phi_{10}^* \phi_{20}$ is an even function whereas $\underset{\sim}{\boldsymbol{\mu}}_1$ is odd.) Similarly, for the excited states,

$$\text{State A}^+: \quad \langle \Psi_{A^+} | \underset{\sim}{\mathbf{H}} | \Psi_{A^+} \rangle = (1/2) \langle \phi_{1a}\phi_{20} + \phi_{10}\phi_{2a} | \underset{\sim}{\mathbf{H}}_1 + \underset{\sim}{\mathbf{H}}_2 + \underset{\sim}{\mathbf{V}} | \phi_{1a}\phi_{20} + \phi_{10}\phi_{2a} \rangle$$
$$= E_a + E_0 + V_{12}$$

$$\text{State A}^-: \quad \langle \Psi_{A^-} | \underset{\sim}{\mathbf{H}} | \Psi_{A^-} \rangle = (1/2) \langle \phi_{1a}\phi_{20} - \phi_{10}\phi_{2a} | \underset{\sim}{\mathbf{H}}_1 + \underset{\sim}{\mathbf{H}}_2 + \underset{\sim}{\mathbf{V}} | \phi_{1a}\phi_{20} - \phi_{10}\phi_{2a} \rangle$$
$$= E_a + E_0 - V_{12} \quad \text{(7-53b)}$$

Thus the two singly excited states of the dimer no longer have the same energy. We say that the perturbation $\underset{\sim}{\mathbf{V}}$ has split the excited state of the dimer into two states. The energy splitting is $2V_{12}$. As indicated in Equation 7-51, it is a function of the distance between the two monomers and also of their relative orientation.

The absorption frequencies of the two transitions $\Psi_0 \rightarrow \Psi_{A^+}$ and $\Psi_0 \rightarrow \Psi_{A^-}$ are

$$\nu_+ = (1/h)(E_a + E_0 + V_{12} - 2E_0) = \nu_{0a} + V_{12}$$
$$\nu_- = (1/h)(E_a + E_0 - V_{12} - 2E_0) = \nu_{0a} - V_{12} \quad \text{(7-54)}$$

So the frequency splitting between the two absorption bands is just $2V_{12}/h$. This is called *exciton splitting* because it is equivalent to similar optical phenomena first noticed in crystals where a collective excitation takes place. In such systems one cannot speak of a particular chromophore being excited. Instead the excitation appears delocalized over a particular region.

Next let us consider the intensities of the two transitions at ν_+ and ν_-. Once again, wavefunctions are used to calculate dipole strengths. We can evaluate the D_{0A^+} and D_{0A^-} simultaneously, keeping a plus sign throughout for the first and a minus sign for the second.

$$D_{0A^\pm} = |\langle \Psi_0 | \underset{\sim}{\boldsymbol{\mu}}_1 + \underset{\sim}{\boldsymbol{\mu}}_2 | \Psi_{A^\pm} \rangle|^2$$
$$= |(1/\sqrt{2}) \langle \phi_{10}\phi_{20} | \underset{\sim}{\boldsymbol{\mu}}_1 + \underset{\sim}{\boldsymbol{\mu}}_2 | \phi_{1a}\phi_{20} \pm \phi_{10}\phi_{2a} \rangle|^2$$
$$= (1/2) |\langle \phi_{10} | \underset{\sim}{\boldsymbol{\mu}}_1 | \phi_{1a} \rangle \pm \langle \phi_{20} | \underset{\sim}{\boldsymbol{\mu}}_2 | \phi_{2a} \rangle|^2$$
$$= (1/2)[|\langle \phi_{10} | \underset{\sim}{\boldsymbol{\mu}}_1 | \phi_{1a} \rangle|^2 + |\langle \phi_{20} | \underset{\sim}{\boldsymbol{\mu}}_2 | \phi_{2a} \rangle|^2 \pm 2(\langle \phi_{10} | \underset{\sim}{\boldsymbol{\mu}}_1 | \phi_{1a} \rangle \cdot \langle \phi_{20} | \underset{\sim}{\boldsymbol{\mu}}_2 | \phi_{2a} \rangle)]$$
$$\text{(7-55)}$$

The first two terms are just the monomer dipole strength, D_{0a}. Furthermore, because $|\langle\phi_{10}|\underset{\sim}{\mu}_1|\phi_{1a}\rangle| = |\langle\phi_{20}|\underset{\sim}{\mu}_2|\phi_{2a}\rangle| = \sqrt{D_{0a}}$, we can write the dimer dipole strengths in terms of observed monomer spectral properties:

$$D_{0A^\pm} = D_{0a} \pm D_{0a}\cos\theta \tag{7-56}$$

where θ is the angle between the transition dipoles involved in the individual electronic absorption of monomer 1 and monomer 2. Equation 7-56 is very useful. It says that the relative intensities of the two absorption bands depend on the geometry of the dimer in a very simple way. It also shows that the total integrated spectral intensity of the dimer is not a function of the geometry. The total intensity is $D_{0A^+} + D_{0A^-} = 2D_{0a}$, or just twice the monomer intensity.

How to analyze a dimer spectrum

First, let us recapitulate the results of the calculations of the last section. Each single absorption band of a monomer is split into two bands in a dimer. The frequency difference between them is a function of the distance between the monomers and of their relative orientation. The relative intensities, D_{0A^+} and D_{0A^-}, are a function only of the angle between the transition dipoles of the two monomers. These intensities can be expressed in terms of the intensity of a single isolated monomer, D_{0a}.

Physically, the two states Ψ_{A^+} and Ψ_{A^-} can be given a simple explanation. In going from Ψ_0 to Ψ_{A^+}, both monomers are excited in phase. That is, transition dipoles on monomers 1 and 2 oscillate together. In going from Ψ_0 to Ψ_{A^-}, both monomers are excited out of phase. The transition dipoles point in opposite directions with respect to the coordinate system of each individual monomer. The overall spectrum depends on the angle between the transition dipoles (determined by the molecule geometry of the dimer) *and* on the relative phase of their excitation (depending on whether state Ψ_{A^+} or state Ψ_{A^-} is involved). This can be confusing, and it is best understood by considering the spectra of dimers in a few simple geometries (Fig. 7-18).

In the first two cases shown (a and b), θ is 0°, so cos θ is 1; therefore, D_{0A^+} is $2D_{0a}$, and D_{0A^-} is zero.[§] All the absorption intensity is associated with the transition $\Psi_0 \rightarrow \Psi_{A^+}$ (see Eqn. 7-56). The frequency at which this transition occurs is determined by V_{12} (see Eqn. 7-54). For a linear head-to-tail array of monomers (case a), V_{12} is negative. The monomers attract each other, just as classical dipoles would in this orientation. Equation 7-54 shows that the frequency ν_+ will be lower than the monomer frequency ν_{0a}. Therefore, the absorption is red-shifted. For a parallel stacked array (case b), V_{12} is positive, and the ν_+ band occurs at a frequency higher than that of the monomer.

[§] Note that θ is zero for both A^+ and A^- states. The angle θ refers to monomer transition dipole directions. It is a function of molecular geometry, but not a function of the choice of linear combination of the two monomer states.

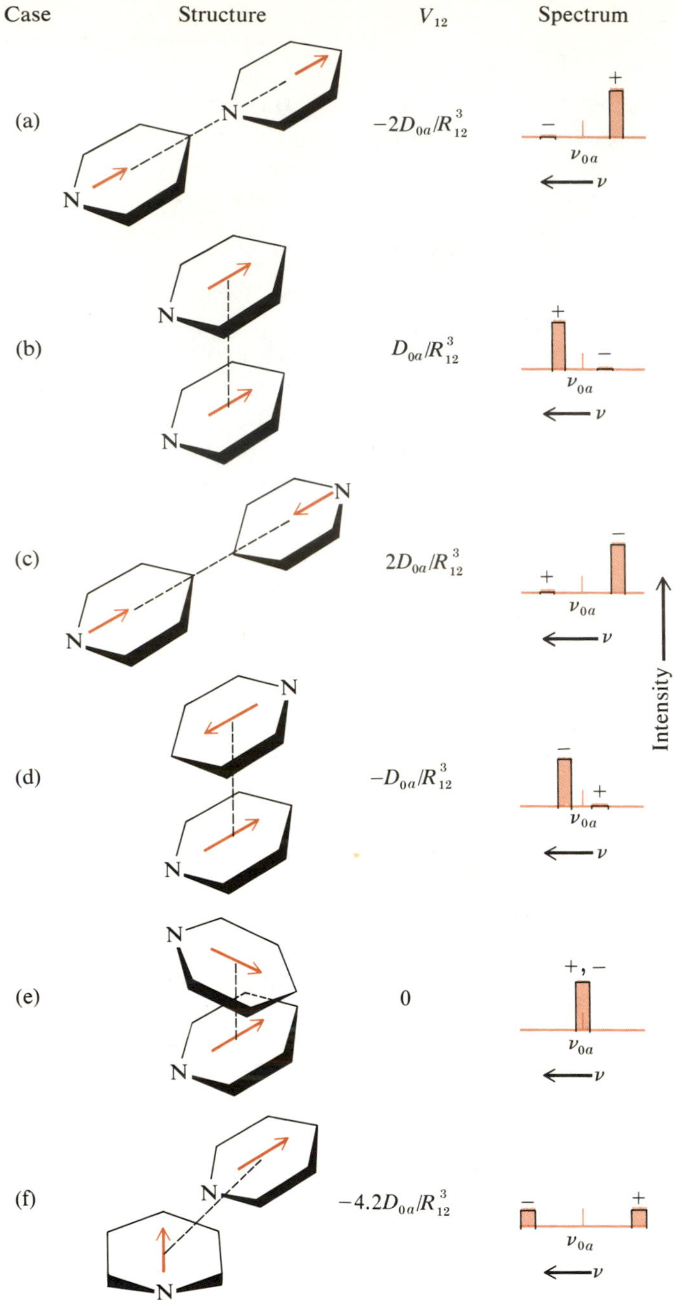

Figure 7-18

Monomer-interaction energies, pair interaction potentials, and schematic spectra for six possible geometries of a dimer of pyridine. The potential V_{12} is expressed in terms of the dipole strength of the monomer ($D_{0a} = |\langle \phi_0 | \underset{\sim}{\boldsymbol{\mu}} | \phi_a \rangle|^2$). The interchromophore distance, R_{12}, is shown as a dashed line; transition dipole orientations are shown as arrows for the transition $\Psi_0 \rightarrow \Psi_{A^+}$ (if the transition $\Psi_0 \rightarrow \Psi_{A^-}$ were shown instead, one of the arrows would point toward the nitrogen instead of away from it). Note that, for a dimer, the intensity of a band can be computed by adding the two individual transition dipoles vectorially and computing the square of the length of the resulting vector.

In the next two cases (c and d), $\theta = 180°$, so $\cos\theta = -1$; therefore $D_{0A^-} = 2D_{0a}$, and $D_{0A^+} = 0$. The monomers are excited out of phase. All intensity is associated with the transition $\Psi_0 \rightarrow \Psi_{A^-}$. A head-to-head structure (case c) has a positive (repulsive) V_{12}, whereas an antiparallel stack (case d) has an attractive V_{12}. The ν_+ band appears at higher frequency than the monomer for case c, and at lower frequency than the monomer for case d. Thus the spectrum of case c is indistinguishable from that of case a, and the spectrum of case d is indistinguishable from that of case b, even though the state that is doing the absorption is different. Note that, in all these cases, no spectral splitting is observable, but only a shift of the absorption maximum. However, for any angle other than 0° or 180°, it usually is possible to observe two spectral bands.

For case e, a dimer with monomers in a stack at 90°, $\cos\theta$ is zero, and transitions from Ψ_0 to both states Ψ_{A^-} and Ψ_{A^+} have equal probability. However, V_{12} is zero in this geometry. Therefore, with no exciton splitting, the spectrum is identical to that of two isolated monomers. In case f, the two monomers are at 90° in the same plane. Whereas $\cos\theta$ is zero in this geometry also, V_{12} is negative in this case. Therefore, two equal bands are seen, with ν_+ occurring at low frequency, and ν_- at high frequency. In a favorable case such as this, one can measure the splitting and the relative intensities. These in turn yield both the distance between the chromophores, (R_{12}) and the angle θ. An actual experimental example is given in Figure 7-19, which

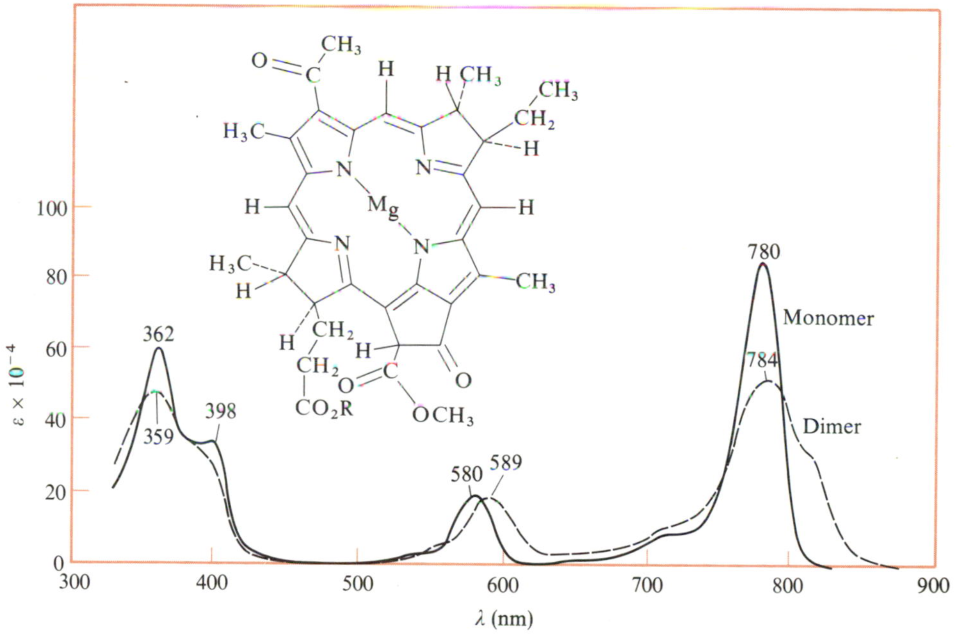

Figure 7-19

Monomer and dimer spectra for solutions of bacteriochlorophyll. A pronounced splitting of the longest-wavelength band in the dimer is visible. [After K. Sauer, J. R. L. Smith, and A. J. Schultz, *J. Am. Chem. Soc.* 88:2681 (1966).]

shows the absorption spectra of monomers and dimers of bacteriochlorophyll. From the splitting of the largest wavelength band in the dimer, the angle between the transition moments giving rise to this band was calculated to be 82°. In practice, the splitting between two bands (V_{12}) often is very small relative to the bands' width. This means that, instead of actually showing separation between two transitions, the result is an apparent spectral shift.

Numbers of bands and chromophores

What happens in more complicated molecules? Consider the case of oligomers composed of identical monomer chromophores. A trimer would be expected to have three absorption bands for each monomer band, a tetramer four, and so on. By the time one reaches a polymer, it might appear that the spectrum would become horrendously complicated. In some cases it does, but fortunately there are simplifications. Note from Figure 7-18 that, even for a dimer, there are some cases in which essentially only one band is seen. The intensity of the other is zero or very small.

As a polymer becomes long enough that we can neglect end effects, if the structure is a regular helix, virtually all the absorption bands have zero intensity. In the case of the polypeptide α helix, only two distinct allowed transitions remain. One has a net transition dipole parallel to the helix axis. The other (really two transitions with the same energy) has a transition dipole perpendicular to this axis. The splitting between these two bands is quite large as evidenced by the spectrum of α-helical poly-L-glutamic acid shown in Figure 7-5. The 190 nm peptide band shows a slight shoulder near 208 nm. When polarized absorption spectra are measured on oriented samples, it is clear that the 208 nm shoulder preferentially absorbs light polarized parallel to the helix axis. Therefore this is the $\pi \rightarrow \pi^*$ polymer transition with a net transition dipole parallel to the helix axis. The perpendicular transition occurs at shorter wavelengths (Fig. 7-5). A second example of splitting in polymer spectra is shown in Figure 7-9, where the spectra of poly-L-lysine in α-helix, β sheet, and random coil are compared. The α-helical sample shows a shoulder at about 205 nm. The wavelength of maximum absorption is shifted to the blue relative to that of the random coil sample. Note that there is no evidence for splitting in either the random coil or the β sheet.

The kinds of electrostatic interactions we have been discussing are not restricted to arrays of identical chromophores. When two chromophores possessing fairly similar transition energies are in close proximity, exactly analogous effects exist. These however are mathematically more complex to analyze (Tinoco, 1972). In nucleic acids, exciton optical effects are very complicated. There are numerous different nearest-neighbor base interactions between bases on the same strand and on opposite strands of a double helix. In absorption spectra, the results are largely averaged out, and no splittings or shifts are visible in the near-UV bands of DNA. However, the exciton effects show up more strongly in optical activity measurements, as we show later.

Hypochromism in chromophore aggregates

The characteristic feature of the exciton effects is that the overall integrated spectral intensity remains constant. However if you examine carefully the spectra of different poly-L-lysine conformations (Fig. 7-9), you will notice that the α helix definitely has a weaker overall absorption intensity than does the coil. This is a manifestation of another optical effect, hypochromism, which simply means "less color." The effect is quite pronounced in DNA (Fig. 7-14). The intact double helix absorbs roughly 30% less than does a mixture of the component monomers. One possible explanation is an exciton effect in which the interaction is so strong that one band has simply shifted far enough to be out of sight. However, calculations of the expected exciton splittings in DNA indicate that they should be very small. In fact, DNA shows no spectral shifts or broadening relative to monomer spectra. Exciton effects are occurring, but they cause insignificant alternations of the absorption spectrum.

Another simple explanation can also be dismissed easily. This is the "shadow effect." Consider a polymer containing chromophores that are absorbing light so intensely that all the incident light is already absorbed before the center of the molecule is reached. The absorbance of a solution of this polymer will be less than that calculated on the basis of the total chromophore concentration, because those chromophores inside the molecule cannot do their fair share of absorbing. Using Beer's law and a typical chromophore extinction of 10^4, one can easily show that most macromolecules are too small for any appreciable shadowing to take place. A satisfactory understanding of hypochromism eluded scientists for a while, but the currently accepted theory (which successfully explains the effect) was finally developed by Ignacio Tinoco and by William Rhodes in the late 1950s. The explanation is not simple.

With the exciton effects described in the previous section, we considered just the lowest excited electronic state of each chromophore. As long as the chromophores were separated far apart, each had the same intrinsic intensity. In an ordered structure, the chromophores interact. There are still as many excitation bands as there are chromophores, but now all do not have equal intensity. One can say that some excited-state transitions are borrowing intensity at the expense of others. In addition, energies of interaction between the chromophores lead to shifts in the energies of the excited states.

Hypochromism has some similarities to the exciton effect, but there are also substantial differences. The origin of hypochromism is the interaction between one particular electronic excited state of a given chromophore and different electronic states of the neighboring chromophores. To keep things simple, we shall consider just the interaction of the lowest excited electronic state of one chromophore with all higher electronic excited states of the neighboring chromophores. Suppose the system is illuminated with light of frequency ν_{0a}, which can excite only the lowest electronic transition. In the chromophore that absorbs, there will be a light-induced dipole, $\boldsymbol{\mu}_{0a}$. The magnitude of this dipole determines the absorption intensity.

How is this effected by the other chromophores nearby? A quantum mechanical treatment must describe how the wavefunction of one chromophore is altered by the presence of its neighbors. The interactions can be described (just as we did for exciton effects) by coupling between transition dipoles of one chromophore and those of its neighbors. Coupling the same transitions (0 + a) on two or more chromophores just leads to the exciton effect, which cannot change the overall intensity in the region near frequency ν_{0a}. But if we couple transition $0 \rightarrow a$ on one chromophore with all higher excited transitions ($0 \rightarrow b$, $0 \rightarrow c$, etc.) on its neighbors, the overall intensity can be altered.

The actual quantum mechanical expressions are very complicated, and it is easier to seek a classical analogy. Even if neighboring chromophores do not *absorb* light at frequency ν, they are affected by its oscillating electromagnetic field. Because molecules are polarizable, a dipole is induced in the presence of a field:

$$\boldsymbol{\mu}_{\text{ind}} = \underset{\sim}{\boldsymbol{\alpha}}(\nu) \cdot \mathbf{E}(\nu) \tag{7-57}$$

Here $\underset{\sim}{\boldsymbol{\alpha}}$ is the electric polarizability tensor of the molecule at frequency ν, and $\mathbf{E}$ is the electric field vector of the light. These induced dipoles do not dissipate the energy of the light, because they simply reradiate. This elastic-scattering phenomenon is responsible for the refractive index of transparent materials. The connection between this classical description and the quantum mechanical description is that $\underset{\sim}{\boldsymbol{\alpha}}$ is a function of the transition dipoles of all of the excited states of the chromophore:[§]

$$\underset{\sim}{\boldsymbol{\alpha}}(\nu) = (2/h) \sum_b [(\nu_b \langle \Psi_0 | \underset{\sim}{\boldsymbol{\mu}} | \Psi_b \rangle \langle \Psi_b | \underset{\sim}{\boldsymbol{\mu}} | \Psi_0 \rangle)/(\nu_{0b}^2 - \nu^2)] \tag{7-58}$$

Depending on the frequency (ν), the induced dipole moment $\boldsymbol{\mu}_{\text{ind}}$ can be in phase or out of phase with the exciting light, $\mathbf{E}$. That is, α can be positive or negative (Fig. 7-20). Each individual absorption band $0 \rightarrow b$ contributes an in-phase component to the polarizability at $\nu < \nu_{0b}$ and an out-of-phase component at $\nu > \nu_{0b}$, as you can see from the form of Equation 7-58. In the special case where ν is at the extreme low-frequency edge of the absorption spectrum, $\boldsymbol{\mu}_{\text{ind}}$ will always be in phase. Because ν_{0a} is at lower frequency than any other absorption band, each must contribute a positive component to the polarizability at $\nu = \nu_{0a}$. Note that we do not have to consider the contribution of $\langle \Psi_0 | \underset{\sim}{\boldsymbol{\mu}} | \Psi_a \rangle$ to the polarizability in computing hypochromism. This contribution is already taken care of when exciton effects are calculated, and (as shown earlier) coupling of transition dipoles of the same states on two chromophores cannot produce hypochromism.

The transition dipole moment of the absorbing chromophore, $\langle \Psi_0 | \underset{\sim}{\boldsymbol{\mu}} | \Psi_a \rangle$, will interact with the induced dipoles ($\boldsymbol{\mu}_{\text{ind}}$) of the neighboring chromophores. Figure 7-21 shows the result for two schematic molecular structures. In a parallel stack, all the dipoles are aligned in a mutually repelling fashion. This makes it more difficult

[§] The product $\langle \Psi_0 | \underset{\sim}{\boldsymbol{\mu}} | \Psi_b \rangle \langle \Psi_b | \underset{\sim}{\boldsymbol{\mu}} | \Psi_0 \rangle$ in Equation 7-58 is a tensor product. It can be expressed as a matrix by considering the i,jth element.

$$\hat{\mathbf{i}} \cdot \langle \Psi_0 | \underset{\sim}{\boldsymbol{\mu}} | \Psi_b \rangle \langle \Psi_b | \underset{\sim}{\boldsymbol{\mu}} | \Psi_0 \rangle \cdot \hat{\mathbf{j}} = \langle \Psi_0 | \underset{\sim}{\boldsymbol{\mu}}_x | \Psi_b \rangle \langle \Psi_b | \underset{\sim}{\boldsymbol{\mu}}_y | \Psi_0 \rangle$$

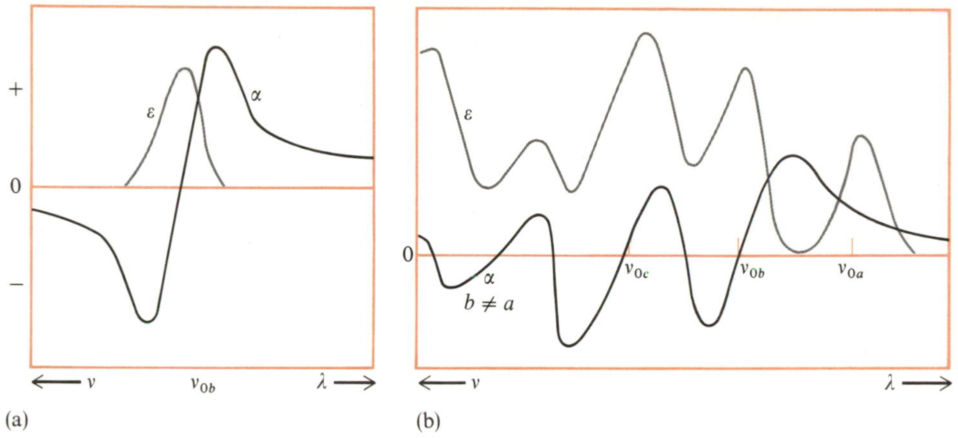

Figure 7-20

The relationship between extinction coefficient and polarizability. **(a)** The contribution of a single absorption band to the polarizability. **(b)** Polarizability as a function of frequency for a molecule with many absorption bands. Shown is the polarizability contributed by all bands except the longest-wavelength transition at ν_{0a}. Note that this polarizability is always positive near ν_{0a}.

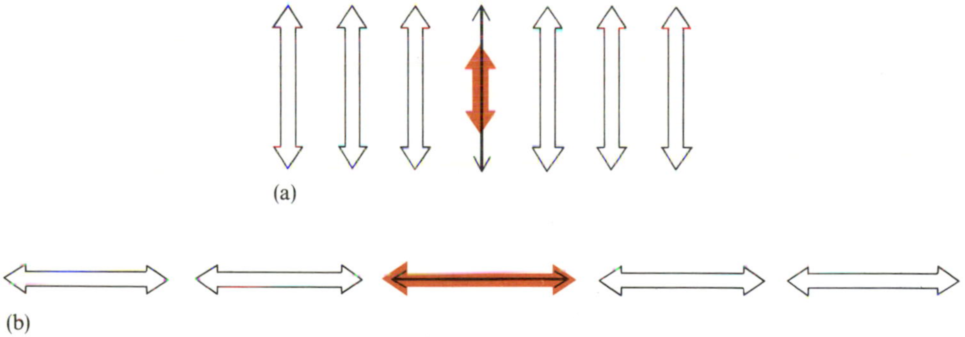

Figure 7-21

Schematic diagram showing origin of hypochromism and hyperchromism. **(a)** This alignment of induced dipoles (*unshaded*) and transition dipole (*black*) produces hypochromism (*shaded*). **(b)** This alignment of induced dipoles and transition dipole produces hyperchromism.

to create a transition dipole on the absorbing chromophore. The result will be a shorter dipole, which means less absorption—that is, hypochromism. The opposite result occurs in an end-to-end arrangement because in this case the induced dipoles attract the transition dipole. The transition dipole lengthens, and hyperchromism occurs.

These results are fairly general for aromatic chromophores because the polarizability tensor has large components in the plane of the chromophore, and the transition dipoles are also in this plane. However, it is important to remember that these predictions are valid only when one is considering the lowest electronic absorption transition. A basic principle of spectroscopy, the Kuhn–Thomas sum rule, states that the overall intensity integrated over all transitions must remain constant and is unaffected by interactions between states. Therefore, if the lowest-energy higher

Box 7-3 THEORY OF HYPOCHROMISM

The discussion of hypochromism in the text is too qualitative to satisfy any reader who likes to see physical effects expressed in mathematical terms. Here is an oversimplified sketch (after Bush, 1974) of how the Tinoco theory can be used to calculate the hypochromism of a dimer. The wavefunctions used earlier in the text were constructed from linear combinations of *unperturbed* monomer wavefunctions. These cannot lead to hypochromism. More accurate wavefunctions must be used. Greater accuracy is achieved by allowing the wavefunction for each state to contain small amounts of wavefunctions of highly excited states. Perturbation theory describes how to calculate the coefficients used to include these contributions.

For the ground state, we allow singly and doubly excited configurations to mix in. The first-order perturbation wavefunction is

$$\Psi_0^1 = \phi_{10}\phi_{20} + \sum_b \sum_{b'} \phi_{1b}\phi_{2b'} V_{1b,2b'}/(E_{b'} + E_b - 2E_0)$$

For the first excited state, only configurations with one additional excitation are considered. The first-order perturbation wavefunctions of the two dimer states are

$$\Psi_{A\pm}^1 = (1/\sqrt{2})(\phi_{1a}\phi_{20} \pm \phi_{10}\phi_{2a}) + \sum_b \phi_{10}\phi_{2b}[V_{1a,2b}/(E_b - E_a)] + \sum_{b'} \phi_{1b'}\phi_{20}[V_{1b',2a}/(E_{b'} - E_a)]$$

The coupling energies $V_{1c,2d}$ have the general form $\langle \phi_{10}\phi_{2d}|\underset{\sim}{\mathbf{V}}|\phi_{1c}\phi_{20}\rangle$ used before, where $\underset{\sim}{\mathbf{V}}$ contains the electrostatic interaction between the two monomers and usually is taken as a dipole–dipole term (Eqn. 7-45). Therefore, these energies represent the interaction between transition dipoles for the $0 \rightarrow c$ transition on monomer 1 with the $0 \rightarrow d$ transition on monomer 2.

The perturbed wavefunctions are used to calculate the dipole strengths of the two dimer transitions, transition has less intensity, some higher-energy transitions must have more, and vice versa. However, for any given absorption band, hypochromic effects of the order of 10% to 50% frequently are observed in ordered macromolecular structures.

The strength of the hypochromism depends on the inverse cube of the distance between the interacting chromophores. (See Box 7-3.) This means that hypochromism is observed only when chromophores are packed fairly close together, and then the effect is dominated by near neighbors. The exact geometric structure also affects the hypochromism. Consider two stacked planar chromophores. If the angle between the planes varies, the effect will be marked. However, if they are kept in parallel planes, rotation of one chromophore in its plane will have a much smaller effect on the hypochromism. This is because the polarizability of planar chromophores is usually relatively isotropic in the plane.

$$D_{0A^+} = |\langle \Psi_0^1 | \underset{\sim}{\mu} | \Psi_{A^+}^1 \rangle|^2$$

$$D_{0A^-} = |\langle \Psi_0^1 | \underset{\sim}{\mu} | \Psi_{A^-}^1 \rangle|^2$$

just as we did in the text. Then the dipole strength of a dimer is computed by summing the strength of the two transitions. The result is

$$\begin{aligned} D_{0A} &= D_{0A^+} + D_{0A^-} \\ &= 2D_{0a} + \sum_b [4E_b/(E_b^2 - E_a^2)](\langle \phi_{20} | \underset{\sim}{\mu}_2 | \phi_{2b} \rangle \cdot \langle \phi_{10} | \underset{\sim}{\mu}_1 | \phi_{1a} \rangle V_{1a,2b} \\ &\quad + \langle \phi_{10} | \underset{\sim}{\mu}_1 | \phi_{1b} \rangle \cdot \langle \phi_{20} | \underset{\sim}{\mu}_2 | \phi_{2a} \rangle V_{1b,2a}) \end{aligned}$$

D_{0A} depends on R_{12}^3 because that quantity enters into the terms $V_{1a,2b}$ and $V_{1b,2a}$.

Quantitatively, hypochromism usually is defined in terms of oscillator strengths (Eqn. 7-41) rather than dipole strengths. The percentage of hypochromism ($\bar{h}$) is

$$\bar{h} = (1 - f_{0A}/f_{0a})100$$

where f_{0A} is the dimer or polymer oscillator strength *per residue* and f_{0a} is the monomer oscillator strength. Sometimes it is convenient to define the percentage of hypochromicity (H) at a particular wavelength λ. By analogy to the equation above, it is

$$H(\lambda) = [1 - \varepsilon_p(\lambda)/\varepsilon_m(\lambda)]100$$

where ε_p is the polymer extinction coefficient per residue, and ε_m is the monomer extinction coefficient.

In practice, hypochromic effects are rarely used to analyze structural details. They are a powerful indication of stacked or end-to-end structures. Beyond this, the effect is usually employed as a simple quantitative index of ordered structure formation or disappearance in response to environmental perturbations. Many examples of this will be seen in later chapters, where conformational changes in nucleic acids are discussed.

Determination of chromophore orientation by linear dichroism

For all protein and nucleic acid spectra discussed thus far, the incident light used was nonpolarized, and the molecules were arranged at random in solution. Therefore, the absorption intensity $|\langle\Psi_0|\underset{\sim}{\boldsymbol{\mu}}|\Psi_a\rangle \cdot \mathbf{E}|^2$ actually measured is an average over all orientations. This is proportional to $|\langle\Psi_0|\underset{\sim}{\boldsymbol{\mu}}|\Psi_a\rangle|^2|\mathbf{E}|^2$. When oriented samples and polarized light are used, one can measure linear dichroism as described earlier. Here we consider a few practical examples.

Consider a sample of DNA oriented so that all double helixes are aligned (Fig. 7-22). From our knowledge about the electronic properties of DNA, it is easy

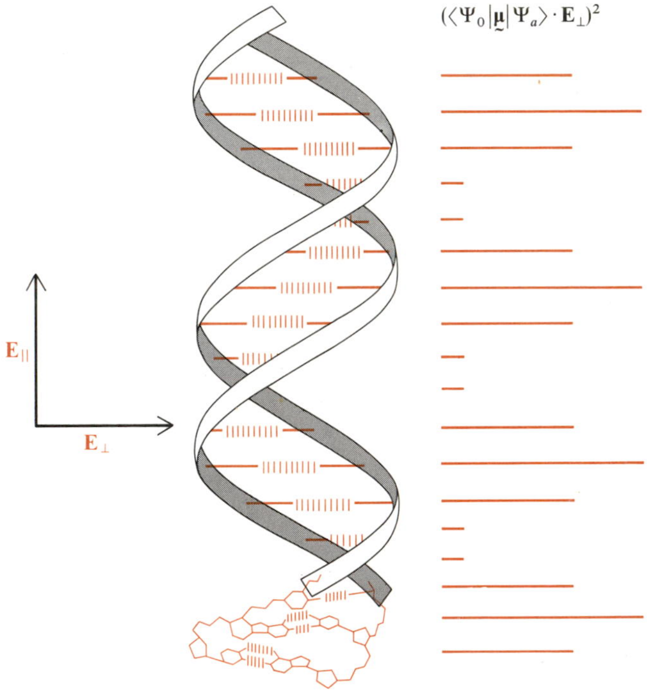

Figure 7-22

Linear dichroism expected for the B-form DNA double helix, when aligned as shown relative to polarized incident light. Because $\langle\Psi_0|\underset{\sim}{\boldsymbol{\mu}}|\Psi_a\rangle$ is in the plane of the base pairs, it is always perpendicular to $\mathbf{E}_{||}$. The intensity of absorption will be periodic along the helix because the angle between $\langle\Psi_0|\underset{\sim}{\boldsymbol{\mu}}|\Psi_a\rangle$ and $\mathbf{E}$ varies with each 36° rotation of successive base pairs. [DNA structure after A. Kornberg, *DNA Synthesis* (San Francisco: W. H. Freeman and Company, 1974).]

to predict that the chances of absorbing light polarized parallel or perpendicular to the helix axis are very different. Remember that, for the main 260 nm π–π^* transition of the nucleic acid bases, the transition dipole lies in the plane of the bases. This means that $\langle\Psi_0|\underset{\sim}{\mu}|\Psi_a\rangle$ is perpendicular to $\mathbf{E}_{||}$. Thus, this polarized component can be only weakly absorbed. In contrast, $\langle\Psi_0|\underset{\sim}{\mu}|\Psi_a\rangle$ is in a plane parallel to $\mathbf{E}_{\perp}$, so substantial absorption occurs. The result is that the dichroic ratio $d = (A_{||} - A_{\perp})/(A_{||} + A_{\perp})$ is less than zero. The observed negative dichroism of DNA is a strong piece of evidence that, in solution, the structure of the double helix must resemble the B form with base pairs in planes approximately perpendicular to the long axis of the molecule.

Another example of linear dichroism is shown in the two $\pi \rightarrow \pi^*$ exciton-split bands of the poly-L-glutamic acid α helix (Fig. 7-5). The peptide transition dipole is skewed at an angle with respect to the helix axis. Exciton interactions between peptide monomers lead to only two intense absorption bands. One is polarized parallel to, and the other perpendicular to, the helix axis. Thus, in an oriented sample, each band exhibits strong linear dichroism. The dichroism of the two bands has opposite signs: the parallel polarized component shows positive dichroism, the perpendicular shows negative.

Summary

The basic principles of molecular electronic spectroscopy are derived from quantum mechanical considerations. These same principles apply, essentially unchanged, to all types of spectroscopy. The primary sources of structural information contained in a spectrum are the wavelength of a transition, which reflects the energy difference between the two states involved, and the intensity, which is proportional to the square of the light-induced dipole moment (the transition dipole).

In proteins the peptide chromophore dominates far-UV absorption, whereas aromatic amino acid side chains make the major contribution to the near-UV spectrum. In nucleic acids, the bases are the dominant absorbing moieties.

Several different effects render the spectrum of a polymer different from the sum of the spectra of its monomeric constituents. Simple environmental effects alter the relative energy of ground and excited states, and this alteration causes spectral shifts. Electronic interactions between chromophores lead to two phenomena: exciton splitting and hypochromism. Exciton splitting arises from the interaction of two identical or near-identical monomer excited states. In a molecule with two chromophores, it results in doubling the number of absorption bands. The intensity of each pair of dimer bands, relative to the corresponding monomer band, depends on the angle between the two chromophores. The energy separation between the pair of exciton bands depends on orientation, but it also decreases as the cube of the distance between the two chromophores. In a polymer with n residues, exciton effects, in principle, split each monomer transition into n bands, but usually all but a few of these have no intensity.

Hypochromism arises from the interaction of a particular electronic excited state of one chromophore with different electronic excited states on other chromophores. For a parallel stacked array, it always results in a diminution of the intensity of the lowest-energy electronic absorption band.

In oriented polymers, the absorption intensity of plane polarized light depends on the angle between the polarization direction and excitation direction. This effect, called linear dichroism, can be used in favorable cases to determine how chromophores are oriented within a macromolecule.

Problems

7-1. Consider a trimer with three identical chromophores. Suppose that the monomers within are oriented such that $V_{12} = V_{23} = V$, and $V_{13} = 0$, where V_{12} is given by Equation 7-51, and V_{23} and V_{13} are given by analogous expressions in which index 1 or 2 is replaced in each term by index 2 or 3, respectively. The wavefunctions for such a trimer in terms of monomer ground- and excited-state wavefunctions are

$$\text{Ground:} \quad \Psi_0 = \phi_1\phi_2\phi_3$$

$$\text{Excited:} \quad \Psi_{A^0} = (1/\sqrt{2})(\phi_{1a}\phi_{20}\phi_{30} - \phi_{10}\phi_{20}\phi_{3a})$$

$$\Psi_{A^+} = (1/2)(\phi_{1a}\phi_{20}\phi_{30} + \sqrt{2}\phi_{10}\phi_{2a}\phi_{30} + \phi_{10}\phi_{20}\phi_{3a})$$

$$\Psi_{A^-} = (1/2)(-\phi_{1a}\phi_{20}\phi_{30} + \sqrt{2}\phi_{10}\phi_{2a}\phi_{30} - \phi_{10}\phi_{20}\phi_{3a})$$

a. Show that these wavefunctions are orthogonal and normalized.

b. Show that these wavefunctions are eigenfunctions of the Hamiltonian

$$\underset{\sim}{H} = \underset{\sim}{H}_1 + \underset{\sim}{H}_2 + \underset{\sim}{H}_3 + \underset{\sim}{V}_{12} + \underset{\sim}{V}_{23} + \underset{\sim}{V}_{13}$$

Compute the energy of each state.

c. Show that the frequencies of the three spectral transitions in the trimer are

$$\nu_{0\to A^0} = \nu_{0a}$$

$$\nu_{0\to A^\pm} = \nu_{0a} \pm \sqrt{2}V/h$$

d. Show that the three excited-state wavefunctions are stationary states of the Hamiltonian given in part b.

e. Show that the dipole strengths of the three trimer absorption bands are

$$D_{0A^0} = D_{0a}(1 - \cos\theta_{13})$$

$$D_{0A^+} = D_{0a}[1 + (\sqrt{2}/2)(\cos\theta_{12} + \cos\theta_{23}) + (1/2)\cos\theta_{13})]$$

$$D_{0A^-} = D_{0a}[1 - (\sqrt{2}/2)(\cos\theta_{12} + \cos\theta_{23}) + (1/2)\cos\theta_{13})]$$

where θ_{ij} refers to the angle between transition dipoles on monomers i and j, and D_{0a} is the monomer dipole strength.

f. Show that exciton effects lead to no hypochromism in the trimer.

g. Show that, if $V_{13} \neq 0$, then the three excited-state wavefunctions no longer are stationary states of the Hamiltonian.

7-2. Consider two monomers in a stacked dimer with an angle of 60° between the transition moments. Compute the linear dichroism expected for the ν_{0A^+} and ν_{0A^-} transitions, assuming that $\mathbf{E}_{\parallel}$ is parallel to the transition dipole of one of the monomers, and $\mathbf{E}_{\perp}$ is perpendicular to it but still located in the plane of the monomer.

7-3. Suppose you have a solution of molecules with radius 10 Å; the solution has an absorbance of 1.0 at 260 nm. The molar extinction coefficient is 10^5 cm^2 M^{-1}. Spontaneously, all of the molecules aggregate to form one spherical particle that contains cubic-closest-packed small molecules. If this particle is still in the light path of your spectrometer, estimate the absorbance of the aggregate. You may assume that the light actually illuminates all of a $1 \times 1 \times 1$ cm^3 sample cell.

7-4. Assume you have a stacked linear dimer, and an equivalent tetramer,

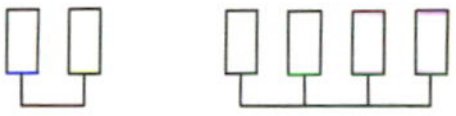

dimer tetramer

in which the distance between adjacent bases is the same as in the dimer. Suppose the dimer is 10% hypochromic. Estimate the hypochromism of the tetramer, using the inverse-cube distance dependance of the interaction that produces the hypochromism.

7-5. A molecule absorbs light with $\lambda_{max} = 280$ nm in cyclohexane. Assume that, in water, the only solvent effect is due to formation of a hydrogen bond between water and the excited state (but not the ground state) of the molecule. If the hydrogen bond has 5 kcal of energy, predict the λ_{max} of the molecule in water.

References

GENERAL

Birks, J. B. 1970. *Photophysics of Aromatic Molecules*. New York: Wiley-Interscience.

Eyring, H., J. Walter, and G. E. Kimble. 1944. *Quantum Chemistry*. New York: Wiley. [Old, but still one of the best available treatments of spectroscopy.]

Freifelder, D. 1976. *Physical Biochemistry*. San Francisco: W. H. Freeman and Company. [A good elementary introduction.]

Jaffe, H. H., and M. Orchin. 1962. *Theory and Application of Ultraviolet Spectroscopy*. New York: Wiley.

Kauzmann, W. 1957. *Quantum Chemistry*. New York: Academic Press. [Good comparison of classical and quantum treatments of optical properties.]

Merzbacher, E. 1970. *Quantum Mechanics*, 2nd ed. New York: Wiley. [A good treatment of time-dependent perturbation theory.]

SPECIFIC

Bush, C. A. 1974. Ultraviolet spectroscopy, circular dichroism and optical rotatory dispersion. In *Basic Principles in Nucleic Acid Chemistry*, ed. P. O. P. Ts'o (New York: Academic Press), p. 92.

Donovan, J. W. 1973. Ultraviolet difference spectroscopy: new techniques and applications. Spectrophotometric titrations of the functional groups of proteins. In *Methods in Enzymology*, vol. 27, ed. C. H. W. Hirs and S. N. Timasheff (New York: Academic Press), p. 497, p. 525.

Gratzer, W. B. 1967. Ultraviolet absorption spectra of polypeptides. In *Poly-L-amino Acids*, ed. G. D. Fasman (New York: Marcel Dekker), p. 177.

Hofrichter, J., and W. A. Eaton. 1976. Linear dichroism of biological chromophores. *Ann. Rev. Biophys. Bioengin.* 5:511.

Holzwarth, G. 1972. Ultraviolet spectroscopy of biological membranes. In *Membrane Molecular Biology*, ed. C. F. Fox and A. Keith (Stamford, Conn.: Sinauer Assoc.), p. 228.

Kasha, M. 1963. Energy transfer mechanisms and the molecular exciton model for molecular aggregates. *Radiation Res.* 20:55.

Timasheff, S. N. 1970. Some physical probes of enzyme structure in solution. In *The Enzymes*, 3d ed., vol. 2, ed. P. D. Boyer (New York: Academic Press), p. 371.

Tinoco, I., Jr. 1962. Theoretical aspects of optical activity, part two: Polymers. *Adv. Chem. Phys.* 4:113.

———. 1965. Absorption and rotation of polarized light by polymers. In *Molecular Biophysics*, ed. M. Weissbluth (New York: Academic Press), p. 269.

8
Other optical techniques

8-1 OPTICAL ACTIVITY

Nearly all molecules synthesized by living organisms are optically active. In fact, optical activity is such a pervasive element of life (as we know it) that it is one of the major criteria used to judge, for example, whether meteorites contain evidence of life elsewhere in the universe. The optical activity of small molecules arises from their lack of symmetry—particularly from the presence of asymmetric carbon atoms and from the effect these atoms have on any nearby chromophores. Here we are less interested in such questions of molecular configuration. Rather, our main focus is to examine how the conformation of a macromolecule can affect its optical activity. Just as with absorption spectra, we shall find that interactions between neighboring chromophores play a most important role.

Experimental detection of optical activity

There are at least four ways that an optically active sample can alter the properties of transmitted light: optical rotation, ellipticity, circular dichroism, and circular birefringence. In Figure 8-1, one is looking along the direction of propagation of a light wave. Suppose one starts with plane-polarized light (Fig. 8-1a). An observer will see the electric vector **E** oscillate sinusoidally in a plane: $\mathbf{E} = \hat{\mathbf{i}}E_0 \sin \omega t$, where $\hat{\mathbf{i}}$ is a unit vector in the x direction, and $\omega = 2\pi\nu$ is the circular frequency of the light.

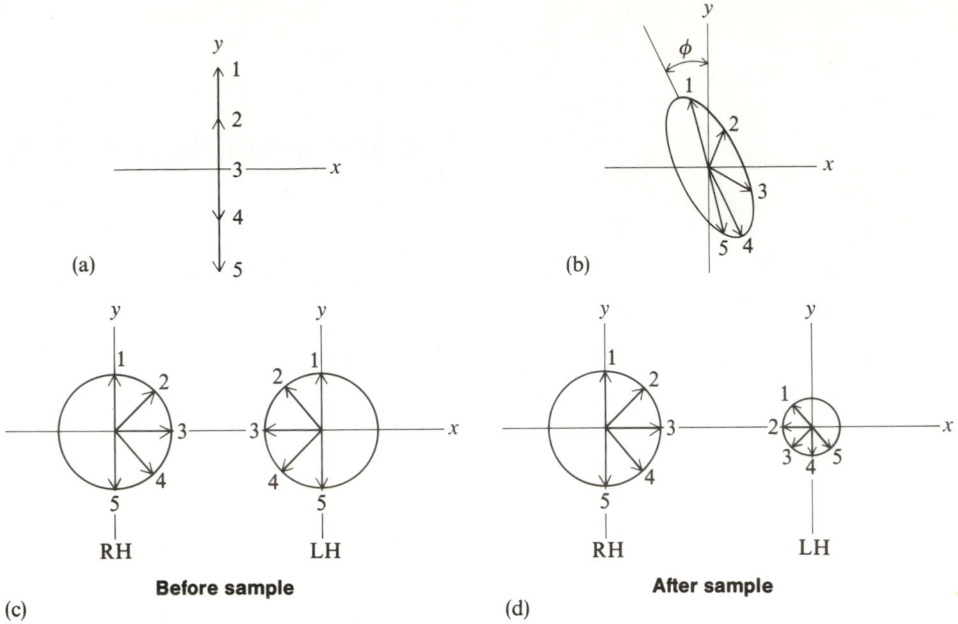

Figure 8-1

Effect of an optically active absorbing sample on incident linearly polarized light. All drawings show the electric field vector viewed along the direction of light propagation. Points 1 through 5 correspond to equal (increasing) time intervals. **(a)** Incident linearly polarized light. **(b)** Elliptically polarized light produced by passing the incident light through an optically active sample. **(c)** Resolution of linearly polarized light into individual right-hand and left-hand circularly polarized components. **(d)** Effect of an optically active sample on the two circularly polarized components. The sum of measurements made with these two separate components must be identical to the result obtained in part b.

After passing through an optically active absorbing sample, the light is changed in two aspects. The maximal amplitude of **E** is no longer confined to a plane; instead it traces out an ellipse (Fig. 8-1b). (Actually, **E** is an elliptical screw but, because we are concerned only with the projection of **E** onto the x–y plane perpendicular to the propagation direction, all that we see is an ellipse.) The ellipticity of the light is one measure of optical activity. Ellipticity is defined as the arc tangent of the ratio of the minor axis to the major axis of the ellipse. For example, an ellipse with a minor/major axial ratio of 1/100 will have an ellipticity of 0.57 degree.

The orientation of the ellipse is the second indication of optical activity. Note that the major axis of the ellipse is not parallel to the polarization direction of the incident light. If the sample absorbed virtually no light, the ellipse would have such a small axial ratio that, for all practical purposes, it would still be equivalent to plane-polarized light. In this case, one could simply say that the plane of polarization had been rotated. Thus the orientation of the ellipse corresponds to optical rotation.

The optical rotation as a function of wavelength is called the optical rotatory dispersion (ORD).

Instead of considering plane-polarized light, we can resolve **E** into two equal-amplitude components of opposite circular polarization:

$$\mathbf{E}_R = \tfrac{1}{2}(\hat{\mathbf{i}}E_0 \sin \omega t + \hat{\mathbf{j}}E_0 \cos \omega t)$$

$$\mathbf{E}_L = \tfrac{1}{2}(\hat{\mathbf{i}}E_0 \sin \omega t - \hat{\mathbf{j}}E_0 \cos \omega t)$$

where $\hat{\mathbf{j}}$ is a unit vector in the y direction (Fig. 8-1c). If these two components are added at each time point, the result is simply plane-polarized light. It is convenient to examine the effect of an optically active medium on each component separately. It turns out that in such a medium the absorbance of left circularly polarized light (A_L) is different from the absorbance of right circularly polarized light (A_R). After passing through the sample, each component is still circularly polarized, but the radii of the circles traced out by the electric vector of each are now different. This is the phenomenon called circular dichroism (CD). When these two opposite circularly polarized light waves are combined, the result will be elliptically polarized light because the two components have different amplitudes (Fig. 8-1d). Thus, CD is equivalent to ellipticity.

Figure 8-1b shows a fourth and final manifestation of optical activity. If the two circularly polarized components are absorbed to different extents at *any* wavelength, then it turns out that the sample will also have a different index of refraction (n) for the two components at virtually *all* wavelengths.[§] This means that one will propagate more rapidly than the other through the medium. The result is a phase shift between the two components, proportional to the refractive index difference, $n_L - n_R$. This effect is called circular birefringence. When the two components are combined, the phase shift results in a permanent rotation of the long axis of the elliptically polarized light. Thus circular birefringence is equivalent to optical rotation.

It is a relatively straightforward algebraic exercise to derive actual expressions for the relationships between CD and ellipticity (θ) and between the circular birefringence and the optical rotation (ϕ). The results for a sample of length l are

$$\phi = 180l(n_L - n_R)/\lambda \text{ degrees} \tag{8-1}$$

$$\theta = 2.303(A_L - A_R)180/4\pi \text{ degrees} \tag{8-2}$$

In practice, usually only two of these quantities are measured experimentally (see Box 8-1). The circular birefringence usually is a very small number, so it is far more

[§] Figure 7-20a illustrates the fact that an absorption band at a particular frequency influences the polarizability (and thus the refractive index) at all frequencies.

convenient to measure ϕ directly. For typical protein or nucleic acid solutions at 10^{-4}M chromophore concentrations, the plane of polarized light will be rotated by 0.01 to 0.1 degree for a 1 cm sample. Current instruments can detect rotations as small as 10^{-4} degree.

CD can be measured easily be exposing a sample alternately to left-hand and right-hand circularly polarized light and detecting just the differential absorption. This difference typically is about 0.03% to 0.3% of the total absorption, a difference

Box 8-1 TYPICAL MAGNITUDES IN OPTICAL ACTIVITY MEASUREMENTS

Optical rotation and circular birefringence. Suppose that a sample of 1 cm pathlength rotates 300 nm polarized light by 0.01 deg. The corresponding circular birefringence (from Eqn. 8-1) will be

$$n_L - n_R = \phi\lambda/180l = (0.01)(3 \times 10^{-5}\text{ cm})/(180)(1\text{ cm}) = 1.67 \times 10^{-9}$$

This value is almost undetectably small. If such a sample were at 10^{-4} M concentration, the molar rotation (from Eqn. 8-3) would be

$$[\phi] = 100\phi/Cl = (100)(0.01)/(10^{-4}\text{ M})(1\text{ cm}) = 10^4\text{ deg M}^{-1}\text{ cm}^{-1}$$

This is a typical value for an individual optically active chromophore. Residue rotation for a monomer imbedded in a helical polymer often is ten times larger.

Ellipticity and circular dichroism. Suppose that a sample of 1 cm pathlength has an ellipticity of 0.01 deg. Because arc tan $a/b \approx a/b$ radians for small a/b, the axial ratio (a/b) of the ellipse will be

$$a/b = (2\pi/360)\theta = (2\pi/360)(0.01) = 1.76 \times 10^{-4}$$

Such an ellipse, drawn to any reasonable scale, would differ imperceptibly from a straight line. The circular dichroism corresponding to an ellipticity of 0.01 deg (from Eqn. 8-2) will be

$$A_L - A_R = 4\pi(0.01)/(2.303)(180) = 3.03 \times 10^{-4}$$

This value is easily measurable. If the sample is at 10^{-4} M concentration, the molar ellipticity (from Eqn. 8-4) will be

$$[\theta] = 100\theta/Cl = 100(0.01)/(10^{-4}\text{ M})(1\text{ cm}) = 10^4\text{ deg M}^{-1}\text{ cm}^{-1}$$

This is a typical value for one isolated monomer residue. The per-residue ellipticities in a polymer are often ten times larger. The value of $\Delta\varepsilon$ is easily calculated from $(A_L - A_R)/Cl = [\theta]/3{,}300 = 3.303$.

that can be determined quite accurately with modern instrumentation. The ellipticity of the light that emerges from an optically active sample usually is very small and would be difficult to measure accurately. However, even though the actual measurements usually are made of $A_L - A_R$, it is common practice to convert the data to ellipticity using Equation 8-2. Because ϕ and θ have the same units (degrees), this conversion facilitates comparisons between the two sets of measurements.

To compare results from different samples, it is necessary to compute optical activity on a molar or residue basis. The following equations define molar ellipticity ($[\theta]$) and molar rotation ($[\phi]$) in terms of the measured quantities, the concentration of the sample (C) in moles per liter, and the path length (l).

$$[\phi] = 100\phi/Cl \tag{8-3}$$

$$[\theta] = 100\theta/Cl \tag{8-4}$$

Circular dichroism sometimes is reported as $\Delta\varepsilon = \varepsilon_L - \varepsilon_R$. Using Equations 8-1 and 8-4, and the Beer–Lambert law, one can show that

$$[\theta] = 3{,}300\,\Delta\varepsilon \tag{8-5}$$

If, as is usual for macromolecules, C is expressed in moles of residues, then these three equations also define the residue ellipticity and residue rotation. (Typical values for all these quantities are described in Box 8-1.)

Relation between ORD and CD

We have already reduced four manifestations of optical activity to two: $[\phi]$ and $[\theta]$. It turns out that these two are not independent either. They are related by a set of integrals called the Kronig–Kramers transforms. These integrals are very general statements about the response of a system to a perturbation such as light. The Kronig–Kramers transforms can be written as

$$[\phi(\lambda)] = \frac{2}{\pi} ⨍_0^\infty \frac{[\theta(\lambda')]\lambda'}{\lambda^2 - \lambda'^2}\, d\lambda' \tag{8-6}$$

$$[\theta(\lambda)] = -\frac{2\lambda}{\pi} ⨍_0^\infty \frac{[\phi(\lambda')]}{\lambda^2 - \lambda'^2}\, d\lambda' \tag{8-7}$$

where the center bar on the integral signs implies how the infinity at $\lambda' = \lambda$ should be treated.

These equations imply that, if the ORD is known at all wavelengths, then one can rigorously calculate the CD (ellipticity), and vice versa. In practice, even if data are available over only a limited wavelength range, it still is possible to approximate the integrals needed for Kronig–Kramers transforms. The resulting transformed data usually are about as accurate as real experimental data. Thus, the information available from experimentally measured $[\theta]$ and $[\phi]$ is redundant. Under most circumstances, it is sufficient to measure only one quantity. For reasons that will be apparent in a moment, CD usually is the measurement of choice.

The ORD and CD of a sample depend strongly on the wavelength of light used to perform the measurement. For absorbing samples, one typically determines the ORD or CD over the same wavelength range used to record an absorption spectrum. The resulting optical activity spectra are called ORD and CD spectra. If the sample contains only strongly allowed electronic transitions (such as $\pi \rightarrow \pi^*$), the shape of the CD spectrum (often called a Cotton effect) is related to the absorption spectrum in a very simple way (Fig. 8-2). Outside the regions of absorption, $[\theta] = 0$. This is reasonable because $[\theta] \propto A_L - A_R$, and both A_L and A_R are zero. In absorbing regions, $[\theta]$ has the same shape as the absorption spectrum. However, it can have a positive or a negative sign. Both the sign and the integrated intensity of each CD band are sensitive functions of molecular structure. Thus CD provides two additional experimental parameters for each electronic transition of a macromolecule.

The ORD spectra in Figure 8-2 look at first glance like derivatives of the CD spectra. There is a point of zero optical rotation (the crossover point) that is coincident with each CD maximum or minium. However, the ORD spectra die off quite slowly at wavelengths outside the absorption band. The limiting form, called a Drude equation, is

$$[(\phi(\lambda)] = A_0/(\lambda^2 - \lambda_0^2) \tag{8-8}$$

where A_0 is a constant related to the intensity of the corresponding CD spectrum, and λ_0 is the crossover wavelength. The form of the Drude equation shows why molecules such as sucrose can demonstrate substantial optical rotation with visible light, even though they cannot absorb this light.

In the early days of optical activity measurements on proteins and nucleic acids, CD instrumentation was unavailable, and ORD could be measured only at wavelengths far away from any absorption bands. Therefore, experimental data had to be fit to the Drude equation. Conclusions about molecular structure and conformation were derived from the values of the parameters A and λ_0. In some cases. more than one Drude term (or similar terms) had to be used to fit data. Although this approach was extremely useful, it has now been superseded by methods that will be described later. These methods make use of the whole range of data currently accessible experimentally.

Physical origins

Suppose the absorption spectrum of a molecule is known. Because CD and absorption bands usually have the same shape, one could predict the CD spectrum from the

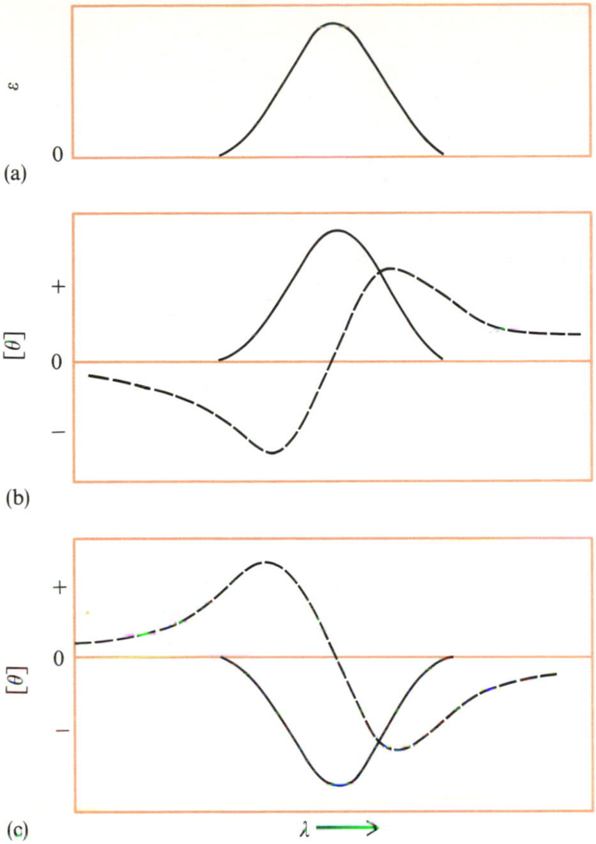

Figure 8-2

Schematic spectra for an allowed transition. **(a)** Absorption spectrum. **(b)** CD (*solid line*) and ORD (*dashed line*) spectra for a positive Cotton effect. **(c)** CD and ORD spectra for a negative Cotton effect.

absorption spectrum, just by computing the area under each CD band and noting the sign. These quantities together are defined as the rotational strength (R_{0a}) of the $0 \rightarrow a$ transition. The rotational strength is analogous to the dipole strength used as a measure of absorption intensity (see Chapter 7).

$$R_{0a} = (3hc/8\pi^3 N_0) \int \{[\theta(\lambda)]/\lambda\} \, d\lambda \tag{8-9}$$

where h is Planck's constant, c is the speed of light, and the integral is taken over only the CD produced by the transition from state 0 to state a.

L. Rosenfeld first showed in 1928 how the rotational strength can be computed from quantum mechanical principles and a knowledge of the wavefunctions of the ground (Ψ_0) and excited (Ψ_a) states of an asymmetric molecule. The result is

$$R_{0a} = \text{imag}(\langle\Psi_0|\underset{\sim}{\mu}|\Psi_a\rangle \cdot \langle\Psi_a|\underset{\sim}{m}|\Psi_0\rangle) \tag{8-10}$$

where $\underset{\sim}{\mu}$ is an electric dipole operator, $\underset{\sim}{m}$ is a magnetic dipole operator, and "imag" means the imaginary part of the expression that follows. For each electron, $\underset{\sim}{m}$ can be written as

$$\underset{\sim}{m} = (e/2mc)(\underset{\sim}{r} \times \underset{\sim}{p}) \tag{8-11}$$

where e and m are the charge and mass of the electron, $\underset{\sim}{p}$ is the momentum operator, and $\underset{\sim}{r}$ is the position operator of the electron. The term $\underset{\sim}{r} \times \underset{\sim}{p}$ is simply orbital angular momentum of an electron, so the magnetic dipole operator $\underset{\sim}{m}$ really corresponds to a circulation of charge. (See Box 8-2 for a review of the properties of the cross product.) Physically, the magnetic transition dipole $\langle\Psi_a|\underset{\sim}{m}|\Psi_0\rangle$ can be regarded as a light-induced current loop, in the same way that the electric transition dipole is a light-induced oscillating dipole (Fig. 8-3). The linear momentum operator is $\underset{\sim}{p} = (h/i)(\nabla)$.

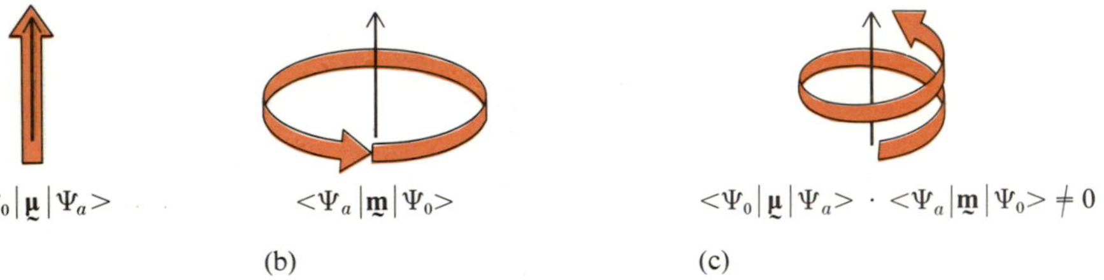

Figure 8-3

Schematic diagram of light-induced charge displacement in a molecule. **(a)** Pure electronic absorption. **(b)** Pure magnetic absorption. **(c)** Optical activity.

Because this is a pure imaginary operator, the magnetic transition dipole also is imaginary. However, R_{0a} is the imaginary part of the product between a real number and an imaginary number. Therefore, R_{0a} itself is a real number—as it must be because it is a physically observable quantity.

The actual derivation of Equation 8-10 is a lengthy and complex algebraic process. A reader with a good background in quantum mechanics and electromagnetic theory is invited to look elsewhere (Schellman, 1975). Here we shall just explore the properties of R_{0a} in the hope of gaining some physical insight into the origins of optical activity. First recall the simpler expression for absorption intensity that we

Box 8-2 VECTOR PRODUCTS

There are four commonly defined products of vectors that are used in this text. These are the dot product, the cross product, the scalar triple product, and the tensor product. They are defined as follows, for vectors $\mathbf{A} = \hat{\mathbf{i}}A_x + \hat{\mathbf{j}}A_y + \hat{\mathbf{k}}A_z$, and $\mathbf{B} = \hat{\mathbf{i}}B_x + \hat{\mathbf{j}}B_y + \hat{\mathbf{k}}B_z$.

The dot product:

$$\mathbf{A} \cdot \mathbf{B} = A_xB_x + A_yB_y + A_zB_z$$

The dot product is a scalar, and $\mathbf{A} \cdot \mathbf{B} = \mathbf{B} \cdot \mathbf{A}$. The dot product of a vector with itself is the length of the vector squared. In matrix notation,

$$\mathbf{A} \cdot \mathbf{B} = (A_xA_yA_z)\begin{pmatrix} B_x \\ B_y \\ B_z \end{pmatrix}$$

The cross product:

$$\mathbf{A} \times \mathbf{B} = \hat{\mathbf{i}}(A_yB_z - A_zB_y) + \hat{\mathbf{j}}(A_xB_z - A_zB_x) + \hat{\mathbf{k}}(A_xB_y - A_yB_x)$$

It is most convenient to remember this definition as a determinant:

$$\mathbf{A} \times \mathbf{B} = \begin{vmatrix} \hat{\mathbf{i}} & \hat{\mathbf{j}} & \hat{\mathbf{k}} \\ A_x & A_y & A_z \\ B_x & B_y & B_z \end{vmatrix}$$

In this form, it is clear that $\mathbf{A} \times \mathbf{B} = -\mathbf{B} \times \mathbf{A}$. The cross product of two vectors is a third vector perpendicular to the plane formed by the two vectors.

The scalar triple product: $\mathbf{A} \cdot \mathbf{B} \times \mathbf{C}$. This product can be put into explicit form by combining the expressions for the dot and cross products. This product is a scalar, and it obeys the relationship

$$\mathbf{A} \cdot \mathbf{B} \times \mathbf{C} = -\mathbf{A} \cdot \mathbf{C} \times \mathbf{B} = \mathbf{C} \cdot \mathbf{A} \times \mathbf{B} = -\mathbf{C} \cdot \mathbf{B} \times \mathbf{A} = \mathbf{C} \times \mathbf{B} \cdot \mathbf{A} = \cdots$$

The scalar triple product of **A**, **B**, and **C** is the volume of a parallelipiped generated by the three vectors when they are placed at a common origin.

The tensor product: **AB**. For three-component vectors, the tensor product is a 3×3 matrix formed as follows:

$$\mathbf{AB} = \begin{pmatrix} A_x \\ A_y \\ A_z \end{pmatrix}(B_xB_yB_z) = \begin{pmatrix} A_xB_x & A_xB_y & A_xB_z \\ A_yB_x & A_yB_y & A_yB_z \\ A_zB_x & A_zB_y & A_zB_z \end{pmatrix}$$

Note that $\mathbf{BA} = (\mathbf{AB})^{\dagger}$, where the superscript $\dagger$ denotes the transpose of a matrix.

can write as the dipole strength,[§]

$$D_{0a} = \langle \Psi_0 | \underset{\sim}{\mu} | \Psi_a \rangle \cdot \langle \Psi_a | \underset{\sim}{\mu} | \Psi_0 \rangle$$

This expression says that absorption is proportional to the square of the light-induced electric dipole. For optical activity, one electric dipole matrix element was replaced by a magnetic dipole matrix element. Because R_{0a} is still a function of $\langle \Psi_0 | \underset{\sim}{\mu} | \Psi_a \rangle$, a molecule that cannot absorb light also cannot show CD, because in this case $\langle \Psi_0 | \underset{\sim}{\mu} | \Psi_a \rangle$ will be zero. In order for R_{0a} to be nonzero, $\langle \Psi_a | \underset{\sim}{\mathbf{m}} | \Psi_0 \rangle$ also must be nonzero.

A light wave contains both oscillating electric and oscillating magnetic components. The electric component usually dominates pure absorption properties, because magnetic effects are so much smaller than electric effects. However, optical activity is a phenomenon that involves combined electric and magnetic interactions. The critical feature of Equation 8-10 is the dot product. This dot product means that, for a molecule to be optically active, $\langle \Psi_a | \underset{\sim}{\mathbf{m}} | \Psi_0 \rangle$ must have a component parallel to $\langle \Psi_0 | \underset{\sim}{\mu} | \Psi_a \rangle$. Therefore, light must induce a helical charge circulation around the direction of $\langle \Psi_0 | \underset{\sim}{\mu} | \Psi_a \rangle$ (Fig. 8-3). For this situation to occur, the molecule must be asymmetric. Otherwise, there could be no preferred helical direction. This observation is the origin of the familiar rules for optically active molecules. It also explains why many helical macromolecules have very intense optical activity: the helical structure facilitates a helical flow of charge.

Calculation of the CD of a dimer

We are interested not so much in the optical activity of isolated chromophores as in the *changes* induced in the optical activity when a set of chromophores is assembled into a macromolecular structure. To keep things as simple as possible, consider a dimer with two identical chromophores. The same wavefunctions described in Chapter 7 for exciton effects in a dimer still apply. There will be two singly excited states (Eqn. 7-52). We must evaluate the electric and magnetic transition dipoles involved in the transition between states Ψ_0 and Ψ_A. We will simultaneously calculate results for states Ψ_{A+} and Ψ_{A-}, keeping the upper sign in all expressions for the $+$ state. The dimer electric transition dipole can be evaluated in terms of the transition dipoles of the two monomers:

$$\begin{aligned} \langle \psi_0 | \underset{\sim}{\mu} | \psi_{A\pm} \rangle &= \langle \phi_{10}\phi_{20} | \underset{\sim}{\mu}_1 + \underset{\sim}{\mu}_2 | (1/\sqrt{2})(\phi_{1a}\phi_{20} \pm \phi_{10}\phi_{2a}) \rangle \\ &= (1\sqrt{2})(\langle \phi_{10} | \underset{\sim}{\mu}_1 | \phi_{1a} \rangle \pm \langle \phi_{20} | \underset{\sim}{\mu}_2 | \phi_{2a} \rangle \end{aligned} \tag{8-12}$$

[§] Here we explicitly show D_{0a} as the product of a transition dipole and its complex conjugate (cf. Eqn. 7-40). Although $\langle \Psi_0 | \underset{\sim}{\mu} | \Psi_a \rangle = \langle \Psi_a | \underset{\sim}{\mu} | \Psi_0 \rangle$, not all operators obey such a relationship. For example, because $\underset{\sim}{\mathbf{m}}$ is imaginary, $\langle \Psi_a | \underset{\sim}{\mathbf{m}} | \Psi_0 \rangle = -\langle \Psi_0 | \underset{\sim}{\mathbf{m}} | \Psi_a \rangle$.

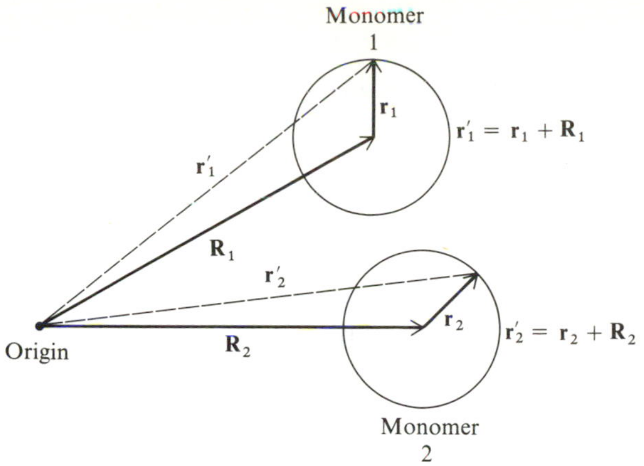

Figure 8-4

Coordinate system used to define the magnetic dipole operator of a dimer.

The magnetic dipole operator of the dimer is more complex. It will be the sum of the magnetic dipole operators of the two monomers, but it will also depend on the distance between the two monomers. This is clear if one chooses the same origin for both monomers (Fig. 8-4). Using the coordinate system of Figure 8-4, the magnetic dipole operator can be written (starting from Eqn. 8-11) as

$$\begin{aligned}\underset{\sim}{\mathbf{m}} &= (e/2mc)\underset{\sim}{\mathbf{r}}'_1 \times \underset{\sim}{\mathbf{p}}_1 + (e/2mc)\underset{\sim}{\mathbf{r}}'_2 \times \underset{\sim}{\mathbf{p}}_2 \\ &= (e/2mc)\underset{\sim}{\mathbf{r}}_1 \times \underset{\sim}{\mathbf{p}}_1 + (e/2mc)\underset{\sim}{\mathbf{r}}_2 \times \underset{\sim}{\mathbf{p}}_2 + (e/2mc)\underset{\sim}{\mathbf{R}}_1 \times \underset{\sim}{\mathbf{p}}_1 + (e/2mc)\underset{\sim}{\mathbf{R}}_2 \times \underset{\sim}{\mathbf{p}}_2 \qquad (8\text{-}13)\end{aligned}$$

Recognizing that the first two terms no longer depend on the choice of origin, we can express $\underset{\sim}{\mathbf{m}}$ as

$$\underset{\sim}{\mathbf{m}} = \underset{\sim}{\mathbf{m}}_1 + \underset{\sim}{\mathbf{m}}_2 + (e/2mc)\underset{\sim}{\mathbf{R}}_1 \times \underset{\sim}{\mathbf{p}}_1 + (e/2mc)\underset{\sim}{\mathbf{R}}_2 \times \underset{\sim}{\mathbf{p}}_2 \qquad (8\text{-}14)$$

where $\underset{\sim}{\mathbf{m}}_1$ and $\underset{\sim}{\mathbf{m}}_2$ are the one-electron magnetic dipole operators for each monomer. They are defined relative to the center of each monomer. $\underset{\sim}{\mathbf{R}}_1$ and $\underset{\sim}{\mathbf{R}}_2$ are position operators that define the location of the center of each monomer; $\underset{\sim}{\mathbf{p}}_1$ and $\underset{\sim}{\mathbf{p}}_2$ are electron momentum operators defined relative to the center of each monomer. The dimer magnetic transition dipole is calculated using Equation 8-14. The result can be expressed in terms of the properties of the individual monomers:

$$\begin{aligned}\langle\Psi_{A\pm}|\underset{\sim}{\mathbf{m}}|\Psi_0\rangle &= (1/\sqrt{2})\langle(\phi_{1a}\phi_{20} \pm \phi_{10}\phi_{2a})|\underset{\sim}{\mathbf{m}}|\phi_{10}\phi_{20}\rangle \\ &= (1/\sqrt{2})(\langle\phi_{1a}|\mathbf{m}_1|\phi_{10}\rangle \pm \langle\phi_{2a}|\underset{\sim}{\mathbf{m}}_2|\phi_{20}\rangle + (e/2mc)\mathbf{R}_1 \times \langle\phi_{1a}|\underset{\sim}{\mathbf{p}}_1|\phi_{10}\rangle \\ &\qquad \pm (e/2mc)\mathbf{R}_2 \times \langle\phi_{2a}|\underset{\sim}{\mathbf{p}}_2|\phi_{20}\rangle) \qquad (8\text{-}15)\end{aligned}$$

For exact wavefunctions, it is possible to replace a momentum operator expectation value $\langle\phi_{1a}|\underset{\sim}{\mathbf{p}}|\phi_{10}\rangle$ by an electric dipole expectation value.[§]

[§] This theorem is easy to prove if one is familiar with the handling of commutators such as $[\underset{\sim}{\mathbf{H}},\underset{\sim}{\mathbf{r}}]$. (See almost any quantum mechanics text—for example, Mertzbacher, 1961, p. 453.)

$$(e/2mc)\langle\phi_{1a}|\underset{\sim}{\mathbf{p}}|\phi_{10}\rangle = (i\pi/\lambda_{0a})\langle\phi_{1a}|\underset{\sim}{\boldsymbol{\mu}}|\phi_{10}\rangle \tag{8-16}$$

where $\lambda_{0a} = c/\nu_{0a}$ is the wavelength of the absorption transition between states 0 and a in the unperturbed monomer. Using this equation and the equivalent one for $\langle\phi_{2a}|\underset{\sim}{\mathbf{p}}|\phi_{20}\rangle$, we can rewrite Equation 8-15 as

$$\begin{aligned}\langle\Psi_{A^\pm}|\underset{\sim}{\mathbf{m}}|\Psi_0\rangle = (1/\sqrt{2})(\langle\phi_{1a}|\underset{\sim}{\mathbf{m}}_1|\phi_{10}\rangle \pm \langle\phi_{2a}|\underset{\sim}{\mathbf{m}}_2|\phi_{20}\rangle \\ + (i\pi/\lambda_{0a})(\mathbf{R}_1 \times \langle\phi_{1a}|\underset{\sim}{\boldsymbol{\mu}}_1|\phi_{10}\rangle \pm \mathbf{R}_2 \times \langle\phi_{2a}|\underset{\sim}{\boldsymbol{\mu}}_2|\phi_{20}\rangle))\end{aligned} \tag{8-17}$$

All that remains is to evaluate the rotational strength of the two transitions $\Psi_0 \rightarrow \Psi_{A^\pm}$ in the dimer. From Equation 8-10 (and using Eqns. 8-12 and 8-17), we have

$$\begin{aligned}R_{0A^\pm} &= \text{imag}(\langle\Psi_0|\underset{\sim}{\boldsymbol{\mu}}|\Psi_{A^\pm}\rangle \cdot \langle\Psi_{A^\pm}|\underset{\sim}{\mathbf{m}}|\Psi_0\rangle) \\ &= \tfrac{1}{2}\,\text{imag}\big((\langle\phi_{10}|\underset{\sim}{\boldsymbol{\mu}}_1|\phi_{1a}\rangle \pm \langle\phi_{20}|\underset{\sim}{\boldsymbol{\mu}}_2|\phi_{2a}\rangle) \\ &\qquad \cdot [\langle\phi_{1a}|\underset{\sim}{\mathbf{m}}_1|\phi_{10}\rangle \pm \langle\phi_{2a}|\underset{\sim}{\mathbf{m}}_2|\phi_{20}\rangle \\ &\qquad + (i\pi/\lambda_{0a})(\mathbf{R}_1 \times \langle\phi_{1a}|\underset{\sim}{\boldsymbol{\mu}}_1|\phi_{10}\rangle \pm \mathbf{R}_2 \times \langle\phi_{2a}|\underset{\sim}{\boldsymbol{\mu}}_2|\phi_{20}\rangle)]\big)\end{aligned} \tag{8-18}$$

$$R_{0A^\pm} = \tfrac{1}{2}\,\text{imag}(\langle\phi_{10}|\underset{\sim}{\boldsymbol{\mu}}_1|\phi_{1a}\rangle \cdot \langle\phi_{1a}|\underset{\sim}{\mathbf{m}}_1|\phi_{10}\rangle + \langle\phi_{20}|\underset{\sim}{\boldsymbol{\mu}}_2|\phi_{2a}\rangle \cdot \langle\phi_{2a}|\underset{\sim}{\mathbf{m}}_2|\phi_{20}\rangle) \tag{8-19a}$$

$$\begin{aligned}&\pm \tfrac{1}{2}\,\text{imag}(\langle\phi_{10}|\underset{\sim}{\boldsymbol{\mu}}_1|\phi_{1a}\rangle \cdot \langle\phi_{2a}|\underset{\sim}{\mathbf{m}}_2|\phi_{20}\rangle \\ &\qquad + \langle\phi_{20}|\underset{\sim}{\boldsymbol{\mu}}_2|\phi_{2a}\rangle \cdot \langle\phi_{1a}|\underset{\sim}{\mathbf{m}}_1|\phi_{10}\rangle)\end{aligned} \tag{8-19b}$$

$$\begin{aligned}&+ (\pi/2\lambda)(\langle\phi_{10}|\underset{\sim}{\boldsymbol{\mu}}_1|\phi_{1a}\rangle \cdot \mathbf{R}_1 \times \langle\phi_{1a}|\underset{\sim}{\boldsymbol{\mu}}_1|\phi_{10}\rangle \pm \langle\phi_{10}|\underset{\sim}{\boldsymbol{\mu}}_1|\phi_{1a}\rangle \cdot \mathbf{R}_2 \\ &\qquad \times \langle\phi_{2a}|\underset{\sim}{\boldsymbol{\mu}}_2|\phi_{20}\rangle \pm \langle\phi_{20}|\underset{\sim}{\boldsymbol{\mu}}_2|\phi_{2a}\rangle \cdot \mathbf{R}_1 \times \langle\phi_{1a}|\underset{\sim}{\boldsymbol{\mu}}_1|\phi_{10}\rangle \\ &\qquad + \langle\phi_{20}|\underset{\sim}{\boldsymbol{\mu}}_2|\phi_{2a}\rangle \cdot \mathbf{R}_2 \times \langle\phi_{2a}|\underset{\sim}{\boldsymbol{\mu}}_2|\phi_{20}\rangle)\end{aligned} \tag{8-19c}$$

The third term of this complex result (Eqn. 8-19c) can be simplified considerably. We mentioned earlier that $\langle\phi_{1a}|\underset{\sim}{\boldsymbol{\mu}}_1|\phi_{10}\rangle = \langle\phi_{10}|\underset{\sim}{\boldsymbol{\mu}}_1|\phi_{1a}\rangle$, with similar results for monomer 2. Any triple scalar product $\mathbf{A} \cdot \mathbf{B} \times \mathbf{C}$ will be zero if two or more of the vectors are the same. Therefore, the first and fourth subterms of Equation 8-19c vanish. The others can be combined by recognizing that, for the triple scalar product, any permutation of adjacent vectors changes the sign. Thus $\mathbf{A} \cdot \mathbf{B} \times \mathbf{C} = -\mathbf{B} \cdot \mathbf{A} \times \mathbf{C} = \mathbf{B} \cdot \mathbf{C} \times \mathbf{A}$, etc. Therefore, Equation 8-19c becomes

$$+(\pi/2\lambda)(\pm\mathbf{R}_2 \cdot \langle\phi_{20}|\underset{\sim}{\boldsymbol{\mu}}_2|\phi_{2a}\rangle \times \langle\phi_{10}|\underset{\sim}{\boldsymbol{\mu}}_1|\phi_{1a}\rangle \pm \mathbf{R}_1 \cdot \langle\phi_{20}|\underset{\sim}{\boldsymbol{\mu}}_2|\phi_{2a}\rangle \times \langle\phi_{10}|\underset{\sim}{\boldsymbol{\mu}}_1|\phi_{1a}\rangle) \tag{8-19d}$$

Now, using the fact that $\mathbf{R}_{12} = \mathbf{R}_2 - \mathbf{R}_1$ is the vector between the centers of the two monomers, Equation 8-19d can be further simplified to

$$\pm(\pi/2\lambda)\mathbf{R}_{12} \cdot \langle\phi_{20}|\underset{\sim}{\boldsymbol{\mu}}_2|\phi_{2a}\rangle \times \langle\phi_{10}|\underset{\sim}{\boldsymbol{\mu}}_1|\phi_{1a}\rangle \tag{8-19e}$$

Although the result is complicated, it has a fairly simple physical meaning. There will be two CD bands in the dimer, one corresponding to each of the two exciton absorption bands $\Psi_0 \rightarrow \Psi_{A^{\pm}}$. The three terms of Equation 8-19 contribute to each band. The first term (Eqn. 8-19a) is just the sum of the CD of the two separate monomers; this is called the one-electron term, and it usually is small. Actually, if we had used a more accurate set of wavefunctions, the first term would contain additional contributions due to perturbation of both monomers by the static field of the dimer.

The second term (Eqn. 8-19b) is called electric–magnetic coupling. It can be very important when one monomer has a small magnetic transition dipole and the other has a small electric transition dipole. The most frequently encountered chromophores, in fact, all have large electric transition dipoles and small magnetic transition dipoles, so the second term usually is small. This means that the CD of a dimer is dominated by the third term (Eqn. 8-19c,e), called the coupled oscillator or exciton term. The value of this term depends on the distance between the two chromophores and on the geometry of the molecule. Note that, even if the monomeric chromophores were optically inactive, this term (Eqn. 8-19e) could still be large, given the appropriate geometry.

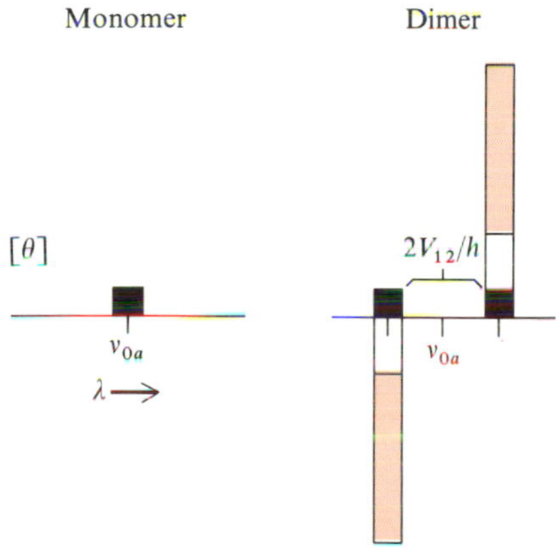

Figure 8-5

Contributions to the CD of a dimer predicted by Equation 8-19. In the monomer, there is a single CD band at ν_{0a} produced by a term like Equation 8-19a. In the dimer, there are two bands at $\nu_{0a} \pm V_{12}/h$, which arise from the one-electron term (Eqn. 8-19a; *dark shading*), from electric–magnetic coupling (Eqn. 8-19b; *unshaded*), and from coupled oscillator terms (Eqn. 8-19e; *colored*).

The contribution of all three terms to the CD of a dimer is illustrated schematically in Figure 8-5. When Equation 8-19e dominates, the two CD bands of the dimer will have equal rotational strengths of opposite sign (as you can see by the $\pm$ in front of this term). The integrated rotational strength, $R_{A^+} + R_{A^-}$ will be zero, and such a spectrum is called conservative. It is illustrated in Figure 8-6 for two different dimer geometries. Note that, although the rotational strengths of the two individual bands can be very large, the actual measured CD is much smaller. The two bands are separated only by a narrow splitting ($2V_{12}$) as shown in Equation 7-54. Because of this narrow splitting, much of the CD intensity of the exciton bands is canceled.

Equation 8-19e predicts that the rotational strengths of the two individual bands will increase linearly as the distance between the two chromophores grows. However,

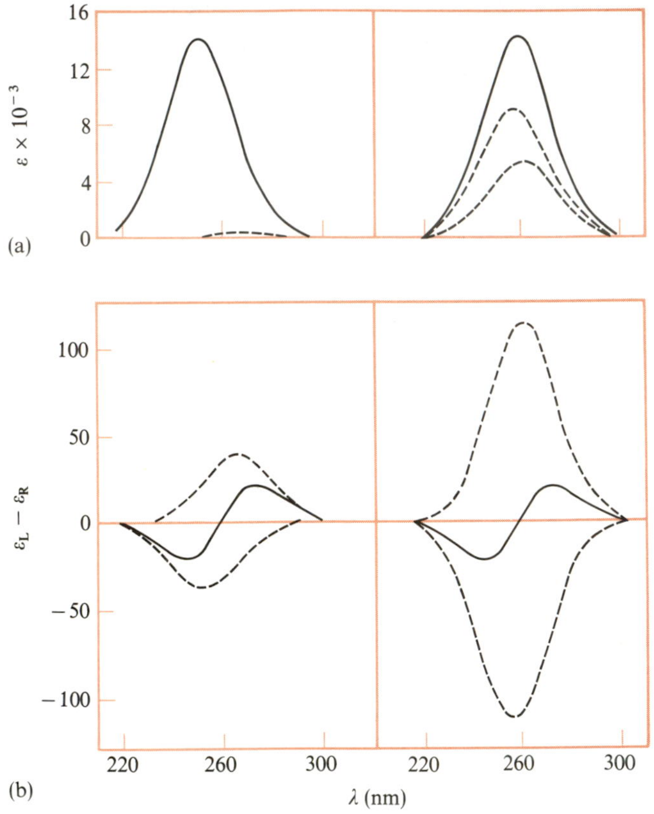

Figure 8-6

Calculated spectra for dimers with two identical chromophores. **(a)** Absorption spectra. **(b)** CD spectra. Solid lines show the expected observed spectra; dashed lines show individual contributions to the spectra from the two exciton-split bands. The dimer illustrated at left has two monomers at a 15° angle; the one at right has the monomers at a 75° angle. Note that very different geometries can yield virtually identical final spectra, despite large differences in individual components. Also note that, in the 15° dimer absorption spectrum, one exciton band has negligible intensity. [After I. Tinoco, Jr., and C. R. Cantor, *Methods Biochem. Anal.* 18:81 (1970).]

because the splitting decreases as R_{12}^{-3}, the actual observable CD will diminish roughly as the square of the distance. Because the exciton splitting is very small, it is not easily detected in a typical absorption measurement. CD magnifies the effect enormously because the two spectral bands produced are of opposite sign. However, because the magnitude of CD depends on both V_{12} and $R_{0A^\pm}$, it may not be easy to sort out the contributions of the two effects. This combination of effects is shown for simulated spectra in Figure 8-6.

Earlier we stated the general rule that CD and absorption spectra have the same shape for allowed transitions. Now you can see that chromophore interactions, such as exciton or electromagnetic coupling, lead to a violation of this rule. The integrated CD in a wavelength region dominated by either of these effects is close to zero. Such a spectrum usually is a good indication that one is dealing with a set of interacting chromophores.

Equation 8-19e contains a triple scalar product; thus, for many conformations of the dimer, the CD vanishes. It will be zero if both chromophores are coplanar, or if they are oriented parallel or perpendicular to one another. Figure 8-7 illustrates the CD calculated for six different geometries of a dimer, considering only the exciton term (Eqn. 8-19e). These results indicate that the CD of an assembly of chromophores is indeed quite sensitive to the molecular conformation.

Frequently, monomeric chromophores have small (but not negligible) CD when compared with a dimer or polymer. In these cases it usually is convenient to compute the difference CD:

$$[\theta]' = [\theta]_{\text{dimer}} - [\theta]_{\text{monomers}} \tag{8-20}$$

As long as the splitting is small and there are no static field contributions to the one-electron term (Eqn. 8-19a), then $[\theta]'$ will be a good approximation to the electric–magnetic and exciton terms. It should be a conservative spectrum, even though the observed dimer CD may not be conservative.

The calculation of the CD of an actual biological oligomer or polymer proceeds along very similar lines to the simple model we have just considered. Cases where the exciton term (Eqn. 8-19e) dominates are especially favorable for conformational analysis. No knowledge of magnetic transition dipoles is required, because the monomer contribution can be subtracted just as in Equation 8-20. The length of each monomer electric transition dipole can be determined from the absorption spectrum: $|\boldsymbol{\mu}_{0a}| = (D_{0a})^{1/2}$. The directions in many cases can be determined by polarized absorption spectroscopy, as described earlier. Therefore, the only quantity that must be calculated from wavefunctions is the splitting, $2V_{12}$.

In any real case however, the actual procedure is much more complex. Each monomer will have more than one absorption band. One must compute coupled oscillator and electric–magnetic coupling terms between each band on one monomer and all the bands on the others. The simplest case of real biological interest is polypeptides. In a molecule such as poly-L-alanine, the side chain contributes little to the optical properties. So for satisfactory analysis of the near-UV CD, one must account

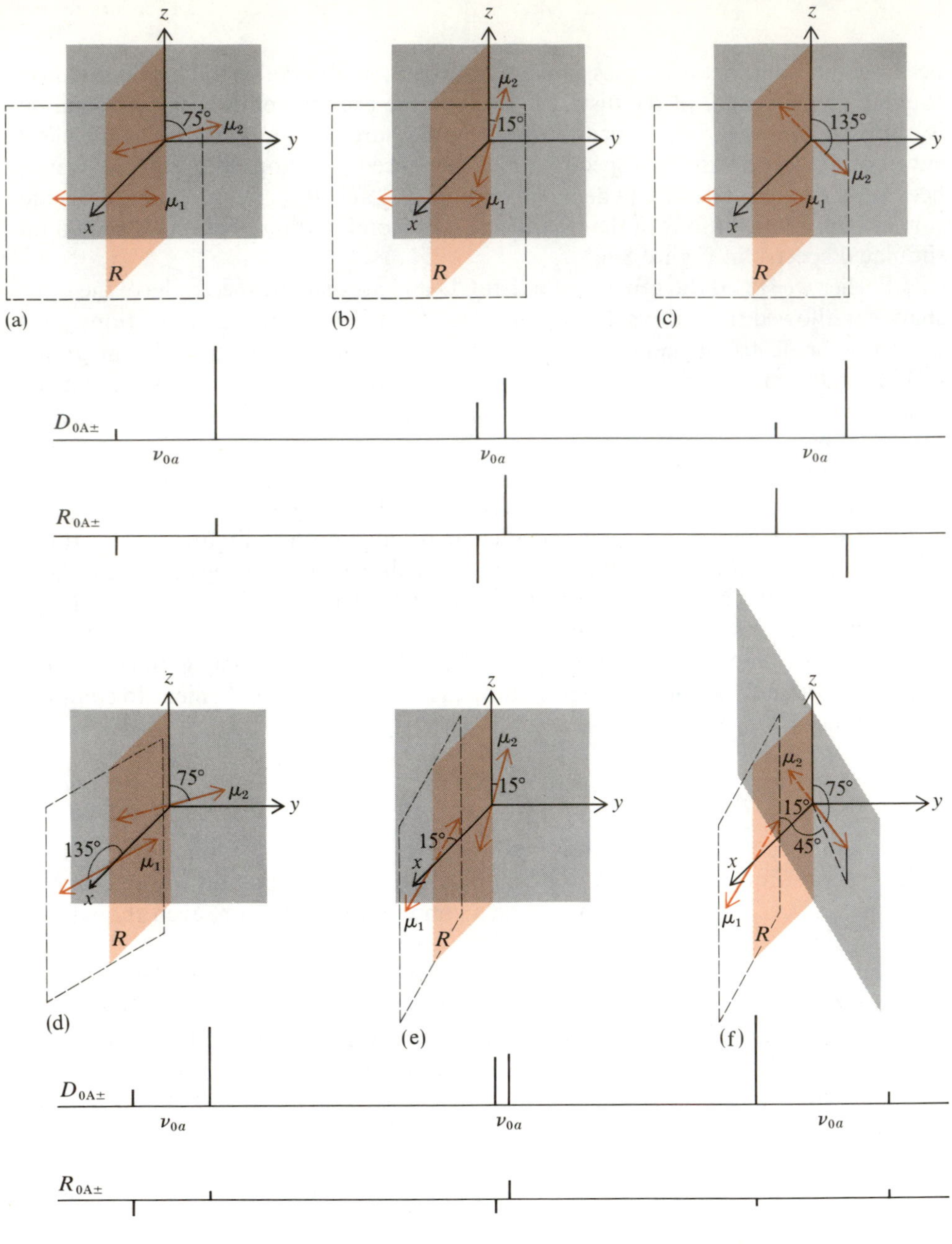

Figure 8-7

Six different geometries of a dimer with two identical chromophores, showing the corresponding absorption dipole strengths and CD rotational strengths. Only the exciton term (Eqn. 8-19e) has been used in computing the CD rotational strengths. **(a,b,c)** In the first three dimers, both transition dipoles are parallel to the *y–z* plane and thus are perpendicular to the line (*R*) between the two chromophores. **(d, e, f)** In the other three dimers, the chromophores are not in a parallel stack. In every case, the monomer frequency is halfway between the two bands. [After I. Tinoco, Jr., *Radiation Res.* 20:133 (1963).]

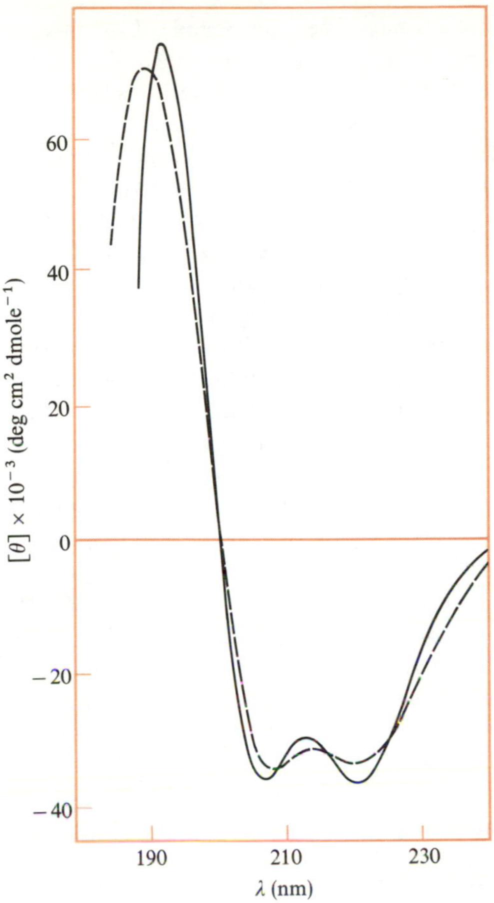

Figure 8-8

*CD spectrum of poly-*L*-alanine in an* α*-helical conformation.* Calculated (*dashed line*) and observed (*solid line*) spectra are shown. [After R. W. Woody, *J. Chem. Phys.* 49:4797 (1968).]

for only the $\pi \rightarrow \pi^*$ and $n \rightarrow \pi^*$ electronic transitions of the peptides. Figure 8-8 shows the results of such a calculation for α-helical poly-L-alanine, compared with experimental results. The agreement is impressive and demonstrates the potential power of optical activity in conformational analysis.

Semiempirical computation of protein optical activity

In practice, the quantum mechanical calculation of the CD of most bioploymers is still a formidable task. Chromophores such as the nucleic acid bases and aromatic amino acids are far more complex than the peptide group. Many more electronic states are involved, and transition moment directions and intensities are rarely known for enough of them with sufficient precision. However, a large body of experimental optical activity data exists for polypeptides, proteins, and nucleic acids with known three-dimensional structures. A number of ways have been developed for using such data to predict or analyze the optical properties of less-well-characterized systems. We shall discuss proteins and nucleic acids separately because the methods that have been successful for the two classes of compounds are somewhat different.

For proteins, the major objective has been to deduce the average secondary structures of the peptide chain from measured CD or ORD spectra. Unless an unusual fraction of aromatic amino acids is present, optical activity in the region of the spectrum between 190 nm and 230 nm is dominated by the peptide backbone. Numerous experiments have shown that, at least qualitatively, the nature of particular aliphatic side chains does not markedly affect the CD spectrum in this region. Therefore, as an approximation, one considers a protein simply as a linear combination of backbone regions with α-helical, β-sheet, or random coil structures. Estimates for the spectra of each type of region can be obtained by using the measured CD of homopolypeptides known to be pure α helices, β sheets, or random coils. Such a set of basis spectra is similar to the one shown in Figure 8-9. If the actual fractional composition of secondary structure types (χ_α, χ_β, χ_r) were known for a particular protein, one could calculate the residue CD at each wavelength:

$$[\theta(\lambda)] = \chi_\alpha[\theta_\alpha(\lambda)] + \chi_\beta[\theta_\beta(\lambda)] + \chi_r[\theta_r(\lambda)] \tag{8-21}$$

where $[\theta_\alpha]$, $[\theta_\beta]$, and $[\theta_r]$ are the measured CD of polypeptides in the conformations indicated. An identical equation is used for ORD.

In practice, it is a much more common situation to know the CD spectrum of a protein and to wish to use it to compute χ_α, χ_β, and χ_r, thus obtaining an approximate analysis of the secondary structure. This is done by choosing a set of wavelengths (λ_i) and solving simultaneously a set of corresponding equations such as Equation 8-21. Three wavelengths are the minimal number needed; it is far better to pick a much larger set and use least squares or other statistical procedures to obtain the most reliable values for the fractions of each type of secondary structure.[§] The precision of such a calculation (even assuming that all the assumptions are accurate) depends on the extent to which the basis spectra $[\theta_\alpha]$, $[\theta_\beta]$, and $[\theta_r]$ are linearly independent functions. Fortunately, as is clear from Figure 8-9, the spectrum of each secondary structure is fairly distinctive.

There are at least two potential weaknesses to the approach just described. The CD of an α helix or β sheet is a function of its length. The model compounds used to set up the basis spectra are homopolymeric structures much larger than typical helices or sheets in a globular protein. In principle, this can be corrected for, but it requires the introduction of more parameters. Experimental data in many cases are not

---

[§] A particularly compact way to do this is to note that the measured CD and desired mole fractions of secondary structure can be written as vectors $\boldsymbol{\theta}$ (i wavelengths) and $\boldsymbol{\chi}$ (three components), whereas the CD basis set is an $i \times 3$ matrix $\underset{\sim}{\mathbf{M}}$. Equation 8-21 then is simply $\boldsymbol{\theta} = \underset{\sim}{\mathbf{M}} \cdot \boldsymbol{\chi}$. Solving this equation for $\boldsymbol{\chi}$, one has $\boldsymbol{\chi} = (\underset{\sim}{\mathbf{M}}\underset{\sim}{\mathbf{M}}^\dagger)^{-1}\underset{\sim}{\mathbf{M}}^\dagger \cdot \boldsymbol{\theta}$, where $\underset{\sim}{\mathbf{M}}^\dagger$ denotes the transpose of $\underset{\sim}{\mathbf{M}}$, and $\underset{\sim}{\mathbf{M}}^{-1}$ denotes the inverse of $\underset{\sim}{\mathbf{M}}$. This expression is algebraically equivalent to a least-squares fit.

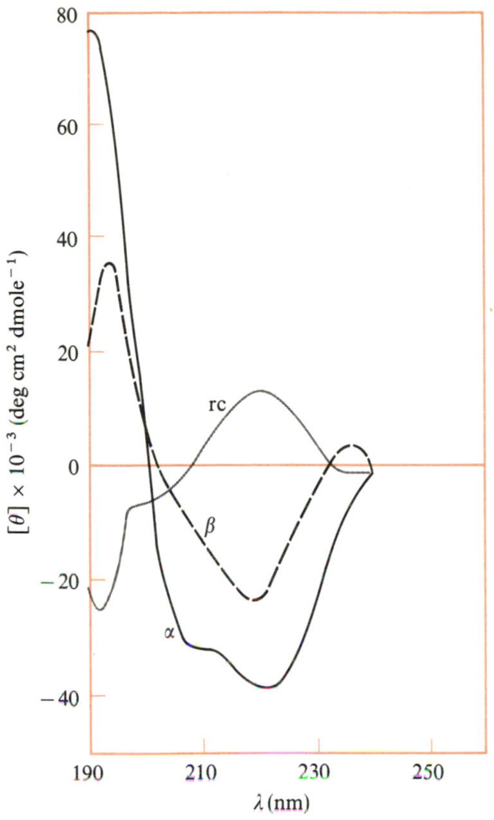

Figure 8-9

CD spectra for α-helix, β-sheet, and random coil conformations, extracted from the spectra of proteins of known three-dimensional structure by the Wetlaufer method. [After V. P. Saxena and D. B. Wetlaufer, *Proc. Natl. Acad. Sci. USA* 66:969 (1971).]

accurate enough to justify this. A second problem is that the tertiary structure of a typical globular protein consists of individual secondary structure regions packed next to each other. Although the interactions responsible for most of the CD of a protein die off as the square of the interchromophore distance, there must surely be some contributions between packed helical and sheet regions that cannot be adequately modeled by extended helical or sheet homopolymers.

A method of semiempircal CD calculation that nicely circumvents both of these problems has been introduced by Donald Wetlaufer. Rather than polypeptides, a set of proteins is chosen to compute basis spectra. Because each protein of this set has a known three-dimensional structure, χ_α, χ_β, and χ_r can be estimated reliably. The experimental CD spectra also are known. Thus, by solving a set of equations corresponding to Equation 8-21, one can extract values for $[\theta_\alpha(\lambda)]$, $[\theta_\beta(\lambda)]$, and $[\theta_r(\lambda)]$. These are approximations for the spectra of each type of secondary structure as they actually occur in typical proteins. They will automatically include average effects of length and of any tertiary interactions. Figure 8-9 shows a typical CD basis set

Table 8-1
Estimates of protein secondary structure from CD measurements

Method	Structure	Protein			
		Carboxypeptidase	α-Chymotrypsin	Myoglobin	Lysozyme
X-ray structure	α Helix	23	8	~68	28
	β Sheet	18	22	0	10
	Random plus other	59	70	~32	62
CD calculation from poly-L-lysine basis set	α Helix	13	12	68	29
	β Sheet	31	23	5	11
	Random	56	65	27	60
CD calculation from protein basis set	α Helix	26	20	—§	—§
	β Sheet	18	20	—§	—§
	Random	56	60	—§	—§

§ Prediction in these cases is not a fair test, because these proteins were included in the original basis set.

derived from protein spectra. The applicability of this experimentally derived basis set can be tested by computing the CD of additional proteins not included in the original calculations. Table 8-1 shows some typical results both from the Wetlaufer method and from the use of a polypeptide basis set. It is clear that both methods are fairly successful.

Semiempirical calculation of nucleic acid optical activity

For nucleic acids, unlike proteins, one cannot ignore spectral differences between different monomeric residues. The bases themselves are directly involved in close interactions in all common secondary structures. Not only the base composition, but in fact some actual sequence information, must be taken into account to explain CD or ORD spectra. The CD spectrum of a dinucleoside phosphate (such as ApG) in aqueous solution is quite different from the sum of the CD spectra of A and G monomers. The base-stacked conformation of even these short oligomers leads to intense contributions to the CD. One can describe the CD of ApG by the contributions from the two monomers and an additional term to account for the base–base interaction:

$$2[\theta_{\mathrm{ApG}}(\lambda)] = [\theta_{\mathrm{A}}(\lambda)] + [\theta_{\mathrm{G}}(\lambda)] + I_{\mathrm{AG}}(\lambda) \tag{8-22}$$

Similarly, for the dinucleoside phosphate GpU, one would have[§]

$$2[\theta_{\mathrm{GpU}}(\lambda)] = [\theta_{\mathrm{G}}(\lambda)] + [\theta_{\mathrm{U}}(\lambda)] + I_{\mathrm{GU}}(\lambda) \tag{8-23}$$

In a trinucleoside diphosphate (such as ApGpU), there would be contributions from three monomers, two interactions between neighboring bases, and a final contribution due to interaction between the next-nearest neighbors, A and U:

$$3[\theta_{\mathrm{ApGpU}}(\lambda)] = [\theta_{\mathrm{A}}(\lambda)] + [\theta_{\mathrm{G}}(\lambda)] + [\theta_{\mathrm{U}}(\lambda)] + I'_{\mathrm{AG}}(\lambda) + I'_{\mathrm{GU}}(\lambda) + I''_{\mathrm{AU}}(\lambda) \tag{8-24}$$

Suppose that the actual geometry of neighboring stacked bases in a dimer were the same as in a trimer. In this case, the optical interactions between neighboring chromophores also would be approximately the same: $I'_{\mathrm{AG}}(\lambda) = I_{\mathrm{AG}}(\lambda)$, and $I'_{\mathrm{GU}}(\lambda) = I_{\mathrm{GU}}(\lambda)$. Because the interactions grow weaker with increasing distance, one ought to be able to neglect the next-nearest-neighbor interaction $I''_{\mathrm{AU}}(\lambda)$ for such structures as stacked helices. Then, simply combining Equations 8-22, 8-23, and 8-24, one has

$$3[\theta_{\mathrm{ApGpU}}(\lambda)] = 2[\theta_{\mathrm{ApG}}(\lambda)] + 2[\theta_{\mathrm{GpU}}(\lambda)] - [\theta_{\mathrm{G}}(\lambda)] \tag{8-25}$$

This equation states that, if we measure the CD spectra of the two dimers ApG and GpU and of the monomer G, we ought to be able to predict the CD of the trimer, ApGpU. Exactly equivalent expressions apply for ORD spectra and also for absorption spectra. An example of such calculations is shown in Figure 8-10. They work fairly well. Note that other calculations and experiments have shown marked sequence dependence of the CD of oligonucleotides. This sensitivity is why such a relatively more complex approach is needed to compute oligonucleotide spectra.

The general approach implicit in Equation 8-25 can be extended to RNA and DNA polymers. At least three types of structural regions must be considered. The spectrum of a random coil can be estimated as simply the average of the properties of the four monomers. That of single-strand stacked helices would contain optical contributions from each of the 16 possible dinucleoside phosphates, weighted by their frequencies of occurrence. Double-strand regions are accounted for in an analogous way, by adding the contributions of each of the 10 possible double-strand dimers (ApG base paired with CpU, and so on). Because these complexes are too unstable for experimental spectra to be measured directly, their optical properties must be extracted from the spectra of larger double-strand molecules. This is done in exactly the same way as was described above for obtaining polypeptide basis spectra from protein spectra.

In practice, a total of up to 30 different spectral contributions must be combined to compute the CD of a molecule such as tRNA that has both single-strand and

[§] Note that the factors of 2 and 3 to the left of Equations 8-22, 8-23, and 8-24 are necessary because measured CD spectra are expressed in residue units. The actual molar ellipticity of a dimer is twice the residue ellipticity.

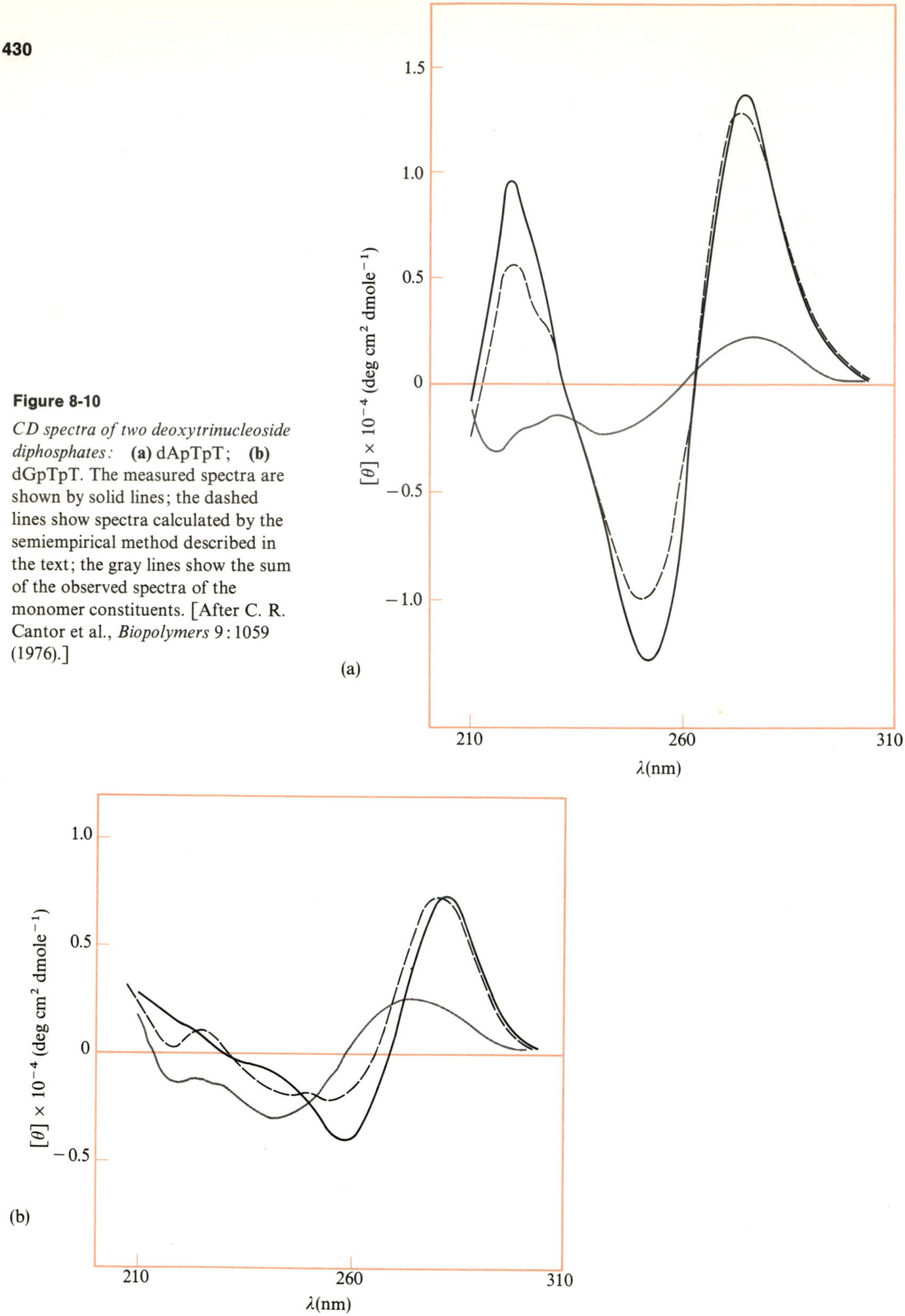

Figure 8-10

CD spectra of two deoxytrinucleoside diphosphates: **(a)** dApTpT; **(b)** dGpTpT. The measured spectra are shown by solid lines; the dashed lines show spectra calculated by the semiempirical method described in the text; the gray lines show the sum of the observed spectra of the monomer constituents. [After C. R. Cantor et al., *Biopolymers* 9:1059 (1976).]

double-strand regions. Such a complete treatment rarely has been possible. One can simplify the problem by factoring all of these contributions into average spectra of single-strand and double-strand regions. This approach is useful for semiquantitative predictions, but it is not precise enough to predict the actual number of base pairs in a tRNA. Actually the problem may be easier for larger RNAs because effects of individual sequence regions are much more likely to average out.

For DNA, a slightly different approach can be used, because the length is great enough that we can neglect the ends. If only double helix is present, there are eight linearly independent basis spectra that can be obtained from measurements on DNAs of regularly alternating sequence. When DNAs containing both purines and pyrimidines on each strand are included in the calculations, quite accurate predictions of the CD of natural DNAs are possible. However, calculations are not very reliable when DNAs such as poly dA·dT are used for the basis set. Such homopurine-homopyrimidine duplexes appear to have average structures different from that of a normal DNA. For example, if all structures were the same, the CD of three synthetic polynucleotides with regularly alternating sequences should be related by the following expression:

CD	$3[\theta_{\text{poly dAAT}\cdot\text{dTTA}}]$	$=$	$2[\theta_{\text{poly dAT}\cdot\text{dTA}}]$	$+$	$[\theta_{\text{poly dA}\cdot\text{dT}}]$
Structure	···AATAATAAT···		···ATATATATA···		···AAAAAAAAA···
	···TTATTATTA···		···TATATATAT···		···TTTTTTTTT···

(8-26)

Equation 8-26 does not apply very well to experimental results, apparently because the structure of dA·dT is different from that of the other helices.

Empirical applications of optical activity

In many cases, not enough is known about the chromophores in an optically active system to permit reliable calculations of the CD spectrum a priori. Nor is a sufficient body of data on homologous systems available for semiempirical calculations. Still, optical activity can be used as a convenient monitor of conformational changes and local environment. In later chapters we deal with kinetic and thermodynamic studies of conformational changes. For such experiments, one typically must follow structural parameters of the system as a function of time, temperature, or other variables. If the system contains several components (or conformations), absorption or optical activity will always be a simple linear average. For example, $[\theta] = \sum_i \chi_i \theta_i$, where χ_i is just the mole fraction of the ith component, and θ_i is the CD of that component. By choosing a set of different wavelengths, in favorable cases, one can follow each component independently.

CD is sensitive essentially only to local interactions, even in cases where complex associations (such as subunit equilibria) are occurring. Thus, changes in CD will primarily reflect changes in local structure, if any occur. The tremendous advantage of CD for monitoring conformational changes is its sensitivity. Even if a detailed structural interpretation is not possible, a change in structure will almost surely show up as a change in CD.

One other useful application of CD is in studying the binding of small molecules to proteins or nucleic acids. This works best with a small molecule that absorbs visible light. An optically active small molecule may show a change in CD upon binding to a macromolecule, either because of electronic interactions with its binding site or because it may undergo a conformational change when it binds. These changes can be detected easily because most common biopolymers have no CD for visible light.

It is especially convenient to study optically *inactive* small molecules. Here an induced CD will usually occur upon binding, for the same reasons just mentioned for optically active small molecules. However, CD measurements in the absorption band of the small molecule will reflect only properties of the bound material. This makes it easier to determine whether there is a class of essentially equivalent binding sites, and to study the number of bound ligands. If all sites are very similar, the induced CD of the small molecule will be given by $\theta = C_b[\theta_b]l/100$, where C_b is the concentration of bound small molecules, $[\theta_b]$ is the CD of bound species (determined by saturation of a sample of small molecules with excess polymer), and l is the path-length. Once $[\theta_b]$ is known, C_b can be calculated from the measured θ. From this, binding constants and (ultimately) apparent thermodynamic data on the binding equilibrium are obtained.

Several factors contribute to the induced CD of a small molecule when it is bound to an asymmetric site on a polymer. The effect of the environment will show up as one-electron-type contributions. In addition, coupled oscillator terms can yield strong induced CD if the energy levels of protein or nucleic acid chromophores are similar to the energy levels of the small molecule. For example, much of the optical activity of bound heme groups in hemoglobins can be accounted for by interaction with nearby tyrosine residues of the protein.

Variations of optical activity measurements

In a magnetic field, optical activity becomes an even more complex phenomenon. Even many optically inactive molecules can show CD and ORD if a magnetic field is present. Optically active molecules show normal CD bands but, in addition, other magnetically induced bands appear. Work in the areas of magnetic ORD and magnetic CD spectroscopy is still in earlier stages of development than that in normal CD. In the future, many useful applications of magnetic CD and ORD are likely to appear.

The operators that lead to normal optical activity are actually tensors, rather than scalars. In normal isotropic solution, the random orientations of molecules average over the tensor properties, and the resulting measured quantity is the trace (sum of diagonal elements) of the optical activity tensor. More information is potentially available about the system if individual elements of the tensor are measured. This can be done by orienting the system. Such experiments are technically very challenging because the oriented system now demonstrates linear dichroism as well as circular dichroism, and the two effects must be sorted out. When all of this is done, a more complete picture of the nature of polymer electronic states is revealed.

If an optically active system is fluorescent (see the next section), two additional aspects of optical activity can be measured. Suppose that circularly polarized exciting light is used, but all emitted light is detected. The intensity of emitted light is proportional to the amount of light absorbed. Thus the difference in emitted intensities when right-hand and left-hand circularly polarized light is used will yield the differential absorbance (that is, the CD) of only the fluorescent chromophores. Thus, for example, if a protein contained two tryptophans, only one of which was fluorescent, normal CD would give the sum of their optical properties. Fluorescence-detected CD would provide the CD of one individual tryptophan; then the CD of the other tryptophan could be obtained by difference. Alternatively, one can use unpolarized exciting light, but measure the difference in the extent to which left-hand and right-hand circularly polarized light is emitted. This approach yields the CD of a chromophore in its excited-state configuration, in distinction to all other CD measurements, which essentially monitor ground-state configurations (see Steinberg, 1978).

One final variation on optical activity is CD in the infrared region of the spectrum. This technique is still in its infancy, but it shows promise as a new structural tool for small and large molecules.

8-2 FLUORESCENCE SPECTROSCOPY

Light emission can reveal properties of biological molecules quite different from the properties revealed by light absorption. The process takes place on a much slower time scale, allowing a much wider range of interactions and perturbations to influence the spectrum.

Basic principles of fluorescence

Consider a hypothetical molecule with two energy levels, S_b and S_a:

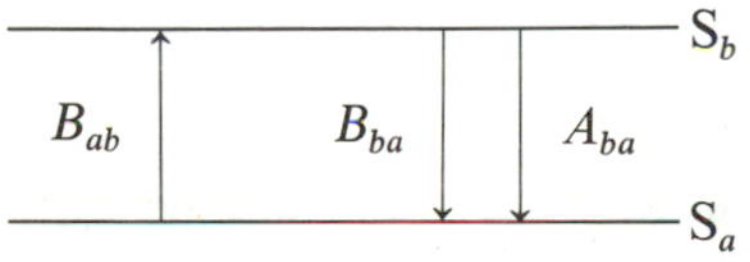

In Section 7-1, we showed that light of radiation density $I(\nu)$ induces the transition from S_a to S_b at a rate of B_{ab} per molecule. The radiation-induced process $S_b \rightarrow S_a$ occurs at exactly the same rate; thus $B_{ab} = B_{ba}$. If the system originally contains n_a molecules in state S_a and n_b in state S_b, then the net rates of conversion are $n_a B_{ab} I(\nu)$ and $n_b B_{ba} I(\nu)$, respectively. At equilibrium, these rates must be equal. Hence $n_a B_{ab} I(\nu) =$

$n_b B_{ba} I(\nu)$, or $n_a = n_b$, independent of radiation density. Unfortunately, this is an absurd result. Without light, virtually all molecules should be in the ground state, S_a. At low light intensities, one hardly expects a major perturbation of the equilibrium values of n_a and n_b. These values are derivable from statistical mechanics. Ignoring degeneracy, a simple Boltzmann factor applies:

$$n_a/n_b = e^{-(E_a - E_b)/kT} = e^{+h\nu/kT} \tag{8-27}$$

where h is Planck's constant.

It was Albert Einstein who first noted that, to resolve this discrepancy, one could postulate a rate of spontaneous emission of a photon from state S_b. The rate of this process (A_{ba}) should be independent of $I(\nu)$. When spontaneous emission is included and the rates of interconversion of S_a and S_b are set equal at equilibrium, the result is

$$n_a/n_b = [B_{ba} I(\nu) + A_{ba}]/B_{ab} I(\nu) = 1 + A_{ba}/B_{ab} I(\nu) \tag{8-28}$$

By setting Equations 8-27 and 8-28 equal, we can evaluate A_{ba}. To do this, we must first insert for $I(\nu)$ the radiation density expected for a black body§ at temperature T:

$$I(\nu) = 8\pi h\nu^3/c^3(e^{h\nu/kT} - 1) \tag{8-29}$$

With this substitution, Equation 8-28 becomes

$$n_a/n_b = 1 + A_{ba}/B_{ab} I(\nu) = 1 + A_{ba}(e^{h\nu/kT} - 1)/8\pi h\nu^3 c^{-3} B_{ab} \tag{8-30}$$

When this value is set equal to $e^{h\nu/kT}$ (from Eqn. 8-27), the result for A_{ba} is

$$A_{ba} = 8\pi h\nu^3 c^{-3} B_{ab} \tag{8-31}$$

Note that this expression depends on the cube of the frequency. This means that, at short wavelengths, A_{ba} is much larger than B_{ab}. Essentially all emission is spontaneous. Earlier (Eqn. 7-29) we showed that $B_{ab} = (2\pi/3\hbar^2)D_{ab}$. Thus we can obtain

$$\boxed{A_{ba} = (32\pi^3\nu^3/3c^3\hbar)D_{ab}} \tag{8-32}$$

Because the dipole strength (D_{ab}) and the frequency (ν) usually can be measured from the absorption spectrum, the rate of spontaneous emission can be determined without performing an emission measurement at all. In the absence of radiation or

§ At equilibrium, the emission and absorption of radiation density must be equal. For an object at temperature T, the radiation density required for equilibrium is given by Equation 8-29. [See R. P. Feynman, R. B. Leighton, and M. Sands, *The Feynman Lectures on Physics*, vol. 1 (Reading, Mass.: Addison-Wesley, 1963), chaps. 41–42.]

any other perturbations or interactions, the rate of deexcitation of molecules initially in state S_b will be

$$dn_b/dt = -A_{ba}n_b \tag{8-33}$$

The solution of this differential equation is $n_b(t) = n_b(0)e^{-A_{ba}t}$, where $n_b(0)$ is the concentration of excited states at zero time. Hence we can define the radiative lifetime of state S_b as

$$\tau_R = 1/A_{ba} \tag{8-34}$$

The dipole strength D_{ab} is a direct measure of the intensity of a spectral absorption. From Equation 8-32, it is clear that D_{ab} and A_{ba} are proportional. Thus Equation 8-34 indicates that the stronger the absorption of a given isolated molecule, the more rapid the emission of fluorescent radiation. Note carefully that Equation 8-34 is valid only so long as the same electronic state that absorbed the radiation is subsequently emitting it. This is not always the case. More elaborate versions of Equation 8-34 exist to handle such complexities as vibronic bands within electronic transitions.

In reality, the actual observed lifetime of an excited singlet state is rarely as long as the radiative lifetime computed by Equation 8-34. This is because the excited state can lose its energy through many other processes besides direct emission of light (Fig. 8-11). We shall discuss each process in some detail because each potentially can yield useful information on biological structures.

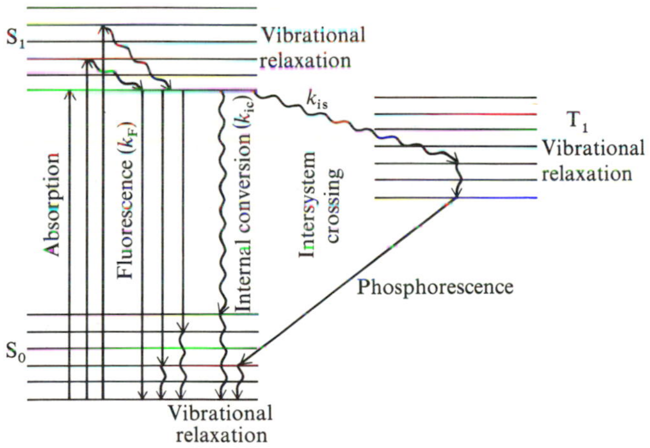

Figure 8-11

Pathways for production and deexcitation of an excited state. [After C. R. Cantor and T. Tao, in *Procedures in Nucleic Acid Research*, vol. 2 (New York: Harper & Row, 1971), p. 31.]

The intrinsic fluorescence rate constant (k_F) can be evaluated from the preceding discussion as

$$k_F = A_{ba} = 1/\tau_R \tag{8-35}$$

as long as stimulated emission is negligible. The nonradiative processes that compete with fluorescence (and the rate constants that characterize them) include internal conversion (k_{ic}), intersystem crossing (k_{is}), and quenching of various types [k_q(Q)]. All of these processes compete directly to depopulate the excited singlet state. Therefore, the fraction of excited singlets that become deexcited through fluorescence (called the fluorescence quantum yield, ϕ_F) is simply

$$\phi_F = k_F/[k_F + k_{ic} + k_{is} + k_q(Q)] \tag{8-36}$$

The fluorescence quantum yield (ϕ_F) is equal to the ratio of photons emitted to photons absorbed by the system.

Factors governing fluorescence intensity

We now examine in more detail the factors that affect the intensity of fluorescence.

1. Internal conversion, occurring at a rate k_{ic}. In this process, excitation energy in S_b is lost by collision with solvent or by dissipation through internal vibrational modes. In general, k_{ic} will increase as the temperature is raised. Therefore, the observed fluorescence will decrease with increasing temperature. This intrinsic temperature dependence is not found in the other spectroscopic phenomena we have discussed. It can seriously complicate attempts to use fluorescence to monitor thermally induced macromolecular conformation changes.

2. Deexcitation resulting from collisions or complexes with solute molecules (Q) capable of quenching the excited state, occurring at a rate k_q(Q). Unlike all the other processes, quenching can (in principle) be a bimolecular step if collisions are involved.

$$S_b + Q \xrightarrow{k_q} S_a + Q \tag{8-37}$$

Because Q usually is in vast molar excess over S_b, the actual observed rate is pseudo first order. The value of k_q can be measured by varying the concentration of quencher, (Q), and then observing the effect on ϕ_F. Aromatic chromophores usually have radiative lifetimes in the range of 1×10^{-9} to 100×10^{-9} sec.

Therefore, quenching processes must be quite effective to compete. Common quenchers, such as O_2 and I^- ion, deexcite essentially every time they collide with an excited singlet. Their rates are limited only by diffusion. At millimolar concentrations of quencher, collisions can approach rates of 10^8 sec^{-1}. Therefore, appreciable quenching is observed.

3. Intersystem crossing, occurring at a rate of k_{is}. In this process, the nominally forbidden spin exchange converts an excited singlet into an excited triplet state. That state can, in turn, convert to the ground singlet state (S_a), either by phosphorescence (emission of a photon) or by internal conversion. The triplet state generally is lower in energy than the excited singlet. Hence, phosphorescence occurs at longer wavelengths and can easily be resolved from fluoresence.

   The intensity for direct singlet–triplet absorption is extremely small. Therefore, triplet states usually can be seen easily only by emission spectroscopy. Another consequence of the low absorption intensity is that (as described by Eqn. 8-34) it will lead to an extremely long radiative lifetime for the triplet state. The lifetime is often in seconds or longer, rather than in the nanoseconds found for singlets. This means that collisions with quenchers or internal conversion can compete all too effectively with phosphorescence. Thus, in solution, phosphorescence rarely is observed. One usually must use rigid glasses at low temperature and must exclude oxygen fairly rigorously in order to see useful phosphorescence intensities. Because these conditions are rather far removed from the biological state, we shall concentrate on fluorescence methods rather than phosphorescence.

Because of all the nonradioactive processes just described, an excited singlet will decay faster than indicated by its radiative lifetime. The kinetic equation describing the decay of the concentration of excited singlets, $(S_b(t))$, is constructed by adding all parallel deexcitation pathways:

$$-d(S_b)/dt = [k_F + k_{ic} + k_{is} + k_q(Q)](S_b) \tag{8-38}$$

This equation has the solution

$$(S_b(t)) = (S_b(0))e^{-t/\tau_F} \tag{8-39}$$

where $(S_b(0))$ is the concentration at time zero, and τ_F is the observed fluorescence decay time:

$$\tau_F = [k_F + k_{ic} + k_{is} + k_q(Q)]^{-1} \tag{8-40}$$

Combining the definitions of τ_F, τ_R, and ϕ_F, we obtain

$$\boxed{\phi_F = \tau_F/\tau_R} \tag{8-41}$$

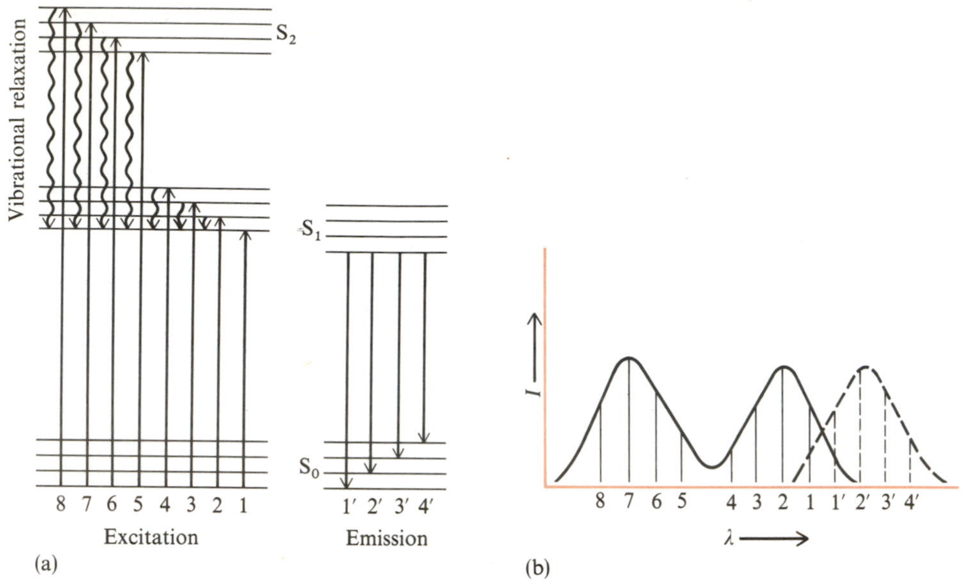

Figure 8-12

Excitation and emission of fluorescence. **(a)** Energy levels. **(b)** Spectra. Note that nonradiative transitions relax S_2 to S_1 much faster than any of the deexcitation processes can return S_1 to the ground state (S_0).

Because τ_R can, in principle, be calculated from the absorption spectrum, a measurement of the fluorescence decay rate (τ_F) is equivalent to a measurement of the quantum yield (ϕ_F).

Thus far we have considered only a single vibronic level of S_a and S_b. Figure 8-12 shows schematically the effect of other vibronic and electronic levels. Internal conversion between higher singlets, such as $S_c \rightarrow S_b$, is much faster than the rates we have considered up to now. Similarly, vibrational relaxation from excited vibrational levels of each electronic state to the ground vibrational level is much faster than photon emission. The result is that all observed fluorescence normally originates from the lowest vibrational level of the lowest excited singlet state. This observation has several important implications.

The spectrum of emitted light should be independent of the excitation wavelength.

Most or all of the fluorescence spectrum will be shifted to lower energies (longer wavelengths) than the longest-wavelength absorption band.

The shape of the emission band will be approximately a mirror image of the longest-wavelength absorption band (Fig. 8-12), providing that the vibronic structures of S_b and S_a are very similar.

The simple arguments shown in Figure 8-12 would predict that the zero–zero transitions in absorption [$S_a(v = 0) \rightarrow S_b(v = 0)$, line 1] and emission [$S_b(v = 0) \rightarrow S_a(v = 0)$, line 1′] should occur at the same wavelength. In solution, this usually is not the case. The absorption process occurs on such a short time scale that the environment can be considered fixed. Therefore, $S_a(v = 0) \rightarrow S_b(v = 0)$ corresponds to the energy difference with the solvent oriented in a favorable way about the ground-state configuration. If, after excitation, the solvent has time to reorient before emission, then $S_b(v = 0) \rightarrow S_a(v = 0)$ corresponds to the energy difference between the states while solvent is oriented favorably about the excited state. There is no reason why these energy differences must be the same and, in fact, either one could be larger. This effect leads to the shift in zero–zero frequencies called the Stokes shift.

Experimental measurements

Figure 8-13 shows a typical experimental arrangement for the measurement of fluorescence. This setup is used in two modes. With the excitation monochrometer M1 fixed, the emission monochrometer M2 can be scanned; this procedure yields an emission spectrum, the wavelength distribution of light emitted by the excited singlet. Alternatively, M2 can be fixed and M1 varied; this procedure produces an excitation spectrum. Because all states within or above S_b rapidly decay to the ground vibronic level of S_b before emission, the excitation spectrum of a pure compound should have exactly the same shape as the absorption spectrum.

With the simple apparatus shown in Figure 8-13, the single-beam arrangement leads to distortions in both excitation and emission spectra. The distortion in excitation spectra occurs because the lamp does not emit equal radiation intensities at all wavelengths. The distortion in emission spectra is mainly the result of frequency-dependent sensitivity of the detector. To correct for these phenomena, many spectrofluorimeters in current use are considerably more complex than that shown in Figure 8-13, but the basic principles remain the same.

One quantity frequently desired is the fluorescence quantum yield, ϕ_F—the fraction of excited singlets that decay by fluorescence. Absolute measurements of ϕ_F are quite difficult, as the following argument demonstrates. The number of excited molecules created by incident light of intensity I_0 is proportional to the number of excited photons. This quantity can be measured at wavelength λ_e by the decrease in light intensity as given by the Beer–Lambert law. Equation 7-33 can be written as

$$I = I_0 e^{-2.303(\lambda_e)Cl}$$

where $\varepsilon(\lambda_e)$ is the extinction coefficient at the exciting wavelength, C is the concentration of absorbing molecules, and l is the path length. For the low absorbances typically

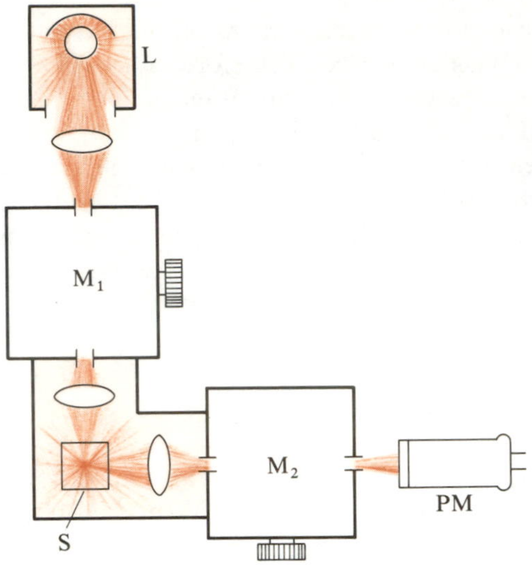

(a)

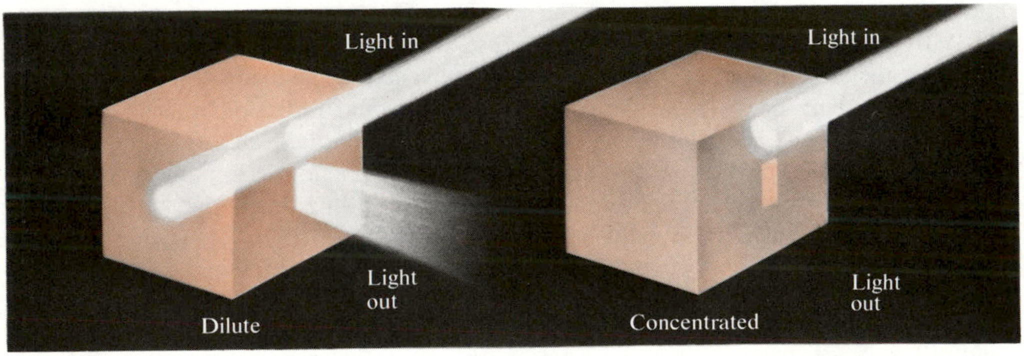

(b)

Figure 8-13

Steady-state fluorescence measurements. **(a)** Schematic diagram of a fluorescence spectrometer. M_1 is the excitation monochrometer, M_2 the emission monochrometer, L the light source, PM a photomultiplier detector, and S the sample. **(b)** An enlarged view of the sample chamber, showing the origins of the inner filter effect.

used in fluorescence, the exponential can be expanded to yield

$$I = I_0[1 - 2.303\varepsilon(\lambda_e)Cl]$$

The concentration of excited molecules will be proportional to the light intensity absorbed:

$$I_0 - I = 2.303\varepsilon(\lambda_e)ClI_0$$

The intensity of emission from one molecule at a particular wavelength λ is governed by the product of three factors: $\phi_F f(\lambda)d$, where ϕ_F is the probability of emitting at all; $f(\lambda)$ is the fraction of total emission that occurs at wavelength λ; and d is the fraction of radiation emitted at λ actually collected by the detector.

The actual observed emission intensity will be given by the product of absorption and emission probabilities:

$$F(\lambda) = 2.303\varepsilon(\lambda_e)ClI_0\phi_F f(\lambda)d = \varepsilon(\lambda_e)\phi_F f(\lambda)CI_0k \tag{8-42}$$

where we have lumped together proportionality constants and the factors l and d that depend on experimental geometry into a single constant k. Integration of Equation 8-42 over λ results in a quantity proportional to the number of excited singlets. However, to determine the quantum yield ϕ_F, we must also know the absolute intensity of the exciting light, I_0, and the constant k.

In practice, rather than face all these complications, one measures relative quantum yields. A standard—such as quinine sulfate in 1 N H_2SO_4 ($\phi_F = 0.70$), or fluorescein in 0.1 N NaOH ($\phi_F = 0.93$)—is used to calibrate the instrument. The integrated fluorescence from Equation 8-42 is compared for the sample and for the standard after adjusting concentrations so that εC for both is the same.

An alternative to quantum yield measurements, suggested by Equation 8-41, is to measure decay rates. Conceptually, the simplest way to do this is to excite the sample with a short ($\sim$1 nsec) pulse of light and directly monitor the emission as a function of time. The intensity of light, $I(t)$, emitted at time t after the pulse will be proportional to the rate of decay of excited singlets and to the fraction of singlets that decay by fluorescence:

$$I(t) \propto \phi_F(S_b)/dt = (S_b(0))(\phi_F/\tau_F)e^{-t/\tau_F} = k_F(S_b(0))e^{-t/\tau_F} \tag{8-43}$$

where we have inserted Equation 8-39 to compute $d(S_b)\,dt$, and have used Equations 8-35 and 8-41 to remove ϕ_F and τ_R. Although it is possible to monitor kinetics directly, indirect methods usually are used to achieve the same effect. Figure 8-14 shows a typical result, obtained by the method of single-photon counting. The single exponential behavior predicted by Equation 8-43 is quite evident.

Suppose that a sample contains more than one fluorescent component. Now the emission spectrum will not be a constant. Rather, it will depend on the choice of

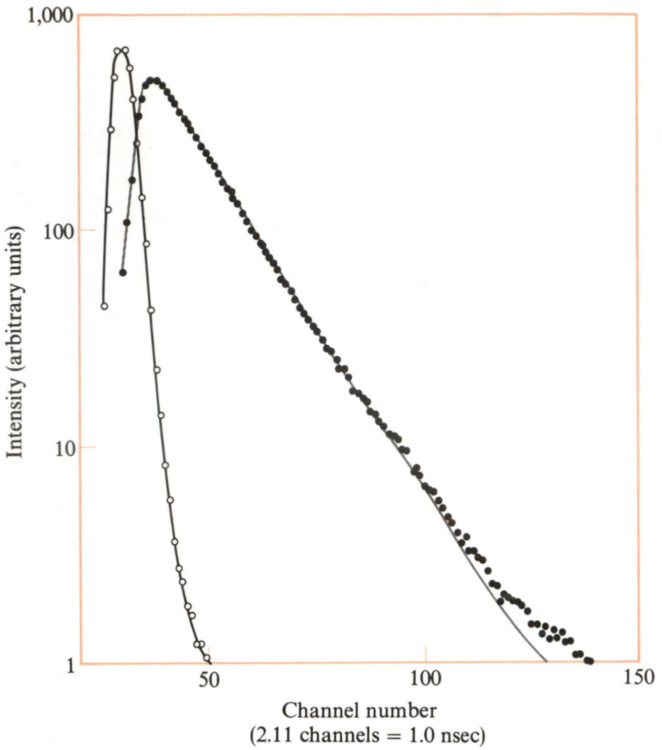

Figure 8-14

Fluorescence decay of the Y base in yeast $tRNA^{Phe}$. The black line shows the exciting pulse. The gray line through the fluorescence observations was generated from a knowledge of the shape of the exciting pulse and the assumption that the excited singlet state decays as a single exponential with $\tau_F = 6.2$ nsec. [After C. R. Cantor and T. Tao, in *Procedures in Nucleic Acid Research*, vol. 2 (New York: Harper & Row, 1971), p. 31.]

exciting wavelengths. For a mixture of two components (1 and 2), we have

$$F(\lambda) = [\varepsilon_1(\lambda_e)\phi_{1F}f_1(\lambda)C_1 + \varepsilon_2(\lambda_e)\phi_{2F}f_2(\lambda)C_2]I_0k \qquad (8\text{-}44)$$

where all quantitites are defined as in Equation 8-42. Similarly, the excitation spectrum will not be simply the sum of the two absorption spectra, because of weighting factors that depend on the quantum yield and wavelength dependence, $f_1(\lambda)$ and $f_2(\lambda)$, of fluorescence of each component. If the emission spectra of the two components overlap, the fluorescence decay will no longer be a single exponential. A sum of exponentials will be observed:

$$I(t) = A_1 e^{-t/\tau_{1F}} + A_2 e^{-t/\tau_{2F}} \tag{8-45}$$

The individual amplitudes (A_1 and A_2) will be weighted by the concentrations of the components.

One critical difference between steady-state and kinetic measurements of fluorescence cannot be ignored. The value of τ_F determined from Equation 8-43 will not be a function of concentration if the sample is a single pure species. In contrast, *Equation 8-42 holds only for optically thin samples* (absorbance at λ_e less than 0.03). In practice, if fluorescence is measured as a function of concentration, the observed intensity in an apparatus such as that shown in Figure 8-13 is linear with C when C is small, then saturates as C increases; and finally (at large enough C), $F(\lambda)$ actually decreases with increasing concentration. The origin of this inner filter effect is shown in Figure 8-13b. All the light is absorbed near the front surface of the cell, and most of the light emitted from this region is not collected by the exit slit.

Properties of typical fluorescent groups

Table 8-2 shows fluorescence characteristics of the chromophores found in proteins and nucleic acids. By and large, quantum yields are low and lifetimes are short. The experimental sensitivity, which will be governed by the product of ϕ_F and ε_{max} (Eqn. 8-42), is low. These are not ideal samples for experimental measurements. Figure 8-15 compares the fluorescence of a typical protein with that expected from

Table 8-2
Fluorescence characteristics of protein and nucleic acid constituents and coenzymes

		Absorption		Fluorescence[§]			Sensitivity
Substance	Conditions	λ_{max} (nm)	ε_{max} $\times 10^{-3}$	λ_{max} (nm)	ϕ_F	τ_F (nsec)	$\varepsilon_{max}\phi_F$ $\times 10^{-2}$
Tryptophan	H_2O, pH 7	280	5.6	348	0.20	2.6	11.
Tyrosine	H_2O, pH 7	274	1.4	303	0.14	3.6	2.0
Phenylalanine	H_2O, pH 7	257	0.2	282	0.04	6.4	0.08
Y base	Yeast $tRNA^{Phe}$	320	1.3	460	0.07	6.3	0.91
Adenine	H_2O, pH 7	260	13.4	321	2.6×10^{-4}	<0.02	0.032
Guanine	H_2O, pH 7	275	8.1	329	3.0×10^{-4}	<0.02	0.024
Cytosine	H_2O, pH 7	267	6.1	313	0.8×10^{-4}	<0.02	0.005
Uracil	H_2O, pH 7	260	9.5	308	0.4×10^{-4}	<0.02	0.004
NADH	H_2O, pH 7	340	6.2	470	0.019	0.40	1.2

[§] Values shown for ϕ_F are the largest usually observed. In a given case actual values can be considerably lower.

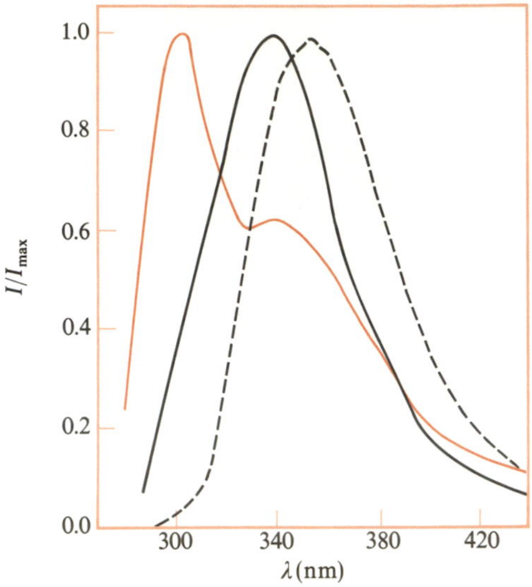

Figure 8-15

Normalized fluorescence emission spectra, produced by excitation at 240–250 nm of human serum albumin (*solid line*), tryptophan alone (*dashed line*), and an 18:1 molar ratio of tyrosine to tryptophan (*colored line*). The latter sample approximates the relative occurrence of these amino acids in the protein. Note that, except for a spectral shift, the spectrum of the protein closely resembles that of pure tryptophan. [After J. W. Longworth, in *Excited States of Proteins and Nucleic Acids*, ed. R. F. Steiner and I. Weinryb (New York: Plenum Press, 1971).]

individual amino acids. In most proteins, the fluorescence is dominated by tryptophan. Although tyrosine is a weaker emitter, one might expect it to contribute significantly because it usually is present in larger numbers. However, tyrosine usually is quenched by any nearby tryptophans because of energy-transfer effects to be discussed later.

The common nucleic acid bases have almost undetectably weak fluorescence—although a few unusual ones, such as the Y base, are intense enough to permit convenient study. This weak intrinsic fluorescence seems discouraging, but it is a blessing in disguise for many studies. The approach is to add a fluorescent probe to the system. This can be done by finding an intense fluorescent chromophore that binds at some specific site, or by covalently attaching such a chromophore. If the probe is properly placed, a variety of types of structural information can be obtained. Usually the probe is chosen so that it can be excited by light that only the probe (not the macromolecule) can absorb. In this case, the macromolecule is invisible, and all information relates to the probe molecule. Table 8-3 summarizes the properties of a few common probes (Fig. 8-16). Notice that the relative experimental sensitivity ($\varepsilon_{max}\phi_F$) shown in the table is much higher than that of most intrinsic protein or nucleic acid constituents.

Sensitivity of fluorescence to the environment

Fluorescence generally is much more sensitive to the environment of the chromophore than is light absorption. Therefore, fluorescence is a most effective technique for following the binding of ligands or conformational changes. The sensitivity of fluorescence is a consequence of the relatively long time a molecule stays in an excited singlet state before deexcitation. Absorption, or CD, is a process that is over in 10^{-15} sec. On this time scale, the molecule and its environment are effectively static.

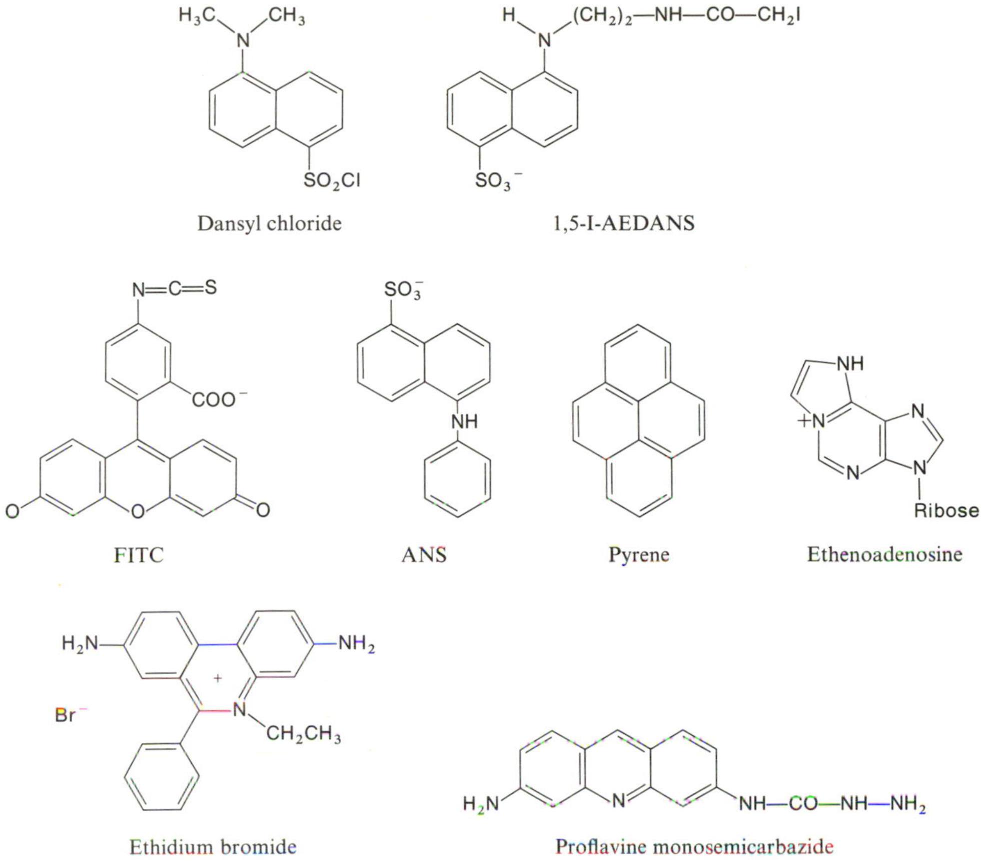

Figure 8-16

Structures of fluorescent probes listed in Table 8-3.

In contrast, during the 10^{-9} to 10^{-8} sec that a singlet remains excited, all kinds of processes can occur, including protonation or deprotonation reactions, solvent-cage relaxation, local conformational changes, and any processes coupled to translational or rotational motion.

A number of fluorescent molecules have a very convenient property: in aqueous solution their fluorescence is very strongly quenched, but in a nonpolar or a rigid environment a striking enhancement is observed. This enhancement can easily be by more than a factor of 20. If the probe can bind to a rigid or nonpolar site on a protein or nucleic acid, the fluorescence spectrum will be dominated by the bound species. Figure 8-17 shows a typical example.

For proteins, the dye 8-anilinonaphthalene sulfonate (ANS) is the most frequently used environmental probe, although several other common ones exist. Ethidium is

Table 8-3
Typical fluorescent probes

		Absorption		Emission[§]			Sensitivity
Probe[#]	Uses	λ_{max} (nm)	$\varepsilon_{max} \times 10^{-3}$	λ_{max} (nm)	ϕ_F	τ_F (nsec)	$\varepsilon_{max}\phi_F \times 10^{-2}$
Dansyl chloride	Covalent attachment to protein: Lys, Cys	330	3.4	510	0.1	13	3.4
1,5-I-AEDANS	Covalent attachment to protein: Lys, Cys	360	6.8	480	0.5	15	34
Fluorescein isothiocyanate (FITC)	Covalent attachment to protein: Lys	495	42	516	0.3	4	116
8-Anilino-1-naphthalene sulfonate (ANS)	Noncovalent binding to proteins	374	6.8	454	0.98	16	67
Pyrene, and various derivatives	Polarization studies on large systems	342	40	383	0.25	100	100
Ethenoadenosine, and various derivatives	Analogs of nucleotides bind to proteins, incorporate into nucleic acids	300	2.6	410	0.40	26	10
Ethidium bromide	Noncovalent binding to nucleic acids	515	3.8	600	~1	26.5	38
Proflavine monosemicarbazide	Covalent attachment to RNA 3′-ends	445	15	516	0.02	—	30

[§] Values shown for ϕ_F and τ_F are near the maximum typically observed in biological samples at ambient temperature. Other (considerably smaller) values often are found.

[#] Structures of these probes are shown in Figure 8-16.

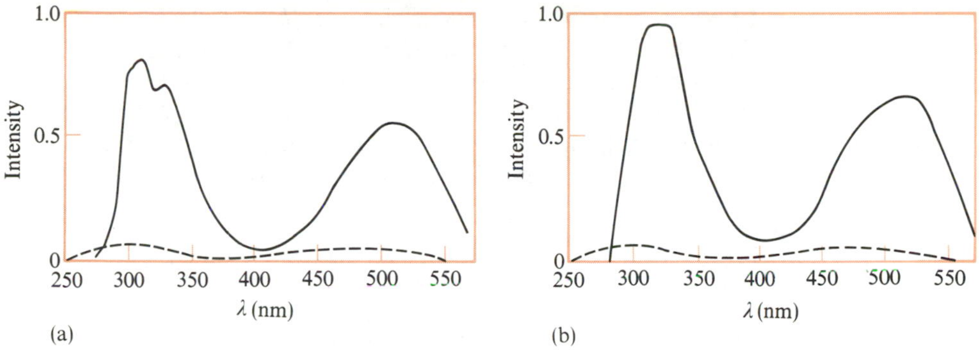

Figure 8-17

Fluorescence excitation spectra of ethidium bromide in aqueous solution (*dashed line*) or bound (*solid line*) to **(a)** double-strand DNA or **(b)** RNA. Emission is monitored at 6.00 nm. [After a figure provided by J. B. LePecq and J. Paoletti.]

most often used for nucleic acids. Ethidium in aqueous solution has very weak fluorescence but, when it intercalates into double-helical regions of a nucleic acid, the fluorescence is very intense. For a fixed total concentration of a fluorescence probe, changes in the number or strengths of binding sites will lead to large alterations in the observed fluorescence. This property obviously can permit accurate measurements of variety of interesting phenomena.

Another environmental effect is the accessibility of a fluorescent chromophore to collisional quenching by solute molecules. Oxygen is a ground-state triplet. Collision with an excited singlet leads to an enhanced rate of singlet–triplet interconversion (intersystem crossing) and thus to quenching of the fluorescence. Any heavy atom (such as iodide or cesium ions) has a similar effect, although the fundamental mechanism is different. A chromophore free in aqueous solution is quite susceptible to such quenching. When incorporated into a macromolecular structure, it may be considerably shielded from the solvent. This shielding can show up as a protection against quenching. Fluorescence, measured as a function of quencher concentration, often can help to discriminate residues on the surface of a protein from those deeply buried. Sometimes these differences are masked by complex effects of the local environment. For example, anionic groups near a chromophore may make it difficult for negatively charged ions such as I^- to approach close enough to quench.

To analyze the effect of a quencher, start with Equations 8-35 and 8-36. If the fluorescence in the presence (F) and in the absence (F_0) of the quencher (or the corresponding quantum yields, ϕ and ϕ_0) are compared, at constant concentration and sample geometry, we have

$$\begin{aligned} F_0/F = \phi_0/\phi &= [k_F + k_{ic} + k_{is} + k_q(\mathrm{Q})]/(k_F + k_{ic} + k_{is}) \\ &= 1 + k_q\tau_0(\mathrm{Q}) \end{aligned} \tag{8-46}$$

where τ_0 is the lifetime in the absence of quencher: $\tau_0 = (k_F + k_{ic} + k_{is})^{-1}$. Because τ_0 can be measured, a plot of F_0/F versus (Q) should yield k_q. Table 8-4 shows some typical values of bimolecular quenching constants. The quenching constant for a free chromophore is of the order of 10^{10} M^{-1} sec^{-1}. Such a rapid rate is characteristic of a diffusion-controlled reaction. However, when chromophores are bound to or incorporated into proteins, k_q sometimes is substantially smaller.

Table 8-4
Quenching of tryptophan fluorescence by collision with small molecules

	Native protein (0.1 M phosphate, pH 7) $k_q \times 10^{-9}$ M^{-1} sec^{-1}		Denatured protein (6 M guanidine-HCl added) $k_q \times 10^{-9}$ M^{-1} sec^{-1}	
Protein	Oxygen	Iodide	Oxygen	Iodide
Tryptophan	12.0	3.9	5.9	1.9
n-Ac-Try-NH_2	11.6	3.8	7.3	2.1
Pepsin	5.7	1.7	4.3	1.8
Trypsinogen	4.3	0.3	6.1	1.2
Carboxypeptidase A	3.8	0.3	3.8	1.1
Carbonic anhydrase	2.6	0.2	4.2	1.0

SOURCE: Adapted from J. R. Lakowicz and G. Weber, *Biochemistry* 12:4171 (1973).

Singlet–singlet energy transfer

A fairly unique feature of fluorescence is the ability of other chromophores quite far away from an excited singlet to cause quenching. In a favorable case, this permits distances between chromophores to be measured up to a range of about 80 Å. Consider a system with just one copy each of two different chromophores, with individual spectra as shown in Figure 8-18. If these chromophores are within a few angstroms, they can interact by the exciton (coupled oscillator) mechanism as described earlier. In this case, splittings and shifts in the apparent locations of spectral bands can occur. This is not the case we want to consider here.

Optical interactions persist between the two chromophores even when they are so far apart that no changes in the shapes of their spectra are observed. This is called the very weak coupling limit. It is convenient to define the component with higher energy absorption as the donor (D), and the other as the acceptor (A). We consider just the ground singlet states (D_a, A_a) and first excited singlet states (D_b, A_b) of each chromophore. Suppose that the donor is excited. It will rapidly lose energy by internal conversion until it reaches the ground vibrational level of the first excited singlet, D_b. If donor emission energies are coincident with acceptor absorption energies, the very weak coupling can permit the following resonance to take place:

$$D_b + A_a \underset{k_{-T}}{\overset{k_T}{\rightleftharpoons}} D_a + A_b \qquad (8\text{-}47)$$

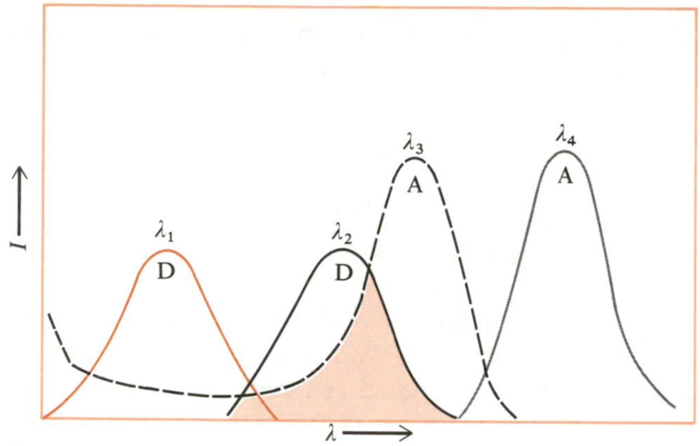

Figure 8-18

Schematic spectra for a donor–acceptor pair suitable for singlet–singlet energy-transfer measurements. Shown are the absorption spectra for donor (*colored line*) and acceptor (*dashed line*), and the emission spectra for donor (*black line*) and acceptor (*gray line*). The spectral overlap (neglecting the factor of v^{-4}) is shaded.

As shown in Figure 8-19, the resulting acceptor singlet (A_b) and donor singlet (D_a) are in excited vibrational states. Vibrational relaxation rapidly converts these to the ground vibrational level. Therefore, even when k_T is an efficient process, the reverse reaction (k_{-T}) is unlikely to occur. The energy-transfer resonance shifts the relative population of excited donors and acceptors. The donor becomes quenched. The acceptor becomes excited, and subsequently it can fluoresce. This process, from the viewpoint of the acceptor, is called sensitized emission. Note that the two chromophores need not necessarily be part of the same molecule. Energy transfer will take place between isolated molecules in solution as long as the concentration is high enough to bring average intermolecular distance within 50 Å or so.

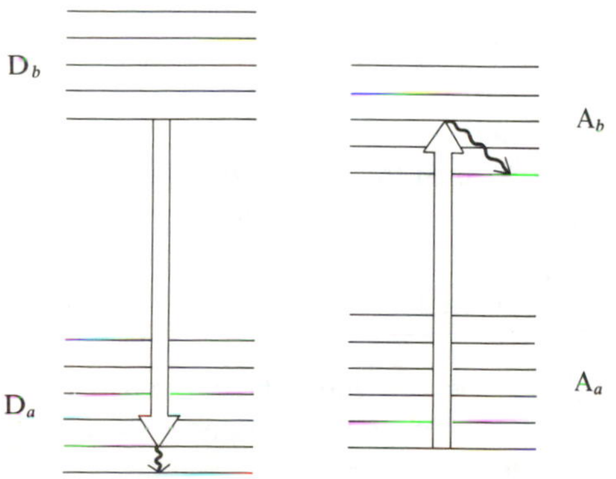

Figure 8-19

Donor deexcitation and acceptor excitation coupled in the resonant interaction that leads to energy transfer. Vibrational relaxation (*wavy arrows*) occurs rapidly and tends to prevent a repeat of the interaction that would relax the acceptor and reexcite the donor.

Singlet–singlet energy transfer can be observed in several ways. Define the efficiency of transfer (E) as the fraction of D_b that is deexcited by transfer to the acceptor:

$$E = k_T/(k_T + k_F^D + k_{ic}^D + k_{is}^D) \tag{8-48}$$

where k_T is the rate of energy transfer, and all other rates refer to donor processes. In Figure 8-18, if the system is excited at λ_1 and observed at λ_2, only emission from D is seen. Two samples must be prepared as identical as possible, except that one contains both donor and acceptor, whereas the other has only donor. When the fluorescence of these two samples is compared, E can be measured. Using Equations 8-36 and 8-48, it is easy to show that

$$\phi_{D+A}/\phi_D = [k_F^D/(k_F^D + k_{ic}^D + k_{is}^D + k_T)] \times [(k_F^D + k_{ic}^D + k_{is}^D)/k_F^D] = 1 - E \tag{8-49}$$

where ϕ_{D+A} is the quantum yield of the donor in the presence of the acceptor.

An alternative experimental arrangement involves keeping the emission wavelength constant at λ_4 and varying the excitation wavelength from λ_3 to λ_1. If the system contains only acceptor, the excitation will just resemble its absorption. But, if donor is present also, there will be an additional peak in the excitation spectrum, corresponding to the absorption maximum of the donor at λ_1. The intensity of this peak will depend on the relative absorbance of donor and acceptor in the region of λ_1, and on the efficiency of transfer (E). At λ_4, only acceptor fluorescence can be observed. For constant experimental geometry and incident light, we can use Equation 8-42 to analyze the result. The relative fluorescence of the sample with acceptor alone (exciting at λ_1 and observing at λ_4) will be

$$F_A \propto \varepsilon_A C_A \phi_A \tag{8-50}$$

whereas the sample with both donor and acceptor will have more fluorescence:

$$F_{D+A} \propto \varepsilon_A C_A \phi_A + \varepsilon_D C_D E \phi_A \tag{8-51}$$

In Equation 8-51, $\varepsilon_D C_D$ is proportional to the number of excited donors; E is the fraction of the excited donors that transfer energy, resulting in an excited acceptor. Combining the results of these two measurements, one can determine E:

$$F_{D+A}/F_A = 1 + (\varepsilon_D C_D/\varepsilon_A C_A)E \tag{8-52}$$

A third independent measurement of E can be obtained by comparing the fluorescence decay times of the donor in the presence ($\tau_{D,A}$) and absence (τ_D) of the acceptor. You should be able to show that

$$\tau_{D,A}/\tau_D = 1 - E \tag{8-53}$$

Measurement of energy transfer by direct observation of lifetimes is quite important because it avoids a potential trivial effect that can mimic transfer.

A requirement for the resonance interaction producing energy transfer is that acceptor absorption must overlap the donor fluorescence (Fig. 8-18). Therefore, in any system capable of energy transfer, an additional process may occur in which the donor emits a photon that is then reabsorbed by the acceptor. This is distinguishable from true singlet–singlet energy transfer, because it leaves the rate of donor emission unchanged. Trivial emission and reabsorption of photons can be detected by comparing the efficiencies derived from Equations 8-53 and 8-49 or 8-52. It can be avoided by working at total chromophore concentrations less than about 10^{-3} M.

Measuring interchromophore distances from energy-transfer efficiencies

To obtain useful structural information from energy transfer, the measured efficiency must be related to the distance R between the two chromophores. This can be done by using the theory developed by Theodore Förster. He computed that the rate of transfer is

$$k_T = (1/\tau_D)(R_0/R)^{-6} \tag{8-54}$$

where τ_D is the lifetime of the donor in the absence of the acceptor. The inverse sixth power comes from the square of the dipole–dipole coupling, which depends on R^{-3} (see Box 8-3). R_0 is called the characteristic transfer distance:

$$R_0 = 9.7 \times 10^3 (J\kappa^2 n^{-4} \phi_D)^{1/6} \text{ cm} \tag{8-55}$$

where

$$J = \int \varepsilon_A(\nu) f_D(\nu) \nu^{-4} \, d\nu \tag{8-56}$$

J is a measure of the spectral overlap between donor emission and acceptor absorption (shaded in Fig. 8-18); f_D is the normalized fluorescence of the donor as defined in Equation 8-42; n is the refractive index of the medium between donor and acceptor; ϕ_D is the quantum yield of donor in the absence of the acceptor; and κ^2 is a complex geometric factor that depends on the orientation of donor and acceptor. If both donor and acceptor are free to tumble rapidly on the time scale of fluorescence emission, κ^2 approaches a limiting value of 2/3. (The origins of all these quantities in Eqns. 8-54, 8-55, and 8-56 are described in Box 8-3.)

Box 8-3 THE FÖRSTER THEORY OF SINGLET–SINGLET ENERGY TRANSFER

Here we sketch an oversimplified derivation of Equations 8-54, 8-55, and 8-56. We start with Equation 7-27, which says that the rate of exciting a molecule is proportional to the square of the expectation value of the interaction causing the excitation. Although we derived this equation for a particular case, it turns out to be a very general result, often called Fermi's Golden Rule. We want to compute the rate at which the state of the chromophore pair D and A changes from $\Psi_{D_b}\Psi_{A_a}$ to $\Psi_{D_a}\Psi_{A_b}$.

We first consider the case where the $a \leftrightarrow b$ transitions of both D and A occur at the same frequency, ν. To describe the weak interaction between D and A, we use the same dipole–dipole coupling described earlier for the somewhat stronger interactions that produce exciton splitting. Therefore, the rate of energy transfer should be proportional to the following expression:

$$k_T(\nu) \propto |\langle \Psi_{D_a}\Psi_{A_b}|\underset{\sim}{\mathbf{V}}|\Psi_{D_b}\Psi_{A_a}\rangle|^2$$

where $\underset{\sim}{\mathbf{V}}$ is given (by analogy to Eqn. 7-48) as

$$\underset{\sim}{\mathbf{V}} = (\underset{\sim}{\boldsymbol{\mu}}_D \cdot \underset{\sim}{\boldsymbol{\mu}}_A)/R^3 - 3(\underset{\sim}{\boldsymbol{\mu}}_D \cdot \mathbf{R})(\mathbf{R} \cdot \underset{\sim}{\boldsymbol{\mu}}_A)/R^5$$

where $\mathbf{R}$ is the distance between donor and acceptor, and $\underset{\sim}{\boldsymbol{\mu}}_D$ and $\underset{\sim}{\boldsymbol{\mu}}_A$ are dipole moment operators.

If all effects of the orientation of D and A are lumped into a parameter κ, then we can rewrite $\underset{\sim}{\mathbf{V}}$ as

$$\underset{\sim}{\mathbf{V}} = \kappa|\underset{\sim}{\boldsymbol{\mu}}_D||\underset{\sim}{\boldsymbol{\mu}}_A|/R^3$$

where the vertical bars indicate that only the lengths of $\underset{\sim}{\boldsymbol{\mu}}_D$ and $\underset{\sim}{\boldsymbol{\mu}}_A$ must be evaluated. Substitution of this expression for $\underset{\sim}{\mathbf{V}}$ into the earlier equation for the transfer rate, k_T, yields

$$k_T(\nu) \propto |(\kappa/R^3)\langle \Psi_{D_a}\Psi_{A_b}||\underset{\sim}{\boldsymbol{\mu}}_D||\underset{\sim}{\boldsymbol{\mu}}_A||\Psi_{D_b}\Psi_{A_a}\rangle|^2$$

Because $\underset{\sim}{\boldsymbol{\mu}}_D$ depends only on the electronic coordinates of the donor group, and $\underset{\sim}{\boldsymbol{\mu}}_A$ depends only on the acceptor group, this expression can be factored to give

$$k_T(\nu) \propto (\kappa^2/R^6)|\langle \Psi_{D_a}|\underset{\sim}{\boldsymbol{\mu}}_D|\Psi_{D_b}\rangle|^2|\langle \Psi_{A_b}|\underset{\sim}{\boldsymbol{\mu}}_A|\Psi_{A_a}\rangle|^2$$

You can see that the transfer rate will depend on the inverse sixth power of the distance. The term in $\underset{\sim}{\boldsymbol{\mu}}_A$ is just the dipole strength of the acceptor, which in turn can be related to an integral over the absorption spectrum using Equation 7-40. In the case we are considering, absorption takes place only at a single frequency ν, and so Equation 7-40 becomes

$$|\langle \Psi_{A_b}|\underset{\sim}{\boldsymbol{\mu}}_A|\Psi_{A_a}\rangle|^2 \propto \varepsilon_A \nu^{-1}$$

The term in $\underset{\sim}{\boldsymbol{\mu}}_D$ is the dipole strength of the donor, which can be related to the fluorescence rate.

Using Equations 8-32 and 8-34, we have

$$D_{ab} \propto \nu^{-3} A_{ba} = \nu^{-3} \tau_{\mathrm{R}}^{-1}$$

From Equation 8-41, we can replace τ_{R}^{-1} by $\phi_{\mathrm{D}}/\tau_{\mathrm{D}}$, where ϕ_{D} is the quantum yield of the donor, and τ_{D} is the lifetime of the donor in the absence of the acceptor. Then

$$|\langle \Psi_{\mathrm{D}_a} | \underset{\sim}{\mu}_{\mathrm{D}} | \Psi_{\mathrm{D}_b} \rangle|^2 \propto \nu^{-3} \phi_{\mathrm{D}}/\tau_{\mathrm{D}}$$

Using all these results, we can write the energy transfer rate as

$$k_{\mathrm{T}}(\nu) \propto (\kappa^2/R^6)(\phi_{\mathrm{D}}/\tau_{\mathrm{D}})\varepsilon_{\mathrm{A}} \nu^{-4}$$

Now we consider the general case where the acceptor absorption occurs over a band of frequencies, so that ε_{A} is a function of ν. Similarly, the fluorescence of the donor occurs over a range of frequencies. Let $f_{\mathrm{D}}(\nu)$ be the fraction of donor fluorescence at frequency ν. The total rate of energy transfer can be computed by integrating the rate expected at each frequency:

$$k_{\mathrm{T}} \propto (\kappa^2/R^6)(\phi_{\mathrm{D}}/\tau_{\mathrm{D}}) \int \varepsilon_{\mathrm{A}}(\nu) f_{\mathrm{D}}(\nu) \nu^{-4} d\nu = (\kappa^2 \phi_{\mathrm{D}}/R^6 \tau_{\mathrm{D}}) J$$

This expression is identical to Equations 8-54, 8-55, and 8-56, except for numerical constants and the appearance of n, the refractive index. The presence of n comes from the fact that $\underset{\sim}{\mathbf{V}}$ as written above applies only in vacuum; in a fluid medium, the true interaction potential at optical frequencies is $\underset{\sim}{\mathbf{V}}/n^2$.

The troublesome part of the Förster theory is the inclusion of the quantity κ^2. Removing the orientation dependence of the donor and acceptor interaction (as we did above) implies that orientations are sampled rapidly during the time the donor is excited. This may not always be true. Unfortunately, κ^2 cannot be measured directly. From the definition of $\underset{\sim}{\mathbf{V}}$, we know that κ^2 must have a value between 0 and 4. Measurements of fluorescence polarization can place narrower limits on κ^2 in each particular case. Uncertainties in the κ^2 value still are a major source of error in distance determinations using energy transfer. Fortunately, the distance R will depend only on $(\kappa^2)^{1/6}$, as you can see by rewriting Equation 8-57 in this form:

$$R = R_0[(1 - E)/E]^{1/6}$$

In actuality, using Equations 8-55 and 8-56, it is more convenient to use spectra measured on a wavelength scale and to express the result in angstroms. Then Equations 8-55 and 8-56 become

$$R_0 = 8.79 \times 10^{-5} (J \kappa^2 n^{-4} \phi_{\mathrm{D}})^{1/6} \text{ Å}$$

$$J = \int \varepsilon_{\mathrm{A}}(\lambda) f_{\mathrm{D}}(\lambda) \lambda^4 d\lambda$$

The efficiency of energy transfer can be calculated by rewriting Equation 8-48 as $E = k_T/(k_T + 1/\tau_D)$. Then, substitution of Equation 8-54 for k_T yields

$$E = R_0^6/(R_0^6 + R^6) \tag{8-57}$$

The result is plotted as the solid line in Figure 8-20 for a particular value of R_0. It is apparent that, for distances near R_0, a measurement of E can yield a fairly accurate determination of the distance. For commonly used pairs of chromophores, R_0 varies from 10 Å to more than 50 Å. Therefore, distances up to about 80 Å are measurable.

The most critical test of the Förster theory came from the work of Lubert Stryer and Richard Haugland (1967). They prepared a set of terminally labeled oligopeptides:

Dansyl—(Pro)$_n$—NH—NH—CO—NH—Naphthyl

The proline residues formed a polyproline type-II helix, which could be confirmed by CD measurements. Because the distance between naphthyl donor and dansyl acceptor was known from the known dimensions of the helix, measured efficiencies could be compared directly with computed values from the Förster theory; the agreement is excellent (Fig. 8-20). Since this pioneering work, energy transfer has been used to measure distances within tRNA, immunoglobulins, rhodopsin, and oligopeptides, and between the protein subunits of assemblies ranging from simple oligomeric proteins to ribosomes. The most difficult part of these measurements is the specific introduction of the two fluorescent chromophores into known portions of the structures. For the full spectrum of measurements, three replica systems must be prepared. They should be as identical as possible, except that one has only donor, one has donor and acceptor, and one has only acceptor.

Energy transfer plays a large role in determining the emission spectrum of normal proteins. The fluorescence of tyrosine is overlapped by the absorption of tryptophan. Although the R_0 for this donor-acceptor pair is only ~9 Å, it is still appreciable compared with average distances expected for tyrosine–tryptophan nearest-neighbor pairs in a globular protein.[§] The result is that the tyrosines are extensively quenched; almost all of the observed emission comes from tryptophan (as we mentioned earlier).

Fluorescence polarization

If plane-polarized light is used to excite a fluorescent system, and if linearly polarized components of the emission are detected, information can be obtained about the size, shape and flexibility of macromolecules. Figure 8-21 shows a typical experimental

[§] A spherical protein of 17,000 mol wt will have a radius (r) of about 17 Å. The root-mean-square distance between random selected pairs of points within a spherical volume is $\sqrt{6/5}\, r$. So even a single tryptophan in such a protein is likely to be near enough to many tyrosines to cause appreciable energy transfer.

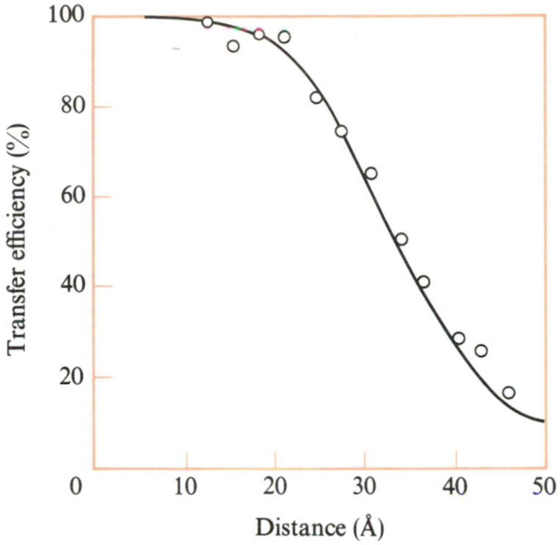

Figure 8-20

Efficiency of energy transfer as a function of distance in dansyl-(L-prolyl)$_n$-α-naphthyl semicarbazide oligomers with $n = 1$ to 12. The curve was fit to the data with Equation 8-57. [From L. Stryer and R. P. Haugland, *Proc. Natl. Acad. Sci. USA* 98:719 (1967).]

arrangement. Light is incident along the x axis and is detected along the y axis. The incident light is polarized along the z axis. Two components of the emitted light are measured: $I_{\parallel}$ is polarized along the z axis, and $I_{\perp}$ is polarized along the x axis. There can be no emission propagating along the y axis that is also polarized along this axis because light, like all electromagnetic radiation, is a transverse wave.

Polarization of rigid systems

First, consider a rigid, isotropic sample. This sample could correspond to a frozen solution of fluorescent molecules, or to chromophores on randomly oriented molecules so large that no appreciable molecular rotations occur on the fluorescence time scale (typically < 100 nsec). The vector $\boldsymbol{\mu}$ defines the orientation of the absorption transition dipole moment ($\langle \Psi_b | \underset{\sim}{\boldsymbol{\mu}} | \Psi_a \rangle$) of one chromophore relative to the laboratory coordinates. The probability that this chromophore will be excited is proportional to $(\boldsymbol{\mu} \cdot \mathbf{E})^2$. Because $\mathbf{E}$ is parallel to the z axis, this probability is $\cos^2 \theta$, where θ is the angle between $\boldsymbol{\mu}$ and the z axis. Therefore, molecules oriented with transition dipoles near the z axis will be preferentially excited. This is the principle called *photoselection.*

To explain fluorescence polarization, we need to calculate the probability of exciting molecules with particular orientations. Then we must compute the probability that these molecules will emit light polarized in certain directions. It is convenient to use spherical polar coordinates (Fig. 8-21). Before excitation, the relative number of molecules with μ oriented at angles between θ to $\theta + d\theta$ and ϕ to $\phi + d\phi$ is $\sin \theta \, d\theta \, d\phi$. (The factor of $\sin \theta$ enters because it is much more probable to find molecules perpendicular to the z axis than parallel to it.) The probability of exciting a molecule oriented at θ and ϕ is proportional to $\cos^2 \theta$, as mentioned earlier. Therefore, the

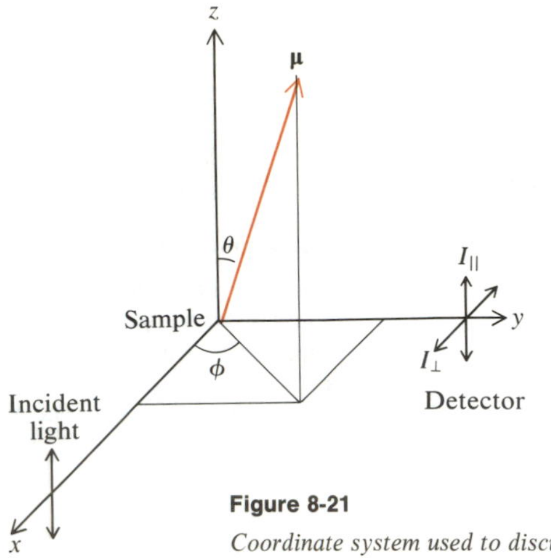

Figure 8-21

Coordinate system used to discuss fluorescence polarization.

relative number of excited molecules oriented at θ to $\theta + d\theta$ and ϕ to $\phi + d\phi$ is

$$P(\theta, \phi)\, d\theta\, d\phi \propto \cos^2\theta \sin\theta\, d\theta\, d\phi \tag{8-58}$$

The fraction of excited molecules oriented at θ to $\theta + d\theta$ and ϕ to $\phi + d\phi$ is

$$W(\theta, \phi)\, d\theta\, d\phi = P(\theta, \phi)\, d\theta\, d\phi \Big/ \int_0^{\pi} d\theta \int_0^{2\pi} d\phi \cos^2\theta \sin\theta\, d\theta\, d\phi \tag{8-59}$$

The integral in Equation 8-59 simply counts all excited molecules. It can be performed by substituting $x = \cos\theta$, $dx = -\sin\theta$, and so on. The result is $4\pi/3$, so Equation 8-59 becomes

$$W(\theta, \phi)\, d\theta\, d\phi = (3/4\pi) \cos^2\theta \sin\theta\, d\theta\, d\phi \tag{8-60}$$

Note that this result depends only on θ. The distribution of excited molecules is cylindrically symmetric about the z axis. (Fig. 8-22 shows a plot of Eqn. 8-60.) Note that excited molecules have transition dipoles oriented preferentially toward the z axis and not along the x axis.

Because the distribution of excited molecules is anisotropic, the resulting fluorescence also will be anisotropic. To calculate this, we must know the relative directions of absorbing and emitting transition dipoles in the chromophore. If the same electronic transition that absorbed does the emitting ($S_a \rightarrow S_b$ followed by $S_b \rightarrow S_a$), then emitting and absorbing transition dipoles are parallel. The probability that emission will occur polarized along the z axis is proportional to $|\boldsymbol{\mu} \cdot \hat{\mathbf{k}}|^2$, where $\hat{\mathbf{k}}$ is a unit vector along the z axis. This expression is proportional to $\cos^2\theta$. To find the relative emission intensity polarized along z, we must multiply the probability

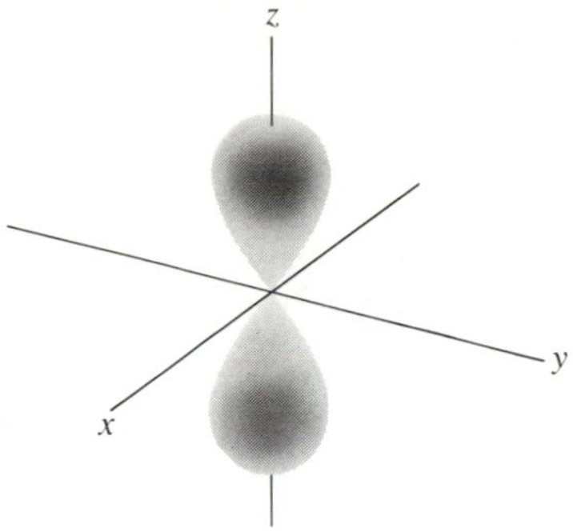

Figure 8-22

Distribution of excited chromophores produced by exciting a sample with z-polarized light propagating along the x axis. The density of the shading is proportional to the probability of finding an excited molecule with its transition dipole at that particular orientation.

of emission by the fraction of excited molecules with each orientation, $W(\theta, \phi)$, and average over all orientations:

$$I_{\parallel} \propto \int_0^{2\pi} d\phi \int_0^{\pi} d\theta \cos^2\theta\, W(\theta,\phi) = (3/4)\pi \int_0^{2\pi} d\phi \int_0^{\pi} d\theta \cos^4\theta \sin\theta = 3/5 \quad (8\text{-}61)$$

The probability of emission polarized along the x axis is proportional to $|\boldsymbol{\mu} \cdot \hat{\mathbf{i}}|^2$, where $\hat{\mathbf{i}}$ is a unit vector along x. This expression is proportional to $(\sin\theta\cos\phi)^2$. To calculate the relative emission intensity, $I_{\perp}$, we again average emission probabilities over the distribution of excited molecules:

$$I_{\perp} \propto \int_0^{2\pi} d\phi \int_0^{\pi} d\theta \sin^2\theta \cos^2\phi\, W(\theta,\phi) = (3/4)\pi \int_0^{2\pi} d\phi \cos^2\phi \int_0^{\pi} d\theta \cos^2\theta \sin^3\theta = 1/5 \quad (8\text{-}62)$$

In practice, what is done is to measure $I_{\parallel}$ and $I_{\perp}$ and compare. Two convenient comparisons of $I_{\parallel}$ and $I_{\perp}$—the polarization (P) and the anisotropy (A)—are defined in Equation 8-63 (see also Box 8-4).

$$P = (I_{\parallel} - I_{\perp})/(I_{\parallel} + I_{\perp}) \qquad A = (I_{\parallel} - I_{\perp})/(I_{\parallel} + 2I_{\perp}) \quad (8\text{-}63)$$

For the totally rigid system we have just described, $P = 1/2$ and $A = 2/5$, as you can see by substituting the results of Equations 8-61 and 8-62 into Equation 8-63. These turn out to be the maximal values of polarization and anisotropy possible under any circumstances.

Box 8-4 POLARIZATION AND ANISOTROPY

The definitions of polarization and anisotropy are not as arbitrary as they seem. P is defined by analogy with the dichroic ratio described in Equation 7-42. It is a relatively simple quantity to measure directly. A is more useful for analysis of experimental data on complex systems. The denominator of A is simply the total light that would be observed if no polarizers were used (see figure). Nonpolarized light incident along x can be resolved into y- and z-polarized

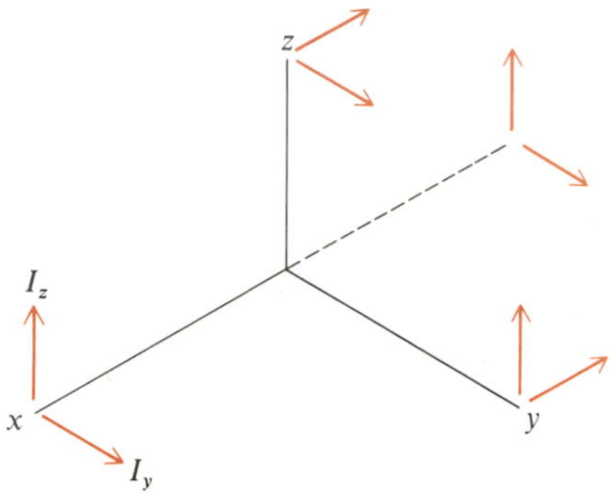

components. The total emission can be found by adding emission along all three Cartesian axes. For z-polarized excitation, there are two perpendicular components propagating along z, and one parallel and one perpendicular component, each propagating along x and y. For y-polarized excitation, there are two perpendicular components propagating along y, and one parallel and one perpendicular component, each propagating along x and z. The result is $8I_{\perp}$ plus $4I_{\parallel}$. Therefore, the total emission has twice as much perpendicular component as parallel component.

Anisotropy has a substantial advantage in many applications because various equations using it are simpler than the corresponding equations for polarization. Another advantage comes in the analysis of mixtures. Multicomponent mixtures of substances with equal fluorescence intensity but variable anisotropy (A_i) show a total anisotropy equal to $\sum_i \chi_i A_i$, where χ_i is the mole fraction of the ith component. Polarization does not obey such simple relationships. However, polarization and anisotropy can be interconverted by the relationship

$$[(1/P) - (1/3)]^{-1} = 3A/2$$

which you should be able to derive from Equation 8-63.

Another common case is a system in which the emission transition dipole is perpendicular to the absorption transition dipole. This will occur in many planar chromophores if absorption takes place into the second excited singlet state, but emission is observed from the first excited singlet. In this case, still for a rigid system, $P = -1/3$ and $A = -1/5$. See if you can derive these values by calculating averages analogous to those worked out for the case just discussed.

The polarization or anisotropy for a rigid system is called the limiting value, and it is denoted by subscript zero. In general, it is given by

$$P_0 = (3\cos^2\xi - 1)/(\cos^2\xi + 3) \qquad A_0 = (3\cos^2\xi - 1)/5 \tag{8-64}$$

where ξ is the angle between absorption and emission transition dipoles. These expressions provide a method of measuring ξ in rigid systems.

Now consider the other extreme case, in which (during the lifetime of the excited state) the chromophore can tumble fast enough to randomize its orientation. In this case, by the time emission occurs, all memory of the original photoselection is lost. Thus $I_{||} = I_{\perp}$, and the polarization and anisotropy are both zero.

Effect of molecular motion

Most macromolecules of biological interest fall between the two extreme cases just described: their rotational motions are *not* negligible on the fluorescence time scale, but neither can they tumble fast enough to achieve random orientation. Suppose that a fluorescent probe is rigidly attached to the macromolecule. The observed polarization will be some intermediate value between Equation 8-64 and zero. To compute this value, we must analyze the relative rates of emission and macromolecular rotational motion. Note that translational motion does not affect fluorescence polarization. Only motion that changes the *orientation* of the transition dipoles can be observed.

First consider what happens if the sample is excited by a pulse of polarized light, and the time dependence of $I_{||}$ and $I_{\perp}$ is measured. If emission and absorption dipoles are parallel, the earliest photons to be emitted are highly likely to be z-polarized because the molecules have not had time to reorient. The last photons to be emitted should have random polarization because by then the system has experienced considerable rotational motion. Therefore, if the polarization or anistropy is measured as a function of the time of fluorescence emission, it will decay from initial values of P_0 or A_0 to final values of zero. The rate of decay is a measure of the rate of rotational motion.

Rotational Brownian motion is described by a diffusion equation quite analogous to that used for translational motion.[§] If $W(\theta, \phi, t)$ is the probability per unit interval of θ and ϕ that a spherical molecule has orientation θ, ϕ at time t, then

$$dW(\theta, \phi, t)/dt = D_{\text{rot}} \nabla^2 W(\theta, \phi, t) \tag{8-65}$$

D_{rot} (the rotational diffusion coefficient) is related to the rotational friction coefficient (f_{rot}):

$$D_{\text{rot}} = kT/f_{\text{rot}} = kT/6V_{\text{h}}\eta \tag{8-66}$$

where V_{h} is the hydrated volume of the molecule, T is the absolute temperature, and η is the viscosity of the solution.

$W(\theta, \phi, t)$ gives the probability that any molecule has a particular orientation at time t. However, in polarization measurements, we observe only those molecules initially excited at time zero. From Equation 8-60, we can write the initial distribution of excited molecules as

$$W(\theta, \phi, 0) = (3/4\pi) \cos^2 \theta \sin \theta \tag{8-67}$$

Solving Equation 8-65 using this boundary condition yields an explicit expression for $W(\theta, \phi, t)$. This expression is a function of the rotational diffusion coefficient just defined. It tells the angular distribution of absorption transition dipoles at any time t.

The population of excited singlets produced at time zero decays with time according to Equation 8-39. Therefore, we must correct $W(\theta, \phi, t)$ to compute the distribution of molecules that are still excited at time t. This expression is

$$W(\theta, \phi, t)e^{-t/\tau_{\text{F}}} \tag{8-68}$$

If emitting and absorbing dipoles are parallel, then the possibility that a photon emitted from an excited molecule oriented at θ, ϕ is polarized along the x or y direction (P_x or P_y) is proportional to $\cos^2 \theta$ or $\sin^2 \theta \cos^2 \phi$, as discussed earlier. In the more general case, the probability of emission along a particular axis is also a function of the angle ξ between absorption and emission dipoles. We can write this as $P_a(\theta, \phi, \xi)$, where the subscript a refers to the axis of the polarized component emitted.

If we now combine $P_a(\theta, \phi, \xi)$ and $W(\theta, \phi, t)e^{-t/\tau_{\text{F}}}$, we have the probability that a molecule excited at time zero will emit a photon at time t polarized along a. Integrating this probability over all orientations gives the expected intensity of polarized emission along a particular axis as a function of time:

$$I_a(t) = \int_0^{2\pi} d\phi \int_0^{\pi} d\theta\, P_a(\theta, \phi, \xi) W(\theta, \phi, t) e^{-t/\tau_{\text{F}}} \tag{8-69}$$

[§] The reader who has had little prior experience with diffusion equations is encouraged to read Sections 10-3 and 12-2 for a justification of the form of Equation 8-65 and a demonstration of some of its properties.

This integral can be evaluated using explicit forms for P_a and W. The results for parallel and perpendicular polarization directions are the following:

$$I_{||}(t) = [(1/3) + (4/15)e^{-6D_{\mathrm{rot}}t}(3\cos^2\xi - 1)/2]e^{-t/\tau_F} \quad (8\text{-}70)$$

$$I_{\perp}(t) = [(1/3) - (2/15)e^{-6D_{\mathrm{rot}}t}(3\cos^2\xi - 1)/2]e^{-t/\tau_F} \quad (8\text{-}71)$$

Thus, although the decay of nonpolarized emission is a single exponential, each of the polarized components decays as the sum of two exponentials. If Equations 8-70 and 8-71 are substituted into the definition of fluorescence anisotropy, $A(t)$, the result is particularly simple:

$$A(t) = [I_{||}(t) - I_{\perp}(t)]/[I_{||}(t) + 2I_{\perp}(t)] = (2/5)e^{-6D_{\mathrm{rot}}t}[(3\cos^2\xi - 1)/2] \quad (8\text{-}72)$$

We define the rotational correlation time τ_c as

$$\tau_c = 1/6D_{\mathrm{rot}} = V_h\eta/kT \quad (8\text{-}73)$$

Then

$$A(t) = (2/5)e^{-t/\tau_c}[(3\cos^2\xi - 1)/2]$$

Note that, in the limit of $t \to 0$, Equation 8-72 simplifies to Equation 8-64.

Thus, the decay of the fluorescence anisotropy of a spherical molecule is a single exponential. A measurement of the decay constant τ_c permits the hydrated molecular volume (V_h) to be calculated if the viscosity is known. Equation 8-73 indicates that the larger the molecule, the slower the decay of the fluorescence anisotropy. This is a reasonable result because larger molecules rotate more slowly. Figure 8-23 shows an example of anisotropy decay data for anthraniloyl chymotrypsin. The measured rotational correlation time is 15 nsec. This is a typical value for a small protein.

If the molecular weight and partial specific volume are known, Equation 10-10 can be used to estimate V_h. The hydrated specific volume of a typical protein is about 1 $cm^3\ g^{-1}$. Therefore, the hydrated volume of a single molecule will be $M/N_0 = (M/6) \times 10^{-23}$ cm^3. The value of η for water at 20°C is about 0.01 centipoise; T is 293 K; k is 1.38×10^{-11} erg deg^{-1}. Thus τ_c can be estimated from Equation 8-73 as

$$\tau_c = (0.01 \times M \times 10^{-23})/(6 \times 293 \times 1.38 \times 10^{-16}) = (M/2.4) \times 10^{-12}\ \mathrm{sec} \quad (8\text{-}74)$$

Roughly, τ_c is 1 nsec for each 2,400 daltons of protein molecular weight, if the protein is approximately spherical. (Equation 8-74 is also approximately correct for globular nucleic acids.) For the example shown in Figure 8-23, the molecular weight is 25,000 daltons; thus τ_c is predicted to be 10 nsec. The difference between predicted and experimental rates arises because chymotrypsin is not a sphere. Elongated shapes rotate more slowly (see Chapter 10).

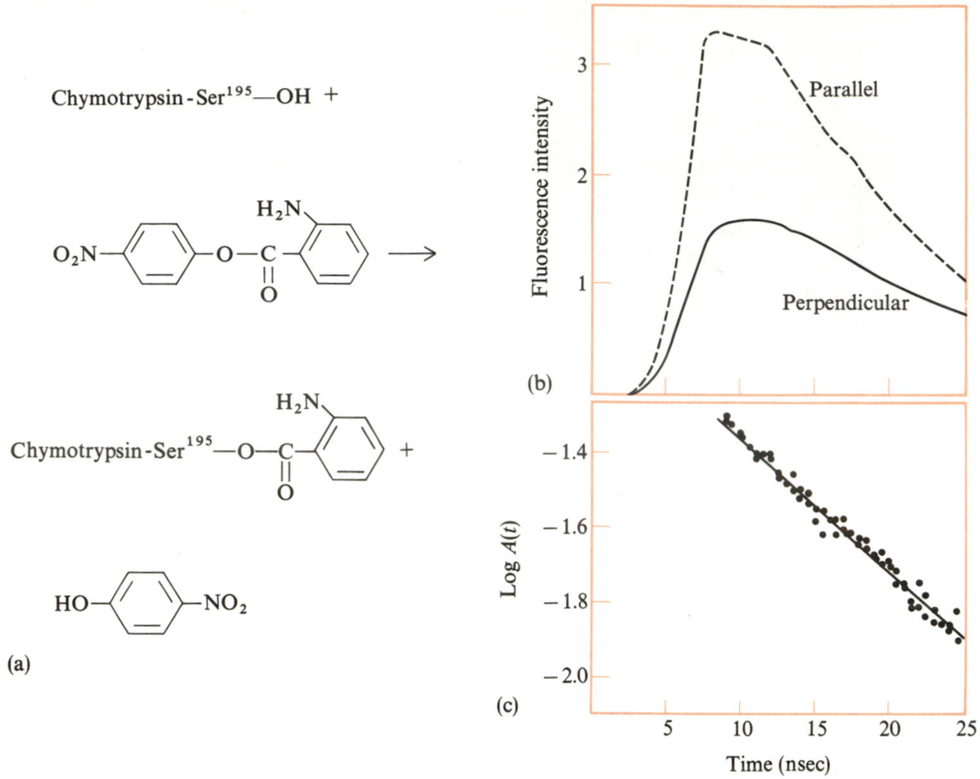

Figure 8-23

Decay of fluorescence anisotropy of anthraniloyl-Ser^{195}-α-chymotrypsin. **(a)** Preparation of this derivative. **(b)** Fluorescence decay data for individual polarized components. **(c)** The anisotropy on a logarithmic scale after deconvolution of the data in order to remove any effects of the shape of the exciting pulse. [After L. Stryer, *Science* 162:526 (1968).]

More often than not, what is measured is the average emission polarized parallel or perpendicular, rather than the time decay. For example,

$$\bar{I}_{||} = \tau_F^{-1} \int_0^\infty I_{||}(t)\,dt \qquad \bar{I}_{\perp} = \tau_F^{-1} \int_0^\infty I_{\perp}(t)\,dt \tag{8-75}$$

These averages values will be observed if constant illumination is employed. Such measurements are called steady-state or static polarization. Equations 8-70 and 8-71 are inserted into Equation 8-75 to compute expected values of the static polarization and anisotropy:

$$\bar{P} = \frac{\bar{I}_{||} - \bar{I}_{\perp}}{\bar{I}_{||} + \bar{I}_{\perp}} = \frac{3}{1 + 10(1 + \tau_F/\tau_c)(3\cos^2\xi - 1)^{-1}} \tag{8-76}$$

$$\bar{A} = \frac{\bar{I}_{||} - \bar{I}_{\perp}}{\bar{I}_{||} + 2\bar{I}_{\perp}} = \frac{3\cos^2\xi - 1}{5(1 + \tau_F/\tau_c)} \tag{8-77}$$

Suppose that measurements are made at such high values of η/T that the macromolecule is unable to rotate during the lifetime of the excited state. In this case, $\tau_F/\tau_c = 0$, and the polarization and anisotropy become

$$\bar{P}_0 = 3/[1 + 10(3\cos^2\xi - 1)^{-1}] = (3\cos^2\xi - 1)/(\cos^2\xi + 3) \tag{8-78}$$

$$\bar{A}_0 = (3\cos^2\xi - 1)/5 \tag{8-79}$$

These values, called the limiting polarization and anisotropy, are precisely the same results as those derived in Equation 8-64 for a rigid system.

The Perrin equations and steady-state polarization measurements

Using the limiting values given for Equations 8-78 and 8-79, we can rewrite Equations 8-76 and 8-77 in a particularly convenient form called the Perrin equations:

$$1/\bar{P} - 1/3 = (1/\bar{P}_0 - 1/3)(1 + \tau_F/\tau_c) = (1/\bar{P}_0 - 1/3)(1 + \tau_F kT/V_h\eta) \tag{8-80}$$

$$\bar{A}^{-1} = \bar{A}_0^{-1}(1 + \tau_F/\tau_c) = \bar{A}_0^{-1}(1 + \tau_F kT/V_h\eta) \tag{8-81}$$

On the right-hand side of the Perrin equations, the rotational correlation time (τ_c) has been expressed explicitly using Equation 8-73. Of all the expressions we have written for polarization, these are the ones most often used in practice. The static polarization or anisotropy of a macromolecular solution is measured as a function of temperature T or viscosity η. When the results are plotted according to Equation 8-80 or 8-81, a straight line should result. The slope will yield the molecular volume (V_h), providing that the fluorescence decay time (τ_F) is known. The intercept at $T/\eta \to 0$ will yield the limiting anisotropy or polarization. Figure 8-24 shows an example of such a Perrin plot for chymotrypsin. The limiting anisotropy (0.3) suggests that the fluorescence probe is nearly rigidly attached. The rotational correlation time (15 nsec) agrees with the value obtained from time-dependent measurements. It is reasonably close to what is expected for a rigid spherical protein, as estimated earlier.

All of Equations 8-65 through 8-81 hold only for spherical molecules. If, instead, a rigid ellipsoid is a more accurate description of the system, the results become much

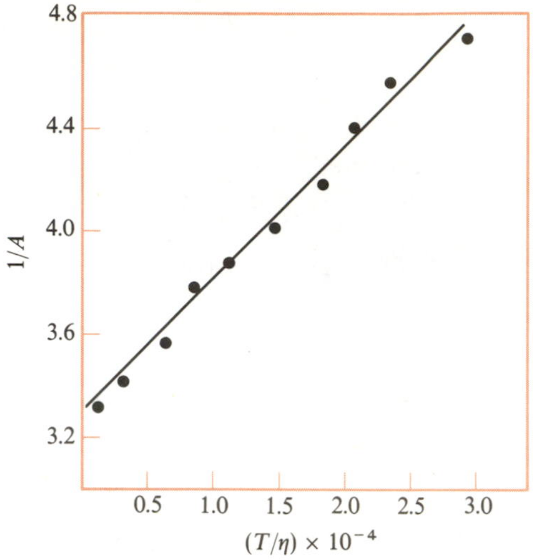

Figure 8-24

Static fluorescence polarization of anthraniloyl-Ser^{195}-α-chymotrypsin (see Fig. 8-23). To generate this Perrin plot, the temperature was maintained constant at 20°C while the viscosity was varied by the addition of glycerol to aqueous solutions of the protein. [After R. P. Haugland and L. Stryer, in *Conformation of Biopolymers*, vol. 1, ed. G. N. Ramachandran (New York: Academic Press, 1967).]

more complicated. The rotational diffusion constant needed for Equation 8-65 now is a tensor. The decay of fluorescence anisotropy, instead of a single exponential as in Equation 8-72, now contains a sum of as many as five exponentials. These exponentials depend not only on the size and shape of the molecules, but also on the orientation of the transition dipoles of the chromophores with respect to the axes of the ellipsoid. In principle, if accurate enough data could be obtained, measurement of anisotropy decay could lead to a detailed picture of the shape of the molecule and the mode of binding of the chromophore. In practice, it is rare that the accuracy of the data justifies fitting to more than two exponentials. In this case, the best one can do is to construct various plausible models for the structure and see how they fit the data. One generalization can be stated: a prolate ellipsoid will tend to have an apparent τ_c larger than that of a sphere. Hence, if the molecular weight is known, some information about axial ratios usually can be obtained.

Suppose that an anisotropy is found to be smaller than that calculated for a spherical molecule of known atomic weight. This is an indication of flexibility of the macromolecule as a whole, or of nonrigid attachment of the chromophore to the macromolecule. Sometimes these effects can be distinguished by carrying out polarization measurements at very high values of T/η. The hydrated volume of a single chromophore is about 1% or less of the volume of a typical macromolecule. If the chromophore is free to move, its rotational correlation time will be about 1/100 that of the macromolecule. At low viscosity, the chromophore will experience all accessible orientations in a period of time that is short compared with the fluorescence lifetime. If these orientations cover a wide range of angles, the observed polarization will be zero. If, however, the rapid motions of the chromophore cover only a limited

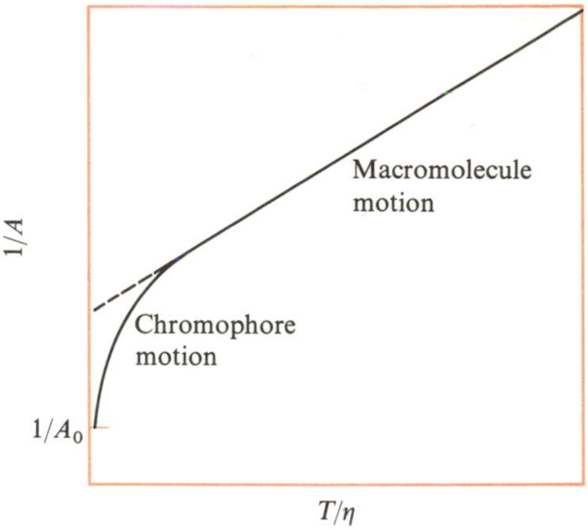

Figure 8-25

Schematic Perrin plot of anisotropy versus temperature or viscosity for a sample in which there is some *limited* flexibility in the attachment of the fluorescent moiety to a macromolecule. Extrapolation of the linear portion to $T/\eta = 0$ (*dashed line*) yields an apparent value for A_0 considerably smaller than the true value. The difference between these values is a measure of the range of angles over which the fluorescent group can move relative to the macromolecule.

range of orientations, some net polarization still will be seen. Macromolecular motion will be required in order for the chromophore to experience all possible orientations. Thus, at low viscosity, the dependence of the residual polarization on T/η will reflect motion of the macromolecule as a whole. Extrapolation of this $\bar{P}$ back to $T/\eta = 0$ on a plot of $\bar{P}^{-1}$ versus T/η allows a measure of the range of angles covered by fast chromophore motion.

At sufficiently high viscosities, the macromolecule is effectively stationary. The chromophore still is free to move, and its motion will be detected by an enhanced dependence of polarization on T/η at high η. The result will be a nonlinear Perrin plot, such as that shown schematically in Figure 8-25. Polarization thus can provide clear evidence of flexibility either near the binding sites of probes or of a whole macromolecular structure. This is important information that often is quite difficult to detect by translational hydrodynamic measurements such as sedimentation.

In addition to monitoring molecular motions, polarization also can be used to detect singlet–singlet energy transfer. Suppose a protein or nucleic acid contains two or more identical fluorescent chromophores. If one of these is excited, energy transfer can lead to migration of the excitation to other chromophores prior to emission. If these chromophores are not oriented parallel to the original absorber, the emitting dipole will no longer be parallel to the absorbing dipole. The polarization observed for the system will be altered as described by Equation 8-64. If the chromophores have fluorescence spectra shifted far to the red of their absorption, these effects are likely to be small because the spectral overlap will be poor, leading to small values of R_0. But many chromophores have fluorescence that substantially overlaps their absorption. In these cases, polarization measurements are an effective way to measure interchromophore distances. However, if energy-transfer effects are large, they will severely complicate efforts to measure molecular size and shape by polarization.

8-3 INFRARED AND RAMAN SPECTROSCOPY

Transitions between vibrational levels of the ground electronic state of a molecule are of much lower energy than are electronic transitions. The absorption bands from these vibrational transitions generally occur between 5,000 cm^{-1} (=2 microns = 2,000 nm) and 200 cm^{-1} (=50 microns = 50,000 nm). The principles that govern vibrational absorption spectra are essentially the same as those discussed previously for electronic spectra. The absorbance is determined by Beer's law. Transition dipoles determine the intensity, and they lead to phenomena such as linear dichroism when oriented samples are observed using polarized light. The spectra of polymers are not the same as those of their monomeric constituents because coupling between nearby absorbing groups can occur in the polymer.

Principles of infrared spectroscopy

In principle, infrared (IR) spectra should yield information of kinds quite similar to that obtained from UV visible spectra. In practice, however, IR measurements have seen much less use. There are several reasons for this neglect of IR spectra. First, IR absorptions typically are far less intense than electronic spectra. For example, IR extinction coefficients of nucleic acid bases are on the order of 10^3, compared to 10^4 for UV extinction coefficients. Thus, much higher concentrations of sample are needed. Second, water absorbs strongly in many of the regions of the infrared where protein or nucleic acid absorption bands are found. This problem can be alleviated, but not eliminated, by the use of D_2O instead of H_2O. Finally, instrumentation (particularly infrared detectors) often is less sensitive than visible or UV instrumentation, making precise quntitative measurements quite a bit more difficult.

Vibrational absorption bands typically arise from transitions that are somewhat localized on a molecule. For example, one talks of bands due to C═O stretching, C—H stretching, particular amides, and so on. These are convenient descriptions, and for some small molecules they are fairly accurate; in general, however, such terminology is quite an oversimplification.

The vibrations of various pairs or sets of atoms are coupled, and the overall molecular vibrations that must be considered in computing the spectrum potentially involve all the atoms. These vibrations must be solved in terms of normal modes, which are the stationary states of the molecular vibrational Hamiltonian. Any specific vibration can be described as some linear combination of the normal modes. Thus, in many cases there is no simple physical picture of a particular vibrational absorption band. Frequently, but not always, a normal mode involves a large displacement of just two bonded atoms, with little motion of any others. In this case, describing the spectral transition as arising from stretching or bending of a particular bond may still be fairly accurate.

By analogy with electronic spectra, the intensity of a vibrational band will depend on the magnitude of the light-induced dipole. To calculate this magnitude, one starts with wavefunctions such as Equation 7-5, in which the Born–Oppenheimer approximation has been used to separate electronic and nuclear motion. A pure vibrational transition can be written as

$$\Psi_0(r, R)\phi_{\mathrm{v}}(R) \rightarrow \Psi_0(r, R)\phi_{\mathrm{v}'}(R) \tag{8-82}$$

where ϕ_{v} and $\phi_{\mathrm{v}'}$ refer to two different vibrational states; r and R are the coordinates of electrons and nuclei, respectively; and Ψ_0 is the ground-state electronic wavefunction. If the nuclei are stationary, the electric dipole operator $\underset{\sim}{\boldsymbol{\mu}}$ depends only on electronic coordinates, and the transition dipole is

$$\langle \Psi_0\phi_{\mathrm{v}}|\underset{\sim}{\boldsymbol{\mu}}|\Psi_0\phi_{\mathrm{v}'}\rangle = \langle \Psi_0|\underset{\sim}{\boldsymbol{\mu}}|\Psi_0\rangle\langle\phi_{\mathrm{v}}|\phi_{\mathrm{v}'}\rangle = 0 \tag{8-83}$$

Equation 8-83 is zero because the first integral over r is odd, and because the second over R is zero since ϕ_{v} and $\phi_{\mathrm{v}'}$ are two different orthogonal eigenstates of the Hamiltonian.

If the nuclei are in motion, then the dipole operator depends implicitly on R because the electron distribution changes with changing R. In this case, $\underset{\sim}{\boldsymbol{\mu}}(r)$ can be expanded in a Taylor series about the equilibrium nuclear position R_0:

$$\underset{\sim}{\boldsymbol{\mu}}(r, R) = \underset{\sim}{\boldsymbol{\mu}}(r, R_0) + \left(\frac{\partial \underset{\sim}{\boldsymbol{\mu}}(r)}{\partial R}\right)_{R_0}(R - R_0) + \cdots \tag{8-84}$$

The expectation value of the first term in the expansion is zero as shown in Equation 8-83, but the second term is nonzero:

$$\langle \Psi_0\phi_{\mathrm{v}}|\underset{\sim}{\boldsymbol{\mu}}|\Psi_0\phi_{\mathrm{v}'}\rangle = \langle \Psi_0|[\partial \underset{\sim}{\boldsymbol{\mu}}(r)/\partial R]_{R_0}|\Psi_0\rangle\langle\phi_{\mathrm{v}}|\underset{\sim}{\mathbf{R}}|\phi_{\mathrm{v}'}\rangle \tag{8-85}$$

The derivative in Equation 8-85 is evaluated at R_0 so, once the nuclear coordinates are specified, it is a function of only the electronic coordinates. This allows the transition dipole to be factored into the product of two terms: an electronic integral and a vibrational integral. The electronic integral is simply the change in *permanent* dipole moment of the ground electronic state that accompanies a change in molecular position. Therefore, just by inspection, one frequently can tell whether particular vibrations are capable of leading to infrared absorption. For example, in the linear molecule CO_2, the symmetric stretch vibration will have no infrared absorption intensity, whereas asymmetric stretch and bending modes are infrared active. The transition dipole will lie along the direction of the changing dipole moment.

symmetric stretch	←O=C=O→	infrared inactive	$\partial\underset{\sim}{\boldsymbol{\mu}}/\partial R = 0$
asymmetric stretch	$\vec{\mathrm{O}}$=$\overleftarrow{\mathrm{C}}$=O→	infrared active	↔$\partial\underset{\sim}{\boldsymbol{\mu}}/\partial R$
bending	O=C=O (C ↑, O ↓, O ↓)	infrared active	↕ $\partial\underset{\sim}{\boldsymbol{\mu}}/\partial R$

Vibrational spectra of biopolymers

For biopolymers, the main concern is how aspects of the conformation affect the normal modes of vibration. As with electronic absorption, it is convenient to focus on how the presence of particular secondary structures alters the spectrum from that of the isolated chromophores. In proteins and polypeptides, three infrared bands have received the most attention. All arise from the peptide backbone and can be assigned to normal modes involving simple atomic groups. The bands are the N—H stretch at about 3,300 cm^{-1}, the C═O stretch at about 1,630 to 1,660 cm^{-1} (amide I), and the N—H deformation at about 1,520 to 1,550 cm^{-1} (amide II). These bands are fairly easy to measure because each peptide residue contributes, and because D_2O does not absorb strongly in this region of the spectrum.

Table 8-5 lists some details about these spectral bands. The C═O stretch and the N—H stretch are polarized essentially parallel to the C═O and N—H bonds, respectively. The amide II band is polarized nearly parallel to the C—N peptide bond, and thus nearly perpendicular to the N—H bond. It turns out that the vibrations leading to this band are roughly equal combinations of a C—N stretching motion and an N—H bending motion (N—H).

Hydrogen bonding shifts the energies of the three peptide vibrations (Table 8-5). The two stretching bands are moved to lower energy. This is easy to understand because the presence of a hydrogen bond makes it easier to stretch the carbonyl oxygen toward the hydrogen-bonding donor, or the amide nitrogen toward the acceptor. The amide II band is shifted to higher energy by hydrogen bonding. It is more difficult to bend the N—H bond because hydrogen bonds are roughly linear, and this effect tends to keep the hydrogen in place between the two electronegative atoms. The results in Table 8-5 also indicate that α helices and β sheets should be resolvable by IR spectroscopy. Actually this is not always possible because of complications to be discussed later.

Table 8-5
Characteristics of principal infrared absorption bands of the peptide group

			Hydrogen-bonded forms				Non-hydrogen-bonded
			α Helix		β Sheet		
Vibration	$\partial\underline{\mu}/\partial R$		Frequency (cm^{-1})	Dichroism	Frequency (cm^{-1})	Dichroism	Frequency (cm^{-1})
N—H stretch	←N—H→	↔	3,290–3,300	‖	3,280–3,300	⊥	~3,400
Amide I (C═O stretch)	←C═O→	↔	1,650–1,660	‖	1,630	⊥	1,680–1,700
Amide II	↑H above N; ←C—N→; ↓ below N	↗	1,540–1,550	⊥	1,520–1,525	‖	<1,520?

SOURCE: Adapted from J. A. Schellman and C. Schellman, in *The Proteins*, 2d ed., vol. 2, ed. H. Neurath (New York: Academic Press, 1962), p. 1.

For oriented polypeptide samples, infrared dichroism is a powerful structural technique. In the α helix, N—H···C=O peptide hydrogen bonds are oriented parallel to the long axis of the molecule. Therefore the N—H stretch and amide I bands should preferentially absorb IR light when the polarization direction is parallel to the helix axis. The amide II band should show just the opposite behavior. Figure 8-26 shows a typical α-helix linear dichroism spectrum. For β sheets, the long axis of the structure is along the extended peptide chains. The peptide hydrogen bonds are perpendicular to this axis (see Fig. 2-23). Therefore, in oriented β sheets, each band will have dichroism opposite to that observed for the corresponding band in an α helix.

Unfortunately, the description just given is a serious oversimplification. In polypeptides or proteins, individual peptide vibrations interact with each other and become coupled. This coupling is exactly analogous to the dipolar coupling we discussed earlier for electronic absorption bands. The implications also are similar. Each monomer vibrational absorption band can, in the polymer, split into a series of several bands. These bands will have different intensities and may also have different polarization directions. The spectrum of the polymer thus is potentially much more complex than that of the monomer. Fortunately, the symmetry of helical polymers leads to a manageable set of bands. For the α-helix-like polymers, T. Miyazawa showed that each monomer transition should lead to two intense polymer transitions: one at a frequency $\nu_{\|}(0)$ polarized parallel to the helix, and two degenerate bands of frequency $\nu_{\perp}(2\pi/n)$ polarized perpendicular to the helix. The latter transitions depend on the number of residues per turn, n. These results are in complete analogy to the electronic spectral bands predicted for the peptide chromophore by the exciton effects. Evidence for splitting in the vibrational spectra of α-helical polypeptides can be seen in the spectra shown in Figure 8-26. Note that the parallel-

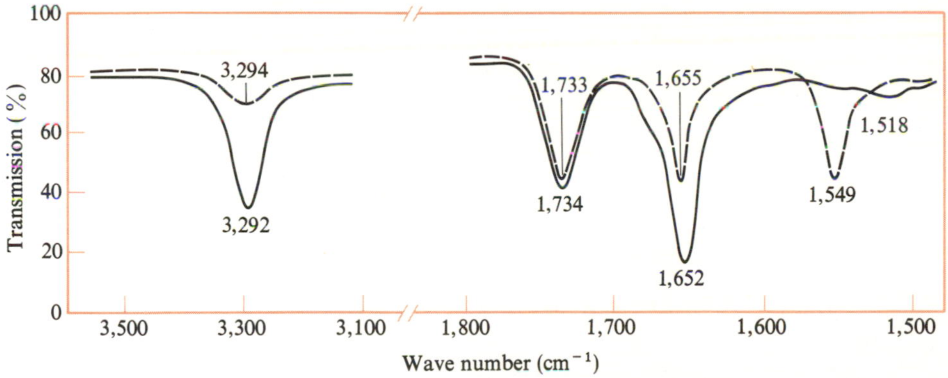

Figure 8-26

Infrared dichroism of oriented films of poly-γ-benzyl-L-glutamic acid. Light was polarized parallel (*solid line*) and perpendicular (*dashed line*) to the fiber axis. [After M. Tsuboi, *J. Poly. Sci.* 59:139 (1962).]

polarized absorption shows peaks at 1,518 and 1,652 cm^{-1}, whereas the perpendicular-polarized absorption has maxima at 1,549 and 1,655 cm^{-1}. These shifts represent the splitting between $\nu_{||}(0)$ and $\nu_{\perp}(2\pi/n)$ amide I and amide II bands, respectively.

For antiparallel β sheets, we must consider the vibrations of the fundamental asymmetric unit (Fig. 8-27). Because four peptides are involved, each monomer vibrational band will be split into four in the antiparallel β sheet. For the amide

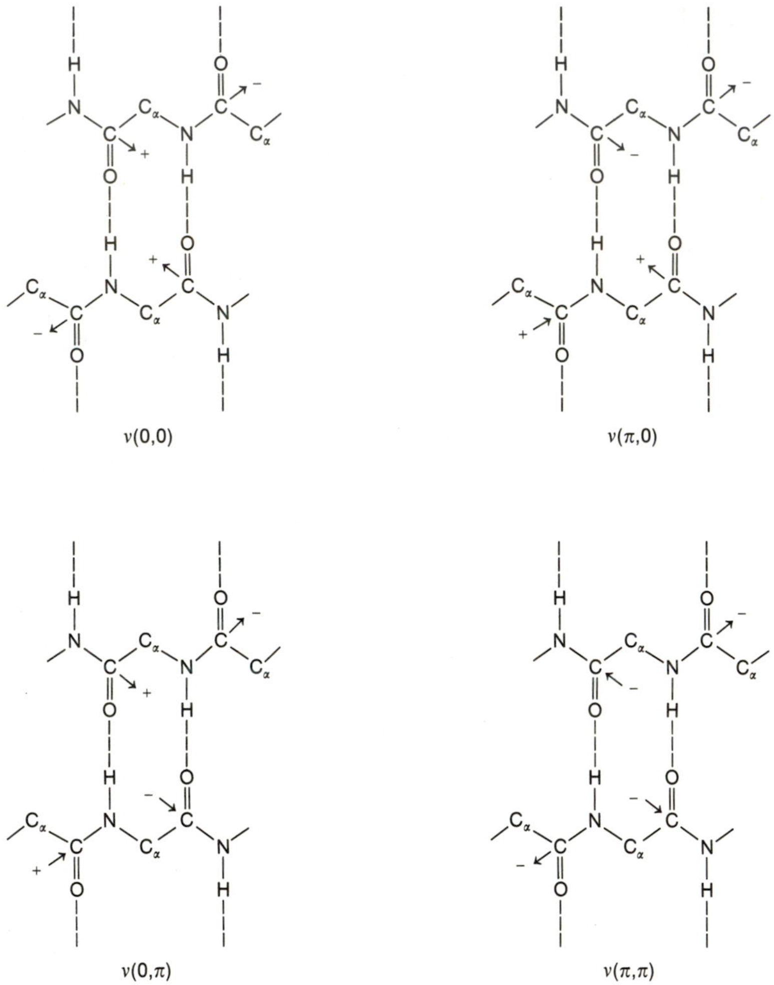

Figure 8-27

Schematic representation of the vibrational modes of the antiparallel β sheet. Arrows represent components of transition moments in the plane of the paper; plus and minus signs represent out-of-plane components. [After T. Miyazawa, *J. Chem. Phys.* 32:1647 (1960).]

Table 8-6
Observed and calculated infrared spectra of polypeptides and proteins

		Amide I		Amide II	
Conformation	Mode	Calculated	Typical observed	Calculated	Typical observed
α Helix	$\nu_{\parallel}(0)$	(1,650)	1,650	(1,516)	1,516
	$\nu_{\perp}(2\pi/n)$	1,647	1,652	1,540	1,546
Antiparallel	$\nu_{\parallel}(0, \pi)$	(1,685)	1,685	(1,530)	1,530
β sheet	$\nu_{\perp}(\pi, 0)$	(1,632)	1,632	1,540	—
	$\nu_{\perp}(\pi, \pi)$	1,668	—	1,550	—
Parallel	$\nu_{\parallel}(0, 0)$	1,648	1,645	1,530	1,530
β sheet	$\nu_{\perp}(\pi, 0)$	1,632	1,630	1,550	1,550
Random coil	ν_0	1,658	1,656	1,535	1,535

NOTE: Frequency values are given in cm^{-1}. Calculated values shown in parentheses were adjusted to equal the corresponding observed values by the choice of parameters.
SOURCE: Adapted from J. A. Schellman and C. Schellman, in *The Proteins*, 2d ed., vol. 2, ed. H. Neurath (New York: Academic Press, 1962), p. 1.

bands, one of the four is infrared inactive, two others are polarized perpendicular, and the fourth is polarized parallel. Table 8-6 compares the results of Miyazawa's normal-mode calculations for α helices and β sheets with experimental results. Agreement is really excellent for the amide I and II bands. Thus, such calculations can form the basis for the rational interpretation of the IR spectra of polypeptides and proteins.

In addition to the three peptide bands we have discussed, there are other peptide bands, as well as bands from various side chains. A detailed discussion of all these bands is beyond the scope of this book.

In nucleic acids, the sugar–phosphate backbone and the various bases lead to a much more complex set of molecular vibrations than those found in peptides. Progress in assigning most of the transitions is at a somewhat less advanced stage. Most attention has been focused on the spectral region between 1,500 and 1,800 cm^{-1}. This region contains vibrations of the carbonyl and of the double bonds of the purine and pyrimidine rings. These vibrations are highly sensitive to base pairing because the atoms involved participate directly in the formation of hydrogen bonds. Figure 8-28 shows a typical example. It is clear from this example that breaking base pairs leads to IR spectral changes that are qualitatively larger than the corresponding UV changes shown in Chapter 7. This advantage helps to compensate for the increased difficulty of making the IR measurements. Also, each of the four common bases has a quite distinct IR spectrum in this spectral region. Thus it is possible to examine G–C and A–U pairs separately. However, far less work has been done with IR spectra on nucleic acid conformational changes than with UV spectra on these changes. The large sample sizes required (and the considerable experimental difficulties) have inhibited what is potentially a very powerful technique.

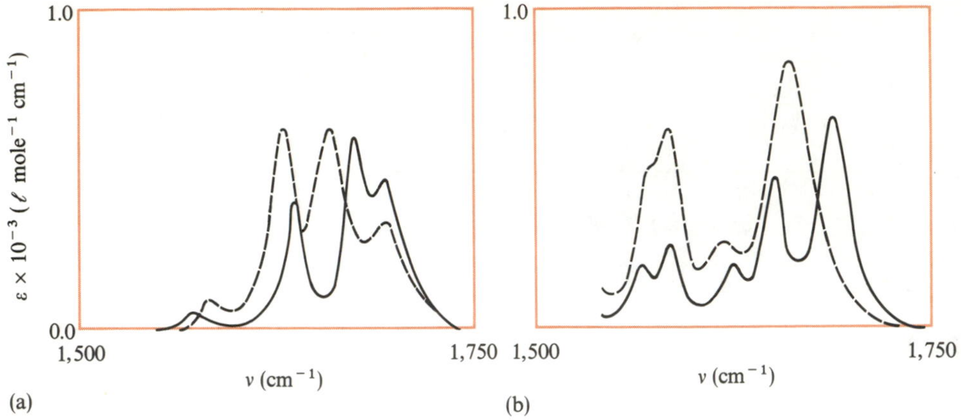

Figure 8-28

Infrared spectra affected by base pairing. **(a)** Double-helical poly rA·rU (*solid line*) and the sum of the spectra of separate samples of poly rA and poly rU (*dashed line*). **(b)** Double-helical poly rG·rC (*solid line*) and the sum of the spectra of G and C mononucleotides (*dashed line*). All measurements in D_2O at pH 7 at temperatures $\leqslant 65°C$. [After G. J. Thomas, *Biopolymers* 7:325 (1969).]

Raman spectroscopy

Raman spectroscopy is an alternative way of gaining information about the vibrational states of molecules. Suppose that a sample is illuminated with an intense beam of light at frequency ν. An oscillating dipole is induced in the molecules, even if ν is a frequency that cannot cause electronic or vibrational excitation. If we describe the time variation of the electric field of the light as $\mathbf{E}(t) = \mathbf{E}_0 \cos 2\pi\nu t$, Equation 7-57 predicts that the dipole induced in the molecule will be

$$\boldsymbol{\mu}(t) = \underset{\sim}{\boldsymbol{\alpha}}(\nu) \cdot \mathbf{E}_0 \cos 2\pi\nu t \tag{8-86}$$

In principle, the polarizability $\underset{\sim}{\boldsymbol{\alpha}}$ is a tensor, but for what follows it is sufficient to ignore this and treat it just as a number, $\alpha(\nu)$.

If the molecule is vibrating with time, its polarizability is not constant. As the nuclei move, the electrons respond, and at each nuclear position a different polarizability results. We can describe this simply by writing

$$\alpha(\nu) = \alpha_0(\nu) + \alpha'(\nu) \cos 2\pi\nu' t \tag{8-87}$$

where α_0 gives the polarizability of the molecule in its equilibrium nuclear configuration, $\alpha'(\nu)$ describes the change in polarizability with nuclear motion, and ν' is a vibrational frequency. Combining Equations 8-86 and 8-87, we have

$$\begin{aligned}\boldsymbol{\mu}(t) &= \mathbf{E}_0[\alpha_0(\nu) + \alpha'(\nu)\cos 2\pi\nu' t]\cos 2\pi\nu t \\ &= \mathbf{E}_0\alpha_0(\nu)\cos 2\pi\nu t + \mathbf{E}_0\alpha'(\nu)\cos 2\pi\nu' t\cos 2\pi\nu t \\ &= \boldsymbol{\mu}(t) + \boldsymbol{\mu}'(t) \qquad (8\text{-}88)\end{aligned}$$

The first term is just the normal induced dipole oscillating at frequency ν. This oscillation gives rise to light scattering and the refractive index of the medium. The second term produces a Raman spectrum. This can be seen if one uses the geometric identity $\cos A \cos B = (1/2)[\cos(A + B) + \cos(A - B)]$ to rewrite it as

$$\boldsymbol{\mu}'(t) = 2\{\mathbf{E}_0\alpha(\nu)[\cos 2\pi(\nu + \nu')t + \cos 2\pi(\nu - \nu')t]\} \qquad (8\text{-}89)$$

A dipole oscillating at a particular frequency results in the emission of radiation at that frequency. The intensity of radiation is proportional to $(\boldsymbol{\mu}')^2$. Equation 8-89 indicates that, when a vibrating sample is illuminated at frequency ν, emission will be observed at frequencies $\nu + \nu'$ and $\nu - \nu'$. The location of the emission peaks (relative to the exciting frequency) permits measurement of the vibrational frequency ν'. The spectral band at lower energy than the exciting light is called the Stokes band, and the band at $\nu + \nu'$ is called the anti-Stokes band. It may seem paradoxical that the sample can emit light at higher energy than the exciting light. A simple way to rationalize this is to view the Raman effect as a two-photon process. If the system takes up two photons of frequency ν, and gives up one each at $\nu + \nu'$ and $\nu - \nu'$, then the overall energy change is zero.

For small molecules, the Raman effect tends to complement infrared spectroscopy. Transitions that produce Raman bands must have a polarizability that changes with nuclear motion, rather than a permanent dipole that changes. Therefore, certain transitions that are low intensity in one technique may have high intensity in the other. For large asymmetric molecules, Raman and IR effects produce essentially the same set of bands. The principle advantage of Raman spectroscopy is that water has a fairly weak Raman spectrum. Therefore, biological samples can be studied conveniently in aqueous solution.

Raman spectra appear to offer particular promise for the study of nucleic acid–protein complexes. A large number of well-resolved spectral bands are visible in the Raman spectrum of samples even as complicated as a whole bacteriophage (Fig. 8-29). Many of these bands can be assigned to specific peptide or side-chain vibrations in the protein moiety, or to specific base or phosphate vibrations in the nucleic acid. Some of these bands are conformationally sensitive.

Table 8-7 shows the Raman bands seen for the amide I and II transitions in several polypeptides as a function of secondary structure. It is evident that clear distinctions among α helices, β sheets, and random coils are available from this technique.

Raman spectra show conformation-dependent shifts and splittings just as infrared spectra do. In addition, there is a Raman hypochromic effect. This is illustrated

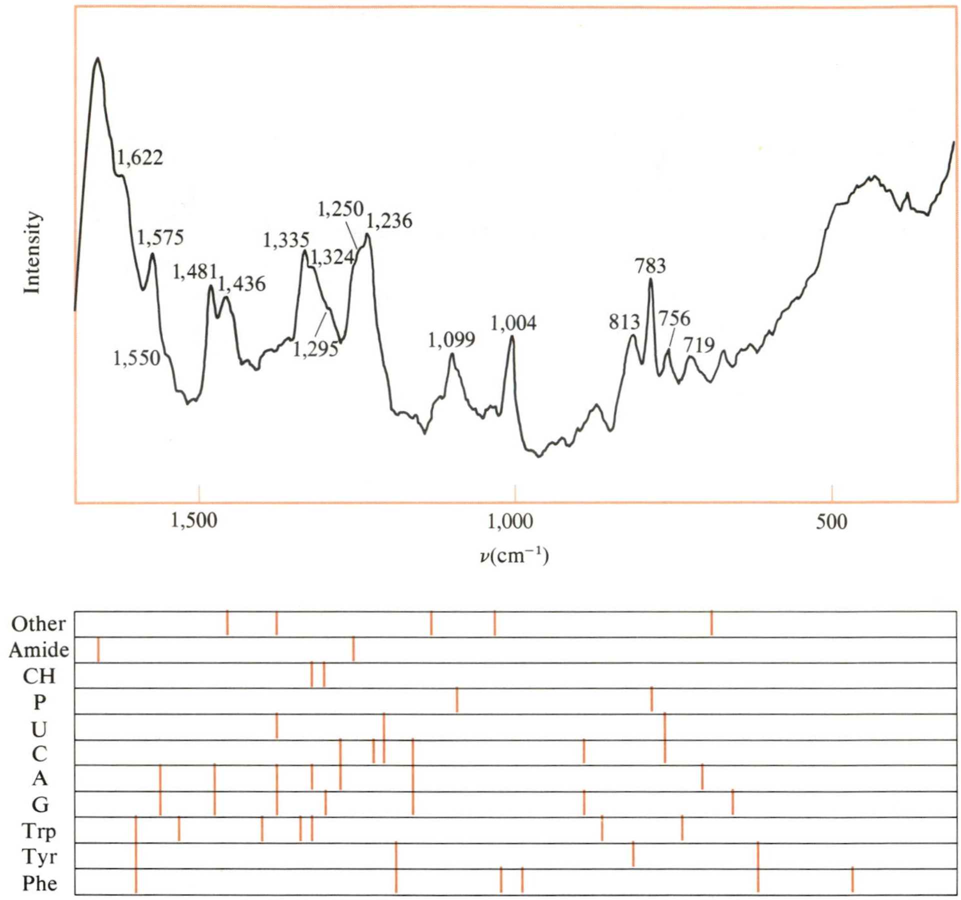

Figure 8-29

Raman spectrum of intact MS2 bacteriophage produced by excitation with an argon-ion laser. Conditions were 60 mg ml^{-1} MS2 in 0.75 M aqueous KCl at 32°C. Expected vibrational bands for various protein or nucleic acid groups are listed on the horizontal axis. [After G. J. Thomas et. al., *J. Mol. Biol.* 102 : 103 (1976).]

in Table 8-8, which shows that various Raman bands increase or decrease in intensity as bases in polynucleotides become more stacked or become hydrogen-bonded.

Until recently, Raman spectroscopy was not often applied to biopolymers because of various experimental difficulties. Suppose that a normal visible light source is used to excite a sample. Even after passage through a monochromator, the band of light that emerges often is contaminated with large amounts of exciting light that partially overlaps the Raman spectrum. Lasers have largely solved this problem because of their much greater spectral purity. As lasers have become available at

Table 8-7
Peptide vibration frequencies seen by raman spectroscopy in aqueous solution

Sample	Amide I			Amide II		
	α Helix	β Sheet	Random coil	α Helix	β Sheet	Random coil
Polyglycine	—	1,674 S	—	—	1,234	—
Poly-L-alanine	1,658 W	1,666	—	1,261 W	1,234	—
Poly-L-glutamate	—	—	1,665		—	1,248
Glucagon	1,658	1,672	—	1,266	1,232	1,248
Poly-L-lysine	1,645	1,670 S	—	1,311	1,240	1,253

NOTE: Frequencies are given in cm^{-1}; S = shoulder; W = weak.
SOURCE: Adapted from W. L. Peticolas, *Biochemie* 57:417 (1975).

Table 8-8
Changes in raman intensity with increasing order (base pairing or stacking) of nucleic acid structure

Base	Vibration frequency (cm^{-1})	Intensity change with increasing order[§]
Adenine	725	Decreases
	1,304	Decreases
	1,380	Decreases
	1,520	Decreases
Uracil	790	Decreases
	1,240	Decreases
Guanine	670	Increases
	1,485	Decreases
	1,580	Decreases
Cytosine	790	Decreases due to frequency shift
	1,256	Decreases at pH < 6.8
	1,547	Decreases at pH < 6.8

[§] Measured at neutral pH, unless otherwise stated.
SOURCE: Adapted from W. L. Peticolas, *Biochemie* 57:417 (1975).

many different frequencies, a new and powerful variant of Raman spectroscopy has become feasible. Suppose that the frequency of light used to excite a sample is near an absorption band of a chromophore. The polarizability at that frequency will be dominated by that chromophore (compare Fig. 7-20a). Thus, as you can see qualitatively from Equations 8-87, 8-88, and 8-89, the Raman spectrum of vibrations localized on that chromophore will be sharply enhanced. This effect is called resonance Raman spectroscopy. It permits the vibrational spectrum of an individual chromophore to be measured selectively, even in the presence of a huge background of other vibrational bands.

Summary

The three types of spectroscopic techniques discussed in this chapter provide different kinds of information about macromolecular structure. Optically active samples show a number of effects, of which the most convenient to measure and analyze is circular dichroism (CD). This is the difference in absorbance of left-hand and right-hand circularly polarized light. CD is strongly affected by the interactions between neighboring chromophores. The effects depend approximately on the inverse square of the distance between the chromophores, and they also depend on the relative orientation of the chromophores. Thus CD is an especially sensitive probe of the types and extents of secondary structure in a protein or nucleic acid. For example, the CD spectra of α helices, β sheets, and random coils are all quite different. By fitting protein spectra against a library of spectra from standard conformations, we can fairly reliably determine the amounts of each type of secondary structure present.

Fluorescence is more sensitive to environment than is absorption, because of the longer time scale of the effect. One can measure either a steady-state spectrum of emitted light or the actual decay kinetics of emission. Energy transfer will occur if two chromophores are near each other in space. Exciting one chromophore can lead to emission from the other. The probability that this will occur depends on the sixth power of the distance between the chromophores. Thus, measurements of energy transfer can yield the distance between specific points in a macromolecule. Excitation of a rigid or frozen sample with polarized light leads to preferential absorption by chromophores with transition dipoles oriented parallel to the polarization direction. In the absence of molecular motion, the fluorescence will show linear polarization. Typical proteins or nucleic acids rotate in solution within times of 10 to 100 nsec, whereas typical fluorescence decay times are in the range of 1 to 30 nsec. Thus the extent of polarization observed in solution will depend critically on the relative rates of fluorescence decay and molecular rotation. It is possible from polarization measurements to determine the size and shape of macromolecules and to explore flexibility of the structure.

Infrared and Raman spectroscopy examine molecular vibrations. Because these vibrations often are somewhat localized within individual sets of atoms, these techniques offer (in principle) considerable detailed information about macromolecular structure. Vibrational spectra are intrinsically quite sensitive to conformation, and they show effects of coupling of neighboring residues quite similar to those seen in electronic absorption spectra. In practice, vibrational spectra have not been used as often as other techniques, because larger samples are required in all except the newest laser Raman methods.

Problems

8-1. A protein has three identical sites arranged in an equilateral triangle. If one is filled with a dye (donor), the measured quantum yield (ϕ_D) is 0.5. Filling one site with a donor dye and a second with an acceptor dye results in a measured donor ϕ_D of 0.25. Predict what will be measured for ϕ_D if one site is filled with donor, and the other two are filled with acceptor.

8-2. The fluorescence anisotropy of a dye-labeled protein was measured first as a function of temperature at fixed viscosity (0.02 poise), then at constant temperature (293°K) at variable viscosity. The resulting Perrin plots were different (Fig. 8-30). Explain as much

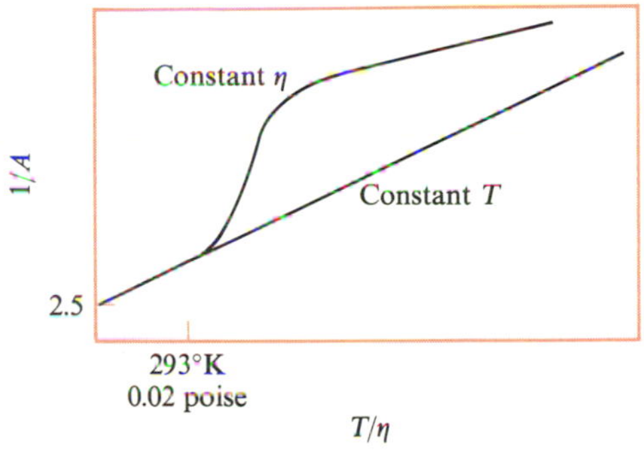

Figure 8-30

Perrin plots for Problem 8-2.

of the experimental evidence as you can, including the different slopes at high T/η values, the different apparent A_0 values when the linear region is extrapolated to $T/\eta = 0$, and the similarity of the data below about 300°K.

8-3. You would like to compare two dinucleoside phosphates, dApdA and rAprA, to see if they have the same base-stacked structure. However, both compounds exist as an equilibrium mixture of stacked and unstacked forms. The equilibrium constants are not known, nor are they easily measurable, but one must assume that they are different for the two samples. Show how CD measurements on monomers and dimers can be used to compare the stacked structures directly, assuming only that the CD of an unstacked form is just the sum of two monomers measured separately.

8-4. Which of the following chromophore pairs will show a CD different from that measured for the separated monomeric chromophores?

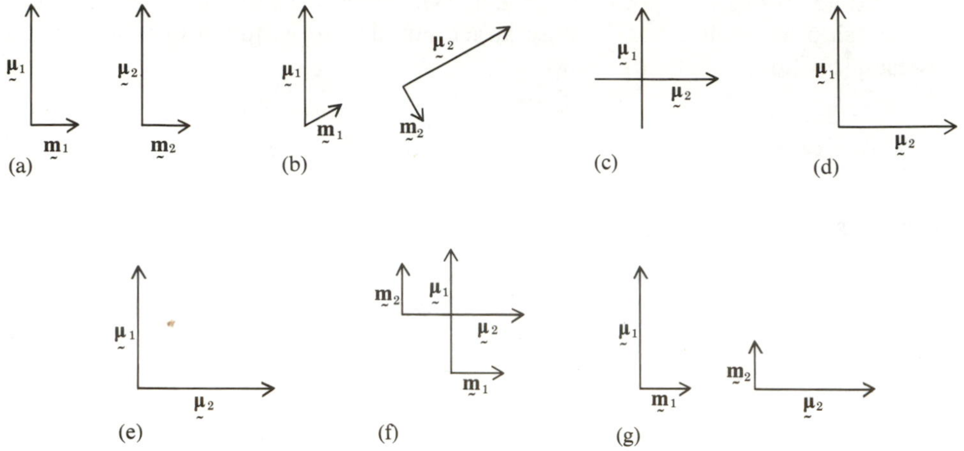

Figure 8-31

Chromophore pairs for Problem 8-4.

a. Fig. 8-31a; all in the same plane.
b. Fig. 8-31b; all in the same plane.
c. Fig. 8-31c; $\mathbf{m}_1$ and $\mathbf{m}_2$ are zero; $\boldsymbol{\mu}_1$ is in a plane parallel to the plane of $\boldsymbol{\mu}_2$, but above it.
d. Fig. 8-31d; otherwise as in part c.
e. Fig. 8-31e; both in the same plane; otherwise as in part c.
f. Fig. 8-31f; otherwise as in part c.
g. Fig. 8-31g; otherwise as in part c.

8-5. How might you use solvent perturbation of IR or Raman spectra to distinguish buried from exposed residues in a protein? What would it mean if in pure D_2O the IR absorption frequencies of several buried peptide residues were at higher frequency than those of exposed residues for a protein with a small fraction of α helix or β sheet?

References

GENERAL

Jirgensens, B. 1973. *Optical Activity of Proteins and Other Macromolecules*, 2d ed., New York: Springer-Verlag.

Pesce, A. J., C. G. Rosen, and T. L. Pasby. 1971. *Fluorescence Spectroscopy: An Introduction for Biology and Medicine*. New York: Marcel Dekker.

Steiner, R. F., and I. Weinryb, eds. 1971. *Excited States of Proteins and Nucleic Acids*. New York: Plenum Press.

Tinoco, I., Jr., and C. R. Cantor. 1970. Applications of optical rotatory dispersion and circular dichrosim to the study of biopolymers. *Methods Biochem. Anal.* 18:81.

SPECIFIC

Belford, G. C., R. C. Belford, and G. Weber. 1972. Dynamics of fluorescence polarization in macromolecules. *Proc. Natl. Acad. Sci. USA* 69:1392. [See this article and that by Chuang and Eisenthal (1972) for the complete treatment of the fluorescence polarization of anisotropic bodies.]

Bush, C. A. 1974. Ultraviolet spectroscopy, circular dichroism and optical rotatory dispersion. In *Basic Principles in Nucleic Acid Chemistry*, ed. P. P. P. Ts'o (New York: Academic Press), vol. 2, p. 91.

Cantoni, G. C., and D. R. Davies, eds. 1971. *Procedures in Nucleic Acid Research*, vol. 2. New York: Harper & Row. [Good articles on nucleic acid fluorescence, CD, infrared and Raman spectroscopy.]

Charney, E. 1979. *The Molecular Basis of Optical Activity*. New York: Wiley. [A complete and useful advanced treatise.]

Chen, R. F., and H. Edelhoch. 1975. *Biochemical Fluorescence*. 2 vols. New York: Marcel Dekker.

Chuang, T. J., and K. B. Eisenthal. 1972. Theory of fluorescence depolarization by anisotropic rotational diffusion. *J. Chem. Phys.* 57:5094.

Fairclough, R., and C. R. Cantor. 1977. The use of singlet–singlet energy transfer to study macromolecular assemblies. In *Methods of Enzymology*, vol. 48, ed. C. H. W. Hirs and S. N. Timasheff (New York: Academic Press), p. 347. [A detailed discussion of energy transfer.]

Fasman, G. D., ed. 1967. *Poly-α-Amino Acids*. New York: Marcel Dekker. [Contains many useful articles.]

Fraser, R. D. B., and E. Suzuki. 1970. Infrared methods. In *Physical Principles and Techniques of Protein Chemistry*, vol. B, ed. S. J. Leach (New York: Academic Press), p. 213.

Holmquist, B., and B. M. Vallee. 1978. Magnetic circular dichroism. In *Methods of Enzymology*, vol. 49, ed. C. H. W. Hirs and S. N. Timasheff (New York: Academic Press), p. 149.

Lehrer, S. S., and P. C. Learis. 1978. Solute quenching of protein fluorescence. In *Methods in Enzymology*, vol. 49, ed. C. H. W. Hirs and S. N. Timasheff (New York: Academic Press), p. 222.

Mandel, R., and G. Holzwarth. 1972. Circular dichroism of oriented helical polypeptides: The alpha-helix. *J. Chem. Phys.* 57:3469.

Mertzbacher, E. L. 1961. *Quantum Mechanics*. New York: Wiley.

Parker, C. A. 1968. *Photoluminescence of Solutions*. Amsterdam: Elsevier. [A solid guide to various types of fluorescence phenomena and experimentation.]

Sears, D., and S. Beychok. 1973. Circular dichroism. In *Physical Principles and Techniques of Protein Chemistry*, vol. C, ed. S. J. Leach (New York: Academic Press), p. 445. [A good treatment of the theory of the CD of polymers.]

Schellman, J. A. 1975. Circular dichroism and optical rotation. *Chem. Revs.* 75:323. [A good discussion of the basic theory of CD.]

Schellman, J. A., and E. B. Nielsen. 1967. Optical rotation and conformation studies on diamide models. In *Conformation of Biopolymers*, ed. G. N. Ramachandran (New York: Academic

Press), vol. 1, p. 109. [How to handle both $n \rightarrow \pi^*$ and $\pi \rightarrow \pi^*$ transitions in dipeptides.]

Shurcliff, W. A., and S. S. Ballard. 1964. *Polarized Light*. Princeton: Van Nostrand.

Steinberg, I. Z. 1978. Circular polarization of luminescence: Biochemical and biophysical applications. *Ann. Rev Biophys. Bioengin.* 7:133.

Stryer, L. 1968. Fluorescence spectroscopy of proteins. *Science* 162:526.

Turner, D. H. 1978. Fluorescence-detected circular dichroism. In *Methods in Enzymology*, vol. 49, ed. C. H. W. Hirs and S. N. Timasheff (New York: Academic Press), p. 199.

Van Wart, H. E., and H. A. Scheraga. 1978. Raman and resonance Raman spectroscopy. In *Methods in Enzymology*, vol. 49, ed. C. H. W. Hirs and S. N. Timasheff (New York: Academic Press), p. 67.

Weber, G. 1972. Uses of fluorescence in biophysics: Some recent developments. *Ann. Rev. Biophys. Bioengin.* 1:553.

9
Introduction to magnetic resonance

9-1 RESONANCE TECHNIQUES AND THEIR APPLICATIONS

As discussed in the previous chapters, the optical properties of biological molecules have been exploited in many ways to give some of our best insights into biological structure and function. In recent years, increasing advantage has also been taken of magnetic properties of biological materials. These properties have become useful through techniques for studying nuclear magnetic resonance (NMR) and electron paramagnetic resonance (EPR).[§]

In many applications, NMR is used as a spectroscopic technique (analogous to infrared, visible, and ultraviolet absorption spectroscopy techniques discussed earlier). NMR techniques monitor the absorption of energy associated with transitions of nuclei between adjacent nuclear magnetic energy levels. The energy is measured as a function of the strength of an externally applied magnetic field or of the frequency of electromagnetic energy supplied by an oscillator. The result is a spectrum of absorption versus field strength or oscillator frequency. Such NMR absorption spectroscopy is valuable because individual nuclei of a given kind within the same molecule (e.g., hydrogen nuclei at chemically distinct positions) commonly absorb at distinct positions in the spectrum. In optical absorption spectroscopy, absorption bands of similar chromophores commonly overlap; in an NMR spectrum, however, hydrogen nuclei within the same molecule often can be resolved, because of the

[§] The terms EPR (electron paramagnetic resonance) and ESR (electron spin resonance) are commonly used interchangeably in the literature.

extreme sensitivity to local environment of the absorption characteristics of a nucleus. Thus, NMR commonly is used to "fingerprint" small molecules; molecules with slightly different structures can have distinct NMR spectra.

Because NMR can resolve nuclei at specific positions in a structure, this technique has been valuable for studying specific loci in biological macromolecules under a variety of conditions – for example, presence or absence of a specific ligand, or different conformational states. If a nucleus is part of a molecule that exchanges between two or more magnetic environments, then it often is possible to obtain quantitative information on the rates of these exchanges. Another attractive feature is the potential for measuring distances between specific nuclei in a complicated structure such as an enzyme–substrate complex.

The basic principles of EPR are analogous to those of NMR: resonance in EPR is achieved by inducing transitions between adjacent magnetic energy levels of a paramagnetic electron. However, EPR depends on the magnetic properties of an unpaired electron; because electrons in most molecules are paired, relatively few molecules produce an EPR signal. In biological applications, this apparent disadvantage of EPR techniques can actually be an advantage. For example, the paramagnetic species in flavins, bacteriochlorophyll, and peroxidases are internal probes that can be studied without interfering effects of absorption bands from the many other groups present. If no internal paramagnetic species is present, a site-specific spin label (a small molecule containing an unpaired electron) can be introduced, and the EPR characteristics of the label then can be used to investigate in detail the structure–function relationships at and near the specific attachment site.

Here we cannot provide an in-depth treatment of NMR and EPR techniques and their roles in biochemistry; entire books are devoted to these subjects. Instead, we consider some basic principles and a few of the many interesting applications, with the hope that we can convey basic physical insights into the techniques and their uses.

9-2 GENERAL PRINCIPLES OF NUCLEAR MAGNETIC RESONANCE (NMR)

The phenomenon of magnetic resonance can be understood (in a complete and accurate sense) only through quantum mechanics. However, many of the basic ideas can be cast into a classical framework. The classical approach has obvious pedagogical benefits in providing good physical insight into many of the elementary ideas.

As an illustration of the mixture of classical and quantum ideas, consider the basic idea of an NMR absorption experiment. According to classical physics, a magnetic moment $\boldsymbol{\mu}_m$ interacts with a steady magnetic field $\mathbf{H}$ such that the interaction energy is $E = -\boldsymbol{\mu}_m \cdot \mathbf{H}$. (In quantum mechanics, this expression is replaced by the Hamiltonian of the interaction.) However, for a nuclear magnetic moment, the values of $\boldsymbol{\mu}_m$ that lie along the field direction are not free to vary at will as in classical physics, but are quantized. Thus the magnetic interaction energy E also is quantized (in units of $\hbar$). That is, the nuclear magnetic moment takes only certain allowed orientations, which give rise to a set of energy levels. In an NMR absorption experiment, the object is to induce transitions between adjacent energy levels that are separated in energy by ΔE. This induction is accomplished by a perturbing field that has a variable frequency. When the frequency (ω) of this field is tuned so that $\hbar\omega = \Delta E$, then absorption of energy occurs as the nuclei go from one state to the adjacent one. In other words, resonance occurs.

However, the phenomenon of resonance also can be understood by considering the interaction of the magnetic fields with magnetic moments, according to equations of classical physics. In this case, a continuous distribution of orientations of magnetic moments in the external field is permitted. The nuclei are viewed as precessing about the external field axis at a characteristic frequency known as the Larmor frequency. A perturbing field is introduced, whose frequency can be varied. When the frequency of this field matches the Larmor frequency of the precessing nuclei, absorption of energy occurs as the nuclei change their net orientation with respect to the field.

In the following discussion, the main features of magnetic resonance are developed from a classical viewpoint. However, of necessity, quantum mechanical considerations are interspersed at appropriate points. The classical treatment is not rigorously correct, but it does give helpful physical insight, and it does explain many aspects of magnetic resonance. With this material as background, the interested reader can consult more rigorous quantum treatments (see references at end of chapter).

Precession of a spinning charged body in a magnetic field

We begin our discussion of NMR by considering a spinning charged body situated in a magnetic field $\mathbf{H}$ (Fig. 9-1a). The body could be a proton.

The movement of charge in the spinning body gives rise to an electric current. This circulating current in turn produces a magnetic moment ($\boldsymbol{\mu}_m$) that interacts with $\mathbf{H}$. The interaction is expressed in terms of a torque $\boldsymbol{\tau} = \boldsymbol{\mu}_m \times \mathbf{H}$. This torque acts to produce a change in the angular momentum ($\mathbf{L}$) of the charged body. The basic electromagnetic principle that relates $\mathbf{L}$ and $\boldsymbol{\tau}$ is

$$d\mathbf{L}/dt = \boldsymbol{\tau} = \boldsymbol{\mu}_m \times \mathbf{H} \tag{9-1}$$

Thus, the direction of $\boldsymbol{\tau}$ and of $d\mathbf{L}$ is at right angles to $\boldsymbol{\mu}_m$ and to $\mathbf{H}$ (Fig. 9-1b).

To understand the motion of the charged spinning body, we must relate $d\mathbf{L}/dt$ to $\mathbf{L}$ and $\mathbf{H}$, rather than to $\boldsymbol{\mu}_m$ and $\mathbf{H}$. For this purpose, we use the following definition (in electromagnetic units) of the magnetic moment $\boldsymbol{\mu}_m$ (see Box 9-1):

$$\boldsymbol{\mu}_m = (1/2) \int \mathbf{r} \times \mathbf{j}\, dV \tag{9-2}$$

where $\mathbf{r}$ is a vector drawn from a fixed point in the body to the middle of a volume element dV, and $\mathbf{j}$ is the current density (charge per unit of time per unit of area).

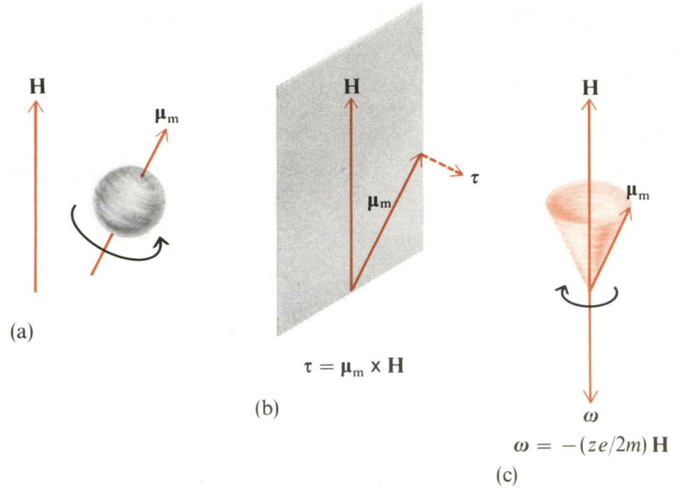

Figure 9-1

A spinning charged body in a magnetic field, illustrated for the case $ze > 0$. **(a)** The spinning charged body with magnetic moment $\boldsymbol{\mu}_m$ is located in a magnetic field **H**. **(b)** The body experiences a torque $\boldsymbol{\tau} = \boldsymbol{\mu}_m \times \mathbf{H}$. **(c)** The torque causes angular precession defined by the angular velocity vector $\boldsymbol{\omega} = -(ze/2m)\mathbf{H}$.

The integral in Equation 9-2 is taken over the entire body. The current density **j** can be written as

$$\mathbf{j} = (ze/m)\rho\mathbf{v} \tag{9-3}$$

where ze/m is the ratio of charge (ze) to mass (m), ρ is the mass density at dV, and **v** is the velocity at dV. Substituting Equation 9-3 into Equation 9-2, we obtain

$$\boldsymbol{\mu}_m = (ze/2m) \int \mathbf{r} \times \rho\mathbf{v}\, dV \tag{9-4}$$

The integral in Equation 9-4 is the definition of total angular momentum **L** of the spinning body, so that

$$\boldsymbol{\mu}_m = (ze/2m)\mathbf{L} \tag{9-5}$$

Thus, there is a direct and simple relationship between $\boldsymbol{\mu}_m$ and **L**. Substituting this relationship into Equation 9-1, we obtain

$$d\mathbf{L}/dt = \mathbf{L} \times (ze/2m)\mathbf{H} \tag{9-6}$$

This is the equation of motion for a constant vector rotating about **H** with an angular velocity $\boldsymbol{\omega} = -(ze/2m)\mathbf{H}$. Therefore, Equation 9-6 becomes

$$d\mathbf{L}/dt = \mathbf{L} \times (ze/2m)\mathbf{H} = -\mathbf{L} \times \boldsymbol{\omega} \tag{9-7}$$

Equations 9-1 and 9-7 state that the change $d\mathbf{L}$ in $\mathbf{L}$ produced by the interaction between $\boldsymbol{\mu}_m$ and $\mathbf{H}$ is at right angles to $\mathbf{L}$ and $\mathbf{H}$. This change in $d\mathbf{L}$ is equivalent to precessional motion at an angular velocity $\boldsymbol{\omega}$ about the field vector $\mathbf{H}$ (Fig. 9-1c). If $ze > 0$, then (looking along $\boldsymbol{\omega}$) the precession is in the clockwise direction ($\boldsymbol{\omega}$ points in the opposite direction from $\mathbf{H}$, so the right-hand rule gives a clockwise rotation); if $ze < 0$, the precession is counterclockwise. The precessional motion is known as *Larmor precession*, and the precessional frequency $\omega_0 = |\boldsymbol{\omega}|$ is the *Larmor frequency*.

Box 9-1 MAGNETIC MOMENTS

Equation 9-2 is a generalization of the simple expression from elementary physics for the magnetic moment of a loop of wire enclosing an area A and carrying a current I. In this case, the magnetic moment is given by $\boldsymbol{\mu}_m = IA\hat{\mathbf{e}}_\perp$, where $\hat{\mathbf{e}}_\perp$ is a unit vector perpendicular to the plane of the current loop (with direction determined by the right-hand rule). To show that Equation 9-2 gives this relationship, consider a loop of cross-sectional area A' and radius r in which the current flow is counterclockwise (see figure). (Recall that, if the current flow is

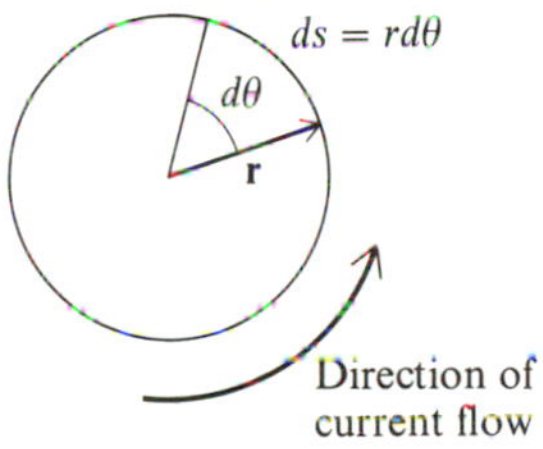

counterclockwise, then the actual flow of electrons is *clockwise*.) In this situation, $\mathbf{r} \times \mathbf{j}$ is perpendicular to (and points out of) the plane of the page; this defines the direction of $\hat{\mathbf{e}}_\perp$. The magnitude of $\mathbf{j}$ is I/A', and $dV = A'\,ds = A'r\,d\theta$. By substituting into Equation 9-2, we thus obtain

$$\boldsymbol{\mu}_m = (1/2)\hat{\mathbf{e}}_\perp \int_0^{2\pi} r(I/A')A'r\,d\theta$$
$$= I\pi r^2\hat{\mathbf{e}}_\perp = IA\hat{\mathbf{e}}_\perp$$

where $A = \pi r^2$.

Nuclear magnetic moments

The factor $ze/2m$ is the ratio of the magnetic moment to the angular momentum; it is known as the magnetogyric ratio. On the atomic scale, angular momentum is quantized in units of $\hbar$. Thus, for example, we expect the angular momentum of an electron to be quantized in units of $e\hbar/2m_e$ (where m_e is the mass of an electron). This fundamental unit is known as the Bohr magneton (β_e); it has the value of 9.27×10^{-21} erg gauss^{-1}. However, quantum mechanical considerations indicate that the relationship between magnetic moment and electron-spin angular momentum is modified by a factor g in Equation 9-5, so that equation becomes

$$\boldsymbol{\mu}_m = -g\beta_e(\mathbf{L}/\hbar) \tag{9-8}$$

where $g = 2.00232$ for the free electron.

An analogous relationship holds for the nuclear magnetic moment and for the nuclear-spin angular momentum. Thus,

$$\boldsymbol{\mu}_m = g_n\beta_n(\mathbf{L}/\hbar) \tag{9-9}$$

where g_n is the nuclear g-factor, and $\beta_n = e\hbar/2m_p$ (where m_p is the mass of a proton). The parameter β_n is called the nuclear magneton, and it has a value of 5.05×10^{-24} erg gauss^{-1}. Equation 9-9 generally is written in a more simple form:

$$\boldsymbol{\mu}_m = \gamma\mathbf{L} \tag{9-10}$$

where $\gamma = g_n\beta_n/\hbar$ is the magnetogyric ratio. This parameter may be positive or negative, depending upon the nucleus. Thus, the nuclear magnetic moment may point either in the same direction as **L** or in the direction opposite to it. For an electron, $\gamma = -g\beta_e/\hbar$, and $\boldsymbol{\mu}_m$ points in the direction opposite to **L**.

Equation 9-10 is analogous to Equation 9-5. Substituting Equation 9-10 into Equation 9-1, we obtain

$$\boxed{d\boldsymbol{\mu}_m/dt = \boldsymbol{\mu}_m \times \gamma\mathbf{H}} \tag{9-11}$$

where $-\gamma\mathbf{H} = \boldsymbol{\omega}$, the angular velocity of precession. This relationship will be of use in Section 9-3 when we consider the Bloch equations.

For the proton, the spin angular momentum is characterized by a spin quantum number $I = 1/2$; a neutron has the same spin quantum number. For a nucleus bigger than hydrogen, the angular momenta of the individual neutrons and protons comprising the nucleus combine to give a resultant in which (depending on the nucleus) $I = 0$, half-integral, or integral. It can be shown that $I = 0$ corresponds to nuclei with even mass numbers and even charge numbers (such as ^{12}C and ^{16}O); I = half-integral corresponds to nuclei with odd mass numbers (such as ^{1}H and ^{19}F); and I = integral corresponds to nuclei with even mass numbers and odd charge numbers (such as ^{2}H and ^{14}N).

The magnitude L of the spin angular momentum is given by

$$L = \hbar[I(I+1)]^{1/2} \tag{9-12}$$

From Equation 9-10 it follows that the magnitude μ_m of the magnetic moment is

$$\mu_m = \gamma\hbar[I(I+1)]^{1/2} \tag{9-13}$$

so that nuclei with $I = 0$ have no magnetic moments. Table 9-1 gives some values of I, μ_m (in units of β_n), γ, and Larmor frequency (ω_0) for selected nuclei. The table also gives the natural abundances and sensitivities for detection of these nuclei. (Sensitivity will be discussed shortly.)

The most prominent naturally occurring nuclei in biological materials are ^{1}H, ^{12}C, ^{16}O, ^{14}N, ^{31}P, and ^{32}S. Of these ^{12}C, ^{16}O, and ^{32}S are of no use from the standpoint of NMR because they have no magnetic moments. ^{14}N also is of little practical value, but for a different reason. This nucleus (like all nuclei with $I > 1/2$) has an electric quadrupole moment (see Box 5-1); rapid switching of the nucleus between quadrupole levels gives rise to lifetimes of nuclear magnetic states (in a magnetic field: see below) that are much shorter than those associated with nuclei having $I = 1/2$. Line widths are very broad for nuclei with very short magnetic state lifetimes (such as those for which $I > 1/2$); consequently, such nuclei are not readily studied.

Table 9-1
Magnetic properties of selected nuclei

Nucleus	Z	I	μ_m/β_n	γ (rad gauss^{-1} sec^{-1})	$\omega_0/2\pi$ (MHz/10^4 gauss)	Natural abundance (%)	Relative sensitivity[§]
^{1}H	1	1/2	4.84	26,753	42.6	99.98	1.000
^{2}H	1	1	1.21	4,107	6.5	0.016	0.0096
^{12}C	6	0	—	—	—	98.89	—
^{13}C	6	1/2	1.22	6,728	10.7	1.11	0.016
^{14}N	7	1	0.57	1,934	3.1	99.64	0.0010
^{16}O	8	0	—	—	—	99.76	—
^{19}F	9	1/2	4.55	25,179	40.1	100	0.834
^{31}P	15	1/2	1.96	10,840	17.2	100	0.066
^{32}S	16	0	—	—	—	95.06	—

[§] Relative sensitivity is given for equal numbers of nuclei in equal external fields. SOURCE: After J. D. Roberts, *Nuclear Magnetic Resonance* (New York: McGraw-Hill, 1959), and R. M. Lynden-Bell and R. K. Harris, *Nuclear Magnetic Resonance Spectroscopy* (London: Thomas Nelson, 1969).

Thus ^1H and ^{31}P are the only prominent nuclei that can be easily studied with NMR techniques in natural biological materials. Of course, ^1H is widely distributed in virtually all biological samples, whereas ^{31}P is largely confined to nucleic acids and phospholipids. Furthermore, the magnetic sensitivity of ^1H is much greater than that of ^{31}P.

Special procedures permit the study of nuclei other than ^1H or ^{31}P. For example, isotopic enrichment makes it possible to investigate carbon centers with ^{13}C (for which $I = 1/2$). (In natural materials, ^{13}C has an abundance of only 1%. This scarcity and its low relative sensitivity make it difficult to study without special enrichment techniques.) In some studies, ^{19}F ($I = 1/2$) is introduced as a site-specific probe that relays information on the magnetic environment of a particular region. ^{19}F is particularly useful for this purpose because of its high sensitivity (due to its relatively large magnetic moment).

Constraints on nuclear magnetic moments

In an external magnetic field, magnetic moments of nuclei have only certain allowed values for their components along the field direction. In an external field **H**, the energy E associated with the interaction between a magnetic moment $\boldsymbol{\mu}_{\mathrm{m}}$ and **H** is

$$E = -\boldsymbol{\mu}_{\mathrm{m}} \cdot \mathbf{H} \tag{9-14}$$

That is, the alignment of lowest energy is that in which the magnetic moment is parallel to the field direction (taken as the z direction). Quantum mechanical constraints limit the values of the component $\mu_{\mathrm{m}z}$ of $\boldsymbol{\mu}_{\mathrm{m}}$ that lies along the field to

$$\mu_{\mathrm{m}z} = m_I \gamma \hbar \tag{9-15}$$

where $m_I = I,\ I-1,\ I-2, \ldots, I-2I$. For ^1H, the allowed values of m_I are $m_I = +1/2,\ -1/2$. Thus, only two possible alignments with respect to the field are allowed:

$$\begin{aligned} E &= -\boldsymbol{\mu}_{\mathrm{m}} \cdot \mathbf{H} = -m_I \gamma \hbar H \\ &= -(1/2)\gamma \hbar H \qquad \text{for } m_I = +1/2 \\ &= +(1/2)\gamma \hbar H \qquad \text{for } m_I = -1/2 \end{aligned} \tag{9-16}$$

The difference in energy between these two alignments is $\Delta E = -\gamma \hbar H$. The ratio of nuclei in the two alignments is governed by the Boltzmann factor $e^{-\Delta E/kT}$. In a field of 10,000 gauss, the ratio of protons lined up with their z components of $\boldsymbol{\mu}_{\mathrm{m}}$ paralleling the field ($m_I = +1/2$) to those antiparallel ($m_I = -1/2$) is 1.000007 at 300°K. Thus, around room temperature, the energy associated with the interaction between magnetic moments and the external field is small compared with randomizing thermal energy; only a minute excess of the nuclei are aligned with the field. However, this small excess is essential for the generation of an NMR absorption signal. If the two states are equally populated, there is no net magnetization of the sample. In such a case, no energy can be absorbed from the oscillator (as we shall see). Furthermore, it is clear that the sensitivity for detection of a nucleus depends on the energy-level spacing; this spacing in turn affects the Boltzmann factor, which determines the population difference between levels. Because the energy-level spacing is determined by γ, it is clear that a larger value of γ corresponds to a greater population difference and a greater sensitivity (Table 9-1). Relative sensitivity is discussed further in Box 9-2.

Physical picture of an NMR experiment: a classical analogy

Consider a collection of nuclei, each of which has a magnetic moment $\boldsymbol{\mu}_m$ (Fig. 9-2). In the absence of an external field, these nuclei and their magnetic moments are randomly oriented. Now suppose that a field $\mathbf{H}_z$ along the z axis is turned on. The nuclei will show some preference to have a net alignment along the axis of the field and to precess about this axis with the Larmor frequency ω_0. The alignment process

Box 9-2 RELATIVE SENSITIVITY

Equation 9-30 gives an expression for the absorption-signal amplitude for the case $I = 1/2$ (e.g., a proton). Under conditions of resonance ($\omega = \omega_0$) and where $\gamma^2 H_{xy}^2 T_1 T_2 \ll 1$, the signal per unit of field is

$$\text{Signal}/H_z = \text{const } N(\mu_{mz}^2/kT)\omega_0\gamma H_{xy} T_2 \qquad \text{for } I = 1/2$$

where we have substituted $\omega_0 = \gamma H_z$. (Resonance is achieved by varying H_z until the nuclear precession frequency and the instrument frequency ω_0 match, as discussed in the text.) For nuclei with $I = 1/2$, we have $\mu_{m(z)}^2 = (1/4)\gamma^2\hbar^2$. Using this relationship for μ_{mz}^2 in the expression above, and denoting a proton with a subscript p, we obtain the relative sensitivity of nuclei with I = 1/2 at given T, ω_0, H_{xy}, and T_2:

$$\text{Relative sensitivity} = (\text{Signal}/H_z)/(\text{Signal}/H_z)_p = \gamma^3/\gamma_p^3 \qquad \text{for } I = 1/2$$

For nuclei with $I = 1$, it is easy to show that

$$\text{Relative sensitivity} = (\text{Signal}/H_z)/(\text{Signal}/H_z)_p = (8/3)(\gamma^3/\gamma_p^3) \qquad \text{for } I = 1$$

These expressions give the values for the relative sensitivities in Table 9-1.

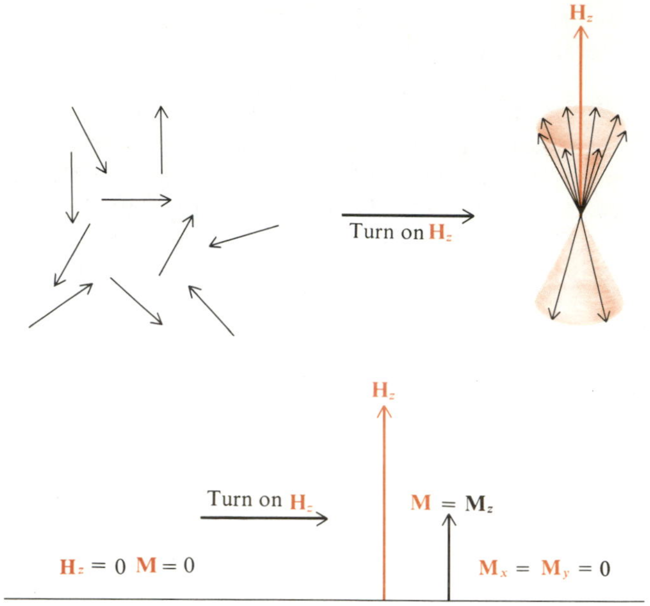

Figure 9-2

Orientation of initially random magnetic moments in an external magnetic field. The external field $\mathbf{H}_z$ causes orientation that produces a net magnetization $\mathbf{M}$ along the field (z) axis.

is accomplished by means of the random motion of the nuclei, through which they establish the Boltzmann distribution with respect to the magnetic interaction energy.

As a result of the net alignment, a total magnetization M_z along the z axis is established. This M_z is simply the sum of the z components of $\boldsymbol{\mu}_{\mathrm{m}}$ for each nucleus. As a consequence of random processes, this magnetization rises to its equilibrium value $\bar{M}_z$ in a simple first-order fashion as a function of the time after $\mathbf{H}_z$ is turned on. Thus,

$$M_z = \bar{M}_z(1 - e^{-t/T_1}) \tag{9-17}$$

where T_1 is the longitudinal relaxation time ("longitudinal" being the direction of the field axis). Hence, T_1 is a measure of the time it takes random processes to establish the equilibrium value $\bar{M}_z$ of the z component of the magnetization.

The net magnetization $\mathbf{M}$ of the sample is given by $\mathbf{M} = \mathbf{M}_z$. The x and y components of $\mathbf{M}$, $\mathbf{M}_x$, and $\mathbf{M}_y$ are zero, because there is no preferential orientation of the nuclei with respect to the x and y axes (Fig. 9-2). Therefore, the nuclei precess around the z axis at a frequency ω_0, but they have no phase relationship among themselves. Imagine now that we place a field $\mathbf{H}_{xy}$ in the x–y plane, and that this field rotates at a frequency ω_0 in the same direction as the precessional motion of the

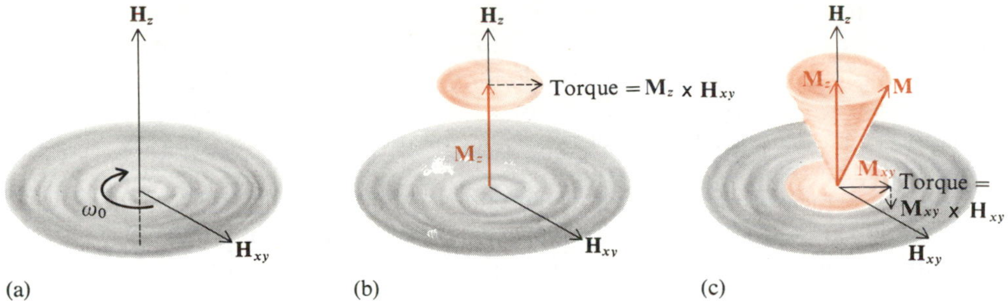

Figure 9-3

Effects of $\mathbf{H}_{xy}$ *on* $\mathbf{M}_z$ *and* $\mathbf{M}_{xy}$. **(a)** The main field $\mathbf{H}_z$ and the rotating field $\mathbf{H}_{xy}$. **(b)** The torque $\mathbf{M}_z \times \mathbf{H}_{xy}$ tips $\mathbf{M}$ toward the x–y plane to generate a component $\mathbf{M}_{xy}$. **(c)** The torque $\mathbf{M}_{xy} \times \mathbf{H}_{xy}$ acts to cause a reversal in direction of $\mathbf{M}_z$.

nuclei (Fig. 9-3a). This field will exercise a torque $\boldsymbol{\tau} = \mathbf{M}_z \times \mathbf{H}_{xy}$ on $\mathbf{M}$ so as to tip $\mathbf{M}$ away from the z axis and give rise to a component $\mathbf{M}_{xy}$ of $\mathbf{M}$ in the x–y plane. As shown in Figure 9-3b, the net component $\mathbf{M}_{xy}$ is 90° out of phase with $\mathbf{H}_{xy}$.

Once the component $\mathbf{M}_{xy}$ is generated, another torque $\boldsymbol{\tau} = \mathbf{M}_{xy} \times \mathbf{H}_{xy}$ comes into play (Fig. 9-3c). This torque acts to drive $\mathbf{M}$ into an antiparallel configuration with respect to $\mathbf{H}_z$. This shift corresponds to an increase in (absorption of) energy. Thus, the component M_z falls below its equilibrium value $\bar{M}_z$.

Imagine now that the field $\mathbf{H}_{xy}$ is suddenly turned off. The z component M_z of $\mathbf{M}$ will spontaneously return to its equilibrium value $\bar{M}_z$ through thermal processes that are governed by the relaxation time T_1 (Fig. 9-4a). In the case of $\mathbf{M}_{xy}$, the value must decay to zero because at equilibrium there is no preferential orientation with respect to the x or y axes. This relaxation also is a spontaneous first-order process characterized by a time constant T_2 such that $\mathbf{M}_{xy} = (\mathbf{M}_{xy})_0\, e^{-t/T_2}$, where t is the time after shutting off $\mathbf{H}_{xy}$, and $(\mathbf{M}_{xy})_0$ is the initial value of $\mathbf{M}_{xy}$ when $\mathbf{H}_{xy}$ is removed (Fig. 9-4b).

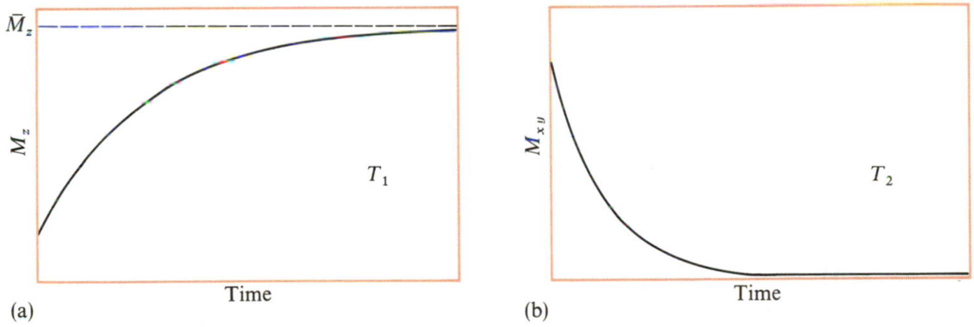

Figure 9-4

Relaxation times. **(a)** The time T_1 is associated with longitudinal relaxation. **(b)** The time T_2 is associated with transverse relaxation.

The parameter T_2 is called the *transverse* relaxation time. Random processes (analogous to those that allow $\mathbf{M}$ to reestablish its equilibrium value $\bar{M}_z$ along the z axis according to T_1) promote the randomization of transverse components to give a vanishing of $\mathbf{M}_{xy}$. However, there are additional processes affecting transverse relaxation that do not affect the longitudinal orientation of $\mathbf{M}$. Therefore, T_2 can never be longer than T_1.

Let us return to the situation depicted in Figure 9-3c. With $\mathbf{H}_z$ turned on and $\mathbf{H}_{xy}$ rotating at the Larmor frequency ω_0, we have noted that the vector $\mathbf{M}_{xy}$ rotates 90° out of phase with $\mathbf{H}_{xy}$, and that there is a torque $\mathbf{M}_{xy} \times \mathbf{H}_{xy}$ acting to flip over the magnetization along the z axis so as to decrease M_z and to give an absorption of energy. However, there is a continuous and spontaneous effort to reestablish the equilibrium value $\bar{M}_z$ of M_z and to randomize the transverse components of the magnetization of the individual nuclei so that $\mathbf{M}_{xy}$ returns to its equilibrium value of zero. As long as these longitudinal and transverse relaxation processes occur efficiently, a steady-state absorption of energy is achieved because the effects of $\mathbf{H}_{xy}$ on $\mathbf{M}$ are continuously "corrected" by the relaxation processes.

It should be noted that the absorption of energy occurs when $\mathbf{H}_{xy}$ is at or near the Larmor frequency ω_0. For protons, the energy difference (ΔE) between magnetic states is (cf. Eqn. 9-16 ff)

$$\Delta E = \gamma \hbar H \tag{9-18a}$$

$$= \hbar\omega_0 = h\nu_0 \tag{9-18b}$$

where $2\pi\nu_0 = \omega_0$. Thus, the $\mathbf{H}_{xy}$ field must oscillate at a frequency that corresponds to the quantized energy gap ΔE between adjacent magnetic states. From a classical standpoint, this can be explained by the necessity for maintaining the 90° phase relationship between $\mathbf{H}_{xy}$ and $\mathbf{M}_{xy}$ in order to generate an NMR signal. The relationship is maintained by having the frequency ω of $\mathbf{H}_{xy}$ close to ω_0. Now imagine the consequences of having ω significantly different from ω_0. At $t = 0$, $\mathbf{H}_{xy}$ acts on the $\mathbf{M}_z$ component of $\mathbf{M}$ to give a torque that generates $\mathbf{M}_{xy}^{(0)}$. At this initial time, $\mathbf{H}_{xy}$ and $\mathbf{M}_{xy}^{(0)}$ are 90° out of phase but, after a time Δt, the phase relationship is 90° + $(\omega - \omega_0)\Delta t$. Moreover, $\mathbf{H}_{xy}$ is continuously acting on $\mathbf{M}_z$; thus, at $t = \Delta t$ for example, it generates a component $\mathbf{M}_{xy}^{(1)}$ that is 90° out of phase with $\mathbf{H}_{xy}$. At this point, $\mathbf{M}_{xy}^{(0)}$ and $\mathbf{M}_{xy}^{(1)}$ are out of phase by $(\omega - \omega_0)\Delta t$. By continuing in this vein, we see that phase coherence is lost, so that $\mathbf{M}_{xy}$ effectively vanishes and, as a result, $\mathbf{H}_{xy}$ cannot act to change the magnetization along the z axis. As a consequence, there is no energy absorption because the equilibrium distribution of the z component of $\mathbf{M}$ is unperturbed.

In an NMR experiment, the effects of $\mathbf{H}_{xy}$ on $\mathbf{M}$ commonly are monitored by following the changes in the y component ($\mathbf{M}_y$) of $\mathbf{M}$. The experimental set-up is shown in Figure 9-5. The main field along the z axis is provided by a large magnet, which is augmented by a secondary, variable field known as the sweep field. The oscillating field is provided by a coil along the x axis, and the receiver coil is placed

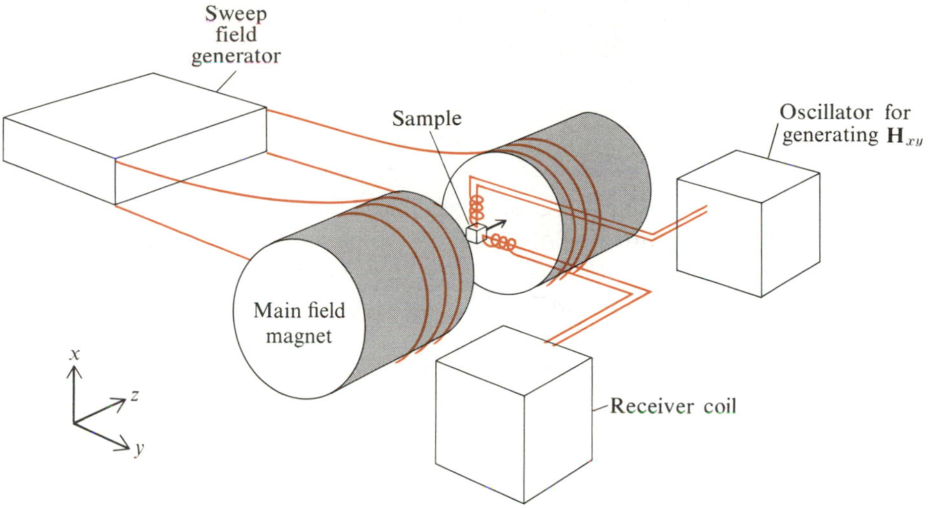

Figure 9-5

Experimental arrangement for an NMR experiment.

along the y axis. We noted earlier that the precessional frequency ω_0 about the z axis is given by $\omega_0 = \gamma H_z$. Therefore, ω_0 can be varied by changing $\mathbf{H}_z$ with the sweep field, a method often used to achieve resonance in an NMR experiment. The field $\mathbf{H}_{xy}$ rotates at a constant frequency ω determined by the instrument, and the field $\mathbf{H}_z$ is varied until ω_0 is equal to ω. At this point, resonance is achieved and, as we shall see, the maximal absorption signal is detected along the y axis. Alternatively, resonance can be achieved by varying the frequency of the oscillating field.

The qualitative description just given can be cast into a quantitative framework to give more accurate insight into some of the essential features of NMR. We accomplish this quantification by considering a simple phenomenological description known as the Bloch equations.

9-3 BLOCH EQUATIONS

Because $\mathbf{M}$ is simply the sum of the $\boldsymbol{\mu}_\mathrm{m}$ vectors for each nucleus, we can write (by analogy with Eqn. 9-11)

$$d\mathbf{M}/dt = \mathbf{M} \times \gamma\mathbf{H} \tag{9-19}$$

where $\mathbf{H} = \mathbf{H}_z + \mathbf{H}_{xy}$ (although $\mathbf{H}_z \gg \mathbf{H}_{xy}$). This equation from classical physics must be augmented, however, to take account of the propensity of individual nuclei to establish their equilibrium distribution of magnetic components through longitudinal and transverse relaxation processes. Bloch assumed that any deviation from the equilibrium value $\bar{M}_z$ of the z component of **M** would result in a first-order relaxation back to equilibrium as described by Equation 9-17, which can also be written in differential form as

$$dM_z/dt = -(M_z - \bar{M}_z)/T_1 \tag{9-20}$$

We must also consider transverse relaxation. As discussed earlier, with an applied field $\mathbf{H}_z$ in the z direction only, there is no preferred orientation of the x and y components of $\mathbf{M}_{xy}$. Any deviations of $\mathbf{M}_x$ and $\mathbf{M}_y$ from zero will tend to relax toward zero in a first-order fashion with a rate given by $1/T_2$. Hence, we may write

$$dM_x/dt = -M_x/T_2 \tag{9-21a}$$

$$dM_y/dt = -M_y/T_2 \tag{9-21b}$$

Combining Equations 9-20 and 9-21 with Equation 9-19, we obtain the complete expression for $d\mathbf{M}/dt$ as

$$d\mathbf{M}/dt = \mathbf{M} \times \gamma\mathbf{H} - (M_x/T_2)\hat{\mathbf{i}} - (M_y/T_2)\hat{\mathbf{j}} - [(M_z - \bar{M}_z)/T_1]\hat{\mathbf{k}} \tag{9-22}$$

where $\hat{\mathbf{i}}$, $\hat{\mathbf{j}}$, and $\hat{\mathbf{k}}$ are unit vectors in the x, y, and z directions, respectively. The magnetic field vector **H** is composed of $\mathbf{H}_z$ and the rotating field $\mathbf{H}_{xy}$. Taking $\mathbf{H}_{xy}$ as rotating in a clockwise direction, we obtain

$$\mathbf{H}_{xy} = H_x\hat{\mathbf{i}} + H_y\hat{\mathbf{j}} \tag{9-23}$$

where $H_x = H_{xy} \cos \omega t$, and $H_y = -H_{xy} \sin \omega t$. The expression for $d\mathbf{M}/dt$ may be written as

$$d\mathbf{M}/dt = (dM_x/dt)\hat{\mathbf{i}} + (dM_y/dt)\hat{\mathbf{j}} + (dM_z/dt)\hat{\mathbf{k}} \tag{9-24}$$

Writing out the cross-product in Equation 9-22 in terms of all the components, we obtain the expressions known as the Bloch equations:

$$dM_x/dt = \gamma M_y H_z + \gamma M_z H_{xy} \sin \omega t - (M_x/T_2) \tag{9-25a}$$

$$dM_y/dt = -\gamma M_x H_z + \gamma M_z H_{xy} \cos \omega t - (M_y/T_2) \tag{9-25b}$$

$$dM_z/dt = -\gamma M_x H_{xy} \sin \omega t - \gamma M_y H_{xy} \cos \omega t - [(M_z - \bar{M}_z)/T_1] \tag{9-25c}$$

Careful inspection of the Bloch equations should provide a physical model of the interaction between each magnetic field component and magnetic moment component to produce a torque that in turn changes each component of **M**. Note that Equation 9-25 reduces to Equations 9-20 and 9-21 when there are no external fields.

In an NMR experiment, a receiver coil is mounted along the y axis so that changes in M_y can be detected (Fig. 9-5). It is useful to focus on $\mathbf{M}_{xy} = M_x\hat{\mathbf{i}} + M_y\hat{\mathbf{j}}$, the rotating magnetic moment vector in the x–y plane. This vector can be expressed as the sum of two components **u** and **v**, where **u** is in phase and **v** is 90° out of phase with $\mathbf{H}_{xy}$ (Fig. 9-6). We anticipate from our previous discussion (and Fig. 9-3) that

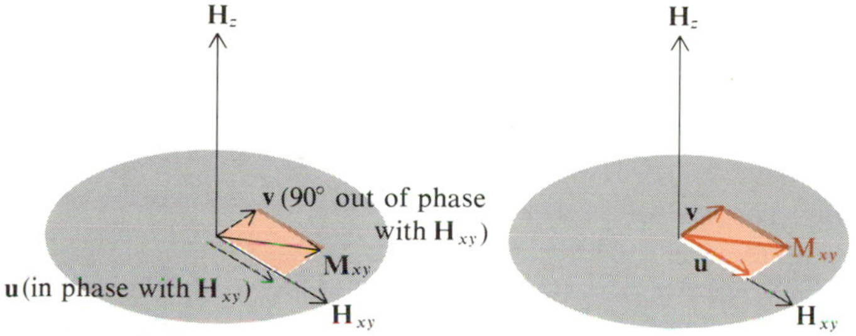

Figure 9-6

Components of $\mathbf{M}_{xy}$. The component **u** is in phase with $\mathbf{H}_{xy}$; the component **v** is out of phase with $\mathbf{H}_{xy}$.

v is the main component of $\mathbf{M}_{xy}$ at resonance ($\omega = \omega_0$). A unit vector $\hat{\mathbf{I}}$ in the direction of $\mathbf{H}_{xy}$ is $\hat{\mathbf{I}} = \hat{\mathbf{i}}\cos\omega t - \hat{\mathbf{j}}\sin\omega t$, so that $u = \mathbf{M}_{xy} \cdot \hat{\mathbf{I}}$, or

$$u = M_x \cos\omega t - M_y \sin\omega t \tag{9-26a}$$

and v is given by

$$\begin{aligned} v &= M_x \cos(\omega + \pi/2)t - M_y \sin(\omega + \pi/2)t \\ &= -M_x \sin\omega t - M_y \cos\omega t \end{aligned} \tag{9-26b}$$

Using the Bloch equations (9-25) in conjunction with differentiation of Equation 9-26, we obtain

$$du/dt = -(\omega_0 - \omega)v - u/T_2 \tag{9-27a}$$

$$dv/dt = (\omega_0 - \omega)u - v/T_2 - \gamma H_{xy}M_z \tag{9-27b}$$

$$dM_z/dt = -(M_z - \bar{M}_z)/T_1 + \gamma H_{xy}v \tag{9-27c}$$

where $\omega_0 = \gamma H_z$, the Larmor precessional frequency.[§]

When the field $\mathbf{H}_z$ is held fixed and has been adjusted so that we are at or near resonance ($\omega \approx \omega_0$), a steady-state signal is received. The steady-state condition means that

$$du/dt = dv/dt = dM_z/dt = 0 \tag{9-28}$$

Using this steady-state condition in Equation 9-27, we can obtain expressions for u, v, and M_z in terms of known quantities. These expressions are easy to derive:

$$M_z = \bar{M}_z[1 + T_2^2(\omega_0 - \omega)^2]/[1 + T_2^2(\omega_0 - \omega)^2 + \gamma^2 H_{xy}^2 T_1 T_2] \tag{9-29a}$$

$$u = \bar{M}_z \gamma H_{xy}(\omega_0 - \omega)T_2^2/[1 + T_2^2(\omega_0 - \omega)^2 + \gamma^2 H_{xy}^2 T_1 T_2] \tag{9-29b}$$

$$v = -\bar{M}_z \gamma H_{xy} T_2/[1 + T_2^2(\omega_0 - \omega)^2 + \gamma^2 H_{xy}^2 T_1 T_2] \tag{9-29c}$$

Note that all of the parameters on the right-hand side of Equation 9-29 are known. For example, $\bar{M}_z$ may be calculated from the Boltzmann distribution.

Equation 9-29 is extremely helpful for understanding an NMR experiment. Note that u vanishes when $\omega_0 - \omega = 0$ (at resonance), so that $\mathbf{M}_{xy}$ is entirely determined by its v component, as we expected. Thus, the vector $\mathbf{M}_{xy}$ is 90° out of phase with $\mathbf{H}_{xy}$. In this case, the maximal possible torque ($\mathbf{M}_{xy} \times \mathbf{H}_{xy}$) can act to produce a change of $\mathbf{M}$ with respect to the z (or field) axis. Changes in the net alignment of $\mathbf{M}$ with respect to the field axis involve a change in the magnetic interaction energy $\mathbf{M} \cdot \mathbf{H}_z$ of the system. According to Equation 9-29c, v is negative (with $\gamma > 0$); with the steady-state condition applied to Equation 9-27c, this means that $(M_z - \bar{M}_z)$ is negative. Thus, the magnetization M_z is pushed below its equilibrium value, meaning that $\mathbf{M}$ has been tilted in a direction so as to be more antiparallel to $\mathbf{H}_z$. This tilt corresponds to an increase in (absorption of) energy for the nuclei. These observations agree with the conclusions of our qualitative discussion of an NMR experiment in Section 9-2.

It is clear that v fixes the absorption of energy arising from a change in M_z with respect to the field axis. A measure of the magnitude of v can be obtained by comparing a component of $\mathbf{H}_{xy}$ with one of $\mathbf{M}_{xy}$ that is 90° apart. For example, at resonance (when $u = 0$, and $\mathbf{H}_{xy}$ is 90° out of phase with $\mathbf{M}_{xy}$), the component H_x is exactly in phase with M_y. Thus, by placing our receiver along the y axis and letting it receive a reference signal of H_x, we obtain a measure of v by measuring the current

[§] In Equation 9-11 and following discussion, we saw that $\boldsymbol{\omega} = -\gamma \mathbf{H}$. That is, for positive γ (which includes most nuclei of interest; see Table 9-1), the direction of $\boldsymbol{\omega}$ is opposite to that of $\mathbf{H}$. As shown in Figure 9-1, if $\mathbf{H}$ points along the positive z axis, then $\boldsymbol{\omega}$ is along the negative z axis, which corresponds to a clockwise rotation. However, in setting up the Bloch equations, we have taken the positive direction of rotation to be clockwise, and therefore the scalar quantity ω_0 is given by $\omega_0 = \gamma H_z$.

induced by M_y superimposed in the receiver on the H_x reference signal. Clearly, when $\omega = \omega_0$, the amplitude of the net signal will be at a maximum because H_x and M_y are in phase; when $\omega \neq \omega_0$, then H_x and M_y will not be in phase, and the resulting amplitude will be lowered. By these means, we obtain a measure of v. Alternatively, we can make equivalent measurements by monitoring the phase relationship between H_y and M_x.

The observed absorption signal will be proportional to the magnitude of v. Figure 9-7 sketches the absorption mode; it is a symmetrical peak with a maximum at $\omega_0 - \omega = 0$. Alternatively, we can measure by a suitable arrangement the u-mode

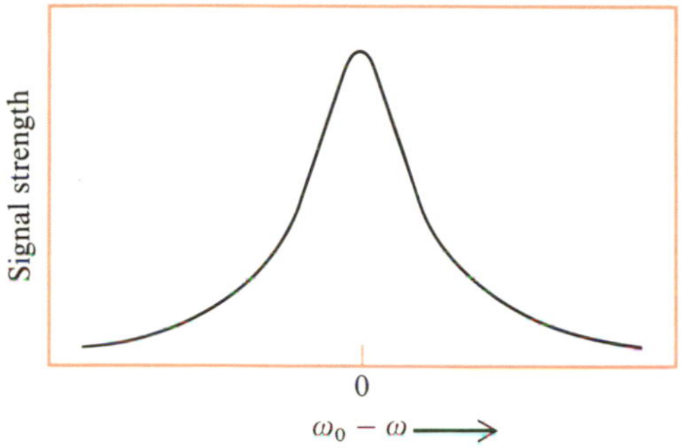

Figure 9-7
Absorption signal (v mode, or absorption mode).

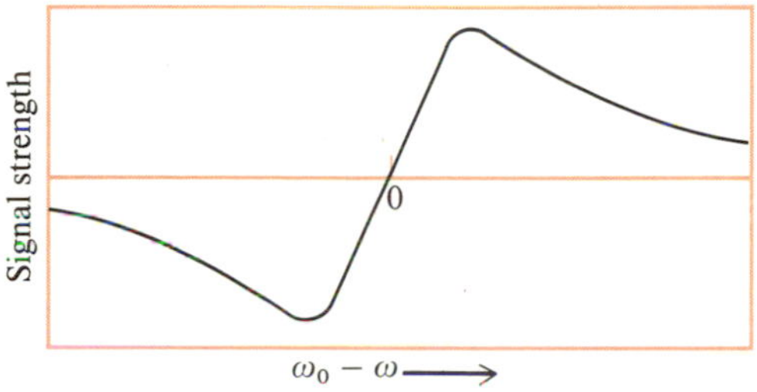

Figure 9-8
Dispersion signal (u mode, or dispersion mode).

(dispersion-mode) signal described by Equation 9-29b (Fig. 9-8). This curve also is symmetrical around $\omega_0 - \omega = 0$, the point at which the component of $\mathbf{M}_{xy}$ in phase with $\mathbf{H}_{xy}$ vanishes.

It will now be useful to consider some of the ramifications of Equation 9-29, and to examine more closely the origins and consequences of T_1 and T_2.

9-4 IMPORTANT IMPLICATIONS OF THE BLOCH EQUATIONS

Signal strength

We expect the absorption signal to be dependent on v. The actual voltage induced in the receiver coil placed along the y axis is proportional to the rate of change of the magnetic moment along the y axis. Under appropriate conditions, this proportionality is given by the expression

$$\text{Signal} \propto \omega_0 v = \gamma H_z v$$

(see Pople et al., 1959). In Equation 9-29c, note that v depends on $\bar{M}_z$. For nuclei with $I = 1/2$, this can be written as $\bar{M}_z = (\bar{N}_{1/2} - \bar{N}_{-1/2})\mu_{mz}$, where $\bar{N}_{1/2}$ and $\bar{N}_{-1/2}$ are the equilibrium numbers of nuclei in the sample with spins in the $+1/2$ and $-1/2$ states, respectively; and μ_{mz} is the value of the magnetic moment along the field axis, which is given by Equation 9-15. From simple Boltzmann statistics (and noting that, usually, $\mu_{mz}H_z/kT \ll 1$), we calculate that

$$\begin{aligned}\bar{M}_z &= N(e^{+\mu_{mz}H_z/kT} - e^{-\mu_{mz}H_z/kT})\mu_{mz} \\ &= 2(H_z\mu_{mz}^2/kT)N\end{aligned}$$

where N is the number of nuclei in the sample volume. Using this expression for $\bar{M}_z$ and substituting Equation 9-29c for v, we obtain

$$\text{Signal} = \text{const } N\left(\frac{\mu_{mz}^2 H_z^2}{kT}\right)\left(\frac{\gamma^2 H_{xy} T_2}{1 + T_2^2(\omega_0 - \omega)^2 + \gamma^2 H_{xy}^2 T_1 T_2}\right) \qquad (9\text{-}30)$$

Analogous equations can be derived for nuclei with other values of I; these differ only in the values of the constant term.

Equation 9-30 shows that the signal strength increases as a quadratic function of H_z. The reader should be able to explain this easily, based on the derivation just given. The inverse dependence of signal on temperature is due simply to less net orientation (along the field direction) accompanying the increased random motion that occurs at higher temperatures. The signal strength and shape also are affected by T_1 and T_2, as we shall discuss shortly.

Finally, it is important to note that the magnitude of the absorption signal is directly proportional to the number of nuclei (N) in the sample volume. As a result, NMR absorption signals are directly related to the concentrations of the absorbing nuclei. This proportionality has obvious practical implications.

Effects of large values of the longitudinal relaxation time (T_1)

Consider the signal strength at the absorption peak ($\omega_0 = \omega$) when T_1 is large. In this situation, Equation 9-30 reduces to

$$\text{Signal} = k/T_1$$

where k is a constant. Thus, under these conditions, the signal is limited by T_1. Similarly, we see from Equation 9-29a that a very large value for T_1 causes M_z to be small. In this situation, the system is said to approach *saturation.* The torque produced by $\mathbf{H}_{xy}$ acting on $\mathbf{M}_z$ has tipped enough nuclei into the antiparallel state so that $\mathbf{M}_z$ eventually vanishes. Because of the large value of T_1, the random processes that act to restore M_z to $\bar{M}_z$ are too sluggish to counterbalance the effects of the torque produced by $\mathbf{H}_{xy}$. Of course, as M_z diminishes, the component v that is 90° out of phase with $\mathbf{H}_{xy}$ also must decrease. In such circumstances, it sometimes is necessary to do experiments under non-steady-state conditions in order to obtain a suitable signal.

Effects of fluctuating local fields and molecular environment on T_1

The environment surrounding the nuclei under study commonly is called the lattice, and T_1 often is called the *spin–lattice relaxation time.* The longitudinal relaxation of the magnetization to its equilibrium value is caused by fluctuating local magnetic fields in the lattice. These are random fields with a distribution of frequency components. One of these components will match the Larmor frequency ω_0 of the nuclei under investigation and will fall in the x–y plane, where it can produce a torque through the "cross-product" action on $\mathbf{M}_{xy}$. This torque, like that produced by $\mathbf{H}_{xy}$, gives rise to a change in the z component of $\mathbf{M}$.

The environment provided by the lattice strongly affects the value of T_1. In solids, where molecular motion is greatly restricted, the randomly generated magnetic fields for the most part have low-frequency components that may fall below ω_0. As a result, T_1 is very large—sometimes on the order of hours. In liquids or gases, where the motion is greater and a wide spectrum of fluctuating local field components are generated, T_1 is very small. For example, in many situations in liquids, $T_1 < 10$ sec.

In liquids, T_1 might be expected to depend on viscosity. Figure 9-9 sketches the effect of viscosity on T_1^{-1}. Note that T_1^{-1} has a maximal value at a particular viscosity and falls to smaller values on either side of the maximum. This behavior is easy to understand from a consideration of the correlation time τ_c, which determines how long it takes an ensemble of oriented magnetic moments to decay to

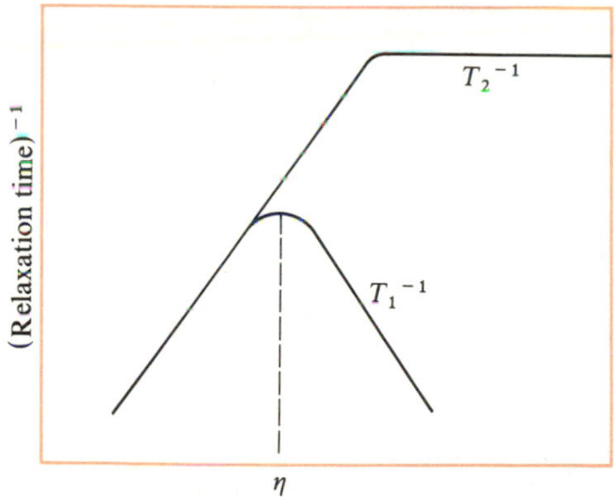

Figure 9-9
Effect of viscosity (η) on relaxation times T_1 and T_2.

within e^{-1} of their statistical equilibrium distribution. For spheres of radius r, this correlation time is [§]

$$\tau_c = 4\pi\eta r^3/3kT \tag{9-31}$$

where η is the viscosity. Clearly, the frequency spectrum of transverse components of fluctuating magnetic fields generated by the lattice nuclei will peak around frequencies close to τ_c^{-1}. Therefore, when η is adjusted so that τ_c^{-1} of the lattice nuclei is comparable to the Larmor frequency of precessing nuclei, an efficient interaction takes place between the transverse components of the fluctuating magnetic fields and those of the precessing nuclear magnetic moments. This interaction permits energy exchange between the spin and lattice systems, so as to establish thermal equilibrium. When η is varied away from this optimal point, τ_c changes so that the peak of the frequency spectrum is shifted away from the Larmor frequency of the spin system. This change gives rise to less efficient energy exchange and to an increase in T_1.

Effects of T_1 and T_2 on line widths

We define the line width $\Delta\omega_{1/2}$ in terms of the value of $\omega_0 - \omega = \pm\Delta\omega_{1/2}/2$ at which the signal drops to half its maximal value (Fig. 9-10). From Equation 9-30, we can write

$$\text{Signal} = \kappa T_2/[1 + T_2^2(\omega_0 - \omega)^2 + \gamma^2 H_{xy}^2 T_1 T_2] \tag{9-32}$$

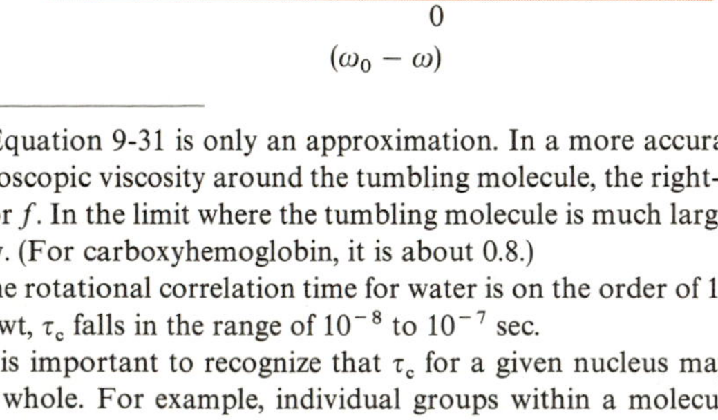

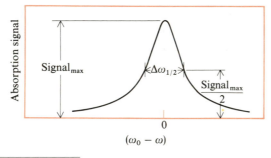

Figure 9-10
Definition of the line width $\Delta\omega_{1/2}$.

[§] Equation 9-31 is only an approximation. In a more accurate treatment, which takes into account the microscopic viscosity around the tumbling molecule, the right-hand side of Equation 9-31 is modified by a factor f. In the limit where the tumbling molecule is much larger than the solvent molecules, f approaches unity. (For carboxyhemoglobin, it is about 0.8.)

The rotational correlation time for water is on the order of 10^{-12} sec. For a protein of 50,000 to 100,000 mol wt, τ_c falls in the range of 10^{-8} to 10^{-7} sec.

It is important to recognize that τ_c for a given nucleus may not be the same as that for the molecule as a whole. For example, individual groups within a molecule may show internal rotation, so that the correlation times for nuclei within those groups are smaller than the correlation time for tumbling of the molecule as a whole.

See Chapter 8 (and cf. Eqn. 8-73) for further discussion of correlation times.

where κ is a constant that lumps together several parameters. The maximal signal is given by

$$\text{Signal}_{\text{max}} = \kappa T_2/(1 + \gamma^2 H_{xy}^2 T_1 T_2) \tag{9-33}$$

and the half-maximal signal is

$$\text{Signal}_{\text{max}}/2 = \kappa T_2/[1 + T_2^2(\Delta\omega_{1/2}/2)^2 + \gamma^2 H_{xy}^2 T_1 T_2] \tag{9-34}$$

Therefore, from Equations 9-33 and 9-34, we conclude that

$$T_2^2(\Delta\omega_{1/2}/2)^2 = 1 + \gamma^2 H_{xy}^2 T_1 T_2 \tag{9-35}$$

and

$$\Delta\omega_{1/2} = (2/T_2)(1 + \gamma^2 H_{xy}^2 T_1 T_2)^{1/2} \tag{9-36}$$

For a given T_1, we see that the line width increases with decreasing values of T_2; with fixed T_2, the line narrows with decreasing values of T_1. Note also that the line width is dependent on H_{xy} but not on H_z.

In many situations, $\gamma^2 H_{xy}^2 T_1 T_2 \ll 1$. This happens whenever one or more of the parameters H_{xy}, T_1, or T_2 is sufficiently small. In particular, it is possible to adjust H_{xy} to a sufficiently small value. In this case, we see from Equation 9-29a that $M_z \approx \bar{M}_z$; that is, the nuclei maintain their Boltzmann distribution with respect to their alignment along the z axis. It follows from Equation 9-36 that

$$\Delta\omega_{1/2} = 2/T_2 \qquad \text{when } \gamma^2 H_{xy}^2 T_1 T_2 \ll 1 \tag{9-37}$$

In this situation, therefore, the line width gives a direct measurement of T_2. Note also, from Equation 9-33, that

$$\text{Signal}_{\text{max}} = \kappa T_2 \qquad \text{when } \gamma^2 H_{xy}^2 T_1 T_2 \ll 1 \tag{9-38}$$

Thus, the signal height also is governed by T_2 in this circumstance. However, it can be shown that the area under the absorption curve is independent of T_2.

The reason for the proportionality between signal height and T_2 (Eqn. 9-38) is not hard to understand. Recall that T_2 governs relaxation of the transverse component of the magnetization ($\mathbf{M}_{xy}$) back to its equilibrium value of zero. Clearly, the smaller the value of T_2, the smaller the steady-state value of M_{xy} (cf. v in Eqn. 9-29c, when $\gamma^2 H_{xy}^2 T_1 T_2 \ll 1$). Hence, the torque $\mathbf{M}_{xy} \times \mathbf{H}_{xy}$ that gives rise to changes in M_z is reduced, and the absorption also is reduced as a result.

The broadening of the line width with smaller T_2 values (Eqn. 9-36) also is not hard to comprehend. At $\omega \neq \omega_0$, the more rapid randomization of the transverse components of the magnetization partially compensates for the fact that $\mathbf{H}_{xy}$ does not match the precessional frequency. With a large value of T_2, the torque $\mathbf{H}_{xy} \times \mathbf{M}_z$ continuously produces components of transverse magnetization that (at $\omega \neq \omega_0$) are out of phase with each other and thereby nullify themselves. With a small value of T_2, the transverse components continuously produced by $\mathbf{H}_{xy}$ rapidly decay, so that at $\omega \neq \omega_0$ they cannot destructively interfere as efficiently with each other. This means that the instantaneous magnetization in the x–y plane can somewhat interact with $\mathbf{H}_{xy}$ even at frequencies where little signal would be received at large T_2. The result is a broader line.

Factors affecting T_2

Several factors determine the transverse relaxation time, T_2. First, we should note that the z-axis field felt by a particular nucleus generally is given not by $\mathbf{H}_z$ but by $\mathbf{H}_z + \Delta\mathbf{H}_{\text{loc}}$, where $\Delta\mathbf{H}_{\text{loc}}$ is a small time-dependent fluctuating perturbation field due to the local environment surrounding that nucleus. If the amplitude of fluctuation is ΔH^0_{loc}, then the nuclei at any instance will have a spread $\Delta\omega$ in their Larmor frequencies of about $\gamma\Delta H^0_{\text{loc}}$. Thus, if they start out in phase and the field $\mathbf{H}_{xy}$ is suddenly switched off, then after $\Delta t = \Delta\omega^{-1}$ the nuclei will be spread over a range of one radian. This effect clearly dissipates the transverse component of magnetization ($\mathbf{M}_{xy}$) and shortens the observed transverse relaxation time.

A second factor influencing T_2 is spin exchange. Neighboring nuclei of oppositely oriented spin can mutually exchange spin. By this process, a nucleus changes its spin state and a neighboring nucleus does also, but in the opposite sense, so that there is no change in net spin. Although this exchange process does not affect the net spin, it does destroy the phase relationship between nuclei, and therefore it causes a decrease in T_2.

Finally, it should be recognized that the longitudinal relaxation process also affects T_2, simply because every change in the nuclear spin energy levels (associated with establishing the Boltzmann distribution with respect to net orientation along the z axis) results in a loss of phase relationships.

The observed rate of transverse relaxation ($1/T_2$) is simply the sum of rates for each of the mechanisms just described. Let $1/T_2'$ be the rate associated with the first two mechanisms (effects of $\Delta\mathbf{H}_{\text{loc}}$ and of spin exchange). Then,

$$1/T_2 = (1/T_2') + (1/2)(1/T_1)$$

The factor 1/2 multiplies $1/T_1$ because the nuclei are actually changing spin states only half as fast as the longitudinal relaxation measured by T_1^{-1}. This is because the difference in the occupation numbers of spin states changes by two when a single nucleus changes from one state to the other.

We can now consider the effect of viscosity (or τ_c) on T_2 (Fig. 9-9). At low values of η, τ_c is so small that the variations in $\Delta\mathbf{H}_{\text{loc}}$ are effectively averaged to zero, and neighboring nuclei spend too short a time in close proximity to undergo spin exchange.

Hence, $1/T_2' \to 0$, and $1/T_2 \approx (1/2)T_1^{-1}$. We see from Equation 9-37 that in this situation the line width $\Delta\omega_{1/2} = T_1^{-1}$; this is known as the *natural line width* or the *extreme narrowing limit.* As η increases, the effects of $\mathbf{\Delta H}_{\text{loc}}$ and of spin-exchange processes are felt, so that $1/T_2'$ rises to finite values. Consequently, T_2 becomes smaller and line broadening occurs (Eqn. 9-37). Eventually, as η increases further, molecular motion reaches the point where $\mathbf{\Delta H}_{\text{loc}}$ and spin-exchange events have their maximal effectiveness. At this point, T_2 reaches a plateau and does not change further with increases in η.

From the preceding discussion, it is clear why line widths are broad in solids, where molecular motion is greatly restricted. Macromolecules in solution also can have broad lines, because of slow tumbling times that do not effectively average out $\mathbf{\Delta H}_{\text{loc}}$.

We should also mention that instrumental limitations can affect the apparent line width. It is not feasible to produce a perfectly homogeneous z-axis field. As a result of field variations across the sample, individual nuclei will precess at slightly different frequencies, thus giving rise to apparent line broadening, known as inhomogeneous broadening. The problem can, in large measure, be eliminated by spinning the sample at a frequency greater than T_2^{-1}, so that the nuclei effectively feel a constant average field. This procedure can also produce "spinning side bands" that occur in an NMR spectrum at integral multiples of the sample spinning frequency.

Measurement of T_1 and T_2 by pulse methods

Special pulse methods commonly are employed to measure transverse and longitudinal relaxation times. These methods use pulses or special pulse sequences of an intense magnetic field $\mathbf{H}_1$ at the Larmor frequency, thus tipping the net magnetization vector $\mathbf{M}$ away from the z axis. These $\mathbf{H}_1$ pulses are applied at a right angle to the main field $\mathbf{H}_z$. Two types of pulses are commonly employed; they are known as the 90° $\mathbf{H}_1$ pulse and the 180° $\mathbf{H}_1$ pulse. (Fig. 9-11). The 90° pulse rotates $\mathbf{M}_z$ 90°, whereas

Figure 9-11

NMR pulse experiment, showing the effects of applying two successive pulses along the x' axis.

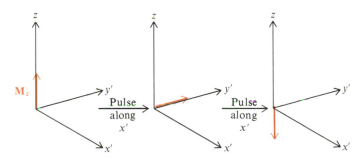

the 180° pulse causes an inversion of $\mathbf{M}_z$. The actual angle of nutation (θ) is determined by $\theta = \gamma H_1 \Delta t$, where γ is the magnetogyric ratio of the nucleus of interest, and Δt is the time period for which the pulse $\mathbf{H}_1$ is applied. Clearly, by proper choice of H_1 and Δt, it is possible to have $\theta = 180°$.

In a rotating coordinate system ($x'y'z$, where x' and y' rotate about fixed z), the $\mathbf{H}_1$ field is applied along the x' axis while the y' component of $\mathbf{M}$ is monitored. The time Δt is commonly on the order of 1 to 200 μsec, which for most situations is sufficiently short that no T_1 or T_2 relaxation occurs during the period of the pulse. After application of the brief pulse, the system is in a nonequilibrium state. As it decays back to its equilibrium value, a transient signal (known as the free induction decay) is observed along y'. This induction signal will decay according to e^{-t/T_2}, where T_2 includes the effects (if any) of magnetic field inhomogeneities. Figure 9-12 shows the free induction decay of the ^{23}Na NMR signal for an NaCl solution.

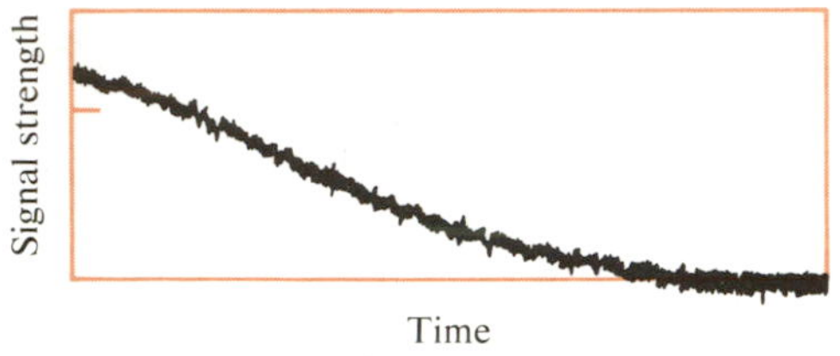

Figure 9-12

Free induction decay. Decay of ^{23}Na signal for an aqueous saturated solution of NaCl. [After T. L. James, *Nuclear Magnetic Resonance in Biochemistry* (New York: Academic Press, 1975).]

Spin–lattice (T_1) and spin–spin (T_2) relaxation times commonly are monitored by means of special pulse sequences. As an example, consider the spin-lattice relaxation time T_1 as measured by a two-pulse sequence of alternate 180° and 90° pulses, with a time interval τ between the two pulses. The 180° pulse inverts the initial value $\bar{M}_z$ to $-\bar{M}_z$; after cessation of the pulse, the z-axis magnetization will relax back to its equilibrium value according to e^{-t/T_1}. After time τ, the difference between M_z and its equilibrium value $\bar{M}_z$ will be reduced by the factor $e^{-\tau/T_1}$; if at this time a 90° $\mathbf{H}_1$ pulse is applied, the magnitude of the resulting $M_{y'}$ will be a measure of how far T_1 relaxation has proceeded. For example, if $\tau \gg T_1$, then $M_z \approx \bar{M}_z$, and the maximal value of $M_{y'}$ will be produced by the 90° pulse. However, if $\tau \leqslant T_1$, then $M_z < \bar{M}_z$, and the component $M_{y'}$ produced by the 90° pulse will be reduced. Hence, by monitoring the amplitude of the signal produced by the 90° pulse as a function of the time τ between the 180° and 90° pulses, a measure of T_1 is obtained. It turns out that, for measurements of T_2, a 90°–τ–180° pulse sequence that gives rise to a "spin echo" is particularly useful.

Absorption spectrum as the Fourier transform of the free induction decay

We have noted that a free induction decay is observed after turning off a pulse $\mathbf{H}_1$ at the resonance frequency. Let the amplitude of the free induction decay be given by $g(t)$. It can be shown that the line-shape function $g(\omega)$ of the absorption spectrum is given by

$$g(\omega) = \text{const} \int_0^\infty g(t) \cos \omega t \, dt \tag{9-39}$$

where the constant is necessary for normalization. [See the right-hand side of Equation 9-30 for a representation of $g(\omega)$.] The sine transform of $g(t)$ gives the dispersion spectrum. Conversely, the Fourier transform of $g(\omega)$ yields the free induction decay $g(t)$. Therefore, measurements of free induction decay by pulse methods and measurements of the absorption and dispersion spectrum by continuous-wave NMR are actually equivalent measurements. However, they are not equivalent procedures from a practical viewpoint, each having its particular advantages.

From a practical standpoint, the major concern is to attain sufficient resolution in the absorption spectrum. With continuous-wave NMR, this is achieved by averaging spectra obtained over a long period of time, commonly using a signal averager called a CAT (computer of average transients). The rationale is that random noise will gradually cancel itself out as more and more spectral scans are averaged together. For N scans, the noise is reduced according to $N^{1/2}$, so that 100 scans can provide a signal-to-noise ratio 10 times larger than that achieved with a single scan. However, the time required to complete just a single scan also depends on the resolution desired. For example, if a resolution of 1 Hz is desired in a spectrum having a width of 10^3 Hz, the time required to scan the entire spectrum will be on the order of 10^3 sec. If many such scans are made using CAT procedures, a large block of time obviously is needed to yield a good spectrum.

In contrast, the Fourier-transform method can use far less time, primarily because in such an experiment the entire spectrum is in effect monitored all at once. That is, by executing the Fourier transform of the free induction decay, one obtains the entire spectrum. For example, with a spectrum 10^3 Hz in width, only about 1 sec is required to obtain a resolution of ~ 1 Hz. As mentioned above, 10^3 sec are required for a single sweep of the same spectrum at that resolution by continuous-wave methods.

The Fourier-transform method has other advantages that will not be discussed here. However, it should be noted that a disadvantage of Fourier-transform NMR is that weak proton signals are more difficult to detect in the presence of water protons, even in situations where D_2O is used, because the effects of residual HDO can be quite serious. Techniques for decoupling H_2O and HDO resonances out of the spectrum can help the situation. On balance, the particular experiment will dictate which procedure is most effective for obtaining the absorption spectrum.

9-5 FEATURES OF NMR SPECTRA

Chemical shifts

We might expect the resonance frequency of a proton to occur at a characteristic frequency γH_z, determined only by the magnitude of the external magnetic field $\mathbf{H}_z$ and the magnetogyric ratio γ. If this were true, NMR would be of limited interest, because protons in various compounds (or different protons in the same compound) would resonate at the same frequency and therefore would be indistinguishable. However, the effective field felt by a proton is not H_z but H'_z, where

$$H'_z = H_z - H_z\sigma = H_z(1 - \sigma)$$

or

$$\sigma = \frac{H_z - H'_z}{H_z} \tag{9-40}$$

In this equation, σ is a constant that describes how much the local environment perturbs the main field to yield the effective field H'_z actually felt by a nucleus. The effective field H'_z is directly proportional to the external field H_z. It is common practice to describe the perturbation in terms of a chemical shift parameter δ given by

$$\begin{aligned} \delta &= [(H_{\text{ref}} - H_{\text{samp}})/H_{\text{ref}}] \times 10^6 \\ &= [(\nu_{\text{ref}} - \nu_{\text{samp}})/\nu_{\text{ref}}] \times 10^6 \end{aligned} \tag{9-41}$$

where H_{ref} is the observed resonance field strength for a nucleus in a reference molecule, and H_{samp} is that for a nucleus of the same kind in the sample of interest. Similarly, ν_{ref} and ν_{samp} are the resonance frequencies ($\nu = \omega/2\pi$) for the reference and sample nuclei, respectively. The factor of 10^6 enters because it is customary to express shifts in parts per million (ppm). By defining δ as a normalized dimensionless parameter, we obviously make its value independent of the operating frequency of the instrument. Equation 9-41 for δ is analogous to that for σ given in Equation 9-40.

At a field strength of 50,000 gauss, most protons resonate over a range of 2,500 Hz centered around 220 MHz. This means that the values of δ typically fall between 0 and 10 ppm.

For protons in aqueous solution, the methyl resonance of sodium 2,2-dimethyl-2-silapentane-5-sulfonate (DSS) is the most widely used reference, whereas tetramethylsilane (TMS) is usually employed in nonaqueous solvents. The resonance frequencies of the methyl protons are almost identical for these two standards. Most other protons resonate at frequencies lower than that of these standards.

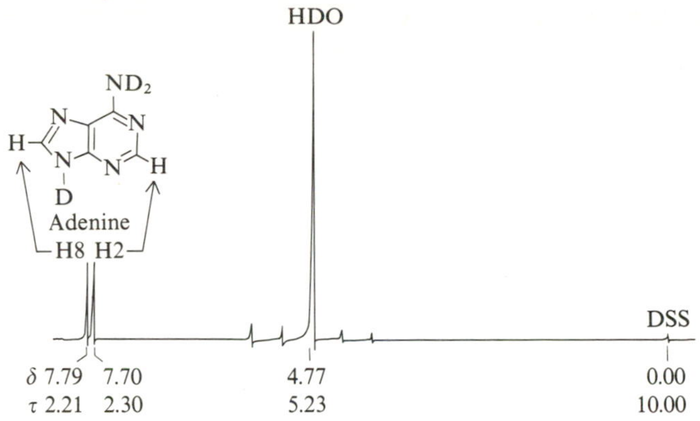

Figure 9-13

The 220 MHz spectrum of adenine in D_2O. Resonances for H-8 and H-2 are indicated. The exocyclic amino protons are exchanged with deuterium. The small peaks flanking the HDO line are due to spinning side bands. The δ and τ values are relative to a DSS standard. [After T. L. James, *Nuclear Magnetic Resonance in Biochemistry* (New York: Academic Press, 1975).]

Figure 9-13 shows the 220 MHz spectrum of adenine in 1 M NaOD (D_2O solvent) and indicates the H-8 and H-2 resonance positions. With DSS as a standard, H-8 is shifted downfield 7.79 ppm and H-2 is shifted 7.70 ppm. Also shown in the figure are the frequently cited τ values. These shift parameters are based on the scale $\tau = 10 - \delta$; thus, the standard is assigned a value of $\tau = 10$ ppm with this system.

The magnitude of the chemical shift is determined by the parameter σ in Equation 9-40. This parameter in turn is determined by three types of effects, so that we may write $\sigma = \sigma_1 + \sigma_2 + \sigma_3$. Local diamagnetic effects determine σ_1. When a field is applied to an atom, circulating electron currents are set up that are at a right angle to the applied field. These currents generate a small magnetic field that is in the opposite direction to $\mathbf{H}_z$. As a result, H_z is reduced (therefore, σ_1 is positive).

The parameter σ_2 usually is determined by mostly paramagnetic effects from neighboring atoms, with a lesser contribution from diamagnetic effects. The parameter σ_3 is determined by interatomic currents. In general, each of these two parameters can be either positive or negative.

It should be clear that, if σ is positive, a stronger external field $\mathbf{H}_z$ will be needed to achieve resonance at a given frequency. The incremental amount of extra field strength is given by σH_z^{res}, where H_z^{res} is the field strength at resonance if $\sigma = 0$.

A special effect occurs in aromatic compounds. This effect is illustrated in Figure 9-14 for benzene. It is known that the protons on the ring resonate in fields of anomalously low strength. This phenomenon is caused by ring currents generated by the six π electrons. The large electron current induced by the external field produces

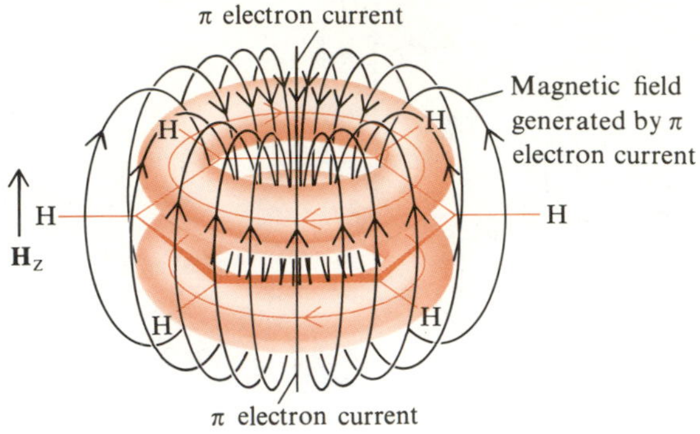

Figure 9-14

The ring-current effect on the local magnetic field felt by aromatic protons. The applied external field is $\mathbf{H}_z$.

a magnetic field, directly above and below the ring, that opposes the main field; but outside of the ring, the ring-current magnetic field acts in the same direction as the main field. Thus, the pendant protons are deshielded and experience a field stronger than $\mathbf{H}_z$. This effect gives rise to resonance in a weaker external field. The effect can be quite sizable; for example, benzene protons are shifted downfield about 1.48 ppm relative to protons on 1,3-cyclohexadiene.

Another important effect in biological systems is the chemical shift associated with hydrogen bonding. In a system such as D—H· · ·A, hydrogen resonance generally is observed to be shifted downfield. Apparently this is due to the effect of the acceptor A on the diamagnetic currents associated with D—H, so that some deshielding results. The resulting shifts can be larger than 1 ppm.

Achieving high resolution

From Equation 9-41, it is clear that the actually observed shifts $H_{samp}-H_{ref}$ and $\nu_{samp}-\nu_{ref}$ are given by δH_{ref} and $\delta \nu_{ref}$, respectively. If a very strong external z field is applied, there must be a proportional increase in the value of the resonance frequency, because the frequency (ν) at which a given nucleus resonates increases according to $\omega = 2\pi\nu = \gamma H$; conversely, if the instrument is operated at a constant high frequency, then a much stronger z field must be applied to achieve resonance. Thus, higher frequencies require stronger external fields, and vice versa. Because the observed chemical shifts are directly proportional to the field strength and to the frequency, lines are farther apart in instruments that use stronger fields and higher frequencies. For this reason, there has been a steady effort to develop NMR machines that operate at higher and higher frequencies; instruments operating at 360 MHz and even higher frequencies are now available. These instruments increase by a factor of six the observed separation between two proton lines that might be too close to visualize separately in the spectrum taken at the once-popular frequency of 60 MHz.

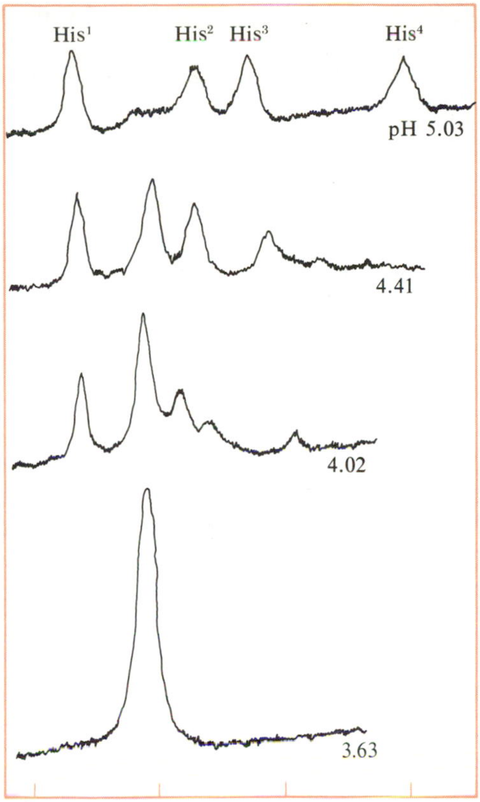

Figure 9-15

The 220 MHz NMR spectrum of staphylococcal nuclease C-2 protons on histidine imidazoles. The histidines are indicated as His[1] through His[4]. [After H. F. Epstein et al., *Proc. Natl. Acad. Sci. USA* 68:2042 (1971).]

The power of high-resolution NMR is well illustrated in Figure 9-15, which shows the behavior in a 220 MHz spectrum of the C-2 protons on the four histidines of staphylococcal nuclease. The protein undergoes a reversible denaturation as a function of pH in the acid region. At pH 5, the protein is in its native form, and all four C-2 protons are clearly resolved. The separate resonances are spread over a region of only 0.5 ppm, but this separation is sufficient to give good resolution at 220 MHz. The distinct resonances mean simply that the native structure generates distinct magnetic environments (different σ values) for each proton. It also is important to note that the area under each peak is approximately the same, denoting equal concentrations of each C-2 proton.[§] As pH is lowered, the protein denatures, and the four peaks coalesce at pH 3.6 to a single peak of 8.44 ppm with the intensity of four protons. Thus, in the unfolded form, all four protons have magnetically equivalent environments.

[§] The relationship between concentration and absorption-peak area follows from Equation 9-30 and the discussion thereafter.

Interestingly enough, in the pH range where the protein is partially denatured, not all C-2 protons have changed in concert; that is, at a given pH, some have lost more intensity than others. This observation indicates that the regions around the four histidines are not totally coupled together in the unfolding process. On the other hand, if the individual intensities are plotted as a function of pH, steep curves are obtained, showing that the unfolding has a strong cooperative component.

These spectra illustrate the highly specific information that can be obtained at 220 MHz. If the same spectra were run at 60 MHz, the line separations would be reduced about fourfold, and resolution of the individual histidines would be impossible.

Spin–spin splitting of resonance lines

Spin–spin splitting arises from the effects of neighboring nuclei on the field felt by the nucleus of interest. Figure 9-16a shows the result of this splitting in the 60 MHz spectrum of ethoxyacetic acid. Note that the methyl group is represented by three lines (a triplet), and the adjacent methylene group shows four lines (a quartet). The spacing J (called the coupling constant) between adjacent components in each multiplet is the same in both the triplet and the quartet.

The line splitting (spin–spin splitting) shown in Figure 9-16a is due to the interactions between adjacent nuclear spins. This interaction is mediated by bonding

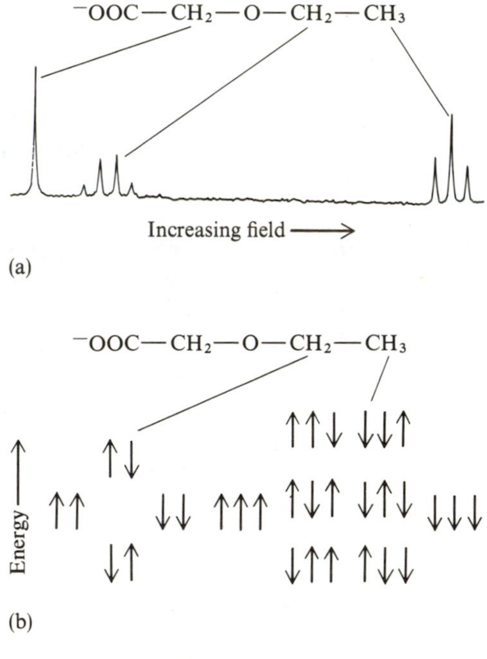

Figure 9-16

The NMR spectrum of ethoxyacetic acid in aqueous solution. **(a)** The 60 MHz spectrum. Lines associated with specific protons are indicated. **(b)** Orientations of proton spins in ethyl group of ethoxyacetic acid. [After T. L. James, *Nuclear Magnetic Resonance in Biochemistry* (New York: Academic Press, 1975).]

electrons; it depends on the distance between nuclei, the type of chemical bond, the bond angle, and the nuclear spin. The number of multiplets and their relative intensities can be predicted for situations in which J is much smaller than the chemical shift frequency $\Delta\nu$ between adjacent groups (typically it is necessary that $J \leqslant 0.1\ \Delta\nu$). For these cases, the multiplets are predicted from the possible orientations of the set of protons ascribed to a particular group. For example, Figure 9-16b shows the orientations for the protons in the methylene group adjacent to the methyl group. There are three distinct orientations, in the ratio of 1:2:1. These provide three distinct nuclear-spin environments for the neighboring methyl group, so that it is split into three lines, about in the ratio 1:2:1. A similar analysis shows that the methyl-group protons have four distinct arrangements, in the ratio of 1:3:3:1, thus accounting for the observed splittings and approximate intensities of the adjacent methylene group. In the case of the methylene next to the COO^- group, there is no splitting, because it is insulated from the protons on the other methylene group by the ether oxygen.

In situations where $J > 0.1\ \Delta\nu$, the spectrum can be much more complex than the largely "first-order" splitting pattern displayed in Figure 9-16a. However, there are methods for simplifying and interpreting complex spectra. For example, spin–spin splitting is not dependent on the external field strength. Because the chemical shifts are dependent on field strength, an increase in field strength will raise $\Delta\nu$ to the point where simple first-order splittings can be seen ($J \leqslant 0.1\ \Delta\nu$). Another approach is to use double resonance. In this case, a second oscillating field is applied at the resonance frequency of the nuclei that are splitting the lines of the ones under study. If this is done to saturate the perturbing nuclei, the multiplet being studied will collapse into a singlet. This procedure is known as spin decoupling.

The coupling constant J contains useful information related to molecular structure. M. Karplus (1959, 1963) has shown that, for an ethanelike molecule, the value of J for hydrogens on adjacent carbons is dependent on the dihedral angle ϕ between

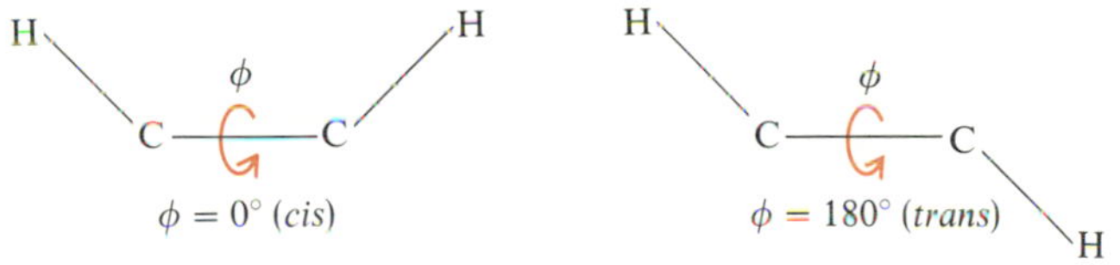

Figure 9-17

The dihedral angle ϕ defining the relationship between two C—H bonds in an ethanelike molecule. (See Chapter 5 for further discussion of dihedral angles.)

the two C—H bonds under consideration (Fig. 9-17; see also the discussion of dihedral angles in Chapter 5). The relationship is

$$J = A + B \cos \phi + C \cos 2\phi \tag{9-42}$$

where A, B, and C are constants. The equation predicts that J has a large value when $\phi = 0°$ or 180° but is sharply diminished when $\phi = 90°$ (an axial-equatorial arrangement).

Various investigators have taken advantage of the relationship between J and angular orientation in order to estimate the conformations of various molecules. Although this procedure is fraught with difficulties (because of lack of detailed knowledge of how the various molecular parameters affect J in different types of chemical species), some useful data have been obtained. For example, in peptides, we can measure the coupling between the hydrogen attached to nitrogen and the adjacent C^{α}-hydrogen

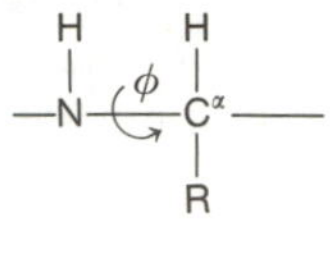

From this coupling measurement, we can estimate the intervening dihedral angle ϕ. A particularly interesting analysis has been carried out by G. Ramachandran and coworkers, who have examined the fairly rigid cyclic peptide alumichrome:

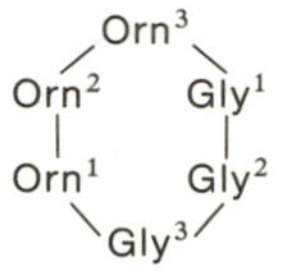

The values of ϕ for the ornithine residues have been estimated from measurements of coupling constants. These values, in turn, have been compared with values determined by an x-ray analysis of ferrichrome A, an iron analog of alumichrome with Gly^1 and Gly^2 replaced by serines. Table 9-2 shows the results of this analysis. It is clear that the values of ϕ determined by NMR are close to those measured by x-ray diffraction. This is good evidence in support of the NMR approach.

Table 9-2

Comparison of dihedral angle ϕ values for ornithine residues computed from coupling constants (J) and those measured by x-ray crystallography

Residue	Average J (Hz; 2 solvents)	ϕ (from J)	ϕ (from x-ray)
Orn^1	7.6	−150°	−145°
Orn^2	5.4	−73°	−77°
Orn^3	8.6	−100°	−104°

NOTE: J is the vicinal H–N–C^{α}–H proton–proton coupling constant of ornithine residues of alumichrome. The value of ϕ "from x-ray" is that for ornithine residues of ferrichrome as determined by x-ray crystallography.

SOURCE: After G. N. Ramachandran, R. Chandrasekaran, and K. D. Kopple, *Biopolymers* 10:2113 (1971).

Much effort has been directed also at examining linear peptides in solution. In these cases, one may assume that a distribution of conformations exists, and that only an average value for ϕ may be obtained. The value of the dihedral angle ψ for the bond between the α-carbon and the carbonyl carbon has been approached by examining hydrogen–nitrogen coupling in systems such as

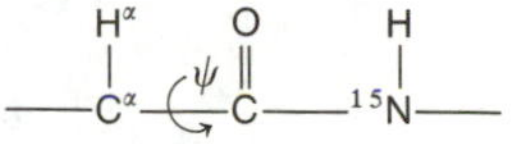

Here, the coupling between H^α and ^{15}N is monitored. Sugar conformations also have been examined by studying vicinal coupling constants. Although these data will not be further considered here, it is clear that the contribution of Karplus (showing that J is sensitive to dihedral angle) has spawned some important applications.

Study of biological complexes with paramagnetic probes

The magnetic moment of an unpaired electron is approximately 10^3-fold larger than that of a nucleus (Section 9-7). Therefore, paramagnetic species can markedly affect the magnetic environment of nuclei, producing paramagnetic shifts in line positions and altered relaxation rates. The paramagnetic shifts can be quite pronounced—as evidenced, for example, by the proton resonance spectrum of paramagnetic Fe^{3+} sperm-whale cyanometmyoglobin, which shows peaks over the wide range of -27 ppm to $+3$ ppm. The effect of paramagnetic species on chemical shifts has led to the development of shift reagents—such as lanthanide chelates—that can be used to simplify NMR spectra.

The large magnetic moment of the unpaired electron enables paramagnetic species to give efficient relaxation mechanisms. The effects of these species on T_1 and T_2 can be exploited to yield useful structural information. For this purpose, paramagnetic Mn^{2+} has been particularly useful because it can replace diamagnetic Mg^{2+} in many systems without loss of biological activity.

In one type of experiment, T_1 values for water protons are measured in the presence of free Mn^{2+} and in the presence of Mn^{2+} bound to a macromolecule (such as an enzyme). From these measurements, the number of water molecules liganded to Mn^{2+} can be calculated. Assuming that Mn^{2+} adheres to strict octahedral coordination (which appears to be the case), and by subtracting the number of coordinated water molecules, one can estimate the number of ligands contributed by the macromolecule. The number of liganded water molecules in a Mn^{2+}–macromolecule complex also gives some idea of the accessibility of water to the metal site. For example, Mn^{2+} complexes of carboxypeptidase A and of pyruvate kinase have one and three liganded water molecules, respectively. This indicates, particularly for carboxypeptidase A, considerable shielding of the bound metal from the aqueous medium.

Another type of experiment takes advantage of the effect of paramagnetic species on T_1 and T_2. The relaxation times of a proton at a distance r from a paramagnetic

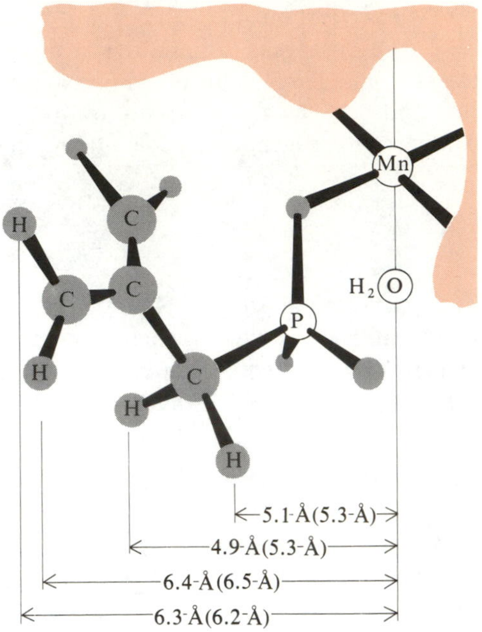

Figure 9-18

Distance relationships in a pyruvate kinase–Mn(II)–α-(dihydroxyphosphinylmethyl) acrylate complex. Distances measured from a molecular model are given; values in parentheses are those obtained from experimental measurement. [After T. L. James and M. Cohn, *J. Biol. Chem.* 249:3519 (1975).]

species are altered in accordance with r^{-6} (James, 1975). Hence, by measuring relaxation rates of known protons in a Mn^{2+} complex, one can estimate distances between the ion and the various protons. Figure 9-18 shows an example of the results obtained with this approach. The structure under study is the pyruvate kinase–Mn^{2+}-α-(dihydroxy-phosphinylmethyl)acrylate complex. Various distances were measured, and then a molecular scale model of the complex was constructed. The scale model could be built so as to yield distances quite close to those measured by NMR. (The distances measured by NMR are given in parentheses in the figure.)

This study by T. L. James and M. Cohn (1975) provides an excellent illustration of the power of paramagnetic probes. Several other systems also have been explored. Particularly interesting are cases in which distances determined by NMR can be compared with those measured by x-ray diffraction on crystals. In a number of cases, good agreement has been obtained for the two types of measurements.

Use of NMR to monitor rate processes

Consider a situation in which a nucleus is mobile and can move between two distinct sites. An example would be a particular proton on a ligand where the ligand binds to its receptor site on a macromolecule. In this case, the free and bound ligands are apt to have very different chemical shifts. In the limit where no exchange occurs, two distinct resonance lines will be evident—one corresponding to each molecular

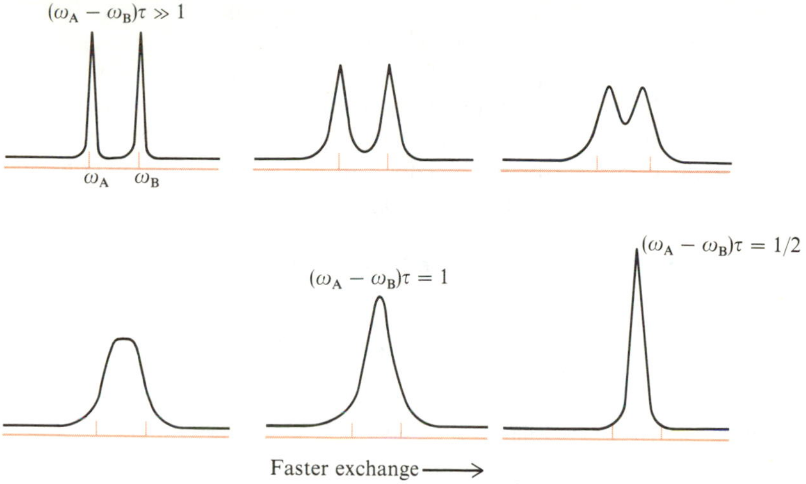

Figure 9-19

Effect of exchange on NMR spectrum. The nuclei exchange between two sites A and B that have characteristic resonance frequencies ω_A and ω_B. We have assumed equal populations in each site ($\hat{\chi}_A = \hat{\chi}_B = 1/2$) and $\tau_A = \tau_B = \tau$. [After T. L. James, *Nuclear Magnetic Resonance in Biochemistry* (New York: Academic Press, 1975).]

environment. If exchange is extremely rapid, only one line will be evident; its position will be between the two originally separated lines.

Figure 9-19 is a schematic representation of the effect of exchange on the spectrum. Let the two chemical environments be designated A and B. The lifetime in environment A is designated τ_A, and τ_B denotes the time spent in environment B. (For a given molecule in state A at $t = 0$, the probability that it will be in state A at time t is e^{-t/τ_A}.) The resonance frequencies of the nucleus in the two environments are ω_A and ω_B. When $\tau_A \gg (\omega_A - \omega_B)^{-1}$ and $\tau_B \gg (\omega_A - \omega_B)^{-1}$, two distinct lines are evident (Fig. 9-19). Under these "slow exchange" conditions, the transverse relaxation time T_{2A} associated with the line at ω_A is given by

$$T_{2A}^{-1} = (T_{2A}^{0})^{-1} + \tau_A^{-1} \tag{9-43}$$

where T_{2A}^{0} is the relaxation time in the absence of exchange. In this situation, the effect of τ_A is simply to decrease T_{2A} and broaden the line at ω_A. Hence, in the slow-exchange limit, the parameter τ_A (and similarly τ_B) is obtained from a measurement of the increased broadening caused by exchange.

When exchange is such that $\tau_A < (\omega_A - \omega_B)^{-1}$ and $\tau_B < (\omega_A - \omega_B)^{-1}$, only a single line is observed (Fig. 9-19). Under these conditions, the transverse relaxation time T_2 associated with the single coalesced line is given by

$$T_2^{-1} = \hat{\chi}_A (T_{2A}^{0})^{-1} + \hat{\chi}_B (T_{2B}^{0})^{-1} + \hat{\chi}_A^2 \hat{\chi}_B^2 (\omega_A - \omega_B)^2 (\tau_A + \tau_B) \tag{9-44}$$

where $\hat{\chi}_A$ and $\hat{\chi}_B$ are the fractions of nuclei in state A and state B, respectively. In this situation, the line width of the coalesced resonance depends on the fractional occupancies of each state and on both lifetimes τ_A and τ_B. This effect is called *exchange broadening*. The parameters $\hat{\chi}_A$ and $\hat{\chi}_B$ often are not difficult to estimate. For example, if proton exchange between water and ethanol molecules is studied in water-ethanol mixtures, the fractions of each state are given by the mole fractions of water and ethanol. If ligand exchange between a free and a bound state is considered, the fractions in each state can be estimated from the equilibrium constant for the binding reaction.

It is of interest to note that, in the limit where τ_A and $\tau_B \to 0$, Equation 9-44 becomes

$$T_2^{-1} = \hat{\chi}_A(T_{2A}^0)^{-1} + \hat{\chi}_B(T_{2B}^0)^{-1} \qquad \text{for } \tau_A \to 0 \text{ and } \tau_B \to 0 \tag{9-45}$$

In this instance, the width of the single coalesced line is independent of the lifetimes τ_A and τ_B.

In certain circumstances, the condition under which two lines coalesce can be used to evaluate τ_A and τ_B. When $\hat{\chi}_A = \hat{\chi}_B = 1/2$, then $\tau_A = \tau_B$, and the point at which the two separate peaks coalesce is given by

$$\tau = 2^{1/2}(\omega_A - \omega_B)^{-1}$$

where $2\tau = \tau_A = \tau_B$, and where it is also assumed that $(T_{2A}^0)^{-1}$ and $(T_{2B}^0)^{-1} \cong 0$. This relationship often is useful.

9-6 NMR SPECTRA OF BIOLOGICAL SYSTEMS

Proton magnetic resonance spectra of proteins

A small protein has on the order of 10^3 or more protons. In the side chains of the free amino acids, these protons resonate at field positions that usually fall within ± 8 ppm of one another. For example, relative to a DSS standard, methyl and methylene resonances of free amino acid side chains generally fall in the range of 0.8 to 3.5 ppm, whereas aromatic, indole, and imidazole protons bonded to carbon fall between 6.5 and 8.0 ppm. Because a huge amount of spectral information is crammed into a relatively small region, it is difficult to interpret the spectrum in terms of specific residues. As we shall see, one of the chief aims of many investigations is to simplify the spectra and to identify resonances of specific residues amidst a complicated background.

The spectra of proteins commonly are obtained in D_2O in order to eliminate the huge H_2O absorption band. Exchangeable protons thus are replaced by deuterium, so that these resonances do not appear. To assure complete exchange of the amide NH protons, it is customary to preheat the sample in D_2O.

We can obtain a rough estimate of the spectrum of a protein simply by summing up the contributions of the various monomeric amino acid side chains, weighting each one according to the amino acid composition. The resulting calculated spectrum is found to correlate reasonably with the spectrum of the random, denatured protein. Figure 9-20a,b illustrates this correlation for the 220 MHz spectrum of lysozyme; the calculated spectrum shows most of the features of that for the denatured form at 80°C. In each instance, the observed resonances fall between about −150 Hz and

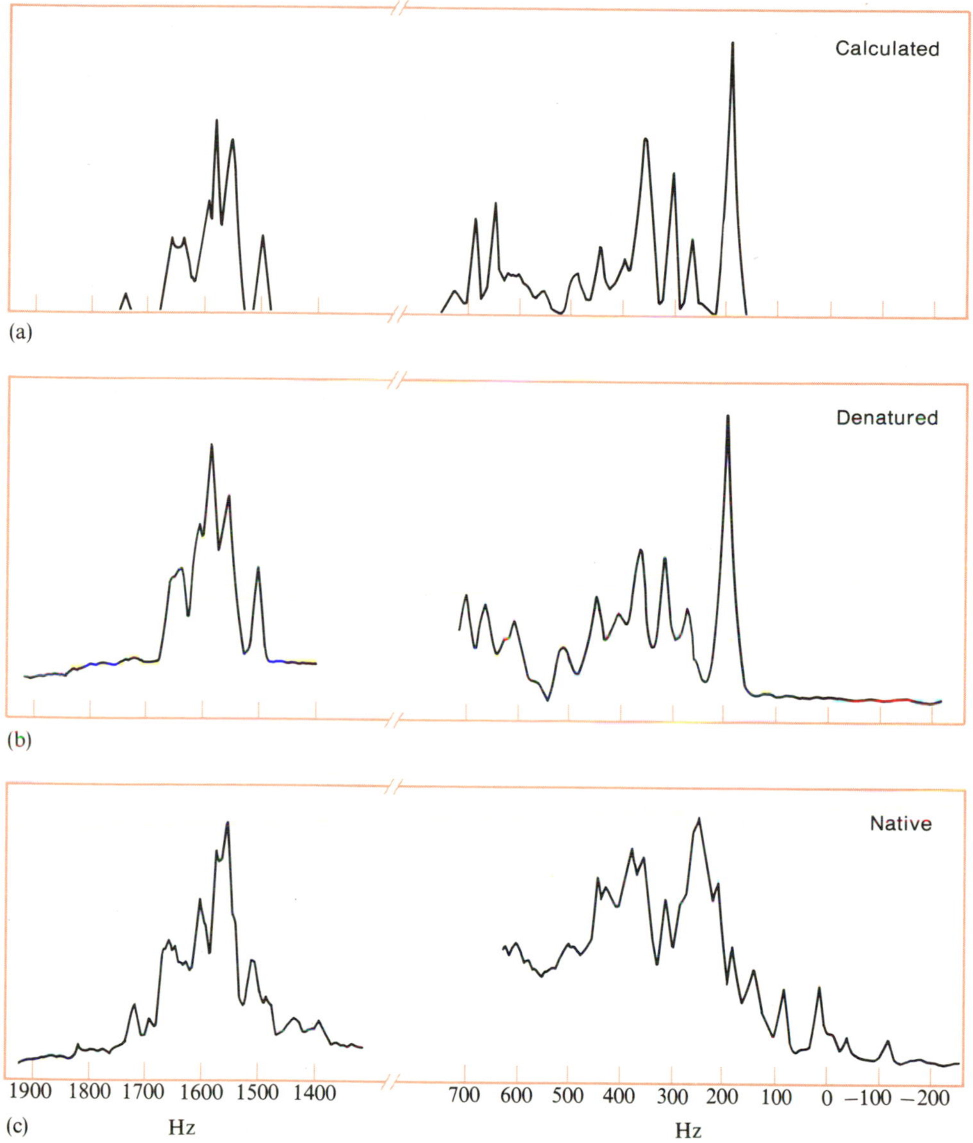

Figure 9-20

Spectrum of lysozyme in D_2O relative to a DSS standard. **(a)** Calculated spectrum, based on amino acid composition. **(b)** Observed 220 MHz spectrum of denatured lysozyme. **(c)** Observed 220 MHz spectrum of native lysozyme. [After C. C. McDonald and W. D. Phillips, in *Fine Structure of Proteins and Nucleic Acids*, ed. G. D. Fasman and S. N. Timasheff (New York: Dekker, 1970), p. 1.]

$-1{,}800$ Hz, relative to a DSS standard. The $-CH_2$ and $-CH_3$ resonances are in the range of -150 Hz to -750 Hz ($\sim$0.7 to 3.4 ppm), whereas aromatic, indole, and imidazole protons bonded to carbon fall in the range of $-1{,}500$ Hz to $-1{,}800$ Hz ($\sim$6.8 to 8.2 ppm).

Although the spectrum of the random coil can be rationalized simply in terms of its constituent amino acids, the native form typically shows a significantly more complex spectrum (Fig. 9-20c). Note that lysozyme in the native state has resonances in upfield positions near and above -100 Hz; these lines are absent in the denatured form. In addition, the downfield part of the spectrum shows many differences in resonance positions and intensities from the spectrum of the random coil.

Clearly, the marked change in the spectrum results from alterations in the magnetic environments around many of the nuclei in the native structure. The spectrum of the native form contains a wealth of potential structural information, but this information is difficult to extract. In certain cases, it has been possible to identify particular residues in the spectrum and to monitor their behavior under various conditions. A good example of this is provided by histidine side chains, for which the C-2 proton resonances can often be monitored because they have the largest chemical shift relative to a DSS standard. (In free histidine, the C-2 proton resonates at 7.91 ppm, or about $-1{,}740$ Hz in a 220 MHz spectrum. A small, well-separated peak, corresponding to the lone histidine in lysozyme, is evident in the simulated spectrum of Fig. 9-20a.)

Ribonuclease A contains four histidine residues. (See Chapter 16 for a discussion of this enzyme.) In the native protein, all four residues can be resolved in a 100 MHz spectrum, because each has a unique magnetic environment. Figure 9-21 shows the four histidine C-2 proton resonances, numbered 1 through 4. By special experimental

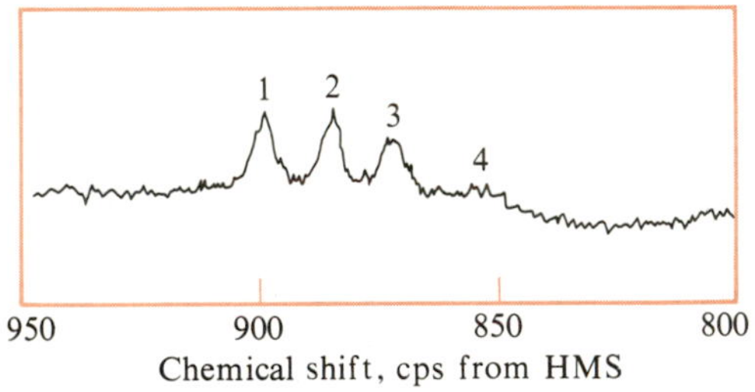

Figure 9-21
The 100 MHz spectrum of the histidine absorption region of ribonuclease A in D_2O. [After D. H. Meadows and O. Jardetzky, *Proc. Natl. Acad. Sci. USA* 61:406 (1968).]

approaches and arguments, it has been possible to assign all four resonances to specific residues in the sequence. As discussed in Chapter 16, this information has provided very valuable insight into the mechanism of the enzyme, because two of the residues (His^{12} and His^{119}) are intimately involved in catalysis.

Carboxymethylation of either His^{12} or His^{119} simultaneously shifts peaks 2 and 3 without altering peaks 1 and 4. Thus, His^{12} and His^{119} must correspond to peaks 2 and 3. Because both peaks are affected when only one of the histidines is chemically modified, it is likely that the two residues are spatially close in the structure. (From other evidence, this is known to be the case.) To resolve which residue is peak 2, advantage was taken of special hydrogen-exchange procedures (described in Chapter 16).

The remaining two histidines are His^{48} and His^{105}. His^{48} is known to be buried in the enzyme structure, so peak 4 (with its anomalous position and broader line) is probably due to this residue. By the process of elimination, therefore, peak 1 is assigned to His^{105}. These assignments also are consistent with the hydrogen-exchange studies described in Chapter 16.

Of course, each protein is a unique problem, and the procedures used to assign specific resonances to particular residues will vary from case to case. Generally, if the protein is small and its sequence and active-site residues are known, one has a good chance of being able to assign one or more of the resonances that are distinct and are not buried among other lines. Also, it is always possible to simplify a complex spectrum by selectively deuterating certain amino acids. This deuteration can sometimes be accomplished in the case of bacterial proteins by isolating them from bacteria grown in a medium containing selected [^{2}H]-amino acids.

^{13}C NMR spectra of proteins

The natural abundance of ^{13}C is only about 1.1% (see Table 9-1). Furthermore, its sensitivity is only 1.6% that of ^{1}H (for equal numbers of nuclei in the same field). Thus it is clear that the signal strength for ^{13}C nuclei is around 10^4-fold weaker than that for protons. Nevertheless, the advantages of ^{13}C NMR are sufficient that effort is expended with signal-enhancing procedures in order to make observation of the ^{13}C resonances feasible.

The advantages of ^{13}C NMR stem from several sources. First, the chemical shifts of ^{13}C in most cases occur over a much broader range than those of ^{1}H. For example, relative to a CS_2 standard, the peptide carbonyl carbon is shifted upfield 16 to 26 ppm, α-carbons tend to fall upfield in the range of 130 to 150 ppm, and the δ-carbon of isoleucine is upfield around 180 ppm. Even among the α-carbons, there is a sizable range. For example, the proline C^α is around 132 ppm, the alanine C^α is around 142 ppm, and the glycine C^α is around 150 ppm. This large range of chemical shifts contrasts with those observed for amino acid protons, where most resonances are within 8 ppm of one another. This wide variation in ^{13}C chemical shifts permits separate visualization of many individual nuclei on various amino acids. Another advantage of ^{13}C NMR (using specimens with natural abundance) is that ^{13}C–^{13}C spin–spin splitting is eliminated, because most nuclei adjacent to a ^{13}C nucleus are nonmagnetic ^{12}C. This effect obviously simplifies the spectrum.

Finally, ^{13}C is sufficiently like ^{12}C that it can be inserted into a molecule at a specific locus to replace ^{12}C. In this way, a site-specific magnetic probe is obtained that does not perturb the structure, as might a more bulky group.

The disadvantage of the low natural abundance of ^{13}C can be offset by preparing samples enriched in this isotope. However, there is another disadvantage to ^{13}C NMR that does offset some of its potential usefulness, even in cases where good spectra can be obtained. Although the ^{13}C chemical shifts for various carbons in amino acid side chains are spread over a wider range than ^{1}H chemical shifts, the ^{13}C shifts are much less sensitive to environment than are ^{1}H shifts. As a result, for example, there is less difference between the ^{13}C spectra of native and denatured proteins than between the corresponding proton magnetic resonance spectra. Therefore, it is more difficult to explore conformational changes, effects of ligand interactions, and so forth. Despite this drawback, there are situations where ^{13}C NMR offers clear advantages. For example, if one wishes to trace the course of a particular carbon fragment in a biochemical pathway, it is convenient to use ^{13}C with observation by NMR.

Some interesting studies of ^{13}C NMR spectroscopy of proteins have appeared. For example, M. W. Hunkapillar et al. (1972) have used selective ^{13}C enrichment of the C-2 of the single histidine in the serine protease α-lytic protease to study the catalytic mechanism. This histidine is a crucial residue because it is part of the Asp–His–Ser triad that is an essential part of the catalytic machinery of the serine proteases.

In another vein, using natural-abundance ^{13}C NMR, K. Ugurbil et al. (1977) have been able to resolve (and to make assignments to) many of the individual carbon centers in the small copper-containing protein azurin. [The interested reader should consult Lévy (1976) for other pertinent applications.]

^{31}P NMR studies

The sensitivity of ^{31}P is only a fraction of that of ^{1}H or ^{19}F (Table 9-1). However, this shortcoming is somewhat offset by the relatively large chemical shifts of ^{31}P, which can extend over a range of 500 ppm. Furthermore, in some biological materials, ^{31}P is usually present in smaller proportions than ^{1}H or ^{13}C. Therefore, fewer resonance lines are observed, thus making assignments somewhat easier.

Several applications of ^{31}P NMR have been made, and many other important studies with this nucleus will be done in the future. One particularly attractive application is the use of ^{31}P NMR to study some phosphate compounds within cells. Such studies are possible because some phosphate compounds (such as ATP) can in certain cases have intracellular concentrations approaching 1 mM or more.

One example of such studies is the work of R. B. Moon and J. H. Richards (1973) on the ^{31}P NMR of 2,3-diphosphoglycerate (DPG) in whole rabbit blood and in hemolysates. Both phosphates of DPG are easily resolved. Because of the sensitivity of the P-2 and P-3 chemical shifts to the states of ionization of the corresponding phosphates, it is possible to obtain a measurement of the intracellular pH from an NMR experiment. And with ^{31}P NMR, Ugurbil et al. (1978) have been able to study glycolysis and some of the bioenergetics of anerobic *E. coli* cells.

^{19}F as a probe for biochemical systems

For a given magnetic field strength, fluorine's sensitivity approaches that of ^{1}H (Table 9-1). Moreover, the fluorine coupling constants and chemical shifts are much larger (10-fold or more) than those of ^{1}H. And ^{19}F has a relatively small size. For these reasons, ^{19}F is ideally suited as a site-specific probe. An illustration of the usefulness and sensitivity of a ^{19}F probe is provided by studies of W. H. Huestis and M. A. Raftery on RNase S,[§] in which Lys^1 and Lys^7 of the S-peptide were reacted to give trifluoroacetyl (TFA) derivatives. This treatment produced a mixture in which the ε-amino groups of Lys^1 and of Lys^7 were attached to TFA, together with some molecules in which the α-amino of Lys^1 also was reacted.

Figure 9-22 shows the spectrum of the TFA-derivatized S-peptide. Three lines

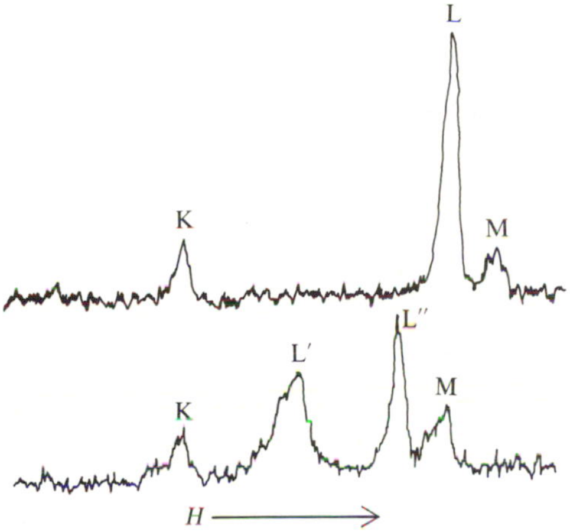

Figure 9-22
The ^{19}F *NMR spectrum of TFA-derivatized S-peptide* (top) *and of S-peptide associated with S-protein* (bottom). [After W. H. Huestis and M. A. Raftery, *Biochemistry* 10:1181 (1971).]

(K, L, and M) are evident. From model compounds, line L is assigned to the ε-amino TFA of Lys^7 and mono-TFA Lys^1, whereas K and M are assigned to the TFA groups on the α- and ε-aminos of the doubly acetylated Lys^1. In RNase S, line L splits into two lines (L′ and L″). Further investigation showed that L′ corresponds to the TFA on the ε-amino of Lys^7, whereas L″ corresponds to the TFA on the ε-amino of the monosubstituted Lys^1. Thus, in the reconstituted protein, the ^{19}F magnetic resonance is sufficiently sensitive to discriminate between two ε-aminos that are separated by only a few residues in the sequence. We should also mention that the ^{19}F-labeled enzyme retains activity, indicating that the probe has not seriously upset the structure.

[§] Ribonuclease S (RNase S) is a modified form of ribonuclease A in which the peptide bond between residues 20 and 21 is broken. The two resulting polypeptides (residues 1–20 and 21–124) are held together by noncovalent forces. The 20-residue moiety is called S-peptide; the longer section is called S-protein. See Chapter 16 for more details.

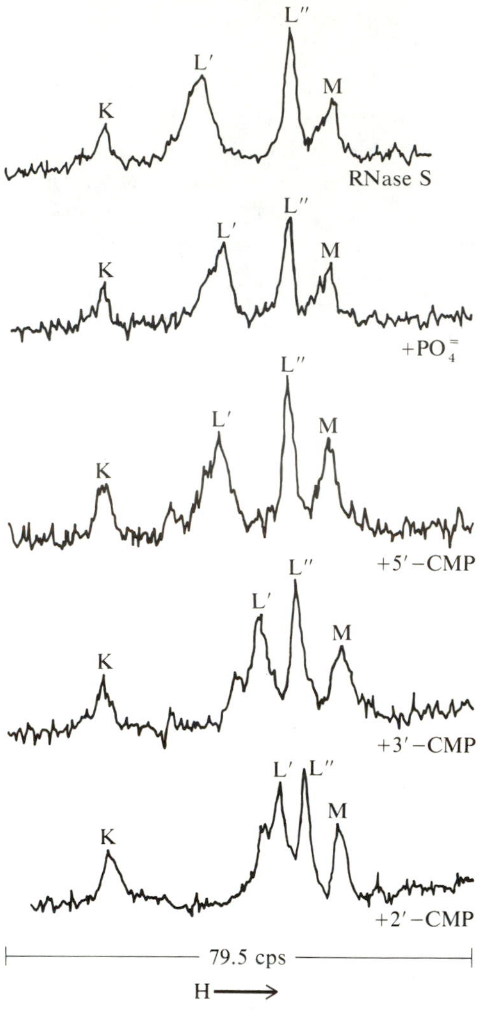

Figure 9-23

The ^{19}F *NMR spectra of TFA-derivatized ribonuclease S in the absence* (top) *and presence of various inhibitors.* [After W. H. Huestis and M. A. Raftery, *Biochemistry* 10:1181 (1971).]

Figure 9-23 shows the effect on the ^{19}F spectrum of adding inhibitors of the labeled RNase S. The line L′ clearly is shifted to higher field and, in certain instances, is split into two components. We will not here pursue the analysis and interpretation of these results; we simply observe that, in this example as in others that have been investigated, ^{19}F is a sensitive and therefore valuable probe.

NMR spectra of nucleic acids

At the mononucleotide and oligonucleotide levels, proton magnetic resonance has provided much information about nucleic acids—information on base stacking, base pairing, ribose–base orientation, and so on. However, in high-molecular-weight chains, the situation is complicated by the fact that nucleic acids are predominantly composed of four nucleotide units, each of which is represented many times in the sequence. Consequently, in the spectrum, a given area of absorption will represent a superposition of the absorption of a particular kind of base from various parts of the structure. In addition, there is the usual problem of serious line broadening that is encountered with high-molecular-weight rigid macromolecules.

On purine and pyrimidine bases, there are two nonexchangeable protons that can be monitored. These are H-2 and H-8 in the case of purines, and H-5 and H-6 in the case of pyrimidines. In addition, there are the protons on the exocyclic methyl group of thymidine.

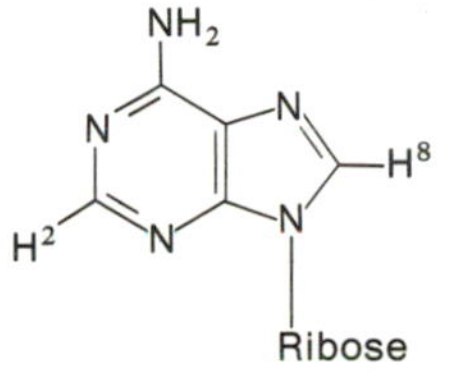

Adenosine

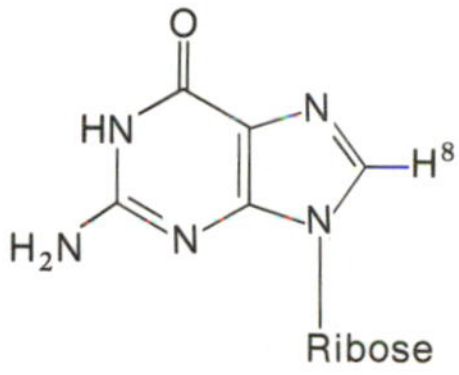

Guanosine

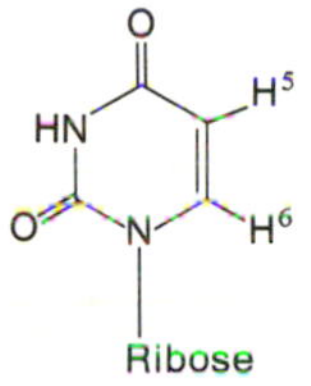

Uridine

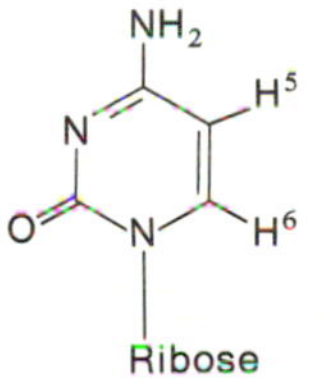

Cytidine

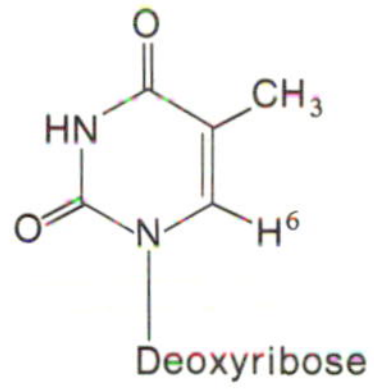

Thymidine

In the free mononucleotides, the purine H-2 and H-8 protons in D_2O are about 8.4 to 9.0 ppm from a TMS standard; pyrimidine H-5 is around 6.3 to 6.6 ppm, and H-6 is at 8.0 to 8.5 ppm; the thymidine methyl protons are around 2.3 to 2.4 ppm. Thus, the nonexchangeable protons fall roughly in the range of 2 to 9 ppm.

Because they exchange, the protons that participate in Watson–Crick hydrogen bonds cannot be observed in D_2O. However, in nonaqueous solvents, it is possible to detect these protons. Figure 9-24 shows the 60 MHz spectrum in dimethylsulfoxide of 1-methylcytosine and of 9-ethylguanine, and of a 1:1 mixture of these two bases. In the case of 1-methylcytosine, the exocyclic amino-group protons are clearly evident; H-5 and H-6 protons each are split into doublets by spin–spin splitting. In the case of 9-ethylguanine, the NH and NH_2 protons are clearly resolved along with H-8. The 1:1 mixture of bases shows pronounced downfield shifts of the 1-methylcytosine NH_2 and of the 9-ethylguanine NH and NH_2. These spectra show good evidence of hydrogen bonding between the bases in solution, with the Watson–Crick geometry. Other studies in nonaqueous solutions have shown A–T pairing and an absence of bonding between pairs such as G–T and A–G.

Although it is difficult to obtain incisive new information from NMR studies of high-molecular-weight nucleic acids in aqueous solution, transfer RNA is a bio-

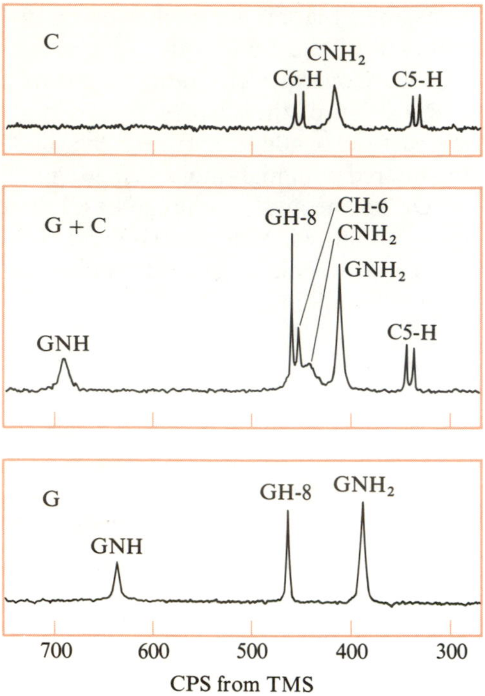

Figure 9-24

The 60 MHz NMR spectra of separated and mixed 9-ethylguanine (G) *and 1-methylcytosine* (C) *in dimethylsulfoxide.* [After R. R. Shoup et al., *Biochem. Biophys. Res. Commun.* 23:194 (1966).]

logically interesting nucleic acid that has been studied fruitfully by NMR methods. One reason for success with this species is its relatively small size (70 to 80 nucleotides). A second reason is that this species proved to provide useful information from spectra run in H_2O rather than the usual D_2O. Although the huge H_2O absorption typically obscures critical parts of a proton magnetic resonance spectrum, in this instance the base-paired hydrogens are shifted downfield sufficiently far that they can be clearly observed.

Figure 9-25 shows the spectrum of yeast $tRNA^{Phe}$ at 35°C. This spectrum is from

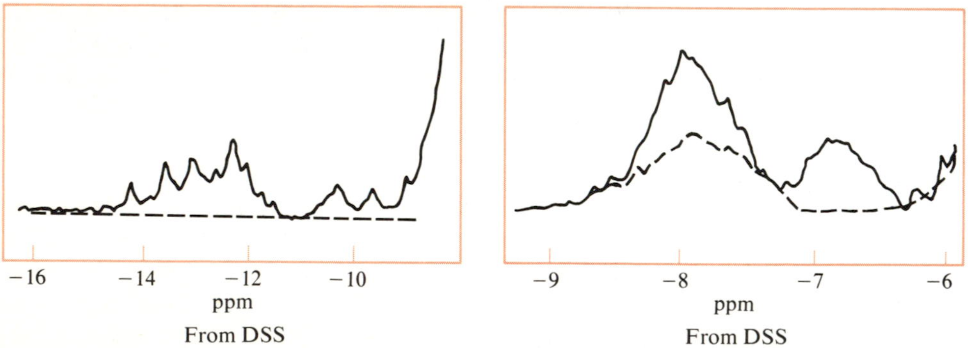

Figure 9-25

The 220 MHz proton NMR spectrum of yeast $tRNA^{Phe}$ at 35°C. The solid line shows the spectrum in H_2O solvent, the dashed line in D_2O solvent. [After Y. P. Wong et al., *J. Mol. Biol.* 72:725 (1972).]

the work of D. R. Kearns, B. R. Reid, and R. G. Shulman—investigators who have been extremely active in the tRNA NMR field. Spectra for both H_2O and D_2O solvents are shown. In H_2O, the region of -11.5 to -15 ppm (DSS standard) arises from the Watson–Crick base-paired hydrogens. This region is completely absent in D_2O, thus demonstrating the exchangeable nature of these paired hydrogens. By systematically examining the low-field region, it has been possible to investigate many aspects of tRNA structure in solution (see Chapter 24).

9-7 ELECTRON PARAMAGNETIC RESONANCE (EPR)

Similarities between EPR and NMR

Electron paramagnetic resonance (EPR)—also called electron spin resonance (ESR)—has assumed an increasingly prominent position in biophysical chemistry during the past decade. The phenomenon is based on the magnetic moment of an unpaired, spinning free electron. In a magnetic field, the unpaired electron—which has a spin quantum number $S = 1/2$ ($m_s = \pm 1/2$)—precesses about the field axis (z axis) with a component of its spin angular momentum either parallel ($m_s = +1/2$) or antiparallel ($m_s = -1/2$) to the z axis. An oscillating magnetic field at right angles to the field axis induces transitions between the two spin states when the frequency of the field is at or near the Larmor frequency of the precessing electron.

From the discussion of Equations 9-1 to 9-11, we know that the angular precessional frequency is $\omega = -\gamma H$, where γ is the magnetogyric ratio. The parameter γ is given by $\gamma = -g\beta_e/\hbar$ (see Equations 9-8 and 9-10, and accompanying discussion). We have already seen that the nuclear magneton (β_n) is over three orders of magnitude smaller than the Bohr magneton (β_e) of the electron. The value of g for the free electron is 2.00232. Thus, for a free electron, $\gamma = -1.7 \times 10^7$ rad gauss^{-1} sec^{-1}, which in absolute value clearly is 10^3-fold or so larger than the absolute values of γ for nuclei (Table 9-1). This means that, in a given magnetic field, the resonance frequency of an unpaired electron typically is 10^3-fold higher than that of the commonly studied nuclei. Therefore, although NMR experiments are carried out in the MHz range, EPR experiments require frequencies in the GHz range (1 GHz $= 10^9$ cycles sec^{-1} $= 10^3$ MHz).

Although there is a difference in the frequency ranges of NMR and EPR, the equations derived earlier for NMR (such as the Bloch equations) and the related discussions apply also to EPR. Thus, factors such as longitudinal (T_1) and transverse (T_2) relaxation times play a key role in determining the nature of an EPR signal. However, there are certain features of EPR that deserve special consideration, and we now turn to these.

Hyperfine interaction

Consider an unpaired electron in an external magnetic field **H**. The environment of the electron is such that local permanent fields $\Delta\mathbf{H}_{loc}$ combine with **H** to give an

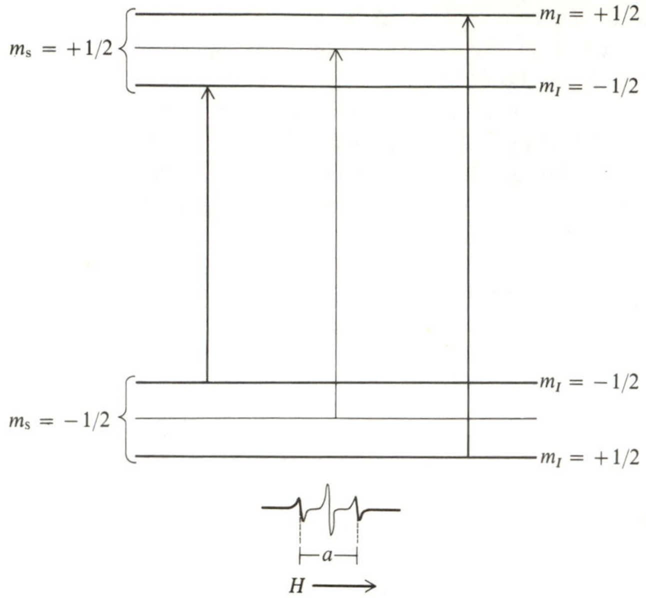

Figure 9-26

Electron spin transitions for the case $I = 1/2$ (dark lines); the lighter line displays the situation for the case $a = 0$. The resulting EPR spectrum is shown at the bottom of the figure.

effective field $\mathbf{H}_{\text{eff}}$. These local fields usually are from the magnetic moments of nuclei. In a case where an electron is essentially localized around one nucleus of spin quantum number I, the value of $\Delta\mathbf{H}_{\text{loc}}$ along the external field axis can take on $2I + 1$ values, corresponding to the $2I + 1$ values of the nuclear spin quantum m_I. Therefore, resonance occurs at $2I + 1$ values of H_{res}, which can be expressed as

$$H_{\text{res}} = H^0_{\text{res}} - am_I \qquad \text{for } m_I = -I, -I + 1, \ldots, I - 1, I \tag{9-46}$$

where a is a constant such that aI is the *magnitude* of the local field, and H^0_{res} is the field strength at which resonance would occur if $a = 0$.

As a particular example, consider the case of a hydrogen atom where $I = 1/2$. In this case, resonance for the unpaired electron occurs at $H^0_{\text{res}} - a/2$ and $H^0_{\text{res}} + a/2$. Figure 9-26 shows the transitions that occur. If $a = 0$, there is a single transition; if a is finite, there are two transitions. The transitions involve a change in the electron spin quantum number ($m_s = \pm 1/2$; $|\Delta m_s| = 1$) at fixed m_I. That is, electron spin transitions effectively occur at fixed nuclear orientations. In the lower part of Figure 9-26, the EPR spectrum is displayed.[§] Two lines are observed, with a separation

[§] It is customary to report EPR spectra as the first derivative of the absorption spectrum. Information extracted from EPR spectra often is obtained by a detailed line-shape analysis and an accurate measurement of hyperfine splitting constants. This kind of analysis is facilitated with first-derivative spectra.

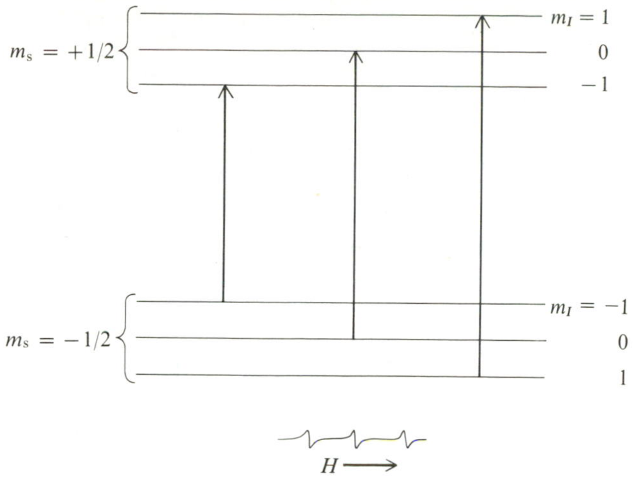

Figure 9-27

Electron spin transitions for the case $I = 1$. The resulting EPR spectrum is shown at the bottom of the figure.

of a. The parameter a is called the *hyperfine splitting constant*, and is expressed in units of gauss. Usually it is much less than the values of the applied field strength (e.g., tens to hundreds of gauss compared to external field strengths of thousands of gauss).

As a further example, Figure 9-27 shows a case of $I = 1$ (as in nitrogen). In this instance, three lines are observed, corresponding to the three values of m_I. In the more general case, the interaction of n equivalent nuclei with a single paramagnetic electron gives $2nI + 1$ lines. Therefore, if 10 equivalent protons interact with an electron, 11 lines are expected. In such a complex situation, the individual hyperfine lines may not be resolved. Instead, a single broad line is observed. This phenomenon is known as inhomogeneous line broadening, and it is encountered in the EPR spectra of molecules in which an unpaired electron is extensively delocalized (e.g., in some aromatic compounds).

Contact interaction

There are two types of hyperfine interactions that give rise to the hyperfine line splittings just discussed. One of these is called *contact* or *Fermi interaction*. This is a quantum mechanical interaction that depends on the probability of the electron being located at the nucleus. This interaction is isotropic (orientation independent) and therefore is sometimes called the isotropic hyperfine interaction. From elementary quantum mechanics, we know that the probability of the electron being located at the nucleus is proportional to the square of the electronic wave function evaluated at the nucleus. This probability is nonzero for electrons in s orbitals. However, the wave functions for electrons in p, d, f, and higher orbitals have a node at the nucleus, so that the probability vanishes at that point, and no contact interaction can take place.

The second type of hyperfine interaction is orientation dependent; it is known as the *anisotropic hyperfine interaction.* It can be explained on the basis of classical physics and is not restricted solely to electrons in *s* orbitals. For a system of rapidly tumbling molecules, the anisotropic contribution averages to zero, leaving only the isotropic (contact) interaction. We shall consider first the situation in which only the contact interaction is of consequence, and then we shall turn to the anisotropic interaction in the next subsection.

For a hydrogen atom, the contact interaction energy can be written simply as

$$\text{Contact energy} = hAm_s m_I \tag{9-47}$$

where m_s and m_I are electron and nuclear spin quantum numbers, respectively, and A is a frequency that can be calculated from the value of the square of the wave function evaluated at the nucleus.[§] For a system of one unpaired electron and a nucleus with $S = 1/2$ and $I = 1/2$, there are four different contact energies, representing the four combinations of $m_s = \pm 1/2$ with $m_I = \pm 1/2$. The energy of just the paramagnetic electron in the magnetic field is given by

$$\begin{aligned} E &= -\boldsymbol{\mu}_m \cdot \mathbf{H} \\ &= -(1/2)g\beta_e H \qquad \text{for } m_s = -1/2 \\ &= +(1/2)g\beta_e H \qquad \text{for } m_s = +1/2 \end{aligned} \tag{9-48}$$

(Compare Equation 9-48 with Equation 9-16, and recall that $\gamma = -g\beta_e/\hbar$.) In a strong magnetic field, these energies are 10^3-fold or more higher than the contact interaction energies. For each state ($m_s = +1/2$ or $m_s = -1/2$), there are two possibilities for the nuclear spin, so the energies of the four levels can be written (in order of decreasing energy) as

$$\begin{aligned} E &= +(1/2)g\beta_e H + hA/4 \qquad \text{for } m_s = +1/2 \text{ and } m_I = +1/2 \\ &= +(1/2)g\beta_e H - hA/4 \qquad \text{for } m_s = +1/2 \text{ and } m_I = -1/2 \\ &= -(1/2)g\beta_e H + hA/4 \qquad \text{for } m_s = -1/2 \text{ and } m_I = -1/2 \\ &= -(1/2)g\beta_e H - hA/4 \qquad \text{for } m_s = -1/2 \text{ and } m_I = +1/2 \end{aligned} \tag{9-49}$$

As stated earlier, when an electron absorbs energy and flips from one spin state to the other, it does so with no change in m_I. (Of course, m_I will change if irradiation is done with an oscillating field at the characteristic frequency of the nucleus.) There-

[§] Equation 9-47 represents the contact energy term H_{con} in the Hamiltonian, at strong magnetic fields. More generally, it can be written as $H_{con} = h A \mathbf{S} \cdot \mathbf{I}$, where **S** and **I** are electron and nuclear spin operators. This relationship can be shown to reduce to Equation 9-47 at sufficiently strong fields (see Swartz et al., 1972).

fore, two transitions occur. These transitions give rise to two lines, which are separated in frequency by A. The parameter A is the hyperfine coupling constant. It is easy to see from Equation 9-49 that the energy change ΔE for the two transitions is

$$\begin{aligned} \Delta E &= g\beta_e H + (1/2)hA \qquad \text{for } m_I = +1/2 \\ &= g\beta_e H - (1/2)hA \qquad \text{for } m_I = -1/2 \end{aligned} \tag{9-50}$$

In an EPR experiment, the frequency ν_0 usually is fixed, and the field strength is varied until resonance is achieved where $\Delta E = h\nu_0$ and $H = H_{res}$. Substituting these relationships in Equation 9-50, we obtain

$$H_{res} = (h\nu_0/g\beta_e) \pm (1/2)(h/g\beta_e)A \tag{9-51}$$

Thus, resonance occurs at two different field strengths. Comparing Equations 9-51 and 9-46 for the case $I = 1/2$, it is clear that $h\nu_0/g\beta_e = H_{res}^0$, and $a = (h/g\beta_e)A$.

Anisotropic hyperfine interaction

As mentioned earlier, the anisotropic interaction can be calculated from classical theory. Consider a nuclear magnetic moment $\boldsymbol{\mu}_{mn}$ separated a distance $\mathbf{r}$ from a paramagnetic electron, where $\mathbf{r}$ makes the angle θ with the field axis or z axis (Fig. 9-28).

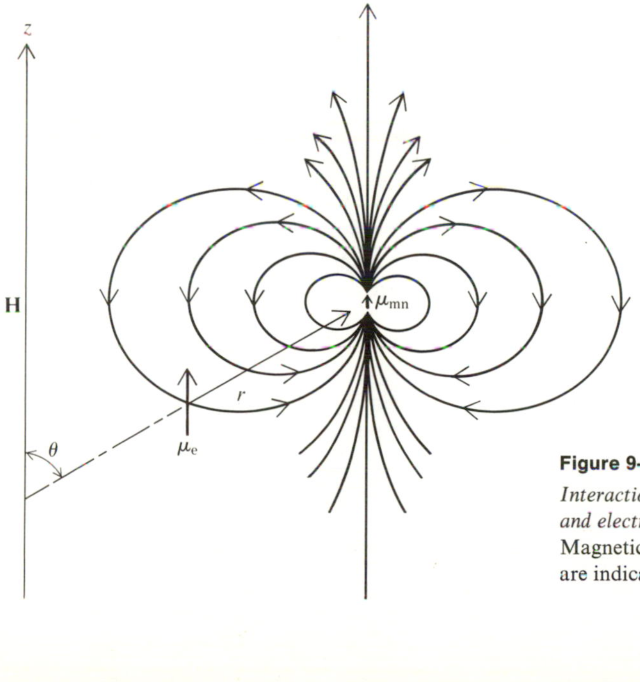

Figure 9-28

Interaction of nuclear magnetic dipole $\boldsymbol{\mu}_{mn}$ *and electron magnetic moment* $\boldsymbol{\mu}_e$. Magnetic flux lines associated with $\boldsymbol{\mu}_{mn}$ are indicated.

Box 9-3 shows that the z component of the local field $\mathbf{\Delta H'_{loc}}$ generated by $\boldsymbol{\mu}_{mn}$ is given approximately by

$$\Delta H'_{loc} = (\mu_{mnz}/r^3)(3\cos^2\theta - 1) \tag{9-52}$$

where μ_{mnz} is the z component of $\boldsymbol{\mu}_{mn}$, and μ_{mnz} can take on $2I + 1$ values. In the absence of an isotropic hyperfine interaction, the hyperfine splitting is determined entirely by the anisotropic contribution so that, for example, $\Delta H'_{loc}$ in Equation 9-52 corresponds to the hyperfine splitting am_I of Equation 9-46.

For the case $I = 1$, three hyperfine lines will be observed, corresponding to the three values of μ_{mnz}. However, the width of a given line will be sensitive to the distribution of values of θ that are adopted. If the paramagnetic species is fixed in a few rigid positions, Equation 9-52 indicates that each of these is associated with a local field ($\mathbf{\Delta H'_{loc}}$) characteristic of the particular angle of orientation (θ). As a result, resonance occurs over a wide range of applied field, as different net fields are experienced by the molecule in each of its various orientations. In an extreme case, a paramagnetic species in a rigid glass or powder experiences all possible orientations and their associated local fields $\mathbf{\Delta H'_{loc}}$. The result is a very broad EPR line. On the other hand, if there is fast tumbling of the molecule, so that all possible values of θ are rapidly generated on a random basis, then $\cos^2\theta$ averages to 1/3, and $\Delta H'_{loc} = 0$. In this case, there is no anisotropic hyperfine interaction, and only the spectral features of the isotropic part of the hyperfine interaction are observed.

To illustrate these ideas, Figure 9-29 shows the spectrum of the di-*t*-butyl nitroxide radical under two different conditions.

Box 9-3 MAGNETIC FIELD NEAR A MAGNETIC DIPOLE

In Box 5-1, we derived an approximate expression for the electric field $\mathbf{E}_A$ at a distance r from an electric dipole $\boldsymbol{\mu}_A$. The z component of this field is (with $\mu_{Ax} = \mu_{Ay} = 0$)

$$[(3z^2/r^5) - (1/r^3)](\mu_{Az}/\varepsilon)$$

where μ_{Az} is the z component of $\boldsymbol{\mu}_A$. By exact analogy, the z component of the magnetic field at a distance r from a magnetic dipole $\mathbf{u}_{mn}$ is approximately (with $\mu_{mnx} = \mu_{mny} = 0$)

$$[(3z^2/r^5) - (1/r^3)]\mu_{mnz}$$

where μ_{mnz} is the z component of $\boldsymbol{\mu}_{mn}$. Because $z^2/r^2 = \cos^2\theta$—where θ is the angle between the z axis and r (see Fig. 9-28)—it follows that the z component of the magnetic field of the dipole $\boldsymbol{\mu}_{mn}$ is given by Equation 9-52.

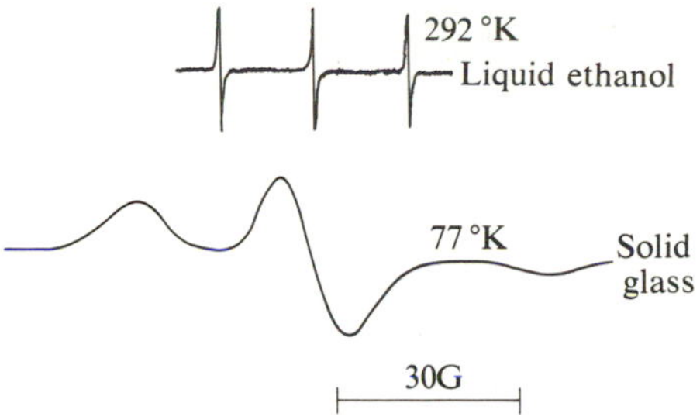

Figure 9-29

*Spectrum of di-*t*-butyl nitroxide radical* in liquid ethanol and in solid glass. [After J. R. Bolton, in *Biological Applications of Electron Spin Resonance*, ed. H. M. Swartz, J. R. Bolton, and D. C. Borg (New York: Wiley–Interscience, 1972), p. 37.]

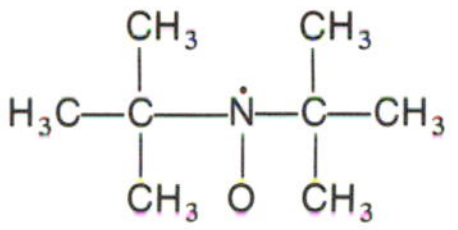

di-*t*-butyl nitroxide

In the top spectrum, the radical is in liquid ethanol at 292°K, so that rapid tumbling occurs and the anisotropic hyperfine interaction averages to zero; this leaves three lines, which come from the isotropic hyperfine interaction of the nitrogen nucleus with the unpaired electron. The lower spectrum is for the same radical in a glass at 77°K. The lines are broadened considerably and poorly resolved, and they are shifted in position somewhat from those of the first spectrum. Clearly, in the glass, the anisotropic contribution to the hyperfine interaction is large and substantially alters the spectrum.

The preceding discussion illustrates the essential idea of the anisotropic hyperfine interaction. The local field generated by the anisotropic interaction can be expressed in terms of a splitting constant a, such as that introduced in Equation 9-46 in connection with the discussion of the isotropic hyperfine interaction. However, in the usual situations of practical interest, the anisotropic hyperfine interaction must be described in terms of a tensor. It often is possible to find an axis system (the principal axis system) such that the tensor representation of the anisotropic hyperfine interaction boils down to a description in terms of three splitting constants: a_x, a_y, and a_z. These constants refer to the anisotropic coupling along the three principal axes (x, y, and z). In radicals that are axially symmetric, the principal x and y axes are equivalent, so that $a_x = a_y \neq a_z$. In this case, there are two anisotropic hyperfine splitting constants $a_{||}$ and $a_{\perp}$, which are given by $\Delta H'_{\text{loc}}$ in Equation 9-52 (with $\theta = 0°$ and $\theta = 90°$, respectively).

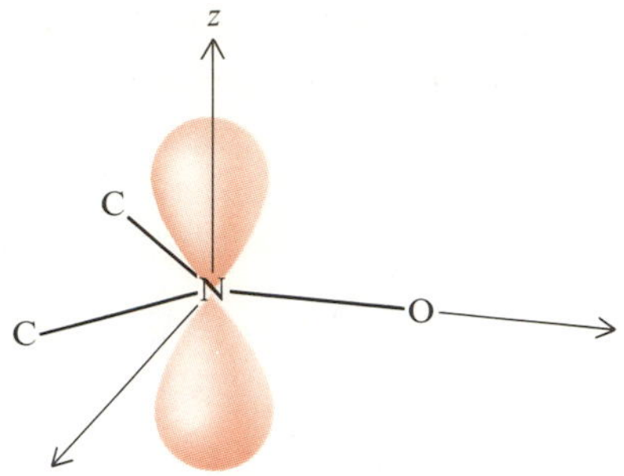

Figure 9-30

Structure of the nitroxide radical, with $2p\ \pi$ orbitals indicated by shading. The two carbons, the nitrogen, and the oxygen are assumed to be coplanar.

As a more detailed illustration of these ideas, consider the structure of the di-*t*-butyl nitroxide radical shown in Figure 9-30. The unpaired electron is largely located in the $2p\ \pi$ orbital that lies along the z axis. Because it is localized in this orbital, there is a strong anisotropic hyperfine interaction that is nearly axially symmetric. The axis system shown in this figure is the principal axis system, and along each axis there is a characteristic nuclear anisotropic hyperfine interaction parameter: a_x, a_y, and a_z. These parameters may be determined from measurements on oriented crystals, with the magnetic field successively pointed along each of the three principal axes of the molecule. Table 9-3 shows the results of such measurements, which clearly

Table 9-3
EPR parameters for the nitroxide radical

g_x	g_y	g_z	a_x	a_y	a_z	g_0	a_0
2.0089	2.0061	2.0027	7.1 ± 0.5	5.6 ± 0.5	32.0 ± 1.5	2.0060 ± 0.0002	15.1 ± 0.5

NOTE: The experimental error for the g values is ± 0.0003. The a values are given in gauss. The values along the axes x, y, and z were measured in single crystals; the values of g_0 and a_0 were measured in solution.
SOURCE: After O. H. Griffith, D. W. Cornell, and H. M. McConnell, *J. Chem. Phys.* 43:2909 (1965).

demonstrate the orientation dependence of these parameters and the especially large differences between values for a along the z axis and those along the x and y axes. In addition, the table gives the value for a_0, the isotropic hyperfine splitting constant. This is the value of the coupling constant for the same radical in solution, where rapid tumbling averages the anisotropic contribution to zero to give

$$a_0 = (1/3)(a_x + a_y + a_z)$$

The experimental values of a_0 in Table 9-3 agree well with the values calculated from this equation using the experimental values of a_x, a_y, and a_z.[§]

g-Factor anisotropy

The energy of an unpaired electron in a magnetic field depends on the parameter g. As mentioned earlier, this parameter is equal to 2.00232 for a free electron. However, local internal fields induced by the external field can cause the value of g for an unpaired electron in a molecule to vary from the value for the free electron. Also, the value of g generally varies with respect to the orientation of the molecule relative to the magnetic field. In general, like the nuclear hyperfine interaction, g must be expressed as a tensor. However, in the principal axis system, only three parameters (g_x, g_y, and g_z) are required to describe the g-factor anisotropy. These correspond to the g factors that are obtained when a magnetic field is successively oriented along the x, y, and z axes of an oriented crystal. In a molecule with true axial symmetry, $g_x = g_y \neq g_z$. In this case, there are only two g factors necessary to account for the anisotropy. These are designated $g_\perp$ (corresponding to g_x and g_y) and $g_\parallel$ (corresponding to g_z).

Table 9-3 also summarizes data obtained on the g factors from the same oriented crystals used to measure a_x, a_y, and a_z. Note that the variation in g factors along the three axes is considerably smaller than that observed in the nuclear anisotropic hyperfine coupling constants. The table also gives the measured value of g_0, the averaged g factor of a rapidly tumbling molecule. This parameter is given by

$$g_0 = (1/3)(g_x + g_y + g_z)$$

and the value of g_0 measured on the radical in solution agrees well with that calculated from the values of g_x, g_y, and g_z measured on the radical in oriented crystals.

EPR study of metal-containing proteins

Many biological molecules contain no unpaired electrons, but some proteins have paramagnetic metal ions that play an essential structure–function role. In these cases,

---

[§] We should note that the parameters a_x, a_y, and a_z determined on oriented crystals (Table 9-3) contain both isotropic and anisotropic contributions to the hyperfine splitting. Using Equation 9-52 to calculate the three *purely* anisotropic parameters, it is easy to show that the sum of these equals zero.

As stated earlier, for a p-orbital electron localized to a single nucleus, the square of the wavefunction vanishes at the nucleus, and there is no isotropic hyperfine interaction. But for an electron in the $2p\ \pi$ molecular orbital of the nitroxide radical, the wavefunction is a linear combination of nitrogen and oxygen atomic orbitals. The square of the oxygen atomic-orbital wavefunction vanishes at the oxygen nucleus but not at the nitrogen nucleus, and vice versa for the nitrogen atomic-orbital wavefunction. As a result, the square of the linear combination of atomic-orbital wavefunctions does not vanish at the nitrogen nucleus, and an isotropic hyperfine interaction occurs with this nucleus ($I = 1$). (NOTE: $I = 0$ for oxygen; see Table 9-1.)

EPR is a particularly attractive way to explore mechanistic features. In the proton NMR of a protein, resonances of a large number of nuclei are compressed into a relatively small region of the spectrum. In the EPR studies, however, only the EPR signal of the metal ion and its environment is monitored, thus avoiding the problem of the superpositioning of many resonances. In effect, the metal ion provides a built-in probe.

A considerable literature is building up on the applications of EPR to metal-containing proteins. As examples, the interested reader is referred to studies by Beinert on flavoproteins and iron–sulfur proteins, and by Vänngard on copper proteins (see Swartz et al., 1972); also see F. L. Bowden (1975) for coverage of some molybdenum-containing enzymes.

Spin-label EPR studies

Most biological macromolecules give no intrinsic EPR signal, because they have no unpaired electrons. Therefore, if EPR is to be used in studying these molecules, one or more radicals (known as spin labels) must be coupled to the molecule under investigation. The spin label thus is an extrinsic probe or reporter group. When such groups are introduced, care must be taken to perturb the host molecule as little as possible. The perturbing effect of the spin label can be checked by measuring the biological activity (e.g., catalytic activity in the case of an enzyme) and other properties with and without the attached probe. If these measurements indicate a negligible effect on the conformation and activity of the system under study, there is some assurance that the spin label will report information that reflects the state of the native system.

For most studies, use is made of spin labels containing the nitroxide radical discussed in conjunction with Figure 9-29. When inserted into the appropriate host, the spin label relays information on structure and dynamics. This information is available because of the sensitivity of the nuclear hyperfine splitting constant and of the g factor to the orientation of the radical with respect to the field. As we have discussed, the line separation and the field strength at which resonance occurs are determined by the splitting constant and g factor. An additional effect comes from the polar character of the nitroxide bond, which causes line-splitting and g-factor parameters to be influenced by the polarity of the environment in which they are situated. The dependence of the EPR signal on orientation and polarity is what enables a spin label to relay structural information about the area around the probe. As we have also discussed, anisotropic hyperfine splittings and g factors are sensitive to motion, so that EPR signals also give data on dynamics.

The spin-label method has been used in tackling many problems (see Berliner, 1976, 1978). Among the most fruitful studies have been those of structure–function relationships in phospholipid bilayers. These applications of the spin-label approach are discussed in some detail in Chapter 25.

Summary

Magnetic resonance spectroscopy is a powerful approach to problems in biophysical chemistry. In an external field, a spinning charge—such as a nucleus (e.g., a proton) or an unpaired electron—will precess about the field axis (z-axis) at a characteristic frequency that is proportional to the external field strength. The z components of the magnetic moments generated by spinning charges are quantized, and magnetic resonance is achieved when they undergo transitions between adjacent quantum levels. This is accomplished by varying the external field until the precessional frequency matches that of an oscillator, which can then exchange energy with the precessing charges.

Among other factors, longitudinal T_1 and transverse T_2 relaxation times play a major role in determining the character of nuclear magnetic resonance (NMR) spectra. These parameters are sensitive to molecular environment and molecular motion, and therefore can be used to obtain structural information and rate constants.

The local environment has a marked influence on the characteristic precessional frequency of a nucleus, so that individual nuclei within the same molecule often can be resolved in a spectrum. In addition, neighboring nuclei can interact through spin–spin interactions, so that the absorption line for a given nucleus is split into two or more lines. The magnitude of the splitting can sometimes be used to calculate internal rotation (dihedral) angles.

Because of their relatively high abundance and good sensitivity for detection, protons are the nuclei most commonly studied in NMR spectroscopy. In the case of biological polymers, proton magnetic resonance spectra are typically complex, owing to the presence of so many groups. However, with ingenuity, resonances sometimes can be assigned to specific residues, so that detailed information on a particular site within a biopolymer is obtained. For some applications, ^{13}C NMR is advantageous, especially because this method may yield better resolution than proton NMR studies, and may do so with fewer complications. In other instances, a ^{19}F probe can be introduced into a specific site as a sensitive indicator of structure and mechanism at a given locus.

The basic theory of electron paramagnetic resonance (EPR) is the same as that of NMR. Additional factors of importance are the nuclear hyperfine interaction and the g factor. The nuclear hyperfine interaction of an unpaired electron with a nucleus of spin quantum number I gives rise to $2I + 1$ lines in the EPR spectrum. This means, for example, that an unpaired electron associated with a nitrogen nucleus ($I = 1$) has a three-line spectrum. Both the hyperfine interaction and the g factor are sensitive to the orientation of the radical with respect to the field, to molecular motion, and to the polarity of the local environment. These factors in turn affect the characteristics of the EPR signal, so that it can be analyzed to obtain information on structure, motion, and polarity.

In many biological systems, there is no group bearing an unpaired electron.

This shortcoming is surmounted by attaching a site-specific probe that bears a radical. The most popular probes, or spin labels, are derivatives of the nitroxide radical. When these are introduced into a biological system, care must be taken that the native state of the system is not significantly perturbed. When such spin labels are successfully inserted, they serve as sensitive reporter groups.

Problems

9-1. Fill in the details necessary to derive the Bloch equations (Equation 9-25) from the relationships immediately preceding them in the text.

9-2. Using the steady-state condition of Equation 9-28, derive Equation 9-29.

9-3. A protein has four histidine residues. The C-2 protons of the four residues can be resolved in a 220 MHz NMR spectrum under conditions where the protein is in its native state. These resonances occur in the range of 8 to 9 ppm from a TMS standard. Addition of a reagent X (e.g., urea) in sufficient amounts is known to denature the protein; under these conditions, only a single C-2 proton peak (at 8.4 ppm) is observed, with area equal to the sum of the areas of the four peaks obtained in the native form. A solution containing native protein is titrated with increasing amounts of the reagent X. At low X concentrations, the four histidine lines split into eight lines, four of which retain the original positions of the native spectrum, and four of which are shifted roughly 0.1 ppm from these positions. As more X is added, the original native structure lines disappear, and the four new lines rise in amplitude. Finally, at high concentrations of X, the four new lines collapse into a single line at 8.4 ppm.

Explain these data in terms of a scheme for unfolding of the protein. Comment on any information you can deduce about the rates of steps in your mechanism.

9-4. A ligand L binds to a protein and, in so doing, affects the protein's 100 MHz NMR spectrum. In particular, proton resonances in the range of -80 to -160 Hz (relative to a DSS standard) are affected most. An affinity label analog L′ of the ligand L is prepared. This label binds (covalently) to the protein. The attachment site proves to correspond to a tyrosine residue.

Are the affinity-labeling results consistent with the NMR results? Explain your answer. If you could run just one more NMR spectrum to check your conclusion, what would you choose for your sample? Explain.

9-5. Consider an unpaired electron that interacts with three equivalent protons. Sketch the electron spin transitions that occur. How many lines does the EPR spectrum have? Prove the statement in the text that, in general, the number of lines is given by $2nI + 1$ for an electron that interacts with n equivalent nuclei of spin I.

References

GENERAL

Abragam, A. 1961. *The Principles of Nuclear Magnetism*. London: Oxford Univ. Press.

Carrington, A., and A. D. McLachlan. 1979. *Introduction to Magnetic Resonance*. New York: Halsted Press.

Feher, G. 1970. *Electron Paramagnetic Resonance with Applications to Selected Problems in Biology* (Les Houches Lectures, 1969). New York: Gordon & Breach.

James, T. L. 1975. *Nuclear Magnetic Resonance in Biochemistry*. New York: Academic Press. [Broad coverage of principles and biological applications.]

Lynden-Bell, R. M., and R. K. Harris. 1969. *Nuclear Magnetic Resonance Spectroscopy*. London: Thomas Nelson. [An introductory treatment.]

Pople, J. A., W. G. Schneider, and H. J. Bernstein. 1959. *High-Resolution Nuclear Magnetic Resonance*. New York: McGraw-Hill. [A classic text on NMR.]

Roberts, J. D. 1959. *Nuclear Magnetic Resonance*. New York: McGraw-Hill. [A popular introductory text with some emphasis on applications in organic chemistry. Although published in 1959, it remains useful for its insight into elementary principles.]

Slichter, C. P. 1963. *Principles of Magnetic Resonance*. New York: Harper & Row.

Swartz, H. M., J. R. Bolton, and D. C. Borg, eds. 1972. *Biological Applications of Electron Spin Resonance*. New York: Wiley–Interscience. [Chap. 1 by J. R. Bolton gives an introduction to EPR theory. Chap. 2 by I. C. P. Smith covers the spin-label method. Other chapters cover a variety of applications.]

Wertz, J. E., and J. R. Bolton. 1972. *Electron Spin Resonance: Elementary Theory and Applications*. New York: McGraw-Hill. [A good introduction to the basics of EPR.]

SPECIFIC

Berliner, L. J., ed. 1976. *Spin Labeling*. New York: Academic Press.

———, ed. 1978. *Spin Labeling II*. New York: Academic Press.

Bowden, F. L. 1975. In *Techniques and Topics in Bio-Inorganic Chemistry*, ed. C. A. McAuliffe (New York: Wiley), p. 205.

Dwek, R. A. 1973. *Nuclear Magnetic Resonance in Biochemistry: Applications to Enzyme Chemistry*. Oxford: Clarendon Press.

Hunkapillar, M. W., S. H. Smallcombe, D. R. Whitaker, and J. H. Richards. 1972. Carbon nuclear magnetic resonance studies of the histidine residue in α-lytic protease: Implications for the catalytic mechanism of serine proteases. *Biochemistry* 12:4732.

Karplus, M. 1959. Contact electron-spin coupling of nuclear magnetic moments. *J. Chem. Phys.* 30:11.

———. 1963. Vicinal proton coupling in nuclear magnetic resonance. *J. Amer. Chem. Soc.* 85:2870.

Levy, G., ed. 1976. *Topics in Carbon-13 NMR Spectroscopy*, vol. 2. New York: Wiley.

McConnell, H. M. 1971. Spin-label studies of cooperative oxygen binding to hemoglobin. *Ann. Rev. Biochem.* 40:227.

Moon, R. B., and J. H. Richards. 1973. Determination of intracellular pH by ^{31}P magnetic resonance. *J. Biol. Chem.* 248:7276.

Roberts, G. C. K., and O. Jardekzky. 1970. Nuclear magnetic resonance spectroscopy of amino acids, peptides, and proteins. *Adv. Protein Chem.* 24:447.

Sykes, B. D., and M. D. Scott. 1972. Nuclear magnetic resonance studies of the dynamic aspects of molecular structure and interaction in biological systems. *Ann. Rev. Biophys. Bioengin.* 1:27.

Ugurbil, K., R. S. Norton, A. Allerand, and R. Bersohn. 1977. Studies of individual carbon sites of azurin from *Pseudomonas aeruginosa* by natural-abundance carbon-13 nuclear magnetic resonance spectroscopy. *Biochemistry* 16:886.

Ugurbil, K., H. Rottenberg, P. Glynn, and R. G. Shulman. 1978. ^{31}P nuclear magnetic resonance studies of bioenergetics and glycolysis in anaerobic *Escherichia coli* cells. *Proc. Natl. Acad. Sci. USA* 75:2244.

10
Size and shape of macromolecules

10-1 METHODS OF DIRECT VISUALIZATION

How big is it? What does it look like? These are two of the most obvious questions to ask about any large biological molecule. There are many techniques available for answering these questions. They provide information at various degrees of reliability and detail. We begin our discussion with techniques that are closely analogous to the optical techniques we use for direct viewing of larger objects.

Molecular electron microscopy

The most informative methods for determining macromolecular size and shape involve working with samples in the solid state. Conceptually, the simplest approach is actually to look at the molecule, and this is possible with the electron microscope. Currently available transmission electron microscopes can provide resolution as high as to 2 to 4 Å. This resolution would be sufficient to resolve many structural features of interest if enough contrast could be obtained between the macromolecular sample and the surface holding it for viewing in the microscope. The scattering of electrons is proportional to the square of the atomic number of the scattering atom. Thus, uranium is almost 10^4 times as intense a scattering center as hydrogen.

Typical biopolymers and the films on which they sit in an electron microscope both consist largely of light atoms. Thus, the contrast available with virtually any biopolymer is insufficient for direct observation. There are exceptions such as ferritin, with its core of Fe_2O_3. Almost always, however, one must resort to staining procedures to visualize the sample. These procedures usually result in a substantial loss in resolution.

Positive stains involve incorporation of heavy atoms into the sample. This can be done by a specific chemical complexation (for example, binding of uranyl ions to DNA, or osmium fixation of lipid bilayers), but a much more common approach is shadowing. Tungsten (or another heavy atom) from a hot wire source is sprayed at the sample. These heavy atoms deposit on the sample and the supporting grid. If the sample is bombarded from a fixed direction, metal atoms pile up in front of the sample, and behind the sample a shadow is formed in which no metal atoms are deposited (Fig. 10-1a). The thickness of the shadow is a key to the height of the sample. Sometimes the sample is rotated while metal is deposited by bombardment at a shallow angle. In this case, no shadow is formed, but the extra metal deposited near vertical features allows them to be visualized.

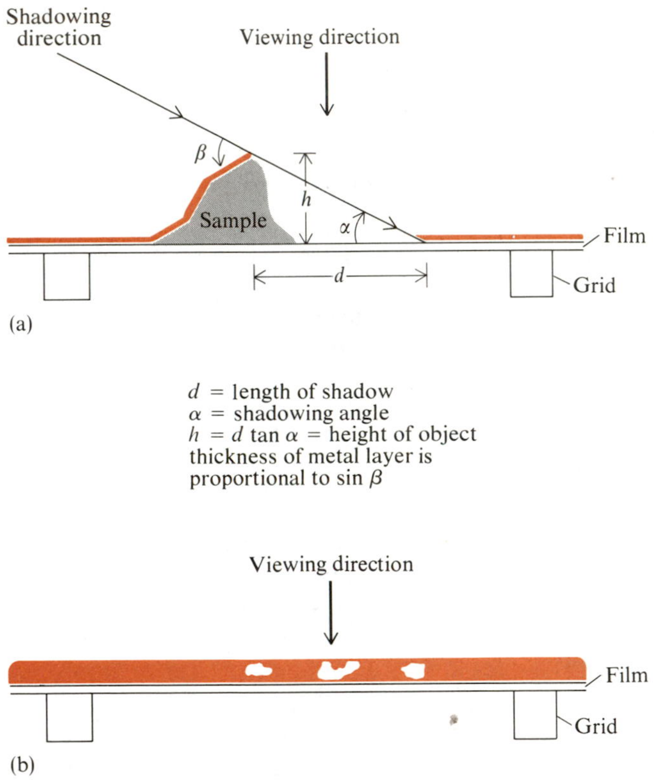

Figure 10-1

Schematic views of samples stained for electron microscopy. In each case, a heavy-atom deposit is shown in color. **(a)** A shadowed sample. **(b)** A negatively stained sample.

A second approach is that of negative staining. Here the sample is flooded with a staining solution containing heavy atoms such as uranyl acetate. These heavy atoms fill all the space around the macromolecule, leaving it as a conspicuous nonscattering center (Fig. 10-1b). Unfortunately, the view afforded by this technique is somewhat indirect, because one really observes where the molecule *isn't*, rather than where it *is*.

All the staining methods suffer from a number of potential difficulties. The sample must be dry and in a high vacuum to be observed in the microscope. The chemical procedures used to introduce heavy atoms are harsh. Thus, there is considerable risk of distorting the structure of the sample. Proteins and nucleic acids contain a considerable amount of bound water under physiological conditions. As discussed earlier, forces dependent on the presence of an aqueous solution play a large role in determining the structure. Clearly, under the high-vacuum conditions of the electron microscope, both hydrophobic forces and hydration must be severely altered. Fixing agents, such as glutaraldehyde, that cross-link the sample are often used to attempt to maintain a structure resembling that in solution. Unfortunately, even these carry the risk of structure perturbation. Finally, the intense electron beam of the microscope can itself damage the sample severely.

Direct comparison between solution structures and electron microscope images would be very useful to assess possible artifacts in the latter. Unfortunately, such comparison is not that simple, because most of the molecules for which high resolution data is available are too small for accurate electron microscopic study. Conversely, most macromolecules that show clear structural detail in the electron microscope are so large that solution techniques are hard-pressed to provide comparable detail. Comparisons are possible with viruses and certain large regular structures such as microtubules and muscle fibers. In general, there is good qualitative agreement, but only fair quantitative agreement, about dimensions or packing of subunits. Much effort is currently being devoted to improvements in electron microscope techniques. For example, the high-resolution scanning microscope developed by A. V. Crewe distinguishes between elastically and inelastically scattered electrons. This provides much better contrast and allows individual heavy atoms to be visualized. As the technique is refined, increased contrast should become available with lighter atoms.

Minimizing drying and shrinking artifacts

Ingenious techniques have been developed to examine samples in which the solvent has not been completely removed. One is freeze etching. A frozen sample is cleaved by striking it with a knife. Solvent is sublimed from the resulting surface to produce a textured terrain, in which nonvolatile macromolecules protrude from a frozen aqueous layer. Then carbon atoms are evaporated onto the surface from a hot source. A thin carbon film ultimately builds up to form a replica of the surface. This replica is carefully lifted off the surface and placed on a normal electron microscope grid. After shadowing with heavy atoms, the replica is viewed normally in the microscope. The appearance of freeze-etched samples or equivalent preparations is often considerably different from that of dried fixed samples. For example, freeze-dried 70S ribosomes show dimensions of 170 × 230 × 250 Å, corresponding to a volume of 5.1×10^6 Å^3. More conventional air-dried preparations typically have dimensions of 160 × 180 × 200 Å, resulting in a much smaller volume, 3.0×10^6 Å^3. The problem is much like comparing a plum with a prune. It is difficult to evaluate the shrinkage caused when the former is dried into the latter.

Shrinkage results from collapse of the structure as liquid is removed. The major cause of this is surface tension. As some of the liquid is removed, the remainder will still cling to the specimen and contract it to keep the liquid pool as an intact unit. Freeze drying can circumvent this problem. Here the sample is frozen and then water is removed by sublimation. No liquid contraction should occur, because there is no liquid. However, the freezing process itself could introduce structural distortions caused by formation of ice crystals.

An alternative technique is critical-point drying. The critical point of a fluid is that temperature and pressure at which there is no detectable transition from liquid to vapor because there is no change in volume. At the critical point, the surface tension of a liquid becomes zero. Hence, the liquid phase can be removed with no shrinkage. Unfortunately, the critical temperature of water is 374°C. Therefore, what is done is to replace the water in a sample (prior to drying) by a liquid with a lower critical temperature. One procedure passes the sample from water to ethanol to amyl acetate to liquid carbon dioxide (critical temperature 36.5°C). This protocol, of course, entails the risk of other artifacts due to the organic solvents. The best strategy is to use several different techniques on the same sample, putting the most trust in those features visualized with all techniques. (See Figs. 10-2 and 10-3 for such comparisons.)

Using symmetry to enhance the electron microscopic image

The resolution and clarity of electron micrographs often can be improved by sample-averaging procedures. Suppose the sample has simple rotational symmetry. An example is the baseplate of bacteriophage T4 (Fig. 10-4a). Superposition of micrographs of rotated samples should cancel out random noise and lead to a clear picture.

A more rigorous and accurate way to accomplish the same end is through mathematical filtering of the image. The photograph produced by the electron

Figure 10-2

Effect of preparation on appearance of liver glycogen samples in electron microscopy. **(a)** Negatively stained sample. **(b)** Air-dried and shadowed sample. **(c)** Sample dried by agar filtration and shadowed. **(d)** Freeze-dried and shadowed sample. **(e)** Freeze-dried sample, exposed to water vapor before shadowing. **(f)** Thin-sectioned glycogen in liver tissue. [Photos courtesy of E. Kellenberger, W. Villiger, and J. Kistler.]

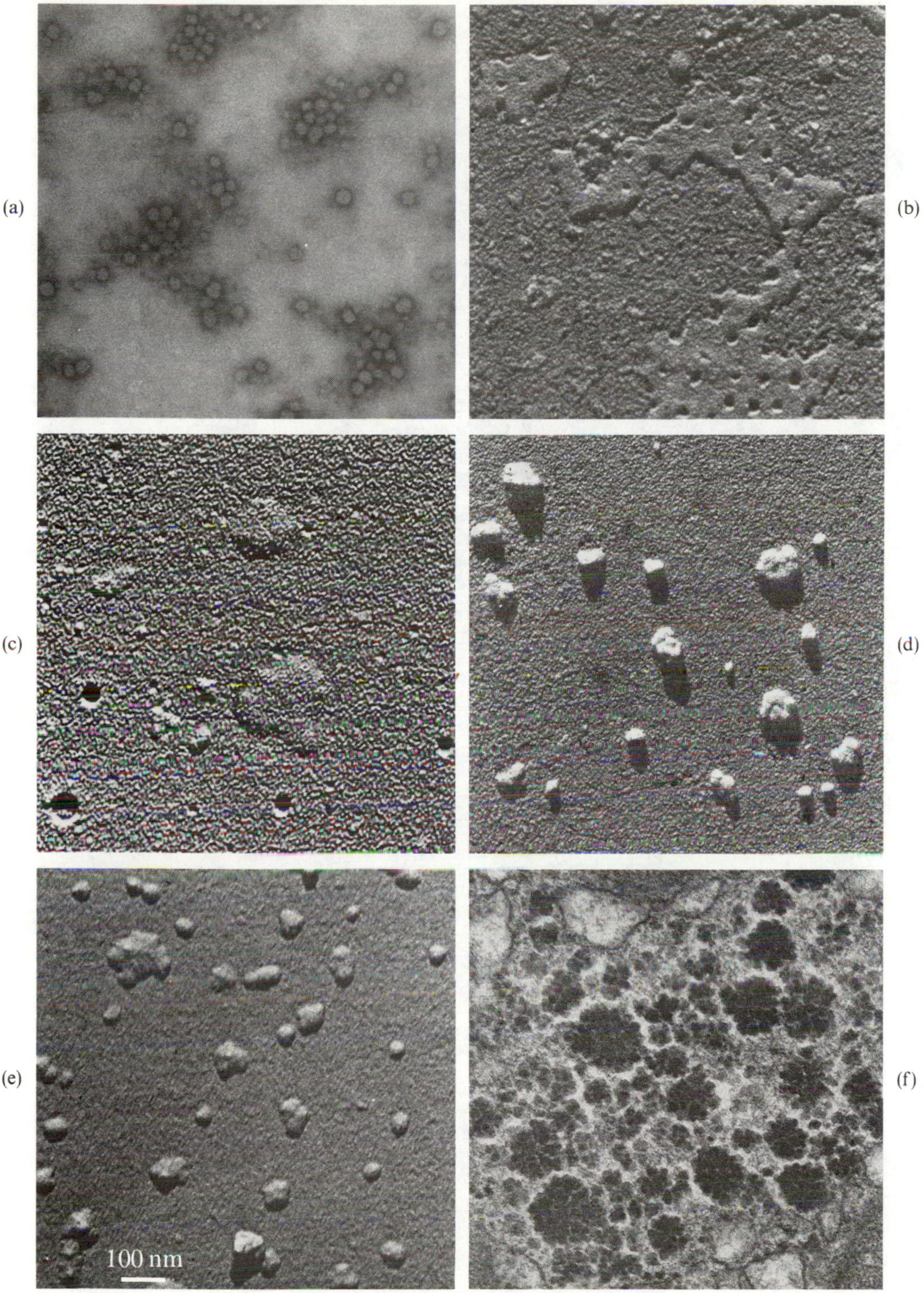
(a)
(b)
(c)
(d)
(e)
(f)
100 nm

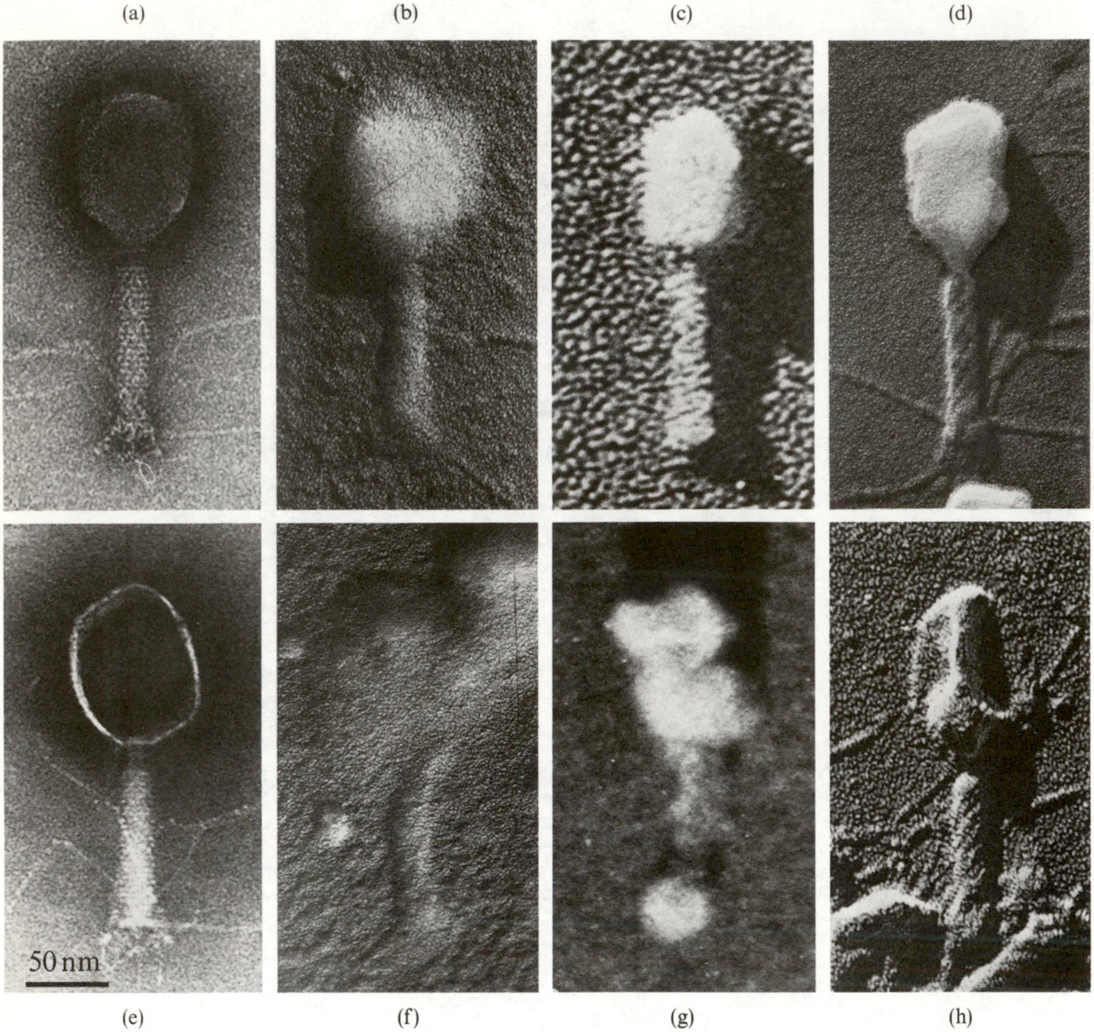

Figure 10-3

Effect of preparation on appearance of T4 bacteriophage samples in electron microscopy. Samples e through h are bacteriophage ghosts, from which the DNA has been removed. **(a,e)** Negatively stained samples. **(b,f)** Air-dried and shadowed samples. **(c,g)** Freeze-dried samples, shadowed according to the microdroplet method (note the collapse of the capsid in g). **(d,h)** Freeze-dried samples, shadowed according to the preadsorption method (note the improved spreading of the tail fibers here in comparison with the other shadowed samples). [Photos courtesy of J. Kistler, B. ten Heggeler, and E. Kellenberger.]

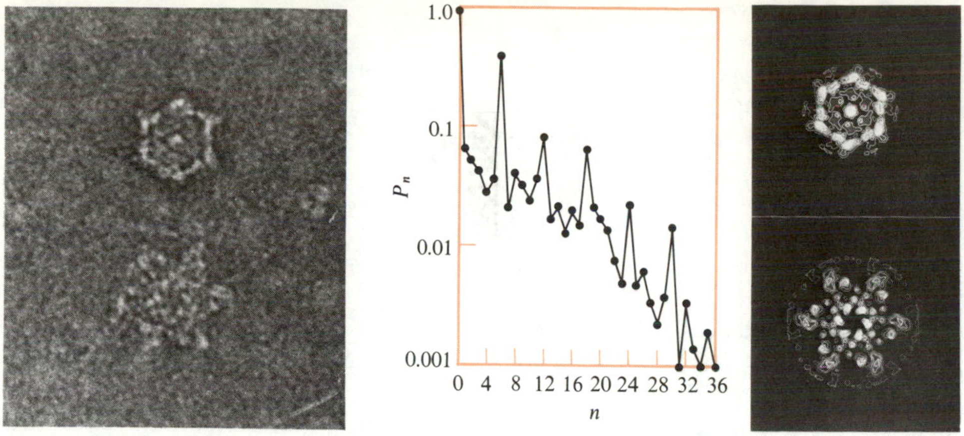

(a) (b) (c)

Figure 10-4

Electron microscopy of the baseplate of bacteriophage T4. An extended baseplate (diameter ~500 Å) is is shown at the top, with a contracted baseplate (diameter ~550 Å) below it. **(a)** Original photomicrograph. **(b)** Rotational power of an image of the extended baseplate. **(c)** Six-fold rotationally filtered images. [From R. A. Crowther, E. V. Lenk, Y. Kikuchi, and J. King, *J. Mol. Biol.* 116:489 (1977).]

microscope is digitized to an array of numbers that represent image density in polar coordinates, $\rho(r, \phi)$. This density is then expanded in cylindrical waves (see, for example, Box 13-2):

$$\rho(r, \phi) = \sum_{n=-\infty}^{\infty} g_n(r) e^{in\phi} \qquad \text{where } g_n(r) = \int_0^{2\pi} \rho(r, \phi) e^{-in\phi}\, d\phi \tag{10-1}$$

Here $g_n(r)$ describes the density with n-fold rotational symmetry at a distance r. Because $\rho(r, \phi)$ is real, it follows that $g_{-n}(r) = g_n^*(r)$, where the asterisk (*) denotes the complex conjugate. Thus, an alternative form of Equation 10-1 is

$$\rho(r, \phi) = g_0(r) + 2 \sum_{n=1}^{\infty} g_n(r) \cos n\phi \tag{10-2}$$

because of the relationships between exponentials and trigonometric functions described in Box 13.1.

The total n-fold rotational component of the structure can be obtained by integrating the density from the center of the image ($r = 0$) to the radius (a). This integral is called the power (ρ_n):

$$\rho_n = \varepsilon_n \int_0^a |g_n(r)|^2 r\, dr \qquad \text{for } n = 0, 1, \ldots, \infty \tag{10-3}$$

The value of ε_0 is one, but all other ε_n are equal to two, because g_{-n} and g_n make equal contributions to the power. A plot of ρ_n versus n then gives a rotational power spectrum of the image (Fig. 10-4b). For this sample (baseplates of phage T4), the power spectrum has clear peaks for values of n that are multiples of six. Thus, a filtered image can be obtained by resynthesizing $\rho'(r, \phi)$, including only these peak terms:

$$\rho'(r, \phi) = \sum_{n=-\infty}^{\infty} g_{6n}(r)e^{i6n\phi} \tag{10-4}$$

The resulting improvement in image is quite striking (compare Fig. 10-4c with the original Fig. 10-4a).

For more extended symmetries, such as those found in crystalline or helical molecules, Fourier transform techniques (see Chapter 13) are used to improve the quality of the images. Here, if the symmetry is displayed over a significant area of the image, a particularly convenient way to perform the Fourier filtering is by optical diffraction rather than numerical calculation (Fig. 10-5a). A photograph of the electron microscope image is used as a grating to produce a diffraction pattern.

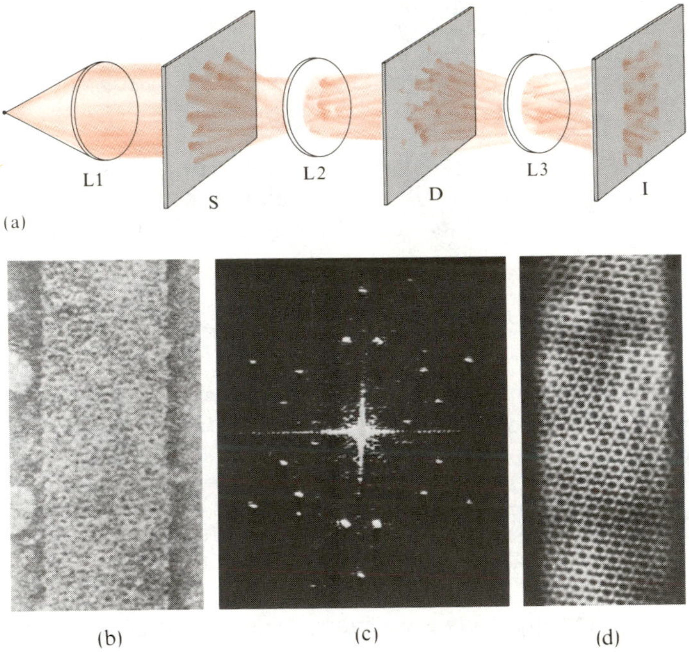

Figure 10-5

Optical filtering of electron micrographs. **(a)** Schematic optical system for Fourier filtering of an electron microscopic image. Light from a point source is transmitted by lenses (L1, L2, L3) through the electron micrograph (S), and focused on the diffraction plane (D) formed by the image at S acting as a diffraction grating, and on the filtered image (I) obtained by recombining light transmitted through D. Filters often are placed between D and L3 to mask out unwanted parts of the image. **(b)** An original electron microscopic image. **(c)** The diffraction pattern formed from b. **(d)** The filtered image formed from b. The sample is a polyhead of bacteriophage T4. This extra-large head arises from defective bacteriophage assembly. The centers of the hexagonal rings of subunits visible in the filtered image are are about 100 Å apart. [Electron microscopic images from R. A. Crowther and A. Klug, *Ann. Rev. Biochem.* 44:161 (1975).]

This pattern can be refocused to give back an image. If the original image has two-dimensional translational periodicity, the diffraction pattern will be spots on a regular lattice (Fig. 10-5c). If a mask is placed in front of the pattern so that only spots on the lattice are transmitted and refocused, the aperiodic noise in the original image will be removed. Sometimes more improvement is possible. The data in Figure 10-5b originate from the head of bacteriophage T4. This is a tubular structure, flattened into a bilayer by drying onto the grid. Thus, the image is from superimposed top and bottom layers. This structure produces diffraction spots on two regular lattices related by axial mirror symmetry. If a mask is used that allows the spots from only a single lattice to be refocused, a very clear filtered image of only one layer results (Fig. 10-5d).

It should be clear that the results of electron microscopy not only are pleasing aesthetically but also offer unusual promise scientifically. The electron microscope is one of the few existing techniques that offer the promise of really revolutionary breakthroughs if further improvements in contrast can be attained. Even with current methods, the technique is extremely useful for the study of very large molecules such as DNA. Some examples will be discussed in later chapters.

High-resolution autoradiography

If a radioactive sample is placed on a suitable photographic emulsion, high-energy particles released by radioactive decay will expose the film. Particular nuclear emulsions have the desirable property that an individual decay event exposes only one or a few silver grains. Thus, after a suitable period of contact with the radioactive sample, the emulsion can be developed to yield a pattern of grains that reveals the shape of the radioactive molecule. A typical example of a molecular radioautogram of DNA is shown in Figure 10-6.

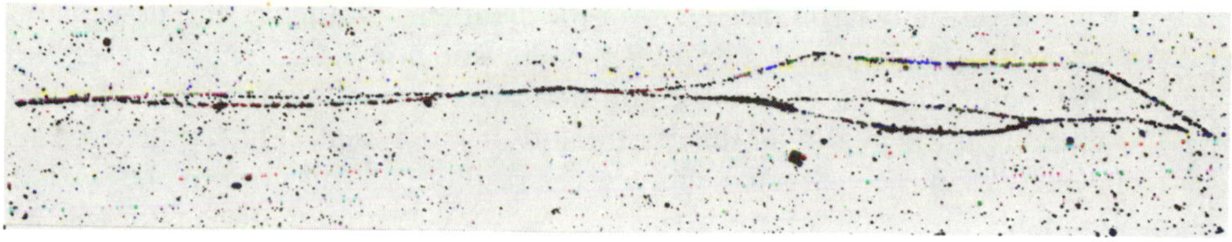

Figure 10-6

Autoradiograph of a replicating DNA molecule from E. coli. About 10% of the original circular duplex has been replicated into two daughter duplexes, forming the extra loop at top right. The length of this single molecule from top to bottom is about 0.65 mm. [Courtesy of Dr. John Cairns.]

The technique is pleasingly simple in some respects, because perturbation of the molecule is limited. However, there are two severe shortcomings. One is that the size of the silver grains and (more significantly) the distance the decay particles can travel limit resolution to hundreds of Å with ^{3}H, thousands of Å with ^{14}C, and even larger distances with ^{32}P. A second and more obvious problem is that quite a few decays per molecule must occur before the sample can be visualized. This means that, unless the investigator has extraordinary patience, a short-lived isotope must be used. Even more serious is the problem that an appreciable fraction of the atoms of the molecule must be substituted with radioisotope.

Suppose one wanted to observe a DNA of the size of *E. coli* DNA (molecular weight $\sim 2 \times 10^9$). A thousand well-spaced decays should clearly define the contour of the molecule. If ^{32}P is used, with a half life of 14 days, and one were willing to wait 14 days, then 2,000 ^{32}P atoms would have to be incorporated per molecule. This works out to a specific radioactivity of about 2.5 curies (mmole P)$^{-1}$. (One curie is 2.2×10^{12} dpm.) This is quite a high density of radiation. The sample would be prepared by allowing *E. coli* to grow on a source of ^{32}P. One must worry not only about radiation damage to the organism but also about damage to the DNA itself, which might break the molecule. Molecular autoradiography is even more difficult with higher organisms, because they are more sensitive to radiation.

In spite of these difficulties, autoradiography has been a powerful adjunct to studies on DNA. One example of its usefulness is that ^{32}P can be preferentially incorporated into newly synthesized regions by pulse labeling. Thus, the structure of regions involved in replication can be specifically examined. Proteins are not as favorable for autoradiographic studies, because the atoms they contain are not easy to substitute extensively with short-lived radioisotopes.

X-ray diffraction

X-ray diffraction is such an important technique that it is treated in detail separately in Chapters 13 and 14. Here it will be sufficient to describe the nature of some of the results obtainable from single-crystal x-ray diffraction. A successful x-ray study provides a picture of the three-dimensional electron density distribution as it repeats through the crystal lattice. One can usually conclude quite a bit about the size and probable shape of the macromolecule from the symmetry of the crystal, its density, and the size of the fundamental repeating element (the unit cell). In many cases, the number of subunits and the symmetries of their arrangement also can be obtained.

Depending on the quality of the crystal and on the extent of data collected, an electron density map is constructed with a given resolution. This governs the level of detail obtainable about the crystal structure. At 10 Å resolution, it usually will be difficult to tell where one molecule in the crystal ends and the next one begins. After all, a diameter of only 30 Å is typical for a small protein, and close contact between neighboring molecules in a macromolecular crystal is not uncommon. It is usually possible to identify the spatial distribution of electron-dense material associated with a single protein or nucleic acid at 6 Å resolution. This yields the detailed overall shape and any particularly striking features such as clefts or holes. An x-ray structure of this resolution is comparable to the very best that can be obtained from Fourier-analyzed electron micrographs of crystalline samples.

Once sufficient x-ray analysis has been done to produce an electronic density map with 4 Å or better resolution, the amount of information available about the structure reaches a level of detail unequaled by any other single technique or even by a combination of many separate techniques (except for the very similar methods of electron or neutron diffraction). At 3 to 4 Å resolution, the nature, length, and orientation of secondary-structure regions become visible. If the primary structure is known, it is usually possible to fit this into the electron density map and start to make reasonable estimates of the spatial location of each individual atom in the structure. By 2 Å resolution, the electron density map is sufficiently detailed that much of the primary structure is revealed directly from the x-ray results.

The enormous effort involved in x-ray crystallography and the need for crystalline samples means that this technique is reserved for substances of particular interest and ready availability. Nevertheless, this technique has taught us more about the structure of large molecules than any other method. It must be considered the benchmark against which the results from less definitive techniques should be tested. X-ray studies are not infallible, nor can they answer all of the questions of interest for any single protein or nucleic acid, but no other method can approach the level of structural detail attainable from a single-crystal x-ray study.

10-2 MACROMOLECULES AS HYDRODYNAMIC PARTICLES

The size and shape of a macromolecule affect its ability to move in fluid solution. Its presence also alters the bulk physical properties of the fluid. A variety of techniques can exploit these phenomena to yield estimates of the molecular weight and shape. Most of these measurements have several features in common. The level of detail inherent in the experimental data is low. Thus, physically plausible models of the molecule must be constructed and used to fit the data. Measurements in isotropic solution usually are blind to molecular asymmetry. Complicated shapes or structural features will not be resolvable. Interactions of the macromolecule with various small-molecule species in solution play an important role in determining the properties of the fluid. Limitations in the knowledge of these interactions will inevitably lead to limitations in the accuracy of the information obtainable about macromolecule structure.

A survey of techniques

The techniques to be discussed can be grouped into several categories. Translational motion dominates several experimental methods. Brownian motion in the absence of an applied force is the origin of diffusion; this can be detected by observing net transport of macromolecules, or by studying fluctuations in the number of macromolecules in a small volume element. External fields can be applied to force additional translational motion. These fields include gravity in simple sedimentation, angular acceleration in velocity sedimentation using the ultracentrifuge, and electric fields in electrophoresis.

Rotational motion dominates other techniques. Fluorescence polarization can examine rotational Brownian motion unperturbed by external fields. EPR and NMR spectra frequently yield information about rotational motions. Here, although an external magnetic field is present, it is not a significant perturbation on the molecular motion. Flow orientation or electrical orientation can alter the normal isotropic distribution of macromolecules in a solution. Rotational motions that relax these orientations can be observed by various techniques including birefringence and dichroism. Viscosity is a measurement of the overall properties of the solution. It is markedly affected by the translational and rotational motions of dissolved large molecules.

Other techniques look essentially at fluctuations in the position or orientation of molecules. All of these contrast the various spatial distributions of macromolecules (or segments of macromolecules) within the surrounding solvent. Light scattering, x-ray scattering, and neutron scattering are fundamentally similar techniques. The kind of information available with each technique differs because of the wavelength of the radiation or particles used and the rules governing the interaction of the scattering object with the incident beam.

Macromolecular volumes and hydration

The molecular volume, V, can be easily estimated for a pure isolated molecule if the molecular weight, M, and the specific volume, $\bar{V}^{\Delta}$, of a pure sample of the substance are known.

$$V = M\bar{V}^{\Delta}/N_0 \tag{10-5}$$

N_0 is Avogadro's number. The specific volume of a pure substance is the inverse of the density. It has units of $\text{cm}^3\ \text{g}^{-1}$.

Standard methods for measuring $\bar{V}^{\Delta}$ have been known since ancient times. The weight of the sample is measured, and the volume is computed from the volume of immiscible fluid displaced when the sample is submerged. Unfortunately, the notion of a pure isolated protein or nucleic acid is not a useful one. Counterions must be present to neutralize the charge of these polyelectrolytes. Even more serious is the possibility of tight associations between normal solvent molecules (such as water) and biopolymers. The hydrated, counterion-containing macromolecule is the species actually present in measurements of aqueous solutions.

In practice, most proteins and nucleic acids are handled in approximately 0.1 M buffered salt solutions. This salt concentration is sufficient to provide ample counterion concentrations. Enormous electrostatic effects occur at much lower ionic strengths; these effects can severely affect hydrodynamic measurements. For example, one could hardly expect to see a charged nucleic acid diffuse away from its positive counterions. The electrostatic energy needed to effect such a charge separation would be enormous. The result is that motions of large molecules and small ions will be coupled, and observed transport properties will become averages of large and small molecule properties.

At high salt concentrations, such a large fraction of the solvent is salt that one must consider a macromolecular solution to have three components: salt, water, and macromolecule. The thermodynamics of such multicomponent systems are quite complex. One must worry about preferential interactions of any two components at the expense of interactions with the third.

The complexities of both high and low ionic strength will be ignored as much as possible in the following discussion. We shall model a macromolecule solution as a two-component system of water (component 1) and macromolecule (component 2). If this were an ideal solution, the volume could be computed by adding the specific volumes of the pure separate components. For a sample with g_1 grams of water and g_2 grams of macromolecule,

$$V_{\text{Tot}} = g_1 \bar{V}_1^{\Delta} + g_2 \bar{V}_2^{\Delta} \tag{10-6}$$

Unfortunately, real solutions do not show such simple behavior. Proteins and nucleic acids interact with substantial numbers of water molecules. These interactions result in nonideal solution behavior. Polar macromolecules can bind appreciable numbers of water molecules fairly tightly. Dry protein films tightly adsorb about 0.4 moles of water per mole of amino acid. One to two moles of water are bound more loosely. Data from many techniques support the idea that strong water–biopolymer interactions exist in dilute solution. For example, NMR shows a small number of very tightly bound water molecules (10 to 50 per entire protein).

Calorimetric studies show that, when a protein solution is frozen, around two moles of water per mole of amino acid (0.4 g H_2O per g protein) remain unfrozen. This water clearly is perturbed by the presence of the protein. NMR and dielectric dispersion studies find that several water molecules per amino acid have moderately restricted motions in a protein solution at room temperature. Presumably, these molecules interact with the protein often enough or tightly enough to slow down the normal rapid tumbling rates of free water molecules. How well these water molecules correspond to those seen by calorimetry is not known. There surely are weaker classes of water–macromolecule interactions that can contribute to the properties of aqueous protein solutions. Bulk water may become trapped in holes or cavities. Although not bound tightly, it will move with the protein and thus contribute to the protein's apparent size. Such water will affect hydrodynamic properties but may not show up as bound water through NMR or thermodynamic techniques.

I. D. Kuntz has developed a particularly simple technique to estimate the amount of tightly bound water. A protein solution is frozen to $-35°C$. All of the water that freezes becomes invisible by NMR spectroscopy. Because dipolar interactions between water molecules cannot be averaged out, the spectrum of solid water is extremely broad. Any water remaining unfrozen is still able to move sufficiently to partially average its environment and give a detectable sharp NMR signal. The number of unfrozen water molecules can be computed from the area under the spectrum. Experiments are carried out at −35°C because this leads to amounts of unfrozen water consistent with what is observable as bound water by other methods. Some examples are shown in Table 10-1.

Table 10-1
Hydration of various biopolymers: A comparison of the results from different methods

Sample	NMR-freezing	Calorimetric	Hydrodynamic[§]	Isopiestic	NMR-calculated[#]
Ovalbumin	0.33	0.32	0.14	0.30	0.37
Bovine serum albumin	0.40	~0.40	0.41	0.32	0.45
Hemoglobin	0.42	0.32	0.63	0.37	0.42
Lysozyme	0.34	~0.30	0.46	0.25	0.36
Myoglobin	0.42	—	0.45	0.32	0.45
DNA	0.59	0.61	—	0.84[¶]	—

NOTE: Values shown are grams bound H_2O per gram anhydrous macromolecule.
§ The average of hydrations computed by viscosity, sedimentation, and diffusion data using known axial ratios or, where these are not available, the average of self-consistent hydrodynamic hydration values such as those shown in Table 12-3.
# Assuming all residues to be fully hydrated, and using values derived for each amino acid residue from model compound studies.
¶ Estimated to be held by short-range interactions.

The variation in bound water in different proteins can be rationalized from the amino acid composition. Kuntz estimated the number of water molecules bound to each amino acid from studies on polypeptides. This value ranges from 6 to 7 for an anion like glutamate to only 1 for nonpolar amino acids. Such variation is reasonable because an ion will generally be tightly solvated by several water molecules, and polar groups should hydrogen-bond strongly to water, whereas a nonpolar residue can bind a water molecule only at the peptide. The amount of water one can expect to be bound to various proteins can be calculated using the hydration values for each amino acid. Agreement with experiment is quite good (as shown in Table 10-1).

Hydration treated thermodynamically

It is impractical to use such a detailed view of water–biopolymer interactions to treat the hydrodynamic properties of proteins or nucleic acids in solution. Also, there is no guarantee that tightly bound water is the sole contributor to hydrodynamic properties. Usually, a very simple model of hydration is constructed. All of the various classes of bound water are merged into a net weight of bound water per weight of macromolecule. It is assumed that this water occupies any internal spaces and also covers the surface of the polymer, smoothing any irregularities. In

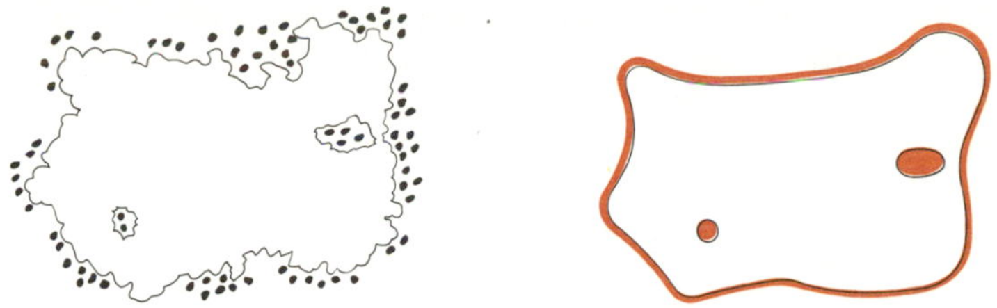

Figure 10-7

Water molecules associated with a protein or nucleic acid. **(a)** In reality, there are many classes of waters bound with various affinities to external and internal regions. The surface of the actual macromolecule has much fine structure. **(b)** In the idealization used to interpret hydrodynamic studies, the molecule's surface is smoothed, and individual bound waters are replaced by a surface film and internal pools.

return, one must admit a certain ignorance about the properties of the bound water and not look too deeply into the significance of the actual amounts needed to explain observed hydrodynamic properties. The approximations made are schematized in Figure 10-7. The real surface of a protein is extremely complicated (see Fig. 2-29).

Suppose a given macromolecule has δ_1 grams of bound water per gram of macromolecule. Let V_1^* be the specific volume of this water. If a solution consists of g_1 grams of water and g_2 grams of anhydrous macromolecule, the total volume of the solution can be computed.

$$V_{\text{tot}} = (g_1 - g_2\delta_1)\bar{V}_1^{\Delta} + g_2 V_2 + \delta_1 V_1^* g_2 \tag{10-7}$$

The first term in this equation is the volume of unbound water. It is presumed to be unperturbed and to have the same specific volume ($\bar{V}_1^{\Delta}$) as pure water. The second term is the volume occupied by macromolecule. The specific volume of the macromolecule in solution (V_2) is not necessarily equal to the specific volume of a pure solid macromolecule ($\bar{V}_2^{\Delta}$). The third term is the volume of bound water (of specific volume V_1^*). Together, the second and third terms give the total volume occupied by the hydrated macromolecules. Because M/N_0 is the weight of a single macromolecule, it is easy to compute the volume V_h occupied by this hydrated molecule:

$$V_h = (M/N_0)(V_2 + \delta_1 V_1^*) \tag{10-8}$$

The difficulty in using this equation is that V_2, δ_1, and V_1^* are not known, nor is there any easy way to measure them. Thermodynamics can be used to resolve this dilemma by replacing these variables with others that are experimentally obtainable.

The partial specific volume of a solute, $\bar{V}_2$, is the change in solution volume when a small increment of solute is added at the limit of infinite dilution. This value is determined by measuring the density of solutions as a function of the weight concentration of solute. The expected result of such measurements can be derived from Equation 10-7. Assuming that the macromolecular component is sufficiently dilute so that the hydration is not a function of concentration, we have

$$\bar{V}_2 = [\partial V_{\text{tot}}/\partial g_2]_{g_1,T} = -\delta_1 \bar{V}_1^{\Delta} + V_2 + \delta_1 V_1^* \tag{10-9}$$

Using this result, we can replace V_2 and V_1^* in Equation 10-8 to obtain

$$V_{\text{h}} = (M/N_0)(\bar{V}_2 + \delta_1 \bar{V}_1^{\Delta}) \tag{10-10}$$

The partial specific volume of a pure substance is equal to the specific volume. Therefore, we can replace $\bar{V}_1^{\Delta}$ in Equation 10-10 by $\bar{V}_1$, the partial specific volume of pure water. So the hydrated volume of a macromolecular solute becomes

$$V_{\text{h}} = (M/N_0)(\bar{V}_2 + \delta_1 \bar{V}_1) \tag{10-11}$$

Because $\bar{V}_1$ is just the inverse of the density of pure water, and $\bar{V}_2$ is measurable independently, the only unknown is the hydration (δ_1) and, as discussed above, there are numerous ways to estimate this.

In practice, partial specific volumes often are tedious to measure because of the large quantity of sample required for conventional density measurements. Accurate microbalances can alleviate these difficulties. In addition, a variety of more specialized techniques exists, including ultracentrifugation in H_2O–D_2O mixtures (Edelstein and Schachman, 1973) and density measurements using mechanical oscillators (Kratky et al., 1973). It is possible to make reasonably accurate estimates of the partial specific volumes of proteins by averaging over the amino acid composition and using estimates of specific volumes of individual amino acid residues in proteins. These estimates range from about 0.60 $\text{cm}^3\ \text{g}^{-1}$ for Asp to 0.90 $\text{cm}^3\ \text{g}^{-1}$ for Leu. A typical protein $\bar{V}_2$ is 0.73 $\text{cm}^3\ \text{g}^{-1}$, and virtually all proteins that do not contain extensive material besides amino acids have $\bar{V}_2$ values between 0.69 and 0.75 $\text{cm}^3\ \text{g}^{-1}$. Nucleic acids are much more dense, with $\bar{V}_2$ values of about 0.50 $\text{cm}^3\ \text{g}^{-1}$ if Na^+ is the counterion, or 0.44 $\text{cm}^3\ \text{g}^{-1}$ if Cs^+ is the counterion.

It is tempting to interpret Equation 10-11 by saying that $(M/N_0)\bar{V}_2$ is the volume of anhydrous macromolecule, and that $(M/N_0)\delta_1\bar{V}_1$ is the volume of bound water. However, this interpretation is hazardous. Both $\bar{V}_2$ and δ_1 contain effects due to water-biopolymer interactions. Partial specific volumes need not be equal to molecular volumes. For instance, the $\bar{V}_2$ of $MgSO_4$ is actually negative. This is because Mg^{2+} and SO_4^{2-} bind water so tightly in their hydration shells that adding $MgSO_4$ to water causes a decrease in total volume. Such electrostriction effects are not usually this dramatic in macromolecules, but they cannot be ignored.

When measured NMR hydrations are used in Equation 10-10, values of the hydrated molecular volume (V_{h}) are predicted that agree (in most cases) with results from hydrodynamic measurements. In these cases, it is probable that $(M/N_0)\bar{V}_2$ is a good estimate of the anhydrous volume. Another procedure is to compute the anhydrous molecular volume from known protein crystal structures. Kuntz has shown that the value for carboxypeptidase is within a few percent of that predicted by $(M/N_0)\bar{V}_2$.

Hydrations of 0.3 to 0.4 g H_2O (g protein)$^{-1}$ are needed to account for the hydrodynamic behavior of typical globular proteins. It is instructive to estimate the added size this water represents. Consider a solid spherical protein of 30,000 mol wt. If $\bar{V}_2$ is 0.72 cm^3 g^{-1}, the anhydrous molecular volume will be 3.6×10^4 Å^3. A hydration of 0.34 g H_2O (g protein)$^{-1}$ will lead to an additional volume of 1.7×10^4 Å^3 because the specific volume of water is very close to 1 cm^3 g^{-1}. Suppose that all of this water exists as a spherical shell on the exterior of the protein. By computing anhydrous (20.5 Å) and hydrated (23.3 Å) radii as simply $(3V/4\pi)^{1/3}$, it is easy to show that the thickness of the spherical hydration shell will be about 2.8 Å. This is equivalent to a shell only one water molecule thick. At least qualitatively, the idea that hydration is surface-bound water has some credence.

Frictional properties of macromolecules in solution

To understand various hydrodynamic measurements—such as sedimentation rates, diffusion, or viscosity—one must be able to explain the motions of the molecules in fluid solution. These motions are much slower than gas-phase motion because of frictional forces between molecules in a condensed phase. In a hypothetical experiment, suppose an external force $\mathbf{F}$ is applied to a molecule with mass m. The equation of motion in a vacuum would be $\mathbf{F} = m\mathbf{a} = m(d\mathbf{v}/dt)$, where $\mathbf{v}$ is the velocity. In a real fluid, however, any moving particle will be subjected to frictional drag. As long as the motion is not fast enough to cause turbulence, the drag is proportional to the velocity. The resulting equation of translational motion is

$$\mathbf{F} - f\mathbf{v} = m(d\mathbf{v}/dt) \qquad (10\text{-}12)$$

Here the frictional force, $f\mathbf{v}$, is shown opposing the applied force. The parameter f is called the translational frictional coefficient. It is a function of the nature of the fluid.

The linear differential equation for motion in a fluid is easy to solve. If the velocity of the macromolecule at time zero is $\mathbf{v}_0$ parallel to the applied force, and $\mathbf{F}$ is constant, the result is

$$\mathbf{v}(t) = (\mathbf{F}/f) + [\mathbf{v}_0 - (\mathbf{F}/f)]e^{-ft/m} \qquad (10\text{-}13)$$

This solution has two important consequences. The particle is accelerated for a short time after the external force is applied. The velocity decays exponentially from the initial value $\mathbf{v}_0$ to a constant final value. This final value, $\mathbf{v}(\infty) = \mathbf{F}/f$, is linear in the applied force.

The rate of approach to a constant velocity is very rapid. For a typical macromolecule of 30,000 mol wt, f is of the order of 5×10^{-8} g sec^{-1}, and m is 5×10^{-20} g. Therefore, the transient $e^{-ft/m}$ term in Equation 10-13 decays on the time scale of 10^{-12} sec. This is the same as the time it takes molecular vibrations to relax.

In the simplest view, frictional forces arise in a fluid because of attractions among the molecules of the fluid. In order to push a solid object through the fluid it is necessary to also move some solvent molecules with respect to others (Fig. 10-8a). Solvent molecules nearest the moving particle will be most perturbed (shown as longest arrows). The disturbance caused by the particle will diminish as one moves away from it in the fluid.

To compute the frictional force, we must calculate the force required to maintain the perturbed velocity distribution of solvent molecules. This force can be phenomenologically related to a property of the fluid called the viscosity. Consider a fluid bounded by two surfaces (Fig. 10-8b). It is reasonable to postulate that the force required to slide one surface past another should be proportional to the area of the sheets (A) and should reflect the difference in velocity. If we consider the surfaces to be sheets of fluid, then for small enough velocity increments (or sheets of fluid close enough together), only the first derivative of the velocity should matter. Then,

$$F = A\eta(dv/dz) \tag{10-14}$$

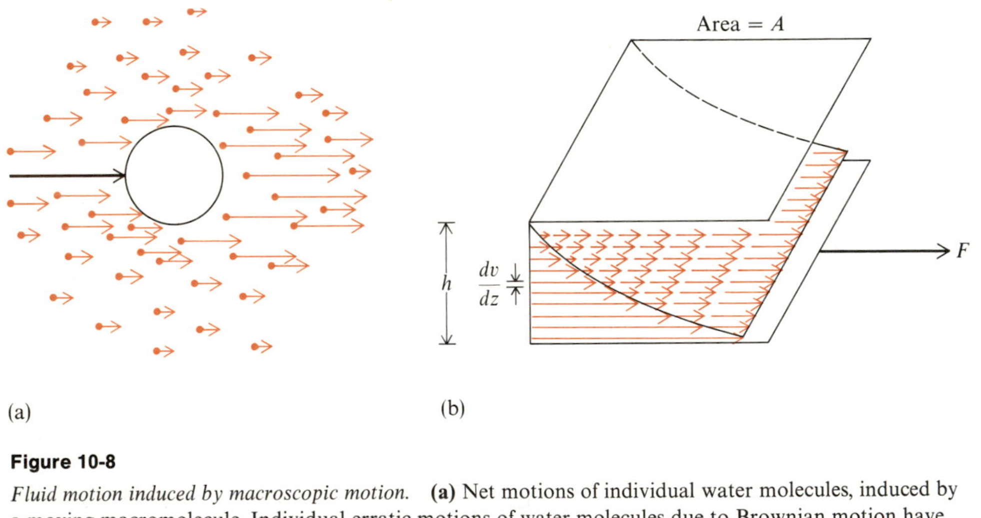

Figure 10-8

Fluid motion induced by macroscopic motion. **(a)** Net motions of individual water molecules, induced by a moving macromolecule. Individual erratic motions of water molecules due to Brownian motion have been neglected. **(b)** Distribution of fluid velocities generated by moving one parallel plane relative to another. The fluid velocity gradient, dv/dz, will be a constant if h is small, or if the relative velocity of the two planes is small. In general, however, the velocity profile need not be linear.

In this equation, η is a proportionality constant called the coefficient of viscosity. It should depend only on properties of the fluid. The velocity gradient normal to the direction of flow, dv/dz, is called the shear.

Relationship between friction and molecular size

We need to find, for the motion of a spherical particle, an equation that relates the coefficient of translational friction, f, to the fluid viscosity, η. The actual derivation is extremely complicated because one must calculate explicitly how particle motion induces velocity gradients in the fluid. Fluid mass must be conserved but, in a real case, particle motion could lead to fluctuations in the density of the fluid. It takes pages of algebra to derive an equation for the frictional coefficient of a spherical particle. However, the form of the equation can be found by dimensional analysis. The only obvious variable for the spherical particle is its radius, r. One can postulate that f is a function of η and r. The simplest possible functional form is

$$f \propto \eta^x r^y \qquad (10\text{-}15)$$

The dimensions of f are g sec^{-1}; from Equation 10-14, η must have dimensions of g cm^{-1} sec^{-1}; r has dimensions of cm. Therefore, in terms of dimensions, Equation 10-15 can be written as

$$\text{g sec}^{-1} = \text{g}^x \text{ cm}^{-x} \text{ sec}^{-x} \text{ cm}^y \qquad (10\text{-}16)$$

The only values of x and y that make Equation 10-16 dimensionally correct are are $x = y = 1$. Thus, $f = \text{const } \eta r$.

The value of the constant cannot be determined by dimensional analysis. It depends on the boundary conditions of fluid flow at the surface of the particle. Two extreme cases usually are considered. If the particle interacts strongly with fluid molecules, it is likely that the layer of fluid immediately adjacent to the particle surface moves at exactly the same velocity as the particle. This case is called stick boundary conditions. The resulting friction equation is called Stokes law.

$$\boxed{f_{\text{sph}} = 6\pi\eta r} \qquad (10\text{-}17)$$

At the other extreme, suppose there is no interaction between particle and fluid molecules. It is reasonable that this case should lead to weaker frictional forces. The fluid simply slips by the particle. These slip boundary conditions yield $f_{\text{sph}} = 4\pi\eta r$.

All of the preceding discussion deals with translational friction. Similar considerations apply for the damping of rotational motions of particles by viscous fluids.

If a constant torque τ is placed on a particle in a fluid, the particle will reach a constant angular velocity, ω, after a transient period. The parameter relating the velocity to the torque is the rotational frictional coefficient, $f_{\text{rot}} = \tau/\omega$. It turns out that, for stick boundary conditions, the rotational frictional coefficient of a sphere is

$$f_{\text{rot}} = 6\eta V \qquad (10\text{-}18)$$

where V is the volume. For slip boundary conditions, f_{rot} of a sphere is zero. This is reasonable, because a rotating sphere will not disturb the fluid if solvent molecules do not interact with it.

A third hydrodynamic case of interest is the effect of spherical particles on the external force needed to maintain a constant velocity gradient (shear) within a fluid. We will not go into a detailed discussion of this because it is mathematically quite formidable. However, it is useful to remember that the critical variable is the volume of the macromolecular particles.

It is interesting to examine the actual distribution of solvent velocities around a dissolved molecule (for translation and rotational motion) and around a molecule suspended in a velocity gradient. Figure 10-9 shows results of detailed hydrodynamic calculations. Here the fluid velocity near the macromolecule is plotted as contours that show the change in velocity from that of the macromolecule to that of unperturbed bulk fluid. The critical thing to note from this figure is that the disturbance caused by the particle dies off very quickly with distance.

For stick boundary conditions, a substantial amount of the disturbance takes place within a thickness corresponding to the dimensions of only a single water molecule. This is a very disturbing result. The surface of a macromole is not described very well by a smooth sphere. It might appear that details of the surface would have to be taken into account for accurate hydrodynamic predictions. Furthermore, it is hard to be comfortable about a macroscopic choice of stick boundary conditions when only a single layer of water molecules is involved. Some slip boundary must be involved also. This will lead to a smaller fluid perturbation, because the difference in velocity between the bulk fluid and the macromolecule surface is smaller.

Fortunately, the effects of the roughness of the surface and the lack of pure stick boundary conditions work in opposite directions. Thus, it is customary to make the approximation summarized schematically in Figure 10-7. Hydration is assumed to add to the volume of a macromolecule without changing its shape. Stick boundary conditions are used, and any roughness of the surface is ignored. Equations 10-17 and 10-18 are used to calculate the frictional coefficients of a sphere, except that one must use a radius or volume appropriate for the hydrated particle (Eqn. 10-11). Thus, for a hydrated spherical macromolecule with a volume V_{h}, the expected frictional coefficients should be $f_{\text{sph}} = 6\pi\eta(3V_{\text{h}}/4\pi)^{1/3}$ and $f_{\text{rot}} = 6\eta V_{\text{h}}$.

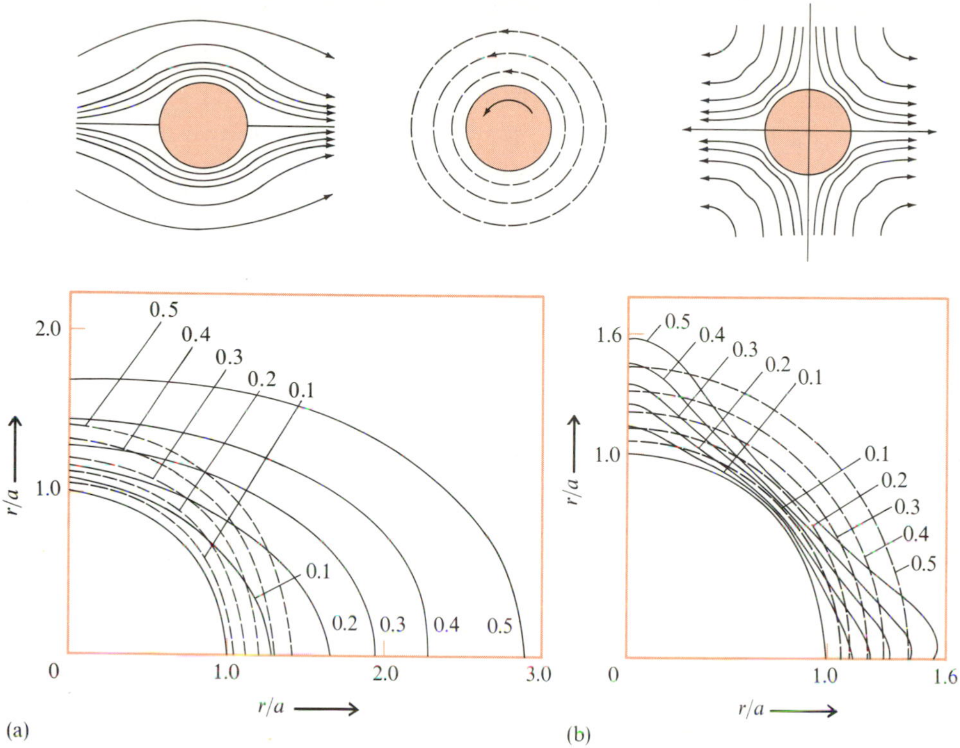

Figure 10-9

Calculated fluid velocity profile generated near a spherical macromolecule obeying stick boundary boundary conditions. In each case, a frame of reference has been chosen to make the macromolecule stationary. A cross section of the fluid, coincident with the center of the macromolecule, is shown at the top; at the bottom, just one quadrant is shown in an expanded view. Fluid velocities are shown as contour lines. Those at the bottom are labeled with the fractional velocity of the unperturbed bulk fluid. The radius of the macromolecule is a, and distance is expressed in units of a. **(a)** Solid lines represent a sphere moving from left to right in the fluid plane. Dashed lines represent a sphere rotating about an axis perpendicular to the fluid plane. **(b)** Dashed lines represent a sphere rotating about an axis perpendicular to the fluid plane. Solid lines represent a sphere in a fluid with a linear velocity gradient oriented at 45°, as indicated at the upper right. [After I. D. Kuntz, Jr., and W. Kauzmann (1974).]

For molecules the size of proteins in aqueous solution, the predictions made by employing these assumptions usually are in fairly good agreement with experiment. Much larger discrepancies can arise for smaller molecules or for less-strongly-interacting solvents. When motions of nonpolar molecules within lipid bilayers are considered, one must begin to pay serious attention to the choice of boundary conditions.

Effects of shape on translational frictional properties

Most macromolecules of biological interest are not spheres. A significant fraction appear to be compact, globular, or irregular rigid bodies. For these, an ellipsoid of revolution is a more realistic model than a sphere. There are two classes of such ellipsoids, both of which are limiting cases of the general ellipsoid with three different axes (see Fig. 10-10). The oblate ellipsoid is a disk shape, generated by rotating an ellipse around its short semiaxis b; the two long semiaxes, a, are identical. A prolate ellipsoid is a rodlike shape, generated by rotating an ellipse around its long semiaxis a; here the two shorter semiaxes, b, are identical. For either kind of ellipsoid, the axial ratio (p_r) is defined as a/b, the ratio of the long to the short semiaxes.

The volume of a sphere is $(4/3)\pi r^3$, whereas that of an ellipsoid is either $(4/3)\pi a^2 b$ (oblate) or $(4/3)\pi ab^2$ (prolate). For equal volumes, the surface area of either ellipsoid will be greater than that of a sphere. It seems reasonable to guess that ellipsoids will have larger frictional coefficients than equivalent spheres, and this guess is confirmed by detailed calculations. Because the volume of a molecule is proportional to the molecular weight (Eqn. 10-11), we see that (for constant mass) the more a molecule deviates from a sphere, the larger its frictional coefficient will become.

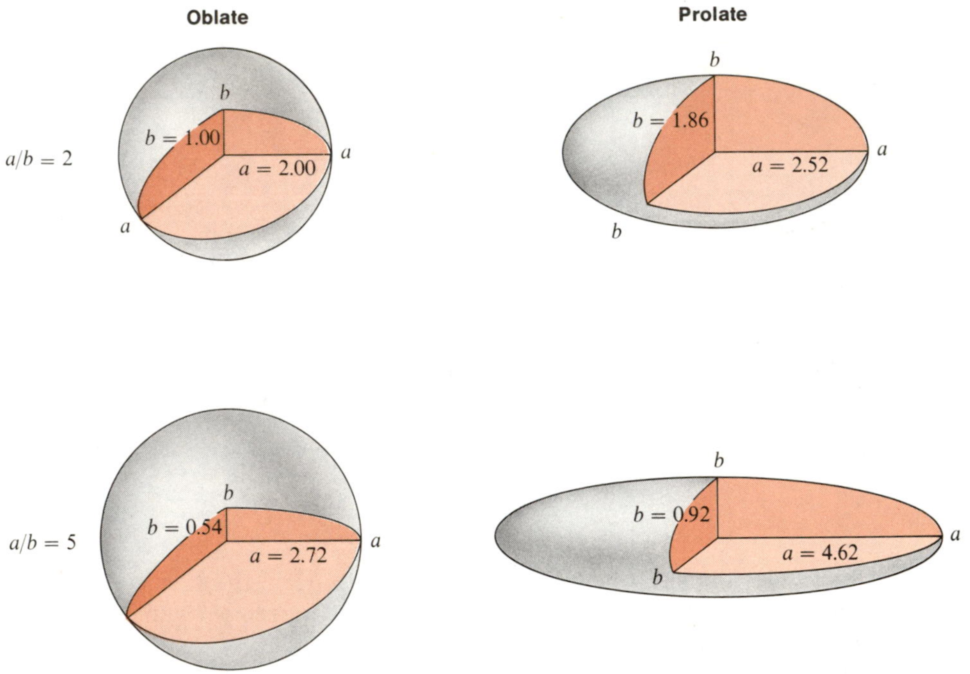

Figure 10-10

Four ellipsoids of revolution, with equal volumes. The near right octant of each ellipsoid has been cut away to show the major (a) and minor (b) axes.

For stick boundary conditions, it is possible to obtain analytical expressions for the dependence of the frictional coefficient of an ellipsoid on the axial ratio. It is useful to express these as the ratio of the frictional coefficient of an ellipsoid (f) to that of a sphere with equal volume (f_{sph}). For translational friction, the results are the following:

$$F = f/f_{sph} = (1 - p^2)^{1/2}/p^{2/3} \ln\{[1 + (1 - p^2)^{1/2}]/p\}$$

for a prolate ellipsoid, where $p = b/a = 1/p_r$ (10-19a)

$$F = f/f_{sph} = (p^2 - 1)^{1/2}/p^{2/3} \tan^{-1}[(p^2 - 1)^{1/2}]$$

for an oblate ellipsoid, where $p = a/b = p_r$ (10-19b)

These equations are not especially convenient to use; therefore, Table 10-2 lists numerical values. The translational frictional coefficient ratio F is often called a shape factor or Perrin factor. Note that the value of these factors increases rather gradually with increasing axial ratio (Fig. 10-11). Prolate ellipsoids always show a higher frictional coefficient than oblate ellipsoids of the same axial ratio. Note that,

Table 10-2
Simha (ν), Perrin (F), and Scheraga–Mandelkern (β) shape parameters for ellipsoids of revolution

Axial ratio	Prolate			Oblate		
	ν	F	$\beta \times 10^{-6}$	ν	F	$\beta \times 10^{-6}$
1	2.500	1.000	2.12	2.500	1.000	2.12
2	2.908	1.044	2.13	2.854	1.042	2.12
3	3.685	1.112	2.16	3.430	1.105	2.13
4	4.663	1.182	2.20	4.059	1.165	2.13
5	5.806	1.250	2.23	4.708	1.224	2.14
6	7.098	1.314	2.28	5.367	1.277	2.14
8	10.103	1.433	2.35	6.700	1.374	2.14
10	13.634	1.543	2.41	8.043	1.458	2.14
15	24.65	1.784	2.54	11.42	1.636	2.14
20	38.53	1.996	2.64	14.80	1.782	2.15
30	74.51	2.356	2.78	21.58	2.020	2.15
40	120.76	2.668	2.89	28.37	2.212	2.15
50	176.81	2.946	2.97	35.16	2.375	2.15
60	242.28	3.201	3.04	41.95	2.518	2.15
80	400.5	3.658	3.14	55.52	2.765	2.15
100	593.7	4.067	3.22	69.10	2.974	2.15
200	2,052.9	5.708	3.48	137.01	3.735	2.15

NOTE: F is calculated by using Equation 10-19; ν and β are calculated as described in Chapter 12 (see Eqns. 12-23 and 12-29).

SOURCE: After H. A. Scheraga, *Protein Structure* (New York: Academic Press 1961).

if there is hydration, Equation 10-19 refers to the ratio of the frictional coefficient of an ellipsoid to that of a sphere with the same *total hydrated* volume.

Effects of shape on rotational frictional properties

The frictional coefficient for nonspherical bodies is a nine-component tensor, $\underset{\sim}{\mathbf{f}}$. It was not necessary to consider this tensor in detail in the equations for the translational frictional coefficient. In most hydrodynamic measurements, one typically observes net mass transport; an explicit knowledge of the orientation of the molecule during transport is not necessary. The magnitude of the frictional coefficient observed (f) is an average over the various components of $\underset{\sim}{\mathbf{f}}$. Thus, instead of the nine parameters needed to describe $\underset{\sim}{\mathbf{f}}$, we required only the single parameter f.

However, measurements of rotational motion must include the detailed distribution of orientations. It is impossible to speak of just a single rotational frictional coefficient. There are two coefficients for ellipsoids: f_a for rotation about the a semiaxis, and f_b for rotation about the b semiaxis. By using stick boundary conditions, J. B. Perrin derived expressions for these coefficients in 1934. The rotational frictional coefficients are similar in spirit to (but much more complex than) the corresponding translational frictional coefficients of Equation 10-19. We give them here in as compact a form as possible. These frictional coefficient ratios are defined relative to the rotational frictional coefficient of a sphere of equivalent volume.

$$F_a = f_a/f_{\text{rot}} = 4(1 - p^2)/3(2 - p^2 S) \qquad (10\text{-}20a)$$

$$F_b = f_b/f_{\text{rot}} = 4(1 - p^4)/3p^2[S(2 - p^2) - 2] \qquad (10\text{-}20b)$$

In these equations, S and p are defined as follows:

Prolate ellipsoid:
$$S = 2(1 - p^2)^{-1/2} \ln\{[1 + (1 - p^2)^{1/2}]/p\} \qquad (10\text{-}21a)$$
$$p = b/a = 1/p_{\text{r}} \qquad (10\text{-}21b)$$

Oblate ellipsoid:
$$S = 2(p^2 - 1)^{-1/2} \tan^{-1}[(p^2 - 1)^{1/2}] \qquad (10\text{-}22a)$$
$$p = a/b = p_{\text{r}} \qquad (10\text{-}22b)$$

Note the similarity of the algebraic form of the parameter S to the translational frictional coefficient ratios for ellipsoids (Eqn. 10-19). In fact $S = 2p^{-2/3}/F$.

Figure 10-11b plots the rotational frictional coefficient ratios for oblate and prolate ellipsoids. Some important qualitative conclusions are evident. Oblate ellipsoids produce roughly the same friction, whether they rotate around the long axis (f_a) or the short axis (f_b). Both motions involve more friction than does the rotation of an equivalent sphere. Prolate ellipsoids rotate around their long axis (f_a)

more easily than an equivalent sphere rotates. However, rotation around their short axis (f_b) is accompanied by extremely great friction. This observation is reasonable, because such motion is bound to perturb the fluid quite considerably. (Just think of the design of a standard magnetic stirrer.)

Each of the rotational frictional coefficients can be related to a rotational relaxation time, τ_r. This variable measures the rate at which an anisotropic distribution relaxes to equilibrium through that particular rotational mode. For a sphere, $\tau_r = f_{rot}/2kT$. For ellipsoids (relative to equivalent spheres),

Prolate ellipsoid:
$$\tau_a/\tau_r = f_b/f_{rot} \tag{10-23a}$$
$$\tau_b/\tau_r = (2/f_{rot})[(1/f_a) + (1/f_b)]^{-1} \tag{10-23b}$$

Oblate ellipsoid:
$$\tau_b/\tau_r = f_a/f_{rot} \tag{10-23c}$$
$$\tau_a/\tau_r = (2/f_{rot})[(1/f_a) + (1/f_b)]^{-1} \tag{10-23d}$$

where τ_a is the relaxation time for orientation about the long axis, and τ_b for orientation about the short axis. You can rationalize these equations if you think about the motion of a prolate ellipsoid. To change the spatial orientation of the long axis (a), you must rotate about one of the two equivalent short axes (b). Rotation about a has no effect. However, to change the orientation of one of the short axes, either rotation about a or b will do, and either should be equally effective, except that the short-axis rotation will involve more friction.

Various measurements of macromolecular rotation yield parameters that are related to τ_a and τ_b but, unfortunately none of the conventional NMR, electric dichroism or birefringence, dielectric relaxation, or fluorescence polarization techniques give τ_a and τ_b directly. Some methods give τ_a and τ_b mixed in with other quantities that may not be known accurately. It is often a formidable task to sort τ_a and τ_b out of results from techniques such as fluorescence polarization. Decay of fluorescence anisotopy mixes τ_a and τ_b with a series of exponentials whose coefficients are weighted by, among other things, the orientation of the fluorescent probe with respect to the principal axes of the ellipsoid (see Chapter 8). Other methods, such as static fluorescence polarization or non-Newtonian viscosity, can yield only the harmonic mean of τ_a and τ_b. For a prolate ellipsoid $\tau_h^{-1} = 1/3[\tau_a^{-1} + 2(\tau_b)^{-1}]$, and the corresponding mean rotational frictional coefficient is $f_h = (2\tau_h)^{-1}$.

It is a shame that it is so difficult to measure τ_a and τ_b independently because, as comparison of Figure 10-11a and 10-11b clearly shows, rotational friction is a much more sensitive measure of shape than is translational friction. However, rotational frictional coefficients are also much more sensitive to the choice of boundary conditions. We have already discussed this dependence for spheres. When slip boundary conditions are used for ellipsoids, translational friction is only moderately altered. Rotation is strikingly affected. With slip boundary conditions, there will be no frictional drag for rotation about the long axis of a prolate ellipsoid or the short

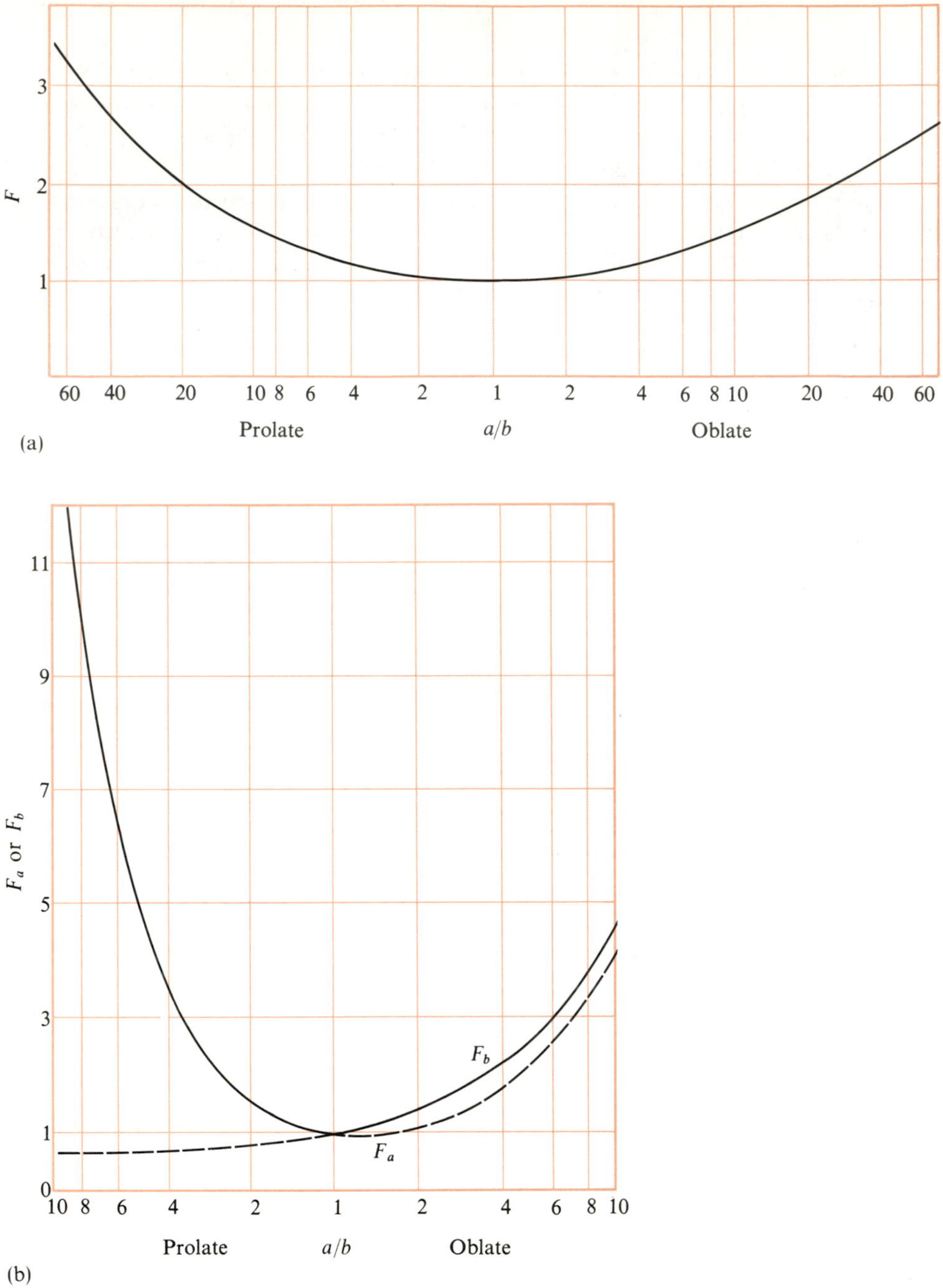

Figure 10-11

Frictional coefficients for ellipsoids, relative to those for spheres of equal volume. **(a)** Translational frictional coefficient ratio F, defined in Equation 10-19. **(b)** Rotational frictional coefficient ratios F_a and F_b, defined in Equation 10-20. The ratio F_b is for rotation around the short axis; F_a is for rotation around the long axis. F_a approaches the value 2/3 as a prolate ellipsoid becomes infinitely long. For oblate ellipsoids, in the limit of large a/b, $F_a = F_b = 3a/4\pi b$. [After S. Koenig, *Biopolymers* 14:2421 (1975).]

axis of an oblate ellipsoid. In contrast, rotation about the other axes involves more and more friction as the axial ratio increases. Robert Zwanzig has calculated that, in the limit of long axial ratios, the frictional coefficients become the same whether stick or slip conditions are used for rotations about the nonunique axes of ellipsoids of revolution (Fig. 10-12).

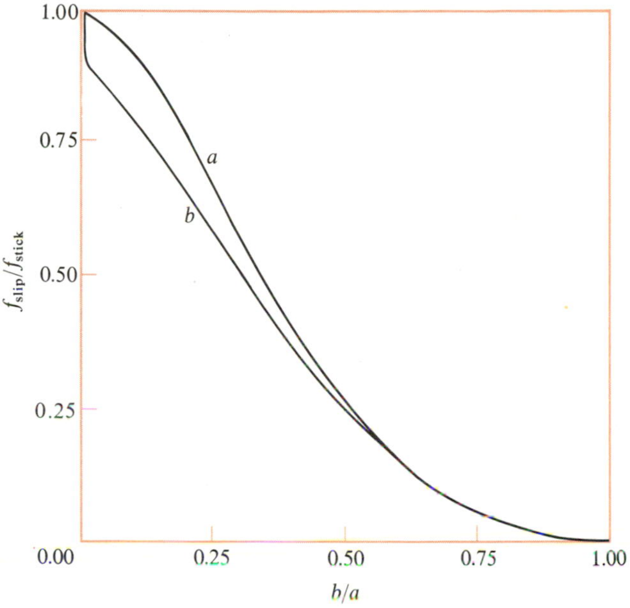

Figure 10-12

Effect of boundary conditions on the rotational frictional coefficients. Curve *a* is for rotation around the long axis of an oblate ellipsoid. Curve *b* is for rotation around the short axis of a prolate ellipsoid. Results are shown as a function of the inverse axial ratio, b/a, to allow extension of the data to infinite axial ratios. [After D. R. Bauer, J. I. Brauman, and R. Pecora, *J. Am. Chem. Soc.* 96:6840 (1974).]

Translation friction of molecules with complex shapes

In many cases, ellipsoids are unsatisfactory models for the shapes of biopolymers. In these cases, one must attempt to predict the hydrodynamic properties of more complex structures. An approximate method for computing the friction of a structure made of identical subunits was developed by J. G. Kirkwood and J. Riseman. There are two major applications of this theory. A linear or coiled polymer can be approximated as a string of beads, each with identical hydrodynamic properties. An oligomeric protein can be modeled as a cluster of identical subunits with the same hydrodynamic properties. In either case, the major problem is to deal with the hydrodynamic interaction between subunits. We will restrict our discussion to the translational motion of a polymer with N segments or subunits. Each subunit, as it moves through the fluid, perturbs the velocity distribution of the fluid nearby. This perturbation is felt by the other subunits.

We need to describe the velocity perturbation, $\Delta\mathbf{v}$, a distance $\mathbf{r}$ away from the point at which a force $\mathbf{F}$ is exerted *on* the fluid by one of the subunits.[§]

$$\Delta\mathbf{v} = \underset{\sim}{\mathbf{T}} \cdot \mathbf{F} \tag{10-24}$$

The tensor $\underset{\sim}{\mathbf{T}}$ that produces the perturbed velocity is called the Oseen tensor and has the form

$$\underset{\sim}{\mathbf{T}} = (1/8\pi\eta r)(\underset{\sim}{\mathbf{I}} + \mathbf{r}\mathbf{r}^{\dagger}/r^2) \tag{10-25}$$

where r is distance, and $\underset{\sim}{\mathbf{I}}$ (the unit tensor) and $\mathbf{r}\mathbf{r}^{\dagger}$ are defined as follows:

$$\underset{\sim}{\mathbf{I}} = \begin{pmatrix} 1 & 0 & 0 \\ 0 & 1 & 0 \\ 0 & 0 & 1 \end{pmatrix} \qquad \mathbf{r}\mathbf{r}^{\dagger} = \begin{pmatrix} x^2 & xy & xz \\ xy & y^2 & yz \\ xz & yz & z^2 \end{pmatrix} \tag{10-26}$$

If the shear in the solution is sufficiently low that no preferential orientation is taken by the polymer, we can average the tensor interaction over all orientations. Such an average is simply the trace of a tensor. For a rigid system, $\langle \mathbf{r}\mathbf{r}^{\dagger} \rangle = (1/3)(x^2 + y^2 + z^2)\underset{\sim}{\mathbf{I}} = (1/3)r^2\underset{\sim}{\mathbf{I}}$, and the average of the Oseen tensor becomes

$$\langle \underset{\sim}{\mathbf{T}} \rangle = (1/8\pi\eta r)[\underset{\sim}{\mathbf{I}} + (1/3)\underset{\sim}{\mathbf{I}}] = \underset{\sim}{\mathbf{I}}/6\pi\eta r \tag{10-27}$$

where the angle brackets denote an average. If the system is flexible, we can use the same equation, but we will have to use an average value $\bar{r}$ instead of the fixed distance r.

Suppose that the bulk fluid is moving with a constant velocity $\mathbf{v}^0$. The local velocity $\mathbf{v}_i$ felt by the ith segment of a dissolved polymer will be different from $\mathbf{v}^0$ because of hydrodynamic interactions with all the other segments (Fig. 10-13a). The jth segment changes the fluid velocity at the ith segment by $\Delta\mathbf{v}_{ij}$. Using Equations 10-24 and 10-27, we find the local velocity to be

$$\mathbf{v}_i = \mathbf{v}^0 + \sum_{j \neq i} \Delta\mathbf{v}_{ij} = \mathbf{v}^0 + (1/6\pi\eta) \sum_{j \neq i}^{N} (1/r_{ij})\mathbf{F}_j \tag{10-28}$$

where r_{ij} is the distance between the ith and jth segments, and $\mathbf{F}_j$ is the force exerted on the fluid by the jth segment. Keep in mind that, if the polymer is flexible or is a distribution of structures, each r_{ij} in this equation will have to be replaced by an average value.

[§] A tensor equation must be used because the velocity perturbation is not necessarily parallel to the force that causes it.

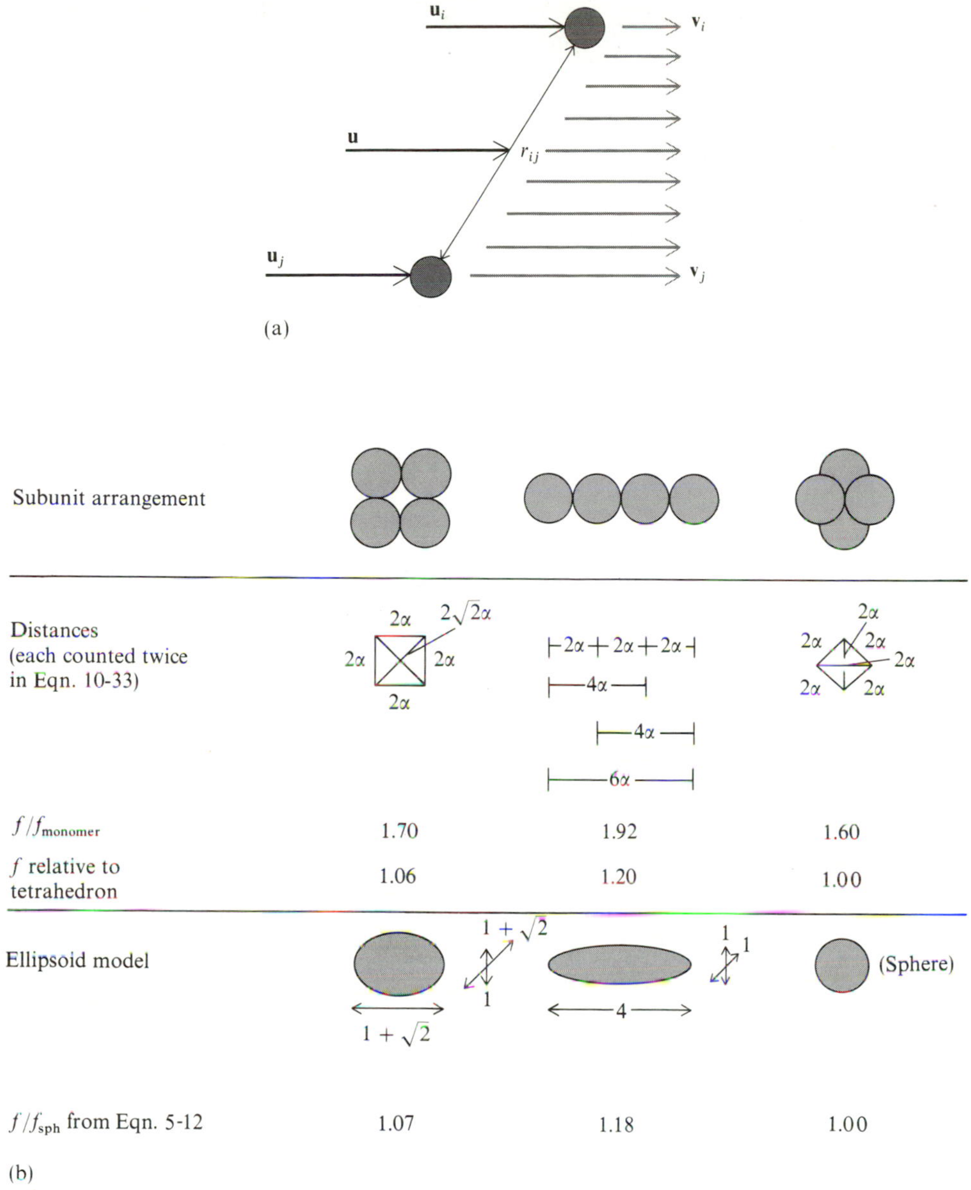

Figure 10-13

Effect of hydrodynamic interactions on frictional properties. **(a)** The hydrodynamic interaction between two segments of a polymer, shown schematically. The two segments are moving with velocities $\mathbf{u}_i$ and $\mathbf{u}_j$. The fluid is moving at velocities $\mathbf{v}_i$ and $\mathbf{v}_j$. The center of mass is moving at velocity $\mathbf{u}$, and r_{ij} is the distance between the two segments. **(b)** Three subunit arrangements for a protein of four identical subunits. The effects of shape on frictional coefficients are calculated by the Kirkwood–Riseman theory (*above*) and by ellipsoidal models (*below*).

The force *on* the fluid from the *i*th segment will be determined by the difference between the velocity $\mathbf{u}_i$ of that segment and the fluid velocity $\mathbf{v}_i$ at the segment:

$$\mathbf{F}_i = \zeta(\mathbf{u}_i - \mathbf{v}_i) \tag{10-29}$$

where ζ is the frictional coefficient of a segment. Equation 10-29 is exactly equivalent to Equation 10-13 after any transient terms have died off. We want to calculate the translational frictional coefficient f for the entire polymer. This is defined by the total force $\mathbf{F}$ exerted by the polymer on the fluid and the velocity ($\mathbf{u}$) of its center of mass relative to that of the bulk fluid ($\mathbf{v}^0$):

$$\mathbf{F} = f(\mathbf{u} - \mathbf{v}^0) \tag{10-30}$$

Simple mechanical considerations lead to two enormous simplifications. The total force exerted by the polymer must be the sum of the forces exerted by the individual segments:

$$\mathbf{F} = \sum_{i=1}^{N} \mathbf{F}_i \tag{10-31}$$

At the hydrodynamic steady state, the average velocity of each segment, $\mathbf{u}_i$, is equal to the velocity ($\mathbf{u}$) of the center of mass. This is true on the average (or else the polymer will deform as it moves), but it need not be true at any given instant. Equation 10-29 becomes

$$\mathbf{F}_i = \zeta(\mathbf{u} - \mathbf{v}_i) \tag{10-32}$$

Equations 10-32 and 10-28 now represent $2N$ linear in 2N unknowns: $\mathbf{v}_i$ and $\mathbf{F}_i$ for $i = 1, 2, \ldots, N$. These equations can be solved in terms of the details of the structure of the polymer, which is contained in all of the distances r_{ij}^{-1}. When the result is combined with Equations 10-30 and 10-31, the final expression for f is surprisingly compact:

$$f = N\zeta \Big/ \left[1 + (\zeta/6\pi\eta N) \sum_i \sum_{j \neq i} r_{ij}^{-1}\right] \tag{10-33}$$

The derivation of Equation 10-33 has been sketched out only roughly. In reality, solving the $2N$ simultaneous equations involves some assumptions that can subsequently be shown to be exact. Furthermore, it is actually the inverse of the friction tensor, $\underset{\sim}{\mathbf{f}}^{-1}$, that must be spatially averaged, rather than $\underset{\sim}{\mathbf{f}}$, because $\underset{\sim}{\mathbf{f}}^{-1}$ is what is actually measured. The interested reader should consult V. Bloomfield et al. (1967) for further details.

Frictional coefficients of oligomers and polymers

Equation 10-33 can be given a relatively simple physical interpretation. Consider the polymer fixed and the fluid moving. If there were no hydrodynamic interactions, the frictional coefficient of a polymer would just be the sum of the frictional coefficients of each segment. For N identical segments, $f = N\zeta$. In the presence of hydrodynamic interactions, the frictional coefficient is less than $N\zeta$, as shown by Equation 10-33. This occurs because, on the average, each segment decreases the fluid velocity near it. The overall result is that each segment is subjected to a smaller fluid velocity. Thus each experiences a smaller frictional force.

Equation 10-33 allows the calculation of the frictional coefficient of any object. To apply the equation to oligomeric proteins composed of N identical spherical subunits of radius R, the segment frictional coefficient ζ is replaced by $6\pi\eta R$ (the frictional coefficient of one subunit, f_m). It is convenient to express Equation 10-33 as the ratio of the frictional coefficient of the oligomer to that of a single subunit. If one measures intersubunit distance (r_{ij}) in units of the subunit radius,

$$f/f_m = N\left[1 + (1/N)\sum_i \sum_{j\neq i} \alpha_{ij}^{-1}\right]^{-1} \tag{10-34}$$

where $\alpha_{ij} = r_{ij}/R$. For example, consider a protein with four identical subunits. Three likely arrangements of these are linear, square planar, or tetrahedral. Figure 10-13b shows relative frictional coefficients for each form, along with the distances needed to calculate these. Notice that the most compact structure, the tetrahedron, has the smallest frictional coefficient. The linear arrangement is most extended and has the largest frictional coefficient. Note though that the differences among the three forms are at most 20%.

For fairly regular structures, results from Equation 10-34 are in good agreement with what is calculated by approximating the shapes as ellipsoids of revolution and using the Perrin factors in Table 10-2. For example we could crudely approximate the tetrahedron as a sphere, the linear array as a 4:1 prolate ellipsoid, and the square as an oblate ellipsoid with an axial ratio of $(1 + \sqrt{2})$:1. Assuming that all three shapes have the same volume gives frictional coefficients *relative* to the tetrahedron within 1% to 2% of those calculated from Equation 10-34. In fact, in the limit of large prolate axial ratios, both Equations 10-34 and 10-19a can be approximated by $N/\ln N$. However, ellipsoid approximations become inaccurate for irregular shapes, and then Equation 10-34 must be used.

All of these results are in the nondraining limit, corresponding to maximal hydrodynamic interactions between segments of a polymer. The opposite limit, called free draining, allows no hydrodynamic interaction. Here, one assumes that each polymer segment interacts with unperturbed bulk fluid velocity. This might be true if, for example, the polymer consisted of widely spaced subunits separated by frictionless linkers. In this case, the Oseen tensor is zero, Equation 10-34 does not hold, and the frictional coefficient of the polymer is simply Nf_{m}. For a compact array of segments, such as an oligomeric protein, subunits are so close that it is clear the nondraining limit must be used. It is not as clear which limit to use for expanded structures, such as random coils. One must calculate how much of the solvent within a coil is trapped and flows with the coil, and how much is free to travel independently, and one must also calculate the expected spatial distribution of polymer segments. Equation 10-34 can be used to derive expressions for the frictional coefficients of coils given equations for the statistical distributions of segments. The result is presented in Chapter 19.

It is important to note that the hydrodynamic interactions described by Equation 10-28 are not restricted to segments of the same polymer. They apply equally well to intermolecular interactions between different polymer molecules in concentrated solutions. These effects become enormously complicated, and they are best handled by attempting to study hydrodynamic properties at concentrations as dilute as possible. A frequent practice is measuring frictional properties as a function of concentration and extrapolating to infinite dilution.

10-3 MACROMOLECULAR DIFFUSION

We now turn from the theory of the hydrodynamics of individual particles to a discussion of several of the experimental methods that can extract the hydrodynamic properties of solutions of macromolecules. Several problems arise with each technique. One must find a way to relate the experimentally observed properties of the solution as a whole to the motions of unseen individual molecules. One must determine the effect of any heterogeneity in the macromolecules on the average properties of the bulk system. One must also deal with the complications caused if concentrations are high enough to allow significant intermolecular interactions.

It is traditional for books of this type to discuss diffusion before any of the other hydrodynamic methods. This is not because diffusion is a powerful or frequently used method. In fact, classical diffusion measurements are extraordinarily difficult for macromolecules. Furthermore, the experimental results derived are not particularly convenient for yielding shapes or molecular weights, unless one of these two quantities is known in advance. However, diffusion is conceptually the simplest of the hydrodynamic techniques and, unlike some other methods, it can be described by equations with convenient analytical solutions. In the past few years, inelastic light scattering has developed as a promising new technique for the rapid and accurate determination of diffusion properties. This method, which could bring a renaissance to the field of diffusion, is discussed briefly in Chapter 14. (For a detailed treatment see Berne and Pecora, 1975.)

Mass transport as flux

In diffusion, and other transport processes such as sedimentation and electrophoresis, the net movement of molecules must be analyzed. The simplest possible systems of biological interest will contain two components, a solvent (component 1), which is usually water, and a dissolved macromolecule (component 2). Most measurements are made on a closed system. Thesefore, any net change in the location of solute within the system must be compensated for by a corresponding change in solvent position. Because these are inextricably coupled, it is usually sufficient to consider the motions of only the solute. The solvent generally can be ignored in the very dilute solutions treated here. This is not always true at higher concentrations. In general, one must assume that the molecular weight of the solute is unknown. This means that the number of solute molecules in a sample is also unknown. Thus, it is most convenient to describe solute transport as mass transport.

The flux, J_2, is defined as the rate of mass transport across a surface of unit area.

$$J_2 = dm_2/dt \quad \text{g sec}^{-1}\ \text{cm}^{-2} \tag{10-35}$$

where m_2 is the number of grams of solute on one side of the surface. If c_2 is the weight concentration of molecules at the surface, and $\langle \mathbf{v}_2 \rangle$ is the average velocity, the flux can be written as

$$\mathbf{J}_2 = c_2 \langle \mathbf{v}_2 \rangle \tag{10-36}$$

A flux can occur only in response to a force on the solution. If only small forces are considered, there should be a linear response of the system.

$$\mathbf{J}_2 = \sum_{i=1}^{N} L'_{2i} \mathbf{X}_i \tag{10-37}$$

Here we allow for a set of generalized forces ($\mathbf{X}_i$) on each component. The L'_{2i} values are phenomenological coefficients that relate the applied force on component i to the resultant flux of component 2. Equation 10-37 implies that a force on component 1 (the solvent or any other component) can affect the motion of component 2, even if the force is not directly felt by component 2. For example a charged solute will move in response to an applied electric field, and this could lead to the motion of a macromolecule even if that macromolecule were uncharged. It is possible to redefine the coefficients L'_{2i} in such a way that the solvent (component 1) can be eliminated from Equation 10-37. This involves choosing a particular frame of reference for the surface across which flow occurs (see Box 10-1). The result is

$$\mathbf{J}_2 = \sum_{i=2}^{N} L_{2i}\mathbf{X}_i \tag{10-38}$$

which for a two-component system is simply

$$\mathbf{J}_2 = L_{22}\mathbf{X}_2 \tag{10-39}$$

Two kinds of forces contribute to each $\mathbf{X}_i$. Electric fields, gravitational fields, and so on are external forces, $\mathbf{F}_i$, These are direct physical forces and can be treated as such. In addition, if the concentration of the solute is nonuniform, there will be a diffusional force $\mathbf{X}_{\mathrm{D}i}$. This describes the tendency of Brownian motion to restore a uniform concentration of all components of a system. The actual force in Brownian motion arises from collisions with solvent. It is simpler, however, to adopt a thermodynamic description of this force.

Any force **F** can be written as the negative gradient of a potential energy function ϕ, so that $\mathbf{F} = -\nabla\phi$, where $\nabla = \hat{\mathbf{i}}\,d/dx + \hat{\mathbf{j}}\,d/dy + \hat{\mathbf{k}}\,d/dz$. The negative sign indicates that the force is always in the direction of moving the system to lower energy. The potential of a component of a solution is simply the chemical potential μ. We shall use solute concentration in units of g cm^{-3} and, therefore, it is convenient to define $\hat{\mu}$ as $\partial G/\partial m$, where G is the Gibbs free energy, and m is the mass of a particular component. Thus, $\hat{\mu}$ is the chemical potential in weight-concentration units—that is, per gram of that particular component in that solution (see Box 10-2).

For solute, $\hat{\mu}_2 = \hat{\mu}_2^0 + (RT/M_2)\ln a_2$, where M_2 is the molecular weight, $\hat{\mu}_2^0$ is the chemical potential in a reference state (defining the energy scale), and a_2 is the activity of the solute. This activity is related to the solute concentration by an activity coefficient, $a_2 = \gamma_2 c_2$. The diffusion force at finite concentrations is

$$\mathbf{X}_{\mathrm{D}2} = -\nabla\hat{\mu}_2 = -(RT/M_2)\nabla \ln a_2 \tag{10-40}$$

The activity coefficient approaches 1 at sufficient dilution, so that $a_2 = c_2$, and then one can write the diffusion force on component 2 as

$$\mathbf{X}_{\mathrm{D}2} = -\nabla\hat{\mu}_2 = -(RT/M_2)\nabla \ln c_2 \tag{10-41}$$

The total flux of component 2 in a multicomponent system is given by

$$\mathbf{J}_2 = \sum_{i=2}^{N} L_{2i}(\mathbf{F}_i - \nabla\hat{\mu}_i) \tag{10-42}$$

Box 10-1 FRAMES OF REFERENCE FOR FLUXES

The surfaces used in the text to describe flow are fixed relative to the cell in which an experiment takes place. The solute flux across such a surface is given by Equation 10-36:

$$(J_2)_c = c_2\langle v_2\rangle_c$$

where the subscript c denotes a cell-fixed frame of reference.

Another possibility is to allow the surface to move with the velocity of the solvent. In this solvent-fixed frame of reference (indicated by a subscript s), the solute flux is

$$(J_2)_s = c_2(\langle v_2\rangle_c - \langle v_1\rangle_c)$$

where $\langle v_1\rangle_c$ is the solvent velocity in the cell-fixed frame. Combining the two equations above, we obtain

$$(J_2)_s = (J_2)_c - c_2\langle v_1\rangle_c$$

A third frame of reference can be constructed by considering the velocity $\langle v\rangle_v$ of the total sample volume at the surface. This is

$$\langle v\rangle_v = \sum_{i=1}^{N} \bar{V}_i c_i \langle v_i\rangle_c$$

where $\bar{V}_i c_i$ is just the volume due to component i, and $\langle v_i\rangle_c$ is its velocity in the cell-fixed frame. The resulting flux of component 2 in this volume-fixed frame is

$$(J_2)_v = c_2(\langle v_2\rangle_c - \langle v\rangle_v) = (J_2)_c - c_2\langle v\rangle_v$$

Note that, if the solution is incompressible and if the volume is unchanged upon mixing the various components, then $\langle v\rangle_v = 0$, and $(J_2)_v = (J_2)_c$.

For each frame of reference, the phenomenological equation for flux (Eqn. 10-37) can be written in terms of coefficients defined for that frame. For a particular frame R,

$$(J_2)_R = \sum_{i=1}^{N} (L_{2i})_R X_i$$

It can be shown that, for the volume-fixed frame or the solvent-fixed frame (Fujita, 1975),

$$(J_2)_R = \sum_{i=2}^{N} (L_{2i})_R X_i$$

Therefore, the true form of Equation 10-38 is

$$(J_2)_c = (J_2)_v = \sum_{i=2}^{N} (L_{2i})_v X_i$$

Here we have recognized the fact that the chemical potential gradient of all i components (not just component 2) can effect a net transport of component 2.

It usually is possible to simplify Equation 10-42 considerably. Most transport properties can be viewed as one-dimensional processes. If the system has only two components, the result is

$$J_2 = L_{22}[F_2 - (\partial \hat{\mu}_2/\partial x)_t] \tag{10-43}$$

We shall use this basic equation to describe various hydrodynamic experiments. Diffusion is particularly simple because there are no externally applied forces:

$$\boxed{J_2 = -L_{22}(\partial \hat{\mu}_2/\partial x)_t} \tag{10-44}$$

Box 10-2 CHEMICAL POTENTIALS

The definition of the chemical potential you have probably seen in the past is a partial molal free energy:

$$\mu = \partial G/\partial n$$

where n is the number of moles of a particular component of a mixture. The chemical potential μ is related to the activity a and the chemical potential in a reference state (μ^0) by

$$\mu = \mu^0 + RT \ln a$$

Here a is a dimensionless quantity, which is related to the molal concentration C as $a = \gamma C$, where γ is an activity coefficient. The reference state by definition in the preceding equation is the state in which the activity is unity. This state usually is chosen as a hypothetical solution of unit molality in which each molecule behaves as if it were at infinite dilution. Thus γ is defined so that it approaches 1 kg mole^{-1} as C approaches 1 mole kg^{-1}. For ideal solution behavior or at low concentration, $a = C$ mole kg^{-1}, so that γ approaches 1 kg mole^{-1}, and $d\gamma/dC = 0$. (See Atkins, 1976, pp. 232–237 and 311–313.)

The chemical potential in weight-concentration units, $\hat{\mu}$, is related to μ simply by

$$\hat{\mu} = \mu/M = \mu^0/M + (RT/M)\ln a = \hat{\mu}^0 + (RT/M)\ln \hat{\gamma}c$$

where M is the molecular weight, c is the concentration in g cm^{-3}, $\hat{\mu}^0$ is the chemical potential in weight-concentration units in a reference state of 1 g cm^{-3}, and the activity coefficient $\hat{\gamma}$ is defined so that it approaches 1 cm^3 g^{-1} as c approaches 1 g cm^{-3}. At low concentration or in ideal solutions, $\hat{\gamma}$ becomes 1 cm^3 g^{-1}, and $d\hat{\gamma}/dc = 0$.

The problem now is to evaluate the coefficient L_{22}. It is easiest to do this by first describing the way in which a diffusion measurement is carried out and then finding a thermodynamic explanation of the results.

Fick's laws of diffusion

Suppose a sample is prepared by layering pure solvent upon an equal volume of a macromolecular solution. If this sample is allowed to stand undisturbed at constant temperature and pressure, the initial sharp discontinuity in solute concentration will broaden gradually until finally the whole sample has a uniform solute concentration (Fig. 10-14). The redistribution of solute is caused by diffusion, and an analysis should yield a value for L_{22}. It is convenient to work in concentration units rather than chemical potentials. At infinite dilution, $(\partial\hat{\mu}_2/\partial x)_t = (RT/c_2M_2)(\partial c_2/\partial x)_t$, so Equation 10-44 becomes

$$J_2 = -(L_{22}RT/M_2c_2)(\partial c_2/\partial x)_t = -D(\partial c_2/\partial x)_t \qquad (10\text{-}45)$$

where we have defined D (the diffusion constant) as $L_{22}RT/c_2M_2$. Usually D is the quantity measured experimentally. First we shall outline how this is done; then we shall show how D (and L_{22}) can be given a molecular interpretation (Eqn. 10-66).

The right-hand side of Equation 10-45 was originally proposed by A. Fick in the nineteenth century by analogy to observations of the heat flow produced by a temperature gradient. It is possible, however, to provide a pretty good justification for the form of Equation 10-45, which is known as Fick's first law of diffusion.

Consider a small zone of fluid bounded by unit-area surfaces at x and $x + dx$ (Fig. 10-15). The thickness, dx, must be chosen such that the concentration of solute varies only slightly from $c_2(x)$ on one side of the zone to $c_2(x + dx)$ on the other. The rate of mass transport from left to right through the zone ought to be proportional to the concentration at the left boundary, $c_2(x)$, and inversely proportional to the thickness of the zone, dx. Transport in the opposite direction is proportional to $c_2(x + dx)/dx$. Define the proportionality constant as D. The net rate of mass transport should be the difference between these two individual rates. Thus, the flux is

$$J_2 = [Dc_2(x) - Dc_2(x + dx)]/dx = -D(\partial c_2/\partial x)_t \qquad (10\text{-}46)$$

in the limit of a thin zone. The result is identical to Equation 10-45 for transport across a surface. Note that, if the concentration increases from left to right with increasing x, then J_2 will be negative, and solute mass will flow from right to left. This is intuitively reasonable.

It often is convenient to measure concentrations rather than fluxes. The same

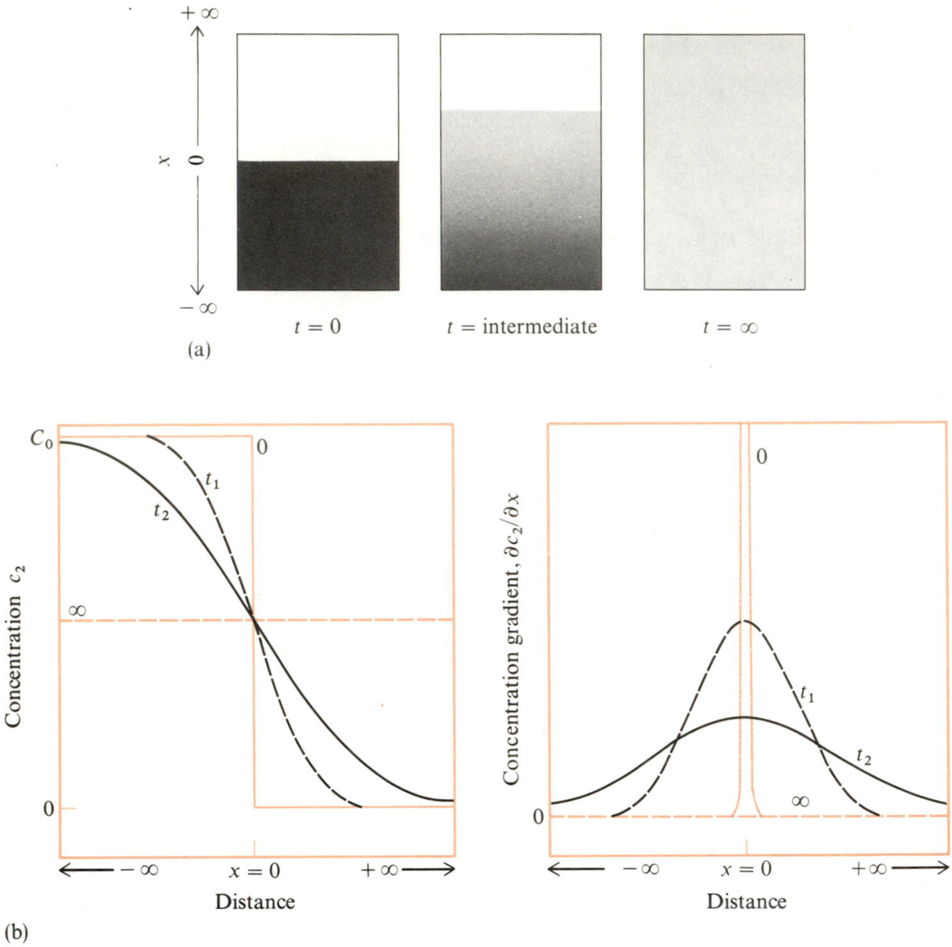

Figure 10-14

Free diffusion. **(a)** Schematic free diffusion measurement. **(b)** Solute concentration and the concentration gradient at times 0, t_1, t_2, and ∞ in a free diffusion experiment starting from an initial sharp boundary like that shown in part a. [After C. Tanford, *Physical Chemistry of Macromolecules* (New York: Wiley, 1961).]

zone (Fig. 10-15) can used to examine how concentrations are changed with time by concentration gradients. Consider the change in the weight of solute in the volume bounded by the two surfaces at x and $x + dx$. The flux across the surface at x from left to right is $J(x)$. The flux across the surface at $x + dx$ from right to left is $-J(x + dx)$. The net rate of accumulation of solute mass m_v within the volume defined by x and dx thus is

$$dm_v/dt = J_2(x) - J_2(x + dx) \tag{10-47}$$

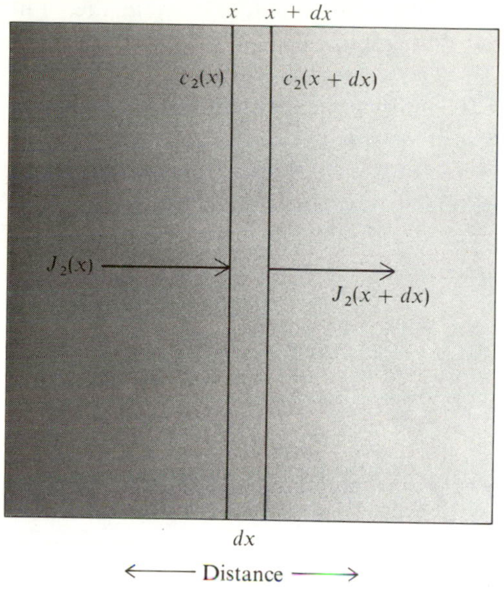

Figure 10-15

Mass flow through a zone in a fluid containing a solute concentration gradient. The flux, J_2, is defined as positive for net transport from left to right. The zone has a thickness dx.

The change in solute concentration c_2 within this volume V is

$$dc_2/dt = (1/V)(dm_v/dt) = (1/dx)(dm_v/dt) \tag{10-48}$$

where the second equality holds because the sides have unit area, and the thickness is dx. When Equation 10-48 is substituted into Equation 10-47, the result is

$$(dc_2/dt)_x = [J_2(x) - J_2(x + dx)]/dx = -(\partial J_2/\partial x)_t \tag{10-49}$$

in the limit of a thin zone.

Equation 10-49 is fundamental to all hydrodynamic studies, and it is true regardless of what forces or gradients are acting on the system. This equation is simply a statement of the conservation of mass. If the fluxes into and out of a volume element are not equal, the concentration within that element must change as a function of time. Equations 10-45 and 10-49 can be combined to yield Fick's second law of diffusion:

$$(dc_2/dt)_x = -\partial(-D\partial c_2/\partial x)/\partial x)/\partial x = D(\partial^2 c_2/\partial x^2) \tag{10-50}$$

Note carefully that the second equality in Equation 10-50 is true only in the limit of very low concentration. As we shall show, D depends on the frictional properties of solute molecules. These are concentration-dependent and, if the concentration of solute varies as a function of x, the value of D could vary also. From Equation 10-50, you can see that D has dimensions of cm^2 sec^{-1}; this has useful implications. By dimensional analysis, we can say that a net distance moved by a diffusing molecule ought to be proportional to $\sqrt{Dt}$. Such a time dependence arises from the fact that solute molecules are executing a random walk in a diffusion process. You will see other examples of random walks when we discuss the conformation of polymer chains in Chapter 18.

Box 10-3 SOLUTION OF FICK'S SECOND LAW BY FOURIER TRANSFORMS

A second common set of boundary conditions for the diffusion process is to start with an initial thin band of solute. This could be a zone of sample in any kind of separation technique, such as gel exclusion chromatography (Chapter 12) or band sedimentation (Chapter 11). In the absence of any external forces, the band will spread by diffusion. If the band is initially at $x = 0$ and contains a total of W_0 grams of solute, the boundary conditions at $t = 0$ are $c_2(x, 0) = W_0\,\delta(0)$, where $\delta(0)$ is the Dirac delta function, whose value is infinite when its argument is zero and is zero everywhere else (see Box 13-3). The diffusion equation of a band is solved with these boundary conditions in a straightforward way using the method of Fourier transforms. By applying the Fourier operator $\int_{-\infty}^{\infty} e^{ikx}\,dx$ to both sides of Equation 10-50, one obtains

$$\frac{d}{dt}\int_{-\infty}^{\infty} e^{ikx}\, c_2(x, t)\, dx = D\int_{-\infty}^{\infty} e^{ikx}\,\frac{d^2}{dx^2}\, c_2(x, t)\, dx \qquad \text{(A)}$$

Integrate the term in d^2c_2/dx^2 by parts twice. The right-hand side of Equation A becomes

$$De^{ikx}\, dc_2/dx\Big]_{-\infty}^{\infty} - (i/k)De^{ikx}\, c_2\Big]_{-\infty}^{\infty} - k^2 D\int_{-\infty}^{\infty} e^{ikx}\, c_2\, dx$$

Using the fact that $c_2(\pm\infty, t)$ and $(dc_2/dx)(\pm\infty, t)$ are 0 for all times, the first two terms are zero. Thus Equation A simplifies to

$$d\overline{c}_2(k, t)/dt = -k^2 D\overline{c}_2(k, t)$$

where

$$\overline{c}_2(k, t) = \int_{-\infty}^{\infty} e^{ikx}\, c_2(x, t)\, dx$$

is the Fourier transform of $c_2(x, t)$ (see Box 13-2). This is now a simple first-order equation.

Solutions to Fick's laws

Solving Equation 10-50 yields the solute concentration everywhere in the sample as a function of time, $c_2(x, t)$. To solve this equation, one must specify boundary conditions. For the free diffusion case shown in Figure 10-14, $c_2(x, 0) = c_0$ if $x < 0$, and $c_2(x, 0) = 0$ if $x > 0$. There are powerful techniques for solving such diffusion equations using Fourier transforms (one example is given in Box 10-3), but it will be sufficient here to outline a less elegant approach.

Because we want a solution of Equation 10-50 for $c_2(x, t)$ in terms of D, which has units of x^2/t, it is reasonable to postulate that c_2 is a function of x^2/t. Define the

Its solution is

$$\bar{c}_2(k, t) = \bar{c}_2(k, 0)e^{-k^2Dt}$$

The initial boundary condition can be used to evaluate

$$\bar{c}_2(k, 0) = \int_{-\infty}^{\infty} e^{ikx}\, c_2(x, 0)\, dx = \int_{-\infty}^{\infty} e^{ikx}\, dx\, W_0\, \delta(0) = W_0$$

Now, to regenerate the concentration as a function of x and t, we must apply the inverse Fourier operator, $(1/2\pi)\int_{-\infty}^{\infty} e^{-ikx}\, dk$ to $\bar{c}_2(k, t)$. The result is

$$c_2(x, t) = (1/2\pi)\int_{-\infty}^{\infty} e^{-ikx}\, W_0 e^{-k^2Dt}\, dk$$

This integral is performed by changing variables to let $y = k\sqrt{Dt} + ix/\sqrt{4Dt}$.

$$c_2(x, t) = (W_0 e^{-x^2/4Dt}/2\pi\sqrt{Dt}) \int_{-\infty + ix/\sqrt{4Dt}}^{\infty + ix/\sqrt{4Dt}} e^{-y^2}\, dy$$

Then the complex integral can be evaluated by standard methods. However, it turns out that it has exactly the same value as the real integral

$$\int_{-\infty}^{\infty} e^{-z^2}\, dz = \sqrt{\pi}$$

so that

$$c_2(x, t) = W_0 e^{-x^2/4Dt}/\sqrt{4\pi Dt} \qquad \text{(B)}$$

As Equation B shows, the shape of the initial thin zone becomes Gaussian. Its width broadens as diffusion proceeds according to $\sqrt{4Dt}$. This is an important consideration in the design of separation procedures. It also allows the diffusion constant to be determined from the width of the band.

variable $y^2 = x^2/t$, and use this relationship to change variables in Equation 10-50.

$$\frac{\partial c_2}{\partial t} = \left(\frac{\partial c_2}{\partial y}\right)_x \frac{\partial y}{\partial t} = -\frac{x}{2t^{3/2}}\left(\frac{\partial c_2}{\partial y}\right)_x = -\frac{y}{2t}\left(\frac{\partial c_2}{\partial y}\right)_x \tag{10-51}$$

$$\frac{\partial c_2}{\partial x} = \left(\frac{\partial c_2}{\partial y}\right)_t \frac{\partial y}{\partial x} = \frac{1}{t^{1/2}}\left(\frac{\partial c_2}{\partial y}\right)_t \tag{10-52}$$

$$\frac{\partial^2 c_2}{\partial x^2} = \left(\frac{\partial}{\partial x}\right)_t \frac{1}{t^{1/2}}\left(\frac{\partial c_2}{\partial y}\right)_t = \left(\frac{\partial}{\partial y}\right)_t \left(\frac{\partial y}{\partial x}\right)_t \frac{1}{t^{1/2}}\left(\frac{\partial c_2}{\partial y}\right)_t = \frac{1}{t}\left(\frac{\partial^2 c_2}{\partial y^2}\right)_t \tag{10-53}$$

The result is that the original equation for Fick's second law can be rewritten as

$$-(y/2)(\partial c_2/\partial y)_x = D(\partial^2 c_2/\partial y^2)_t \tag{10-54}$$

This is now an ordinary second-order differential equation in y. Integrating once yields

$$dc_2/dy = (1/k)e^{-y^2/4D} \tag{10-55}$$

where k is an integration constant to be determined by using the boundary conditions. Changing variables back to x and t,

$$(\partial c_2/\partial x)_t = (1/kt^{1/2})e^{-x^2/4Dt} \tag{10-56}$$

We can integrate this equation from $x = -\infty$ to $x = +\infty$ by assuming that the sample is so long that the concentrations at the top (0) and bottom (c_0) never change from the initial concentrations.

$$\int_{c_0}^{0} dc_2 = -c_0 = -(1/kt^{1/2})\int_{-\infty}^{\infty} e^{-x^2/4Dt}\,dx = -\sqrt{4\pi Dt}/kt^{1/2} \tag{10-57}$$

Using this integral to evaluate k, and substituting the result into Equation 10-56, we have

$$(\partial c_2/\partial x)_t = -(c_0/\sqrt{4\pi Dt})e^{-x^2/4Dt} \tag{10-58}$$

By symmetry, the concentration at $x = 0$, or $c_2(0, t)$, is $c_0/2$ for all times. Equation 10-58 can be integrated again from $x = 0$ to x by using this boundary condition. Using the variable $v^2 = x^2/4Dt$, the result is

$$c_2(x, t) = (c_0/2)\left[1 - (2/\sqrt{\pi})\int_0^{x/\sqrt{4Dt}} e^{-v^2}\,dv\right] \qquad \text{for } x > 0,\ t > 0 \tag{10-59}$$

No analytical form exists for the integral remaining in Equation 10-59. It is called the probability integral, and it is tabulated in many handbooks.

Measurement of diffusion

A variety of experimental methods exists for measuring the concentration of a solute (c_2) or its gradient as a function of its position in the sample. The most direct method is to scan the absorbance of the sample at various times. This yields the concentration from the Beer–Lambert law. For a sample without convenient absorption properties, a Rayleigh interferometer can be used to measure refractive index differences. These can also be converted into a plot of concentration versus distance. Finally, Schlieren optics, which are sensitive to refractive index gradients, yield dc_2/dx directly. These latter two optical systems are elegant but complicated. The interested reader is referred elsewhere for a detailed description of their principles and practical considerations (Van Holde, 1971; Freifelder, 1976). Once c_2 or dc_2/dt is determined, a number of numerical or graphical techniques exist for extracting the value of D from Equation 10-58 or 10-59.

Despite the theoretical simplicity of free diffusion, in practice the technique is plagued with extreme experimental difficulties. The diffusion constants of high-molecular-weight molecules are on the order of 10^{-7} to 10^{-6} cm^2 sec^{-1}. One can estimate (from Equation B in Box 10-3) that an initial thin band will take hours to spread only 1 mm. To measure the spreading accurately, the initial thickness must be much less than 1 mm; creating such bands is not a simple matter. It will take hundreds of hours for the spreading to reach 1 cm. Such a spreading reduces the demands on the sharpness of the original band, but the long time interval introduces other experimental problems. Exactly the same considerations apply to the initial conditions of the step function shown in Figure 10-14. Anything that disturbs the sample during the long time required for the measurements can distort the results. For example, thermal fluctuations in the environment can lead to transitory temperature gradients in the sample; these can cause convective mixing. This mixing produces mass flow much greater than the diffusion. The most common way to prevent thermal gradients is precise thermostatic control of the sample, which usually requires mechanical devices. Any vibrations from these devices can cause mechanical stirring, which again distorts diffusion effects.

Another problem that enters into the design of diffusion experiments is gravitational instabilities. In the experiment shown in Figure 10-14, it is necessary that the macromolecular solution be placed under the less-dense pure solvent. If their positions were reversed, the solution would simply fall through the solvent. A thin horizontal zone will suffer the same fate. A thin vertical zone is not any safer. The unbalanced pressure resulting from the greater density of the zone will lead to immediate collapse by convective mixing. Zones must be gravitationally stabilized by the use of a density gradient to minimize these problems. More will be said about this in Chapter 11 when we discuss sedimentation.

Molecular interpretation of the diffusion constant

It is possible to measure diffusion in a simple two-component system by applying clever experimental design. The diffusion constant of a protein or other macromolecule can be determined from these measurements by using the equations in the preceding section. Now we shall show how to relate this bulk-solution parameter to the frictional characteristics of individual solute molecules.

In a solution, each molecule will undergo Brownian motion. It is accelerated by collisions with other molecules and decelerated by viscous drags. An explicit calculation of diffusion must consider the motion of each molecule. The velocity v_2 of one solute molecule is described by the Langevin equation. For a one-dimensional system, this is

$$m_2(dv_2/dt) = -fv_2 + A(t) \tag{10-60}$$

where m_2 is the mass of the particle, f is the frictional coefficient, and $A(t)$ is a random fluctuating force generated by molecular collisions.

The techniques of statistical mechanics can be used to solve the Langevin equation to predict net mass transport of solute molecules. One must average over the motions of individual molecules. These will depend on the distribution of kinetic energy of individual particles and, therefore, one can predict that the final result must be a function of the absolute temperature of the sample. We shall not attempt to show how the Langevin equation is solved. Instead, we shall use some simple thermodynamic arguments to deduce the relationship between the net flux of solute across a surface and the frictional coefficient of individual macromolecules.

Instead of a distribution of molecular velocities, imagine that all solute molecules are moving across a surface with the same average velocity $\langle v_2 \rangle$. The flux across the surface will be $J_2 = c_2\langle v_2 \rangle$. Molecules moving in this way will feel a total frictional force per molecule of $\langle v_2 \rangle f$. Thus, the force per gram is $N_0\langle v_2 \rangle f/M_2$, where M_2 is the molecular weight. If there is to be no net acceleration, this frictional force must be opposed by other, equal forces. We showed earlier that concentration gradients produce a diffusion force $-(\partial\hat{\mu}_2/\partial x)_t$ that leads to net transport of solute. In a free diffusion experiment with no external forces, we can set this diffusion force equal to the total frictional force:

$$N_0\langle v_2 \rangle f/M_2 = -(\partial\hat{\mu}_2/\partial x)_t \tag{10-61}$$

Equation 10-61 can be used to compute the average velocity, $\langle v_2 \rangle$, that is needed to reproduce the observed flux across a surface due to Brownian motion. Thus,

$$J_2 = c_2\langle v_2 \rangle = -(c_2 M_2/N_0 f)(\partial\hat{\mu}_2/\partial x)_t \tag{10-62}$$

A comparison of Equation 10-62 with Equation 10-44 shows that we have evaluated the phenomenological coefficient L_{22} as c_2M_2/N_0f. However, it is more convenient to rearrange Equation 10-62 so that it can be compared with Fick's laws of diffusion. To do this, recall that

$$\hat{\mu}_2 = \hat{\mu}_2^0 + (RT/M_2) \ln a_2 = \hat{\mu}_2^0 + (RT/M_2) \ln \hat{\gamma}_2 c_2$$

When this substitution is made in Equation 10-62, we obtain

$$\begin{aligned} J_2 &= -\frac{c_2 M_2 RT}{N_0 f M_2}\left(\frac{\partial(\ln \hat{\gamma}_2)}{\partial x} + \frac{\partial(\ln c_2)}{\partial x}\right) \\ &= -\frac{c_2 kT}{f}\left(\frac{\partial(\ln \hat{\gamma}_2)}{\partial x}\frac{\partial(\ln c_2)}{\partial(\ln c_2)} + \frac{\partial(\ln c_2)}{\partial x}\right) \end{aligned} \quad (10\text{-}63)$$

In the second line of Equation 10-63, we have simply multiplied the first partial derivative by $\partial(\ln c_2)/\partial(\ln c_2)$ and substituted the value of Boltzmann's constant k for R/N_0. Rearranging the derivatives in Equation 10-63, and remembering that $\partial(\ln c_2) = c_2^{-1}\partial c_2$, we obtain

$$J_2 = -\frac{kT}{f}\left(1 + \frac{\partial(\ln \hat{\gamma}_2)}{\partial(\ln c_2)}\right)\frac{\partial c_2}{\partial x} \quad (10\text{-}64)$$

Equation 10-64 has exactly the same form as Fick's first law. Setting terms in this equation equal to the corresponding terms in Equation 10-45, we find the diffusion constant to be

$$D = (kT/f)[1 + \partial(\ln \hat{\gamma}_2)/\partial(\ln c_2)] \quad (10\text{-}65)$$

The activity coefficient $\hat{\gamma}_2$ is 1.0 at infinite dilution, $\partial(\ln \hat{\gamma}_2)/\partial(\ln c_2) = 0$, and the diffusion constant becomes

$$D_0 = kT/f \quad (10\text{-}66)$$

where D_0 is the diffusion constant at infinite dilution. Equations 10-65 and 10-66 were first derived by Albert Einstein, and they are called the Einstein–Sutherland equations. A direct solution of the Langevin equation yields the same result (Eqn. 10-65) for the long-time behavior of dilute solutions.

The diffusion constant of a molecule will be a function of the temperature at which the measurement was performed. It will also depend on the solvent viscosity, because this enters into Stokes' law. And solvent viscosities are themselves a function of temperature. Rather than deal with all these variables, it is customary to correct any measured diffusion constant to what would be observed if the measurements were carried out at 20°C in pure water:

$$D_{20,w}/D = (293/T)(\eta_{w,T}/\eta_{w,20})(\eta_{soln,T}/\eta_{w,T}) \tag{10-67}$$

where $D_{20,w}$ is the diffusion constant in pure water at 20°C; T is the actual absolute temperature used; $\eta_{w,20}$ is the viscosity of pure water at 20°C; $\eta_{w,T}$ is the viscosity of pure water at temperature T; and $\eta_{soln,T}$ is the viscosity of the actual solution used at temperature T.

Table 10-3 lists a sampling of representative measured values of $D_{20,w}$. These cover samples that range in molecular weight from less than 10^2 d to more than 10^7 d. The corresponding range of diffusion coefficients and of the frictional coefficients derived from these values by using Equation 10-66 is much narrower.

Table 10-3
Results of diffusion measurements

Sample	Molecular weight	$D_{20,w} \times 10^7$	$\bar{V}_2$ ($cm^3\ g^{-1}$)	f/f_{min}.	Maximum possible δ_1 (g H_2O per g protein)	Maximum possible prolate a/b
Glycine	75	93.3	—	—	—	—
Sucrose	342	45.9	—	—	—	—
Ribonuclease (pancreatic)	13,683	11.9	0.728	1.14	0.35	3.4
Lysozyme (egg white)	14,100	10.4	0.688	1.32	0.89	6.1
Bovine serum albumin	66,500	6.1	0.734	1.31	1.02	6.0
Hemoglobin	68,000	6.9	0.749	1.14	0.36	3.4
Tropomyosin	93,000	2.2	0.71	3.22	23.0	62.0
Fibrinogen (human)	330,000	1.98	0.706	2.35	8.4	31.0
Myosin	493,000	1.10	0.728	3.65	34.7	80.0
Bushy stunt virus	10,700,000	1.15	0.74	1.27	0.78	5.3
Tobacco mosaic virus	40,000,000	0.44	0.73	2.19	6.9	24.0

SOURCE: Data from several sources, including Kuntz and Kauzmann (1974), Tanford (1961), and Van Holde (1971).

Interpretation of measured frictional coefficients

What can be learned from the measured frictional coefficients depends on how much is already known about the macromolecule. In the general case of a hydrated ellipsoid of revolution,

$$\boxed{f = 6\pi\eta r_h F} \tag{10-68}$$

where F is the Perrin shape factor given by Equation 10-19. The hydrated radius, r_h, of the equivalent sphere can be calculated from Equation 10-11:

$$r_h = [(3/4\pi)(M/N_0)(\bar{V}_2 + \delta_1 \bar{V}_1)]^{1/3} \tag{10-69}$$

Assuming that the partial specific volume can be measured or estimated accurately, the frictional coefficient in Equation 10-68 is still a function of three variables: the axial ratio of the ellipsoid, the molecular weight, and the hydration. In most cases, two of these must be known independently for the third to be evaluated. Earlier it was shown that δ_1 can be estimated moderately well or measured. Plenty of methods exist that can yield M. Frequently, therefore, determination of f will allow the calculation of the Perrin factor F. This does not immediately yield the shape, however, unless one knows beforehand whether the particle is prolate or oblate.

A single value of F is consistent with many possible shapes, two of which are ellipsoids of revolution. The ambiguity permitted by this is considerable, as an inspection of Table 10-2 will demonstrate. In the limit of large values of F, one can almost surely assign the shape as prolate, because a disk shape with a Perrin factor greater than 1.5 would have an unrealistically thin minor axis. In the limit of small values of F, the translational frictional properties of oblate and prolate ellipsoids become indistinguishable. This is unfortunate because (as illustrated in Fig. 10-9) a prolate ellipsoid and an oblate ellipsoid, both with axial ratio 2, are significantly different physical objects. Naturally, if the shape is already known independently—say, from electron microscopy—then the frictional coefficient can yield either the molecular weight or the hydration if the other of these quantities also is available independently.

Now suppose you know only the molecular weight and partial specific volume and have measured the diffusion constant. Two variables, δ_1 and F, remain undetermined. There is no way to solve uniquely for them, but limits can be found. Given the anhydrous molecular weight, the minimal frictional coefficient a particle can have, if it is an anhydrous sphere, is $f_{min} = 6\pi\eta r_0$, where $r_0 = (3M\bar{V}_2/4\pi N_0)^{1/3}$. Relative to this minimum value, the observed value will be

$$f/f_{min} = [(\bar{V}_2 + \delta_1 \bar{V}_1)/\bar{V}_2]^{1/3} F \tag{10-70}$$

Pairs of values of F and δ_1 that satisfy this transcendental equation can easily be found graphically. But it probably is more useful to consider limits. If all excess friction is due to the hydration, F is 1, and the molecule is a sphere, then a maximal value of the hydration can be computed as

$$\delta_1^{max} = (\bar{V}_2/\bar{V}_1)[(f/f_{min})^3 - 1] \tag{10-71}$$

If the particle has no hydration, the axial ratio is the maximum possible. In this case, the axial ratio of prolate and oblate ellipsoids is found in tables of the Perrin factors using

$$F_{\max} = f/f_{\min} \tag{10-72}$$

It is rather frustrating to settle for such uncertainty. Note, however, that the value of F that can be derived from Equation 10-70 depends on the cube root of the hydration. This is not too sensitive a dependence. Therefore, what is frequently done is to guess a typical hydration value in order to make a reasonable estimate of the axial ratio. Experience has shown that hydrations of 0.3 to 0.4 g H_2O (g protein)$^{-1}$ are common for most proteins. Table 10-3 lists examples of limiting values of hydrations and axial ratios estimated for several real proteins. We shall see later how these compare with what is known from other techniques.

In the worst case, one could be faced with a diffusion constant for a macromolecule about which nothing is known except the partial specific volume. Because the minimum possible value of F is 1.0, and the minimum value of δ_1 is 0.0, Equation 10-69 will yield an estimate for the maximum possible anhydrous molecular weight of the particle. In practice, this value will be uncertain by up to a factor of two for most systems of interest, and therefore it is not too useful.

Diffusion in multicomponent systems

Most macromolecular solutions contain more than two components. Real proteins or nucleic acid preparations are rarely completely homogeneous; even ignoring this, buffer ions or other salts are inevitably present in aqueous solution. In the case of a three-component system, one can write the analog of Fick's first law by starting with Equation 10-37. If component 2 is macromolecule, and component 3 is some other solute,

$$J_2 = -D_{22}(\partial c_2/\partial x) - D_{23}(\partial c_3/\partial x) \tag{10-72}$$

$$J_3 = -D_{32}(\partial c_2/\partial x) - D_{33}(\partial c_3/\partial x) \tag{10-73}$$

D_{22} and D_{33} are the diffusion coefficients that would be calculated by Stokes' law and Equation 10-66 for the two pure components diffusing in separate two-component solutions. The cross-terms D_{23} and D_{32} describe how the flows of the two components interact. Onsager's reciprocity relationship tells us that D_{23} and D_{32} are related in a simple fashion. Usually these cross-terms become negligible in dilute solutions. This is another reason why diffusion and other hydrodynamic measurements should be extrapolated to zero solute concentration.

One special three-component effect is electrostatic interaction. This does not vanish at low concentrations and must be treated in a different way. So far, our discussion of diffusion has ignored the fact that biologically significant molecules are charged. The equations given above are valid for uncharged molecules, or for charged species only in certain limiting cases. For example, an ion moving in a solution containing charged molecules will experience local electric fields. We must add this effect to the Langevin equation (Eqn. 10-60) to account for the motion of a charged particle:

$$m_2(dv_2/dt) = -fv_2 + A(t) + z_2E(t) \tag{10-74}$$

where z_2 is the charge on the ion, and $E(t)$ is a time-dependent electric field provided by all the other charged particles nearby. Electrostatic forces are strong, and it is impossible for an ion to diffuse a finite distance without counterions also diffusing.

First, consider a solution of a salt that is completely ionized into species A and B. The quantity that is measurable experimentally is the diffusion constant D of the salt. This is related to the diffusion constants D_A and D_B expected by Stokes' law and Equation 10-66 for its individual ions. It can be shown that the solution of Equation 10-74 yields

$$D = q_0^2 D_A D_B/(q_A^2 D_A + q_B^2 D_B) \tag{10-75}$$

where

$$q_A^2 = (4\pi/\varepsilon kT)z_A^2 c_A \tag{10-76a}$$

$$q_B^2 = (4\pi/\varepsilon kT)z_B^2 c_B \tag{10-76b}$$

$$q_0^2 = q_A^2 + q_B^2 \tag{10-76c}$$

Here ε is the dielectric constant of the solution, z_A and z_B are the charges on the two ions, and c_A and c_B are their molar concentrations. Because the solution must be electrically neutral, $z_A c_A + z_B c_B = 0$. If the salt consists of high-molecular-weight polyion with large charge, $|z_A|$, and singly charged counterions, $|z_B| = 1$, use of the electroneutrality relationship allows Equation 10-75 to be simplified to

$$D = (1 + z_A^2)D_A D_B/(D_B + z_A^2 D_A) \tag{10-77}$$

This means that, if the charge z_A varies, the observed diffusion constant for a macromolecule of constant size and shape will change markedly. The expected diffusion constant for the macromolecule, D_A, is much less than that of a counterion, D_B. Thus the measured diffusion, D, will be faster than expected for an uncharged macromolecule. In general,

$$D_A < D < D_B \tag{10-78}$$

This leads to an enormous uncertainty in the meaning of a measured D unless the charge on the macromolecule is known. Unfortunately, it usually is not known, although it is potentially accessible from electrophoretic measurements.

In practice, diffusion and almost all other hydrodynamic measurements of highly charged macromolecules are performed in the presence of a huge excess of low-molecular-weight electrolyte. The result is difficult to analyze because the system is multicomponent. Essentially, all of the added electrolyte of the right charge can effectively act as counterions. Thus the effective counterion concentration c_B is much larger than the macromolecule concentration c_A. Referring to Equation 10-76, this means that $q_B^2 \gg q_A^2$, and so $q_0^2 \cong q_B^2$. Inserting these values into Equation 10-75, we obtain $D = D_A$. Thus, all of the complicated effects of charge on diffusion disappear at sufficiently high ionic strength ($\sum_i z_i^2 c_i/2$), and Stokes' law and Equation 10-66 are valid.

Summary

Electron microscopy allows the size and shape of macromolecules to be visualized. The need to use heavy-atom staining procedures to generate contrast between the molecule and the supporting film usually limits resolution severely. When rotational or translational symmetry exists, averaging procedures can greatly improve the quality of electron microscopic images.

In solution, proteins and nucleic acids contain considerable amounts of associated water; they always behave as hydrated particles. Hydrodynamic techniques examine molecular motions. These techniques can be used to infer information about the size and shape of the hydrated macromolecules. Molecular motions in solution are limited by frictional forces that increase linearly with the velocity. The translational frictional coefficient can be measured by techniques such as sedimentation and diffusion. It increases as the one-third power of the molecular weight, and depends on the shape of a molecule. The frictional coefficient of a prolate ellipsoid is somewhat greater than that of an oblate ellipsoid of equal volume.

Rotational frictional properties are measured by techniques such as fluorescence polarization, and flow and electric birefringence. An ellipsoid of revolution is characterized by two rotational frictional coefficients that correspond to rotations about the long and short axes of the ellipsoid. The frictional coefficient for rotation about the short axis of a prolate ellipsoid is especially sensitive to shape. In general, rotational friction increases linearly with the molecular weight, and thus it is a much more sensitive measure of size than is translational friction.

Diffusion is the simplest hydrodynamic technique to analyze, because the only force acting on the macromolecules arises from spatial variations in their concentration. The rate of mass transport across a surface is proportional to the concentration gradient at that surface. The constant of proportionality (D) is called the diffusion constant. It is inversely proportional to the translational frictional coefficient. D can be measured by observing concentration changes. In practice, direct determinations of D using free diffusion measurements are experimentally quite difficult.

Problems

10-1. Show that the translational frictional coefficient of a dimer containing two spherical subunits in contact is 80% that of a dimer in which the two subunits are held very far apart by a frictionless linker.

10-2. Calculate the rotational relaxation times expected for a protein with a molecular weight of 100,000, assuming that the partial specific volume is 0.74 cm^3 g^{-1}, the hydration is 0.3 g H_2O (g protein)$^{-1}$, and that the protein can be accurately represented by a prolate ellipsoid with an axial ratio of 8. (N.B.: It is easier to use Figure 10-11b than Equation 10-20.)

10-3. Suppose that tropomyosin were an oblate ellipsoid with a hydration of 0.3 g H_2O (g protein)$^{-1}$. Calculate the length of the short axis using the data given in Table 10-3. Comment on the result from your general knowledge about protein structure.

10-4. Calculate the average net velocity of lysozyme molecules (in cm sec^{-1}) due to diffusion in a solution where the concentration changes by a factor of two per mm. (See Table 10-3 for necessary data.)

10-5. Suppose that a protein has an anhydrous molecular weight of 15,000 and a partial specific volume of 0.72 cm^3 g^{-1}. What is the largest diffusion constant the protein could possibly have? What, roughly, is the smallest diffusion constant the protein is likely to have if its hydration is 0.4 g H_2O (g protein)$^{-1}$?

References

GENERAL

Finch, J. T. 1975. Electron microscopy of proteins. In *The Proteins*, 3d ed., ed. H. Neurath and R. L. Hill (New York: Academic Press), p. 412.

Kuntz, I. D., Jr., and W. Kauzmann. 1974. Hydration of proteins and polypeptides. In *Advances in Protein Chemistry*, vol. 28, ed. C. B. Anfinsen, J. T. Edsall, and F. M. Richards (New York: Academic Press), p. 239.

Tanford, C. 1961. *Physical Chemistry of Macromolecules*. New York: Wiley.

SPECIFIC

Atkins, P. W. 1976. *Physical Chemistry*. San Francisco: W. H. Freeman and Company.

Bauer, D. R., J. I. Brauman, and R. Pecora. 1974. Molecular reorientation in liquids: Experimental tests of hydrodynamic models. *J. Am. Chem. Soc.* 96:6840.

Berne, B. J., and R. Pecora. 1975. *Dynamic Light Scattering*. New York: Wiley.

Bloomfield, V., D. Crothers, and I. Tinoco, Jr. 1974. *Physical Chemistry of Nucleic Acids*. New York: Harper & Row.

Bloomfield, V., W. O. Dalton, and K. E. Van Holde. 1967. Frictional coefficients of multisubunit structures, I: Theory. *Biopolymers* 5:135. [See also p. 149.]

Crowther, R. A., and A. Klug. 1975. Structural analysis of macromolecular assemblies by image reconstruction from electron micrographs. *Ann. Rev. Biochem.* 44:161.

Edelstein, S. J., and H. Schachman. 1973. Measurement of partial specific volume by sedimentation equilibrium in H_2O–D_2O solutions. In *Methods in Enzymology*, vol. 27, ed. L. Grossman and K. Moldave (New York: Academic Press), p. 83.

Eisenberg, H. 1976. *Biological Macromolecules and Polyelectrolytes in Solution*. Oxford: Claredon Press. [A useful advanced treatise.]

Fujita, H. 1975. *Foundations of Ultracentrifuge Analysis*. New York: Wiley.

Kratky, O., H. Leopold, and H. Stabinger. 1973. The determination of the partial specific volume by a mechanical oscillator technique. In *Methods in Enzymology*, vol. 27, ed. L. Grossman and K. Moldave (New York: Academic Press), p. 98.

Van Holde, K. E. 1971. *Physical Biochemistry*. Englewood Cliffs, N.J.: Prentice-Hall.

11
Ultracentrifugation

11-1 VELOCITY SEDIMENTATION

Any external force acting on suspended particles can lead to mass transport (as described in Eqn. 10-37). Different types of forces require rather different types of experimental apparatus and involve rather different aspects of molecular structure and properties. In this chapter we treat (in considerable detail) the transport resulting from the response to a particular force: radial acceleration in an ultracentrifuge. This example is chosen because, in practice, ultracentrifugation is by far the most widely used hydrodynamic technique for precise analysis of macromolecular properties.

Sedimentation by gravity or by angular acceleration

Gravitational force has an appealing feature for the determination of particle mass: all particles feel it. The force depends only on the mass and not on charge, shape, or details of chemical composition. Unfortunately, the earth's gravity is very weak, and only the largest subcellular biological structures (such as entire metaphase chromosomes) are really affected by it strongly enough to settle from an aqueous solution. Classical gravity-induced sedimentation is used in the analysis of sizes of large particles (such as sand) and can even be used to separate living cells by size or density.

If an initially homogeneous suspension of uniform large particles is allowed to stand undisturbed, the particles will settle to the bottom of their container, providing

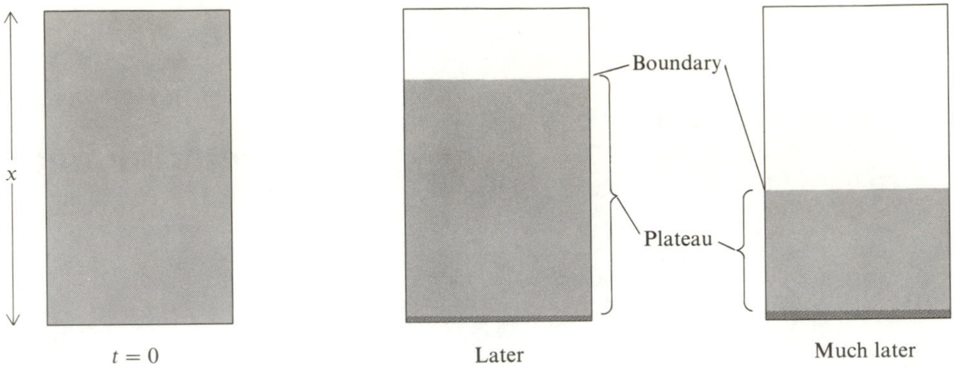

Figure 11-1

Sand settling in a container under the influence of gravity. The initial aqueous suspension is assumed to be uniform. As time proceeds, note the occurrence of a sharp boundary, above which there is only water, and below which there is sand at the same uniform concentration as the original suspension. A second sharp boundary marks the upper limit of the solid sand that has settled out at the bottom.

that they are more dense than the solution (Fig. 11-1). The important thing to note is that a boundary forms as the particles settle. Above this boundary is pure solvent. Below it is a suspension at the same weight concentration as the original homogeneous mixture. This occurs because all particles feel the same gravitational force and therefore move (on the average) with the same velocity. Thus the boundary moves with the same velocity as that of the average sedimenting particle. Note that, if the particles are less dense than the solvent, they will rise rather than sink—just as cream eventually floats out on the top of whole milk.

Now ponder this question: Why do large particles (denser than water) sediment, whereas dense small particles do not? As we shall show, the rate of sedimentation of a particle increases with increased molecular weight, but the rate of diffusion decreases with increased molecular weight. In the limit of large particles such as sand grains, sedimentation is completely dominant and occurs with a very sharp boundary. In the limit of very small particles, diffusion is almost completely dominant, and no boundary forms. However, there is still at least a slight tendency for any particles more dense than the solvent to be found near the bottom of a container rather than near the top. We shall calculate this probability later.

Forces much larger than the earth's gravity are required to cause appreciable sedimentation of typical proteins or nucleic acids. Such forces can be obtained by subjecting particles to an accelerating field. The force is just $F = ma$ for linear acceleration, but it is impractical to sustain such accelerations for extended time periods. The logical alternative is angular acceleration. The radial force produced in a spinning object of mass m is

$$F = m\omega^2 r \tag{11-1}$$

where r is the radius from the center of rotation, and ω is the angular frequency in radians per second. For a mass of 1 g, rotation at 60,000 rpm produces a force of 3.95×10^8 dyn at a radius of 10 cm. This is more than 400,000 times the force of gravity (which is 980 dyn at the earth's surface). Such a force may seem very strong, but it is not sufficient to render diffusion forces insignificant for most proteins or nucleic acids. Thus, the simultaneous effects of sedimentation and diffusion must be considered when we describe motion of macromolecules in the ultracentrifuge.

The ultracentrifuge

Considerable ingenuity has gone into the design of ultracentrifuges, instruments that can spin a sample of solution at up to 70,000 rpm. Figure 11-2 is a schematic diagram of an analytical ultracentrifuge. Samples are held in cells or tubes (Fig. 11-3) in an

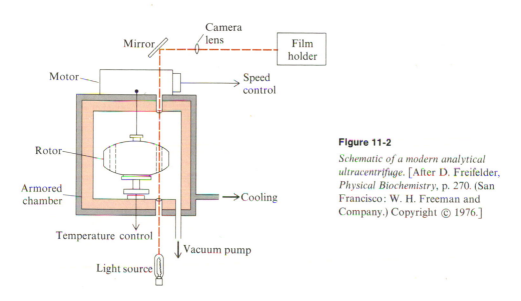

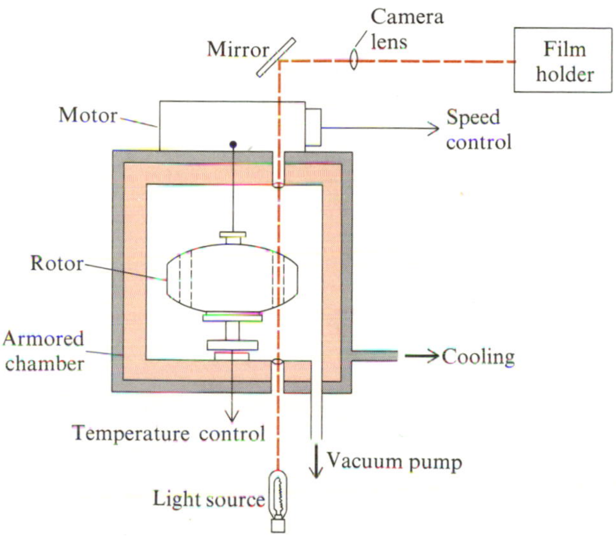

Figure 11-2
Schematic of a modern analytical ultracentrifuge. [After D. Freifelder, *Physical Biochemistry*, p. 270. (San Francisco: W. H. Freeman and Company.) Copyright © 1976.]

aluminum or titanium rotor driven by an electric motor. At typical rotor speeds, friction between the spinning rotor and air would cause intolerable heating. Thus, the chamber in which the rotor spins must be brought to a high vacuum. The temperature of the sample must be regulated precisely, to avoid convective mixing, and this regulation can be difficult. The forces generated within the rotor are enormous, and it is not all that uncommon for a rotor to fly into fragments at high speeds. Thick steel guard rings must surround the rotor chamber to contain these fragments.

Mechanical balance of the rotor is critical. Suppose two sample cells 180° apart on the rotor differ in weight by a milligram. At 400,000 × g, the net force differential on the rotor will be almost a pound. This force is quite considerable and would cause the rotor to wobble on its axis. It is important to balance sample weights, but

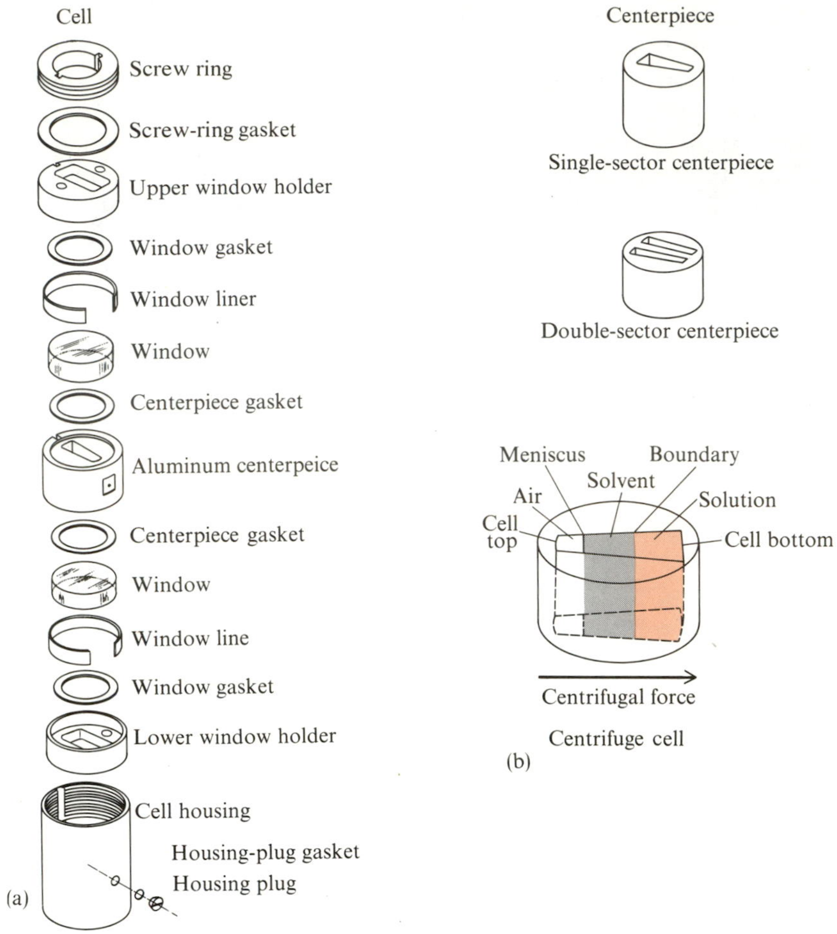

Figure 11-3

Centrifuge cell and centerpieces. **(a)** Exploded view of the cell. **(b)** The centerpiece. [Courtesy of Beckman Instruments.]

they cannot be balanced perfectly. Therefore, a flexible shaft is used in any ultracentrifuge so that the rotor can find its precise center of mass in each case and spin about an axis through that center. This allows a sample mismatch of up to 0.5 g in analytical rotors without adverse effects.

There are two basic applications of ultracentrifugation: preparation and analysis. As a preparative technique, it can be used to purify samples or to separate mixtures for subsequent analysis. Here, the sample is spun at high speed in a cylindrical tube for a fixed time period and then removed from the centrifuge before any measurements are performed. The advantages of this approach are that large sample volumes (5 to 100 ml) can be used, and that a wide variety of biochemical and physical measurements are possible on the recovered sample. The disadvantages are that some stirring or mixing inevitably accompanies slowing the rotor and isolating the sample, and that much information is lost because only a single time point is available.

In analytical ultracentrifugation, the bulk movement of solute molecules is monitored directly as a small sample (0.1 to 1 ml) is spun in the rotor. This is made possible by optical systems that send light through the sample parallel to the rotation axis. The optical system can be absorbance, Rayleigh interference, or Schlieren—the same three systems mentioned for study of diffusion.

A critical feature of the analytical ultracentrifuge is the design of the sample cell. This cell must be sector-shaped when viewed parallel to the rotation axis (Fig. 11-3b). Because the acceleration force is radial, sedimenting molecules will move along radii. If a rectangular cell were used (Fig. 11-4a), the paths of many molecules would lead to collision with the side walls of the cell. They might remain there, or

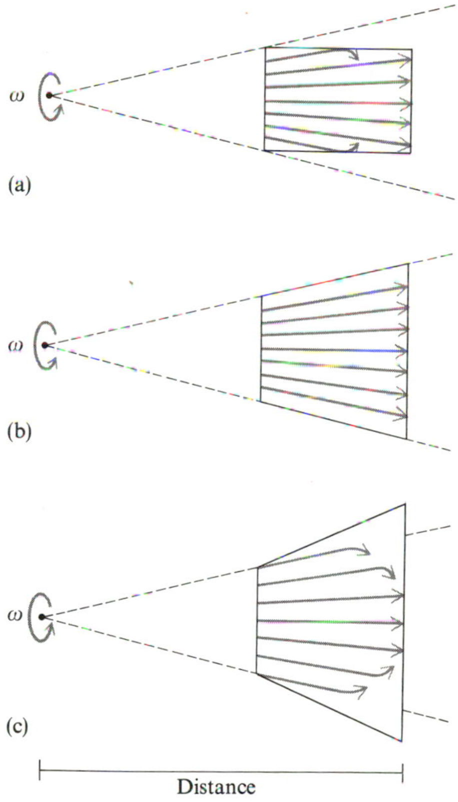

Figure 11-4

Expected solute transport in ultracentrifuge cells of different geometries. In each case, the cells are viewed looking down the axis of rotation. **(a)** In a rectangular cell, the solute collides with the walls. This cell is equivalent to the cylindrical tubes used in all preparative ultracentrifuges. **(b)** In the sector-shaped cell used in all analytical ultracentrifuges, the solute moves freely along radial paths. **(c)** In a cell diverging more sharply than a sector, the solvent collides with the walls and then deflects the solute from the walls.

they might pile up and ultimately fall to the bottom, causing stirring. In either case, the boundary conditions needed to describe the hydrodynamics become very complicated. One might think that the cell shape in Figure 11-4c would circumvent these problems, but it does not. This is because every solute movement must be accompanied by a compensating solvent flow. If macromolecules are sedimenting from left to right in Figure 11-4, then solvent is moving from right to left. In Figure 11-4c, the water will collide with the walls of the cell and be deflected back toward the center, causing stirring.

The sector shape shown in Figure 11-4b makes quantitative sedimentation measurements possible but, as you will shortly see, seriously complicates the mathematical analysis of the experiment. Sample cells in preparative ultracentrifugation traditionally are cylindrical tubes—definitely not sector-shaped. Therefore, some solute molecules are lost on the walls in these tubes—an effect for which correction must be made if accurate quantitation of the results is desired. In practice, not all investigators remember to make the necessary corrections.

Describing transport in the ultracentrifuge: the Lamm equation

Figure 11-5 is an expanded view of a sector-shaped sample cell. Defined are the solvent–air interface, x_m (the meniscus), x_b (the bottom of the cell), ϕ (the angle of the sector shape), and a (the thickness of the cell along the rotation axis). We can treat the hydrodynamics of sedimentation as a one-dimensional problem, with x characterizing the distance from the rotation axis. Because the coordinate x is a radial distance, it defines the location of radial-shaped surfaces.

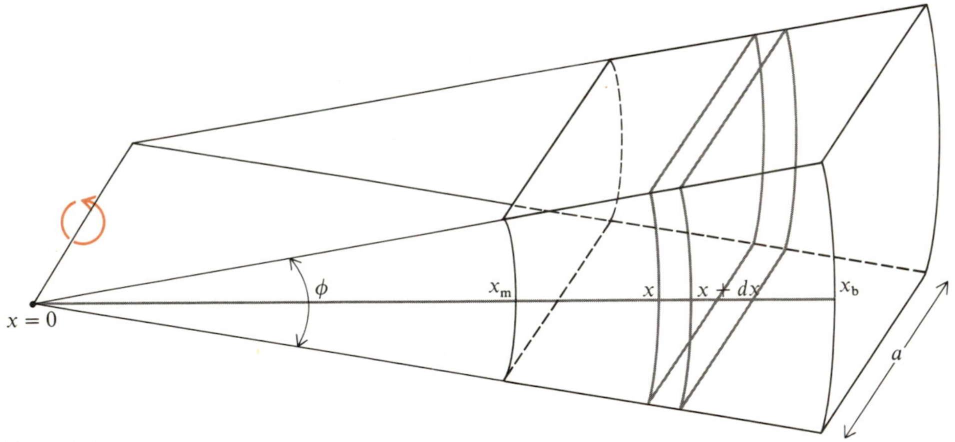

Figure 11-5

Geometry of a typical sector-shaped ultracentrifuge cell. The radial distance is x; the position of the meniscus is x_m (effectively the top of the solution, rather than the actual physical top of the cell); the bottom of the cell is x_b. The rotation axis is at $x = 0$. Mass transport in the zone between x and $x + dx$ is considered in deriving the Lamm equation (Eqn. 11-7).

Consider a system with two components, solvent (1) and solute (2). The phenomenological equation for flux (Eqn. 10-39) immediately yields an equation for the rate of solute mass transport across a surface at radius x in the ultracentrifuge:

$$J_2 = L_{22}[\omega^2 x - (\partial \hat{\mu}_2/\partial x)] \tag{11-2}$$

The first term within brackets accounts for the force due to radial acceleration per gram of macromolecules; the second term describes the diffusion force that will develop in the presence of any chemical potential gradients.

First, we shall use a mechanical description of the sedimentation experiment to identify experimentally measurable parameters. Then, we shall use the thermodynamics of the experiment to connect measured quantities with molecular parameters. The applied force causing flow in the centrifuge is $\omega^2 x$ per unit mass. If there is no diffusion, this force will produce a constant drift velocity (v) in the hydrodynamic steady state. The velocity will be a function of the size and shape of the sedimenting particle, but it must be proportional to the applied force. We define that proportionality constant (s) to be the sedimentation coefficient:

$$s = v/\omega^2 x = (1/\omega^2 x)(dx/dt) \tag{11-3}$$

From Equation 11-3, the units of s are seconds. However, measurements usually are expressed in Svedberg units (1 S = 10^{-13} sec), named after The Svedberg, the inventor of the ultracentrifuge. Keep in mind that s is defined as the velocity per unit field; therefore, any measured value of s should be independent of the angular rotation frequency of the ultracentrifuge.

We can use the definition of s to describe the flux across the surface at radius x in Figure 11-5. Equation 11-2 ensures that contributions due to sedimentation and diffusion can be described separately. If the concentration of solute molecules at x is c_2 g cm^{-3}, and if they all are moving at velocity $v = \omega^2 xs$ due to the applied force, then the flux caused by angular acceleration is $c_2 \omega^2 xs$. If there is a concentration gradient at x, then there will be a flux due to diffusion given by Fick's first law (Eqn. 10-45). Thus, Equation 11-2 becomes

$$J_2 = \omega^2 xsc_2 - D(\partial c_2/\partial x) \tag{11-4}$$

We want to calculate the change in solute concentration in a volume element located a distance x from the rotation axis. First consider the total mass transport across the surface at x. Because J_2 is a flux per unit area, the total mass transport is J_2A. The area A of the surface at radius x is $ax\phi$. Just as we did earlier in discussing diffusion, we must develop an equation for the conservation of mass. The total rate of solute mass change in the volume bounded by surfaces at radii x and $x + dx$ will be

$$dm_2/dt = J_2A(x) - J_2A(x + dx) \tag{11-5}$$

The concentration change is the mass change divided by the volume between the two surfaces:

$$\begin{aligned} dc_2/dt &= (1/\phi xa\,dx)[J_2A(x) - J_2A(x + dx)] \\ &= (-1/\phi xa)(dJ_2A/dx) = -(1/x)[d(xJ_2)/dx] \end{aligned} \tag{11-6}$$

Combining Equations 11-4 and 11-6, we obtain the Lamm equation:

$$\frac{dc_2}{dt} = \left(\frac{1}{x}\right)\frac{d\{x[D(dc_2/dx) - c_2s\omega^2x]\}}{dx} \tag{11-7}$$

This is the most general hydrodynamic description of the mass transport of a two-component system in the ultracentrifuge.

At low solute concentration, the diffusion and sedimentation constants should be independent of concentration. Equation 11-7 then can be simplified:

$$dc_2/dt = D[(d^2c_2/dx^2) + (1/x)(dc_2/dx)] - s\omega^2[x(dc_2/dx) + 2c_2] \tag{11-8}$$

A few of the features of this equation are worth noting. If ω^2 is zero, Equation 11-8 simply describes diffusion. It differs from the form of Fick's second law (Eqn. 10-50) because of the sector-shaped cell. Note that the second diffusion term is undefined at $x = 0$. This is not a problem, because one never places the beginning of a cell at the rotation axis. Even if one did, the volume (and therefore the mass) at that point would be zero anyhow.

Solving the Lamm equation with constant *s* and no diffusion

If D is zero, Equation 11-8 can be solved exactly (see Fujita, 1975). For a sample that at zero time has a uniform solute concentration c_0, the results are

$$c_2(x,t) = 0 \qquad \text{if } x_m < x < \bar{x}$$
$$c_2(x,t) = c_0 e^{-2s\omega^2 t} \qquad \text{if } \bar{x} < x < x_b \tag{11-9}$$

where

$$\bar{x} = x_m e^{s\omega^2 t} \tag{11-10}$$

Here we have ignored the accumulation of material at the bottom of the cell.

Figure 11-6a shows these results schematically. There is a sharp increase in solute concentration from zero to a value that is independent of x at any particular time. The

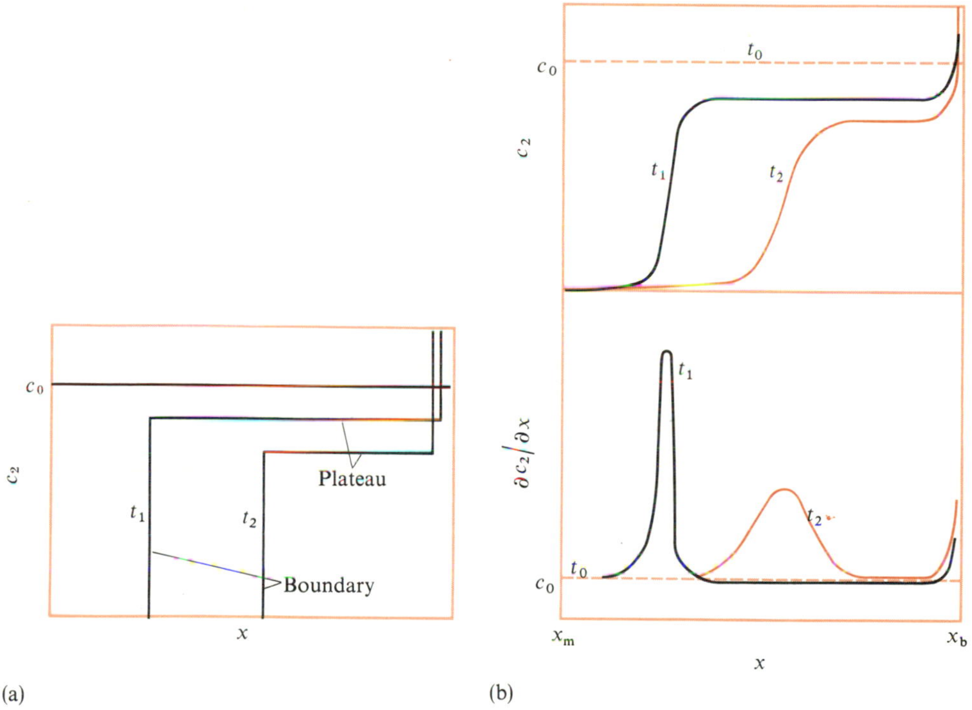

Figure 11-6

Solute concentration profiles during ultracentrifugation of a typical solution of a homogeneous macromolecule. The times shown might be $t_0 = 0$ hr, $t_1 = 1$ hr, and $t_2 = 2$ hr for a large protein at 50,000 rpm. **(a)** If $D = 0$, a sharp boundary is formed. The boundary moves in time. Above the boundary is a flat plateau, whose concentration decreases with time. **(b)** If $D \neq 0$, diffusion causes a broadening of the boundary as a function of time. The concentration gradient $(\partial c_2/\partial x)$ is also shown, as it might be measured by some optical systems used in the ultracentrifuge.

step-function increase occurs at $\bar{x}$, which is called the boundary position. The position $\bar{x}$ changes with time, moving with increasing speed (as shown by Eqn. 11-10). The region $x > \bar{x}$ is called the plateau. It remains "flat" (uniform concentration), but the actual concentration decreases progressively with time (as shown by Eqn. 11-9).

It may seem paradoxical that a region of x-independent solute concentration can be maintained in the cell while the concentration in this region decreases exponentially with time. This occurs because the radial acceleration and the cross-sectional area both increase linearly with x, and their effects are combined. Consider a thin layer of solution at position $x' > \bar{x}$ at time t_1. There molecules are sedimenting with velocity $v' = \omega^2 s x'$. Pick another layer at position $x'' > x'$ (Fig. 11-7). Molecules there

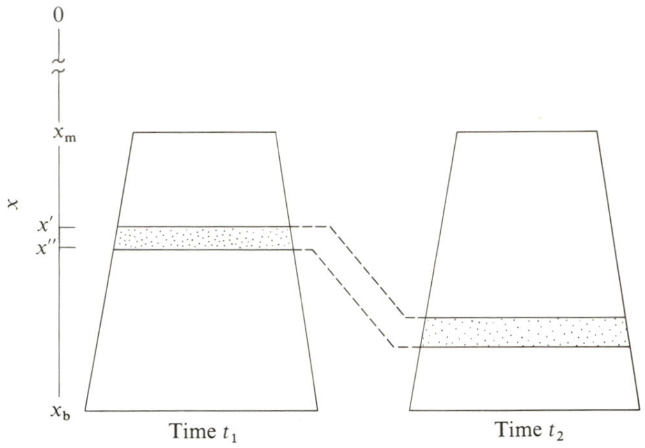

Figure 11-7

Demonstration of radial dilution. A finite zone of solute is followed from an early to a late time during velocity sedimentation. Each dot represents a single macromolecule. The area of the solute zone increases because of the sector shape. Furthermore, the zone thickness increases slightly because the solute at the bottom (x'') is always subjected to a somewhat larger acceleration ($\omega^2 x''$) than is the solute at the top (x'), which feels an acceleration of $\omega^2 x'$, so that the bottom of the zone accelerates away from the top of the zone.

are moving at velocity $v'' = \omega^2 s x'' > v'$. The two layers define a zone as shown in the figure. Now examine the position of the two edges of the zone at some time t_2. The molecules originally at x'' have moved farther than those originally at x', because they have been subjected to stronger radial acceleration at all times. Therefore, the width of the zone increases with time. The area of the zone also increases because of the sector shape. Thus the concentration of solute in the zone drops (because the number of solute molecules inside has remained constant while the volume has increased). This effect is called radial dilution.

To show that a plateau is maintained, we must demonstrate that all zones are subject to the same radial dilution. It is sufficient to prove that the relative volume change of a zone is independent of its position. The volume V of a zone is its thickness $(x'' - x')$ multiplied by its area. The area is the average of the two surfaces: $\frac{1}{2}(\phi x'a + \phi x''a)$. Therefore,

$$V = \tfrac{1}{2}\phi a[(x'')^2 - (x')^2] \tag{11-11}$$

The velocity of the x' surface of the zone is

$$v' = dx'/dt = \omega^2 s x' \tag{11-12}$$

For a surface originally (time t_1) located at $x'(t_1)$, solution of this simple differential equation gives, at time t_2,

$$x'(t_2) = x'(t_1)e^{\omega^2 s(t_2 - t_1)} \tag{11-13}$$

Similarly, for a surface originally located at $x''(t_1)$, we obtain

$$x''(t_2) = x''(t_1)e^{\omega^2 s(t_2 - t_1)} \tag{11-14}$$

Thus the volume of the zone at time t_2 compared with that at time t_1 is

$$\begin{aligned} V(t_2)/V(t_1) &= \{[x''(t_2)]^2 - [x'(t_2)]^2\}/\{[x''(t_1)]^2 - [x'(t_1)]^2\} \\ &= e^{2\omega^2 s(t_2 - t_1)} \end{aligned} \tag{11-15}$$

This volume ratio is independent of x. Therefore, because the number of molecules in each zone remains constant and each zone increases in volume by the same ratio, the concentrations in all the zones must remain equal. Note that we have, in effect, rederived Equations 11-9 and 11-10. Let $t_1 = 0$ in Equations 11-13 and 11-15, and let $x'(0)$ be the top of the cell (x_m). Then Equation 11-13 is identical to Equation 11-10 If the original concentration at $t = 0$ is c_0, then Equation 11-15 shows that the plateau concentration at a later time t_2 will be diluted to $c_0/e^{2\omega^2 s t_2}$, which is identical to Equation 11-9.

Solving the Lamm equation for more realistic cases

The Lamm equation has been solved analytically or numerically only in certain limiting cases (Box 11-1). The results show that there is in fact a boundary and a plateau, so long as sedimentation forces are strong compared with diffusion. In these more realistic cases, the boundary is not perfectly sharp, but it has a breadth that is dependent on D. The plateau concentration does decrease with time (Fig. 11-6b). Note that diffusion forces will be felt only near the boundary where there is a concentration gradient. Thus it is reasonable that a plateau can be maintained, even in the presence of diffusion.

Because we now know that a plateau will exist (assuming that s is constant), we can solve Equation 11-8 at the plateau in a very simple fashion. The solute concentration at the plateau (c_p) is not a function of x. Therefore, both dc_p/dx and d^2c_p/dx^2 are zero at all times, and Equation 11-8 becomes

$$dc_p/dt = -2c_p s\omega^2 \tag{11-16}$$

The solution of this differential equation is trivial, using the boundary condition that the concentration of macromolecules at zero time (c_0) is uniform everywhere in the cell:

$$c_p(t) = c_0 e^{-2s\omega^2 t} \tag{11-17}$$

This result is identical to Equation 11-9. Thus, diffusion does not affect the rate at which the concentration of the plateau decreases with time.

Using Equation 11-17, we can compute the sedimentation constant s from a plot of ln c_p versus t. However, direct concentration measurements in the ultracentrifuge are not always easy. For example, Schlieren optics yield $\partial c_2/\partial x$ rather than c_2. An alternative method for determination of s is based on Equation 11-3, which can be rewritten as $s = \omega^2[d(\ln x)/dt]$. If one could observe an individual particle in the plateau region, a plot of the logarithm of its position as a function of time would yield s. Alternatively, if diffusion is negligible, the boundary between solvent and

Box 11-1 SOLUTIONS OF THE LAMM EQUATION

H. Fujita (1975) summarizes many of the attempts to solve the Lamm equation. W. J. Archibald provided (in 1938) an exact analytical solution for Equation 11-8 with s and D constant. However, the coefficients are so complex that even numerical calculations using the Archibald solutions are too laborious to be of general use in evaluating the results of sedimentation experiments.

H. Faxén (in 1929) showed that it is possible to solve Equation 11-8 in a much simpler form, using special boundary conditions. Instead of the usual sector-shaped cell, he considered an infinite sector: $x_m = 0$, and $x_b = \infty$. His solution can be adapted to predict the results of sedimentation experiments in the limiting case of early times ($\omega^2 st \ll 1$) and weak diffusion ($2D/s\omega^2 x \ll 1$). The results are essentially identical to those shown in Figure 11-6b. Fujita (in 1956) was able to extend Faxén's solution to cover the case where D is constant, but s varies with concentration as $s = s_0(1 - k_s c)$.

However, all these solutions are so cumbersome to use that experimentalists usually have chosen to adapt simple approaches that permit s to be measured from concentration data in only limited regions of the cell. This approach is described in detail in the text.

solution is perfectly sharp, and the velocity of this boundary is the same as the velocity of particles in the plateau. From Equation 11-10, this velocity is $d\bar{x}/dt = s\omega^2\bar{x}$, allowing s to be evaluated as

$$s = \omega^{-2}[d(\ln \bar{x})/dt] \tag{11-18}$$

In a real case, the finite width of the boundary means that molecules at different positions are moving at different net velocities. It is not immediately obvious which position on the boundary to pick to correspond to the motion of the molecules unaffected by diffusion. A common choice is the point of steepest slope, corresponding to a maximal value of $\partial c_2/\partial x$. This choice is made because the point can be measured conveniently; however, it is not a correct choice (although often it leads to only a negligible error).

Obtaining the sedimentation coefficient from the boundary position

We need to find the position on the boundary that moves with the same velocity as that of $\bar{x}$. We can then use this point in Equation 11-18 to compute s. Figure 11-8 shows schematically how the point is located. Suppose we pick some point x_p on the

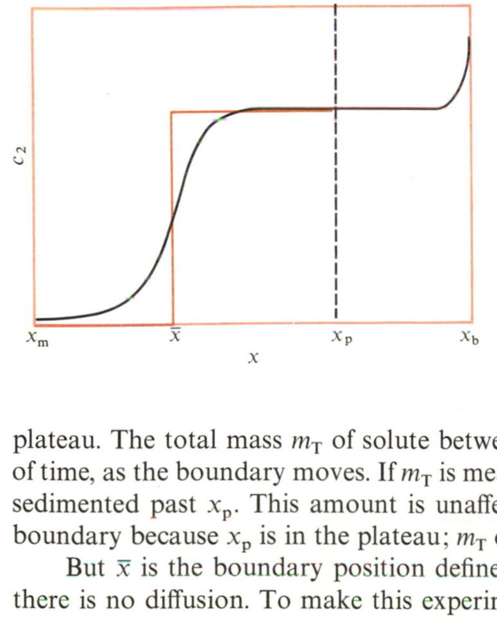

Figure 11-8

Effect of diffusion on sedimentation profile. The sedimentation profile of a real solute (*solid line*) is compared with that expected for a sample of identical sedimentation velocity but no diffusion (*colored line*). A position $\bar{x}$ on the real boundary can be found to move with the same velocity as that of molecules in the plateau region. A position x_p well within the plateau region can be used to measure plateau concentrations, even with the real boundary.

plateau. The total mass m_T of solute between x_m and x_p will decrease as a function of time, as the boundary moves. If m_T is measured, it will reveal how much solute has sedimented past x_p. This amount is unaffected by diffusion or by the shape of the boundary because x_p is in the plateau; m_T depends only on s, ω^2, t, and c_0.

But $\bar{x}$ is the boundary position defined in a hypothetical experiment in which there is no diffusion. To make this experiment correspond to a real experiment (in which the solute has the same value of s), we must ensure that m_T is the same in both cases. The value of $\bar{x}$ that leads to this result is equal to the boundary position in the real experiment that must be used to calculate s.

We find m_T by integrating the concentration $c_2(x, t)$ multiplied by volume $\phi xa\,dx$. Because the concentration is a simple step function in the hypothetical experiment, the integral is easily done in that case:

$$m_T = \int_{x_m}^{x_p} a\phi x c_2(x)\,dx = 0 + c_p\phi a \int_{\bar{x}}^{x_p} x\,dx$$

$$= \frac{1}{2} a\phi c_p(x_p^2 - \bar{x}^2) \qquad (11\text{-}19)$$

In the real experiment, the integral

$$m_T = \int_{x_m}^{x_p} a\phi x c_2(x)\,dx \qquad (11\text{-}20)$$

cannot be performed analytically because $c_2(x)$ is not known explicitly. However, it can be simplified by partial integration:

$$m_T = a\phi c_2(x)x^2/2\Big]_{x_m}^{x_p} - \frac{1}{2} a\phi \int_{x_m}^{x_p} (\partial c_2/\partial x)x^2\,dx \qquad (11\text{-}21)$$

After any finite time in the ultracentrifuge at high speeds, $c_2(x_m) = 0$. The first term in Equation 11-21 then is $\frac{1}{2}a\phi c_p x_p^2$. Equating Equations 11-19 and 11-21 and solving for $\bar{x}$, we obtain

$$\bar{x}^2 = (1/c_p) \int_{x_m}^{x_p} (\partial c_2/\partial x)x^2\,dx \qquad (11\text{-}22)$$

If c_p cannot be measured directly, it can always be obtained as

$$c_p = \int_{x_m}^{x_p} (\partial c_2/\partial x)\,dx \qquad (11\text{-}22a)$$

Thus, $\bar{x}^2$ is simply the second moment of the concentration gradient distribution. This is not the same as the peak position. In actual experimental sedimentation measurements, $\bar{x}$ generally falls at a slightly larger value of x than the peak in $\partial c/\partial x$. Because Schlieren optics give $\partial c/\partial x$ directly, $\bar{x}$ can be evaluated easily, and a plot of $\ln \bar{x}$ versus time yields the correct sedimentation constant.

Equations 11-16 and 11-18 give two independent measures of the sedimentation constant—one actually determined at the boundary, and the other determined at the plateau. These values should agree, and therefore we can write

$$s = (d\bar{x}/dt)/\omega^2\bar{x} = -(1/2\omega^2 c_p)(dc_p/dt) \qquad (11\text{-}23)$$

As the boundary moves from x_m to $\bar{x}$, the concentration at the plateau will go from c_0 to c_p, and therefore

$$\int_{x_m}^{\bar{x}} d(\ln \bar{x}) = -\frac{1}{2}\int_{c_0}^{c_p} d(\ln c_p) \tag{11-24}$$

After integration and rearrangement, we obtain a simple description of radial dilution:

$$c_p/c_0 = (x_m/\bar{x})^2 \tag{11-25}$$

In principle, the detailed shape of the boundary contains information about the diffusion constant of the sedimenting particles (Dishon et al., 1967), and it also can reflect any polydispersity in the sample (Schachman, 1959). In practice, it often is difficult to obtain useful information, because any effects of concentration on s or D are felt very strongly in the boundary region. Before these effects are discussed, it is useful to establish the relationship between s and the properties of individual sedimenting macromolecules. We can do this by simple mechanical considerations.

11-2 ANALYSIS OF SEDIMENTATION MEASUREMENTS

The Svedberg equation

We first consider a mechanical derivation of the relationship between s and molecular properties. There are three forces on a hydrated macromolecule in the plateau region during ultracentrifugation (Fig. 11-9). If M is the anhydrous molecular weight and δ_1

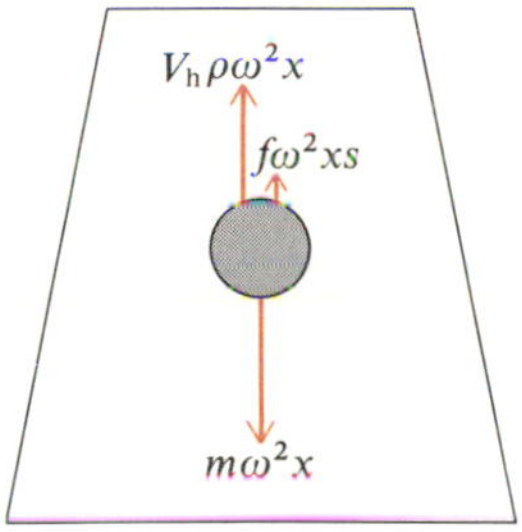

Figure 11-9
Forces on a macromolecule in the ultracentrifuge. The molecule is traveling at the constant velocity $s\omega^2 x$. The acceleration force $m\omega^2 x$ is balanced by the sum of opposing buoyant and frictional forces. Note that the frictional force $f\omega^2 xs$ is actually about one-third the magnitude of the buoyant force $V_h\rho\omega^2 x$ for a typical protein.

is the hydration (in g/g), then the total acceleration force F_a on a single particle is $(M/N_0)(1 + \delta_1)\omega^2 x$. This force is opposed by two forces. The force due to frictional drag is just $F_f = -fs\omega^2 x$. In addition, there is a buoyant force that is proportional to the mass m_s of solution displaced by the macromolecule: $F_b = -m_s\omega^2 x$. The displaced solution has a volume equal to the hydrated volume V_h of the macromolecule, which can be calculated from Equation 10-11. If the solution density is ρ, then

$$F_b = -V_h\rho\omega^2 x = (M/N_0)(\bar{V}_2 + \delta_1\bar{V}_1)\rho\omega^2 x \tag{11-26}$$

The sum of the forces is zero in the hydrodynamic steady state, so

$$(M/N_0)(1 + \delta_1 - \bar{V}_2\rho - \delta_1\bar{V}_1\rho) = sf \tag{11-27}$$

In dilute solution, $\rho^{-1} \cong V_1$, and so the second and fourth terms of the left-hand side of Equation 11-27 cancel, yielding

$$\boxed{s = M(1 - \bar{V}_2\rho)/N_0 f} \tag{11-28}$$

This derivation, although it gives a good mechanical insight, is somewhat fraudulent because of the approximation used to obtain Equation 11-28 from Equation 11-27, and because solution density was used to describe the buoyant force without any real justification. However, the resulting buoyancy term $1 - \bar{V}_2\rho$ is correct, and it does have some interesting consequences. Suppose that $\bar{V}_2 > \rho^{-1}$. This is the case, for example, when lipid-containing samples are placed in aqueous solution. Because the term $1 - \bar{V}_2\rho$ will be negative, in the ultracentrifuge the lipid will float toward the top of the cell rather than sediment toward the bottom.

A more rigorous derivation of Equation 11-28 begins with a thermodynamic description of the flux in a two-component system (Eqn. 11-2). The effect of friction is the only factor connecting flow with the forces due to sedimentation and diffusion. The phenomenological coefficient must be the same as that we evaluated for diffusion alone in Equation 10-61. Thus, $L_{22} = M_2c_2/N_0 f$. Therefore, we can write the flux as

$$J_2 = (M_2c_2/N_0 f)[\omega^2 x - (\partial\hat{\mu}_2/\partial x)] \tag{11-29}$$

In an ultracentrifuge experiment at constant temperature, the thermodynamic variables capable of affecting $\hat{\mu}_2$ are pressure (P) and concentration (c_2). If we consider their effects separately, we have

$$\partial\hat{\mu}_2/\partial x = (\partial\hat{\mu}_2/\partial P)_T(\partial P/\partial x) + (\partial\hat{\mu}_2/\partial c_2)_T(\partial c_2/\partial x) \tag{11-30}$$

The pressure gradient at position x in a solution is simply $\omega^2 x\rho$, where ρ is the *solution* density at x, and $\omega^2 x$ is the accelerating force. The term $\partial\hat{\mu}_2/\partial P$ is defined as the partial specific volume, and this is simply $\bar{V}_2$. The second term in Equation 11-30 was evaluated when diffusion was discussed; for an ideal solution, it is $(RT/M_2c_2)(\partial c_2/\partial x)$. (See Eqns. 10-41 through 10-45). So Equation 11-29 becomes

$$\begin{aligned} J_2 &= (M_2c_2/N_0 f)[\omega^2 x - \bar{V}_2\rho\omega^2 x - (RT/M_2c_2)(\partial c_2/\partial x)] \\ &= [M_2(1 - \bar{V}_2\rho)/N_0 f]c_2\omega^2 x - (kT/f)(\partial c_2/\partial x) \end{aligned} \tag{11-31}$$

When this expression is combined with the mechanical description of flux (Eqn. 11-4), it is apparent that

$$s = M(1 - \bar{V}_2\rho)/N_0 f \qquad \text{and} \qquad D = kT/f \tag{11-32}$$

where we have eliminated the subscript 2 on the solute molecular weight.

Note from Equation 11-28 or 11-32 that the sedimentation constant is a function of both the anhydrous molecular weight and the frictional coefficient. In a two-component system, it is the true anhydrous molecular weight that matters, because any acceleration force due to the extra mass of bound water should be canceled by the buoyant force on that bound water. In a multicomponent system, the situation is more complex. There J_2 will depend not only on the forces on component 2, but also on the forces on the third component: $J_2 = L_{22}X_2 + L_{23}X_3$. In addition, $\hat{\mu}_2$ will be a function of the concentration of the third component (c_3). Thus, Equation 11-30 will have an extra term $(\partial\hat{\mu}_2/\partial c_3)(\partial c_3/\partial x)$. Three-component effects can be significant if sedimentation is carried out to concentrated salt solutions. These effects are considered later in this chapter, when density gradient centrifugation is discussed. (Also see Box 11-3.)

In deriving Equation 11-29, we made the tacit assumption that the frictional coefficients affecting diffusion and sedimentation are the same. This identity is not completely rigorous, especially if we want to compare frictional coefficients measured under different conditions: f_D in passive diffusion, and f_s in the high forces of an ultracentrifuge. For example, a flexible molecule forced to sediment at hundreds of thousands of gravities can distort in shape, thus changing the fractional coefficient. These complications are rare, however. It usually is safe (and always is convenient) to combine the expressions of Equation 11-32 and thus eliminate the frictional coefficient. The result is the Svedberg equation:

$$M = sRT/D(1 - \bar{V}_2\rho) \tag{11-33}$$

Note that the units of R in this equation are cgs: $R = N_0 k = 8.31 \times 10^7$ erg mole^{-1} deg^{-1}.

Determining molecular weights from sedimentation data

The Svedberg equation permits an unambiguous determination of the molecular weight, independent of any details of shape—providing that separate sedimentation, diffusion, and partial-specific-volume measurements can be performed on equivalent samples under conditions as similar as possible. In practice, the conditions used for the three measurements rarely are identical. It is customary to correct sedimentation data to the values that would be observed if the measurements had been performed at 20°C in pure water.

The corrections for diffusion are given in Equation 10-67. The corresponding corrections for sedimentation are calculated by noting (from Eqn. 11-32) that s will depend on solvent and temperature (because $\bar{V}_2$ and ρ are functions of these variables) and on the viscosity η (because f depends on viscosity; see Eqn. 10-68). Therefore, the ratio of the sedimentation constant measured in water at 20°C to that measured under different conditions will be

$$\frac{s_{20,\mathrm{w}}}{s} = \frac{1 - 0.9982(\bar{V}_2)_{20,\mathrm{w}}}{1 - \bar{V}_2\rho}\left(\frac{\eta_{T,\mathrm{w}}}{\eta_{20,\mathrm{w}}}\right)\left(\frac{\eta_{T,\mathrm{soln}}}{\eta_{T,\mathrm{w}}}\right) \tag{11-34}$$

where the subscripts soln and T represent the actual sample solution and temperature used; $\eta_{T,\mathrm{w}}$ is the viscosity of pure water at temperature T; $\eta_{20,\mathrm{w}}$ is 0.01002 poise; 0.9982 g cm^{-3} is the density of water at 20°C; and $(\bar{V}_2)_{20,\mathrm{w}}$ is the partial specific volume measured in water at 20°C.

The sedimentation and diffusion coefficients, although assumed constant in most of the equations derived in this chapter, actually are concentration-dependent in practice. Therefore, it is common (and often is necessary) to measure s and D as functions of concentration and then to extrapolate to infinite dilution to obtain s^0 and D^0; these quantities then are corrected to the values expected at 20°C in pure water at infinite dilution: $s^0_{20,\mathrm{w}}$ and $D^0_{20,\mathrm{w}}$.

Table 11-1 shows some typical measured values of $s_{20,\mathrm{w}}$. What can be done with these values depends on what else is known about the system. If the molecular weight is known, a determination of $s_{20,\mathrm{w}}$ yields the frictional coefficient f, which can be compared to f_{min}, the coefficient expected for an anhydrous sphere. The ratio

Table 11-1
Results of sedimentation measurements

Sample	Molecular weight	$s_{20,\mathrm{w}}$ (S)	$\bar{V}_2$ (cm^3 g^{-1})	f/f_{min} (sedimentation)	f/f_{min} (diffusion)
Ribonuclease A (bovine)	12,400	1.85	0.728	1.29	1.14
Lysozyme (chicken)	14,100	1.91	0.688	1.22	1.32
Serum albumin (bovine)	66,500	4.31	0.734	1.33	1.31
Hemoglobin	68,000	4.31	0.749	1.28	1.14
Tropomyosin	93,000	2.6	0.71	2.65	3.22
Fibrinogen (human)	330,000	7.6	0.706	2.34	2.35
Myosin (rod)	570,000	6.43	0.728	3.63	3.05
Bushy stunt virus	10,700,000	132.	0.74	1.27	1.27
Tobacco mosaic virus	40,000,000	192.	0.73	2.65	2.19

SOURCE: Data from several sources, including Kuntz and Kauzmann (1974), Tanford (1961), and Van Holde (1971).

f/f_{min} can be analyzed in turn to obtain limits on shape and hydration, exactly as described earlier for diffusion coefficients. A comparison of frictional coefficients derived from diffusion and from sedimentation is shown in Table 11-1. The agreement is fairly good in most cases. If both $s_{20,w}$ and $D_{20,w}$ are known, the molecular weight can be computed from Equation 11-33. Then either expression of Equation 11-32 would yield the frictional coefficient. Unfortunately, because sedimentation and diffusion give the same measure of combined shape and hydration, a knowledge of both quantities does not resolve ambiguities as to how much of the excess friction (over f_{min}) is due to hydration. Table 11-2 compares molecular weights obtained using the Svedberg equation with values determined by other techniques. The various methods generally are quite consistent. They should not agree exactly for a number of reasons, including the fact that counterions are weighted to different extents by some of the methods.

Table 11-2
Molecular weights determined by various methods

Sample	Chemical structure	Sedimentation and diffusion	Archibald	Sedimentation equilibrium	Scheraga–Mandelkern	Other
Sucrose	342.3	—	—	341.5	—	—
Raffinose	504.5	—	495	—	—	—
Ribonuclease A (bovine)	13,683	12,400	13,750	13,700	15,200	—
Lysozyme (chicken)	14,211	14,100	—	14,500	12,400	14,100[§]
Serum albumin (bovine)	66,296	66,000	70,000	68,000	59,000	70,000[#]

[§] From small-angle x-ray measurements.
[#] From light-scattering measurements.

Equation 11-33 cannot be used as often as one would like, because diffusion measurements are not that common. It is possible to combine sedimentation data with other hydrodynamic measurements, as will be demonstrated later. All too frequently, a sedimentation coefficient is the only hydrodynamic information available on a system of unknown molecular weight. This is because sedimentation experiments either are faster or require less sample than most other methods. Less than 0.1 mg of a protein (and an order of magnitude less of nucleic acid) can suffice for the determination of $s_{20,w}$ using sensitive photoelectric scanning detection. The actual measurement takes less than a few hours, during which time the boundary moves on the order of a centimeter. This distance can be measured very accurately.

Easy as it is to obtain, a value of $s_{20,w}$ alone requires that some guess be made about either shape or molecular weight in order to estimate the other quantity. If the macromolecule is a sphere, substitution of Equations 10-68 and 10-69 into Equation 11-28 (with the Perrin shape factor $F = 1$) yields

$$s = (M^{2/3}/N_0^{2/3})(1 - \bar{V}_2\rho)\{6\pi\eta[(3/4\pi)(\bar{V}_2 + \delta_1\bar{V}_1)]^{1/3}\}^{-1} \qquad (11\text{-}35)$$

The sedimentation constant for spherical molecules with similar hydration and partial specific volumes is proportional to $M^{2/3}$. This is a very useful result. If, for example, different associated forms of protein subunits are studied, their molecular weights can easily be related by sedimentation measurements, providing that none of the forms is too aspherical. Equation 11-35 is used in a wide variety of cases to estimate molecular weights by guessing values of the hydration. These estimates are fairly valuable for most cases but, in any given case, the prudent investigator will be somewhat cautious.

Shape information from sedimentation data

The shape-dependent predictions of hydrodynamic theory can be tested by using a homologous series of molecules identical in all respects except length. A convenient system, used for this purpose by B. D. Harrison and A. Klug, is tobacco rattle virus (TRV). The infective virus is a cylinder 250 Å in diameter and 1,850 to 1,970 Å long. However, infection also produces shorter particles that can be isolated and studied.

To treat a homologous series, it is convenient to write the frictional coefficient in the form $f = 6\pi\eta(3V_e/4\pi)^{1/3}F$, where V_e is the volume of the equivalent hydrated sphere, and F is a shape factor. The actual volume for a cylindrical rod is $\pi r^2 l$, where l is the length and r is the radius. Using these expressions for f and V_e, we can rewrite Equation 11-28 as follows:

$$M/l = \{[6\pi\eta N_0(3r^2/4)^{1/3}]/(1 - \bar{V}_2\rho)\}(sF/l^{2/3}) \tag{11-36}$$

The molecular weight per unit length (M/l) is constant for a series of homologous rods. Similarly, the first term on the right-hand side is a constant. Therefore, the prediction is that $sF/l^{2/3}$ should be a constant. Figure 11-10a shows a test of this prediction. The prediction is fairly accurate, using Perrin F values for prolate ellipsoids with axial ratios of $l/2r$.

An even better fit is obtained if rods are modeled slightly more realistically. The volume of a prolate 2 ellipsoid is $4\pi ab^2/3$, where a is the major semiaxis, and b is the minor semiaxis. The volume of a cylinder is $\pi r^2 l$. We must set these two volumes equal, to ensure that there are equal numbers of particles in the solution for a given fixed weight concentration.

$$\pi r^2 l = 4\pi ab^2/3 \tag{11-37}$$

Now, selecting an ellipsoid with the same length as the rod; $l = 2a$. Therefore Equation 11-37 can be put in the following form:

$$l/a = 2 = 4b^2/3r^2 = (1/3)(b^2/r^2)(l^2/a^2) \tag{11-38}$$

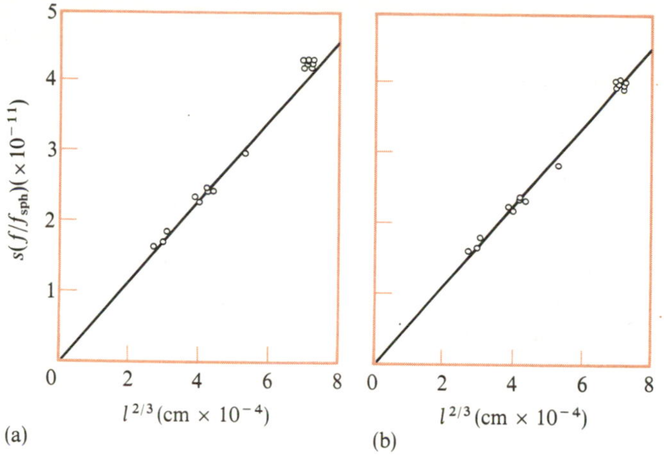

Figure 11-10

Sedimentation coefficient of TRV (tobacco rattle virus) particles as a function of length. Lengths (l) were measured in the electron microscope, and these dimensions were used to calculate f/f_{sph} for prolate ellipsoids. The actual plots used here show (measured length)$^{2/3}$ versus measured s times calculated f/f_{sph}. **(a)** The axial ratio for ellipsoid is calculated as l/d, where d is the diameter of the rod. **(b)** The axial ratio is calculated as $\sqrt{2/3}l/d$. [After B. D. Harrison and A. Klug, *Virology* 30:738 (1966).]

The second and fourth expressions can be rearranged to give

$$a/b = l/6^{1/2}r = (2/3)^{1/2}(l/d) \tag{11-39}$$

where $d = 2r$ is the diameter of the rod. Therefore, an ellipsoid is chosen with an axial ratio slightly smaller than that of the rod. When this is done, the fit of Equation 11-36 to experimental data is nearly perfect (Fig. 11-10b).

The accuracy of sedimentation measurements permits their use in fairly exacting structural discriminations. Hemoglobin has a measured sedimentation constant of about 4.45 S. Each of its four subunits (two α and two β) has an individual sedimentation constant of about 1.77 ± 0.055 S. One can calculate the sedimentation constant of a tetramer from monomer properties using the Kirkwood–Riseman theory (Section 10-2). If both the tetramer (t) and monomer (m) have the same $\bar{V}_2$, one can write (from Equation 11-28)

$$s_t/s_m = (M_t/M_m)/(f_t/f_m) = 4f_m/f_t \tag{11-40}$$

where f_m/f_t is given by Equation 10-34 and must be computed for a specific geometric model of the tetramer. Using the measured monomer sedimentation constant and the values of f/f_m shown in Figure 10-12, we can predict that $s = 3.91 \pm 0.11$ S for a linear tetramer; $s = 4.18 \pm 0.12$ S for a square planar tetramer; and $s = 4.43 \pm 0.13$ S for a tetrahedral tetramer. Thus the data yield sufficient precision to permit the correct deduction that hemoglobin has a tetrahedral arrangement of subunits. Figure 11-11 shows the results of a more complicated example of the use of the Kirkwood–Riseman theory.

To use sedimentation as a monitor of conformational change, it is convenient to make the most precise possible measurements of the difference in sedimentation of two samples of the same substance under slightly different conditions. An effective technique for the purpose is difference sedimentation. The cell used has two identical sector-shaped sample chambers placed side by side (the double-sector counterpiece in Fig. 11-3b). The samples to be compared are placed in the separated sectors and are spun together. Optical systems are used to detect just the differences between the two sectors, in much the same way as difference spectroscopy is carried out. It turns out to be particularly convenient to use Rayleigh optics, which give the differences in solute concentration in the two sectors as a function of position. This technique permits measurement of sedimentation-constant differences as small as 0.01 S with better than 1% precision. However, it places very stringent requirements on the optical system, and these preclude routine use.

Effects of concentration on sedimentation velocity

Everything we have said thus far is applicable to a two-component system (water plus macromolecule) in the limit of great dilution. Formidable difficulties can occur

Model

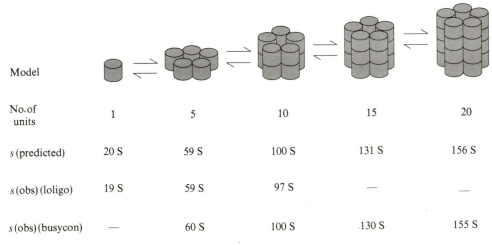

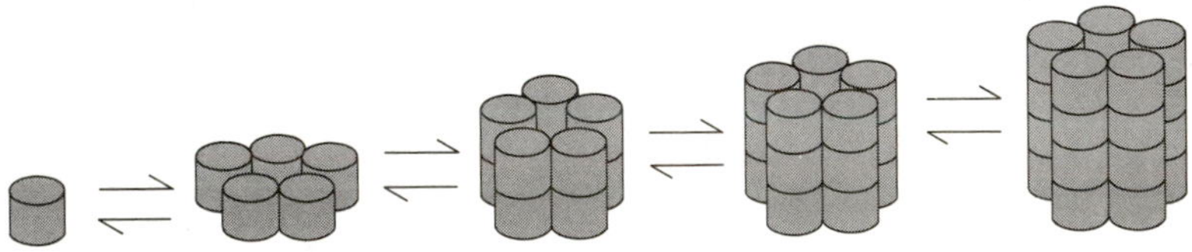

No. of units	1	5	10	15	20
s (predicted)	20 S	59 S	100 S	131 S	156 S
s(obs) (loligo)	19 S	59 S	97 S	—	—
s(obs) (busycon)	—	60 S	100 S	130 S	155 S

Figure 11-11

Measured and predicted sedimentation constants for two species of molluscan hemocyanins (loligo and busycon). The Kirkwood–Riseman theory was used to calculate the predicted values for the oligomeric models shown. [After V. Bloomfield, W. O. Dalton, and K. E. Van Holde, *Biopolymers* 5:135 (1967).]

when these conditions are not met. Here we provide a brief sketch of some of the more commonly encountered effects.

Even in a two-component system, the sedimentation velocity is a function of solute concentration. The dependence of sedimentation on concentration for a single macromolecular component generally can be fit to equations of the form

$$s = s^0(1 + k'c)^{-1} \tag{11-41}$$

or

$$s = s^0(1 - k''c) \tag{11-42}$$

where s^0 is the value at infinite dilution, and k' and k'' are empirically derived constants. Both equations account for a decrease in observed sedimentation velocity with increasing concentration (Fig. 11-12). With small values of c, the equations are

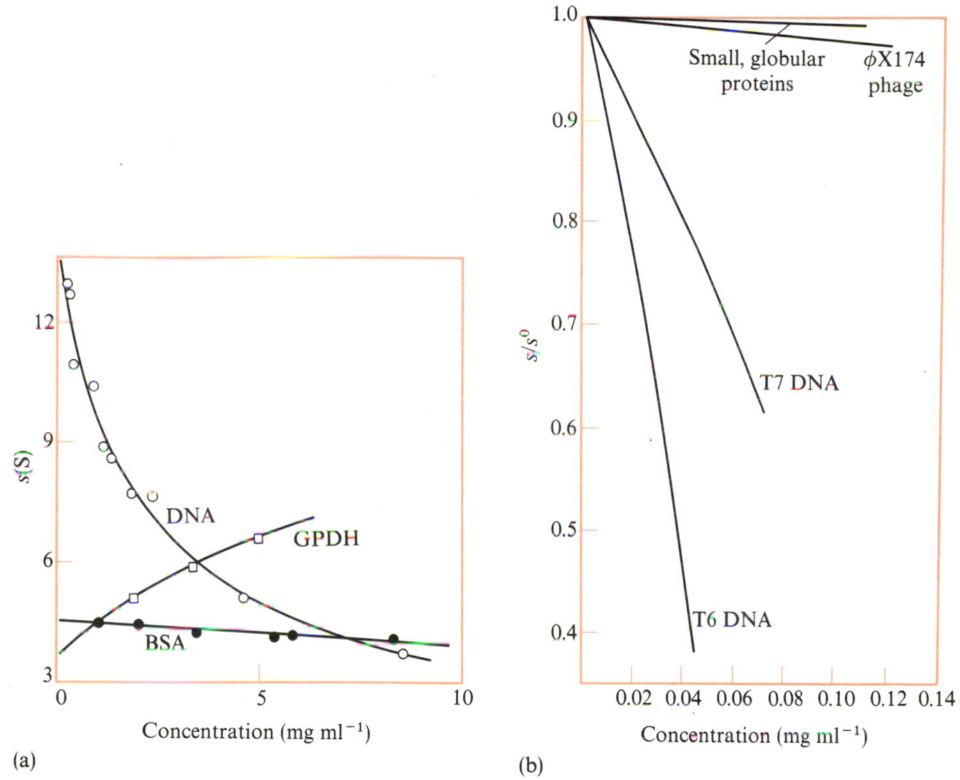

Figure 11-12

Concentration dependence of the sedimentation constant. **(a)** Actual sedimentation data for three macromolecules: DNA of about 3×10^6 mol wt, bovine serum albumin (BSA), and glycerol phosphate dehydrogenase (GPDH). [After C. Tanford (1961), K. E. Van Holde (1971).] **(b)** Relative sedimentation data for several other molecules. Note that the concentration dependence becomes more severe as the molecular weight increases. For example, T7 DNA has a molecular weight of 2.5×10^7 d. [After D. Freifelder, *Physical Biochemistry*, p. 285. (San Francisco: W. H. Freeman and Company.) Copyright © 1976.]

equivalent, because $(1 + kc)^{-1} = 1 - kc +$ (terms of higher order in c). However, the two expressions predict very different behavior at high concentrations.

The physical justification of the two equations also is very different. Equation 11-41 is the form expected if the major effect of solute concentration is to alter the frictional properties of the solution. The sedimentation constant is proportional to f^{-1}, and f is proportional to the solution viscosity. As shown in Chapter 12, η can always be written as a power series in the concentration of solute. The first two terms are $\eta = \eta_0(1 + [\eta]c)$, where η_0 is the viscosity of pure solvent, and $[\eta]$ is called the intrinsic viscosity. Combining this expression with Equation 10-68, inserting the result into Equation 11-28, and ignoring the effects of concentration on density, we obtain Equation 11-41 with $k' = [\eta]$. In practice, however, this equation usually does not fit the observed data very well. Except for DNA, experimentally determined values of k' often are larger than those of $[\eta]$.

Equation 11-42 is the form expected if backward flow of solvent is the major force that slows down sedimentation with increasing solute concentration. For example, a spherical particle drags about four times its own volume of solvent with it. Thus, during the sedimentation of a macromolecule, a considerably larger volume of solvent must flow in the cell in the direction opposite to that of sedimentation. This flow will be proportional to the volume of the sedimenting macromolecules multiplied by their velocity. As viewed by an observer outside the cell, the macromolecules will appear to be moving more slowly. They are, in effect, swimming upstream. As we shall show in Chapter 12, the constant $[\eta]$ in a power expansion of the viscosity is proportional to the volume fraction occupied by solute. Therefore, k' and k'' are related, and they become equivalent in the limit of low concentration. At high concentration, neither Equation 11-41 nor Equation 11-42 can account very well for experimental data, and more complicated expressions are needed.

The effect of solute concentration on the shape of the boundary observed in a sedimentation experiment can be dramatic. When a concentrated DNA solution is centrifuged, molecules in the plateau will sediment slowly (as shown in Fig. 11-12) because they feel the full effect of the high concentration. Suppose there were a broad boundary (Fig. 11-13a). The DNA molecules at the trailing edge are at very low concentration. They should be sedimenting very fast. If they do this, they catch up with the molecules at the leading edge of the boundary. The result is that a broad boundary cannot form, even if diffusion is considerable. Instead, the boundary self-sharpens and remains sharp upon further sedimentation (Fig. 11-13b). This can hide the effect of any heterogeneity of molecular weight in the sample.

Effect of self-association on sedimentation velocity

One of the experimental samples in Figure 11-12 has a sedimentation constant that increases with increasing concentration. From an argument exactly opposite to that used for self-sharpening boundaries, one can predict that the boundary should broaden in this case. Indeed, this is the observed result, but what is its origin? Sup-

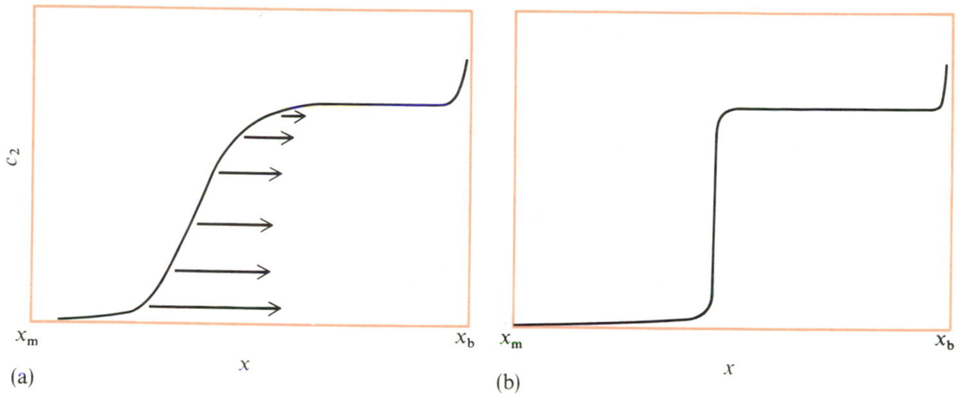

Figure 11-13

Self-sharpening boundary. **(a)** Assume a broad boundary formed in a concentrated solution of DNA. The concentration dependence of s shown in Figure 11-12 would mean that different points of the boundary would move at very different velocities. **(b)** In concentrated DNA, the broad boundary is never actually seen. Instead, a hypersharp boundary forms, stabilized by the effects shown in part a.

pose that a protein monomer P can associate to form oligomers such as $2\ P \rightleftarrows P_2$, or even large aggregates such as $n\ P \rightleftarrows P_n$. If the rates of interconversion are infinitely slow (on the time scale of the sedimentation experiment), then two boundaries should be seen—one corresponding to P and one to P_2 (or P_n).

At faster rates of interconversion, for a monomer–dimer equilibrium, the two boundaries merge into one asymmetric boundary. Even if concentration effects on intrinsic sedimentation rates are ignored, this boundary will have a complex shape that can be analyzed by the theory developed by G. A. Gilbert (see Cann, 1970; Van Holde, 1975). The total solute concentration is near zero at the trailing edge of the boundary, and the equilibrium will favor the monomer. The dimer can be favored if the concentration is high enough at the leading edge. The resulting boundary will be asymmetric, and the position of maximal $\partial c/\partial x$ will be a function of the extent to which the equilibrium favors monomer or dimer.

For a rapid monomer–n-mer equilibrium (with $n > 2$), two boundaries will be seen in certain concentration ranges, and the relative height of the faster-moving boundary will increase with increasing concentration. In the limits of low or high concentrations, a single asymmetric boundary will result. If there are only two species, it sometimes is possible to determine the molecular weight of the aggregate (and even information about the association constant) by analyzing the relative positions and heights of the boundaries. However, in general, an exact analysis of boundary shape in systems with reacting species is a formidable problem that still taxes current research.

Regardless of what is happening at the boundary, some information is available in a straightforward manner from the plateau region. Whether the system is a mixture of interacting or of noninteracting components, one still can describe the flux in the plateau (or in the final plateau, past any and all observed boundaries) from Equation 11-4 as

$$J_p = \omega^2 x \sum_i c_i s_i \tag{11-43}$$

where c_i is the weight concentration of material with sedimentation constant s_i, and diffusion is neglected because $\sum_i(\partial c_i/\partial x_i) = 0$. By analogy to Equation 11-16, the concentration change with time at the plateau then becomes

$$dc_p/dt = d\left(\sum_i c_i\right)\Big/dt = -2\omega^2 \sum_i c_i s_i \tag{11-44}$$

The value of dc_p/dt is an experimentally observable quantity. If one did not know there was a mixture, one would associate it with an apparent sedimentation constant $\overline{s}$:

$$dc_p/dt = -2\omega^2\overline{s}c_p = -2\omega^2\overline{s}\sum_i c_i \tag{11-45}$$

Equating the right-hand sides of Equations 11-44 and 11-45, you can see that the apparent sedimentation constant that will be measured is $\overline{s} = \sum_i s_i c_i / \sum_i c_i$. Because c_i is in weight units, $\overline{s}$ is the weight-average sedimentation constant. For some applications, $\overline{s}$ is useful, but much information about the system is lost in performing the average.

Effects of multiple macromolecular components on sedimentation velocity

If a mixture of more than one macromolecule type is sedimented, more than one boundary generally will be seen. However, the analysis of boundaries in multicomponent systems becomes very complicated. The concentration dependence of sedimentation in a mixture with two or more components is a function of all the components in a given region. Among other things, this means that the backward flow created by a fast component can be felt by a slow component, and in some cases can even overwhelm the sedimentation of the slow component.

To highlight this effect, we consider a somewhat unusual case in which the fast boundary is nearer the rotation axis than is the slow boundary. This effect can be produced at short times in the centrifuge by using a synthetic boundary cell (Fig. 11-14a). This is a double-sector cell made with thin channels connecting the two sectors. At normal gravity, hydrostatic pressure is insufficient to overcome the resistance of these channels to flow. Once in the centrifuge, the hydrostatic forces become

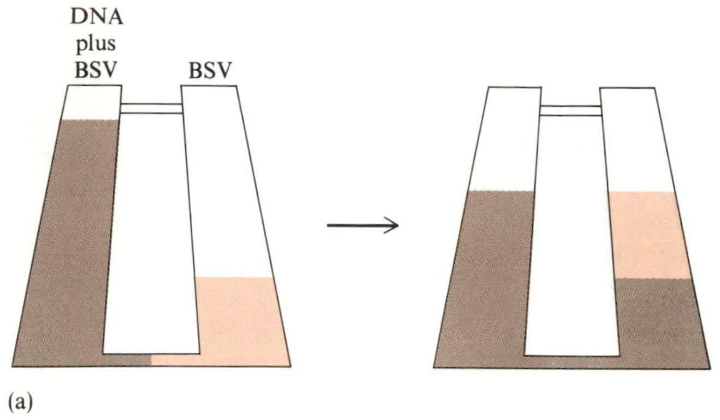

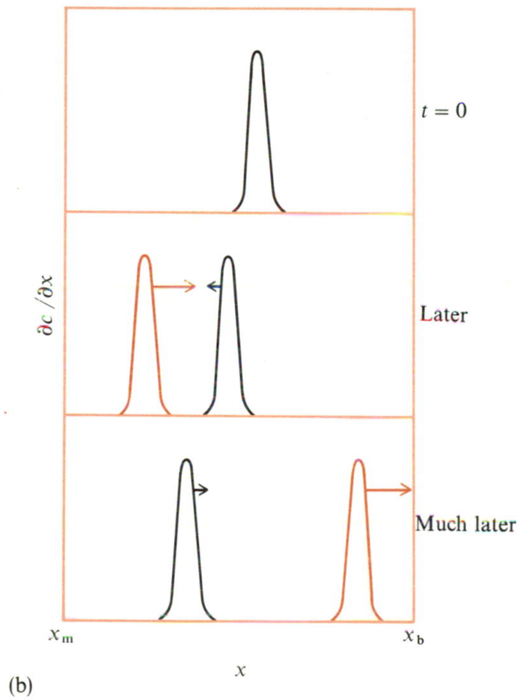

Figure 11-14

Demonstration of solvent backward flow. **(a)** A synthetic-boundary cell is loaded with separate samples of pure bushy stunt virus (BSV) and a virus–DNA mixture. In the ultracentrifuge, hydrostatic forces equalize the fluid levels in the two sectors, resulting in a DNA boundary in the center of the right-hand sector. **(b)** Sedimentation profiles seen in the right-hand sector as a function of time after fluid flow between the sectors. The initial DNA boundary moves up until the BSV boundary passes it. Then it reverses direction and sediments toward the bottom.

enormous, and any differences in the heights of the fluid in the two sectors are eliminated by flow through the channels.

To make a synthetic boundary of slowly sedimenting DNA in the middle of a uniform solution of bushy stunt virus (BSV), one places a small volume of BSV solution in one sector and a large volume of BSV at the same concentration containing added DNA in the second sector. Figure 11-14 shows that, after a very short time, fluid flow produces in one sector a DNA boundary in the midst of a uniform BSV concentration. This sector is used for further examination. The BSV rapidly forms a boundary near the meniscus, and this boundary travels down the cell at about 120 S. When measured separately, the DNA used would move down with a sedimentation constant of about 10 S. However, in this experiment, it actually moves up with a constant of about -10 S in the presence of the sedimenting BSV. Thus, the backward flow experienced by the DNA is about 20 S. The schematic illustration of Figure 11-14b is an oversimplification, however, because it ignores another effect.

The size of a boundary is $\int(\partial c/\partial x)\,dx$; in a one-component system, this size is simply c_p and is a measure of concentration at the plateau. It has long been known that, when mixtures of two components are sedimented, the boundary sizes do not correctly correspond to the fractional composition. This is called the Johnston–Ogston effect. To simplify the explanation of this effect, our discussion here ignores radial dilution. The fast component always sediments in the presence of the slow with a rate s_F (Fig. 11-15a). However, the slow component must sediment at different rates: s_S above the fast-component boundary, where it feels only its own concentration, and $s_{S,F}$ below that boundary, where both fast and slow components are present. Equation 11-41 or 11-42 ensures that $s_{S,F} < s_S$, but this has serious consequences. Imagine

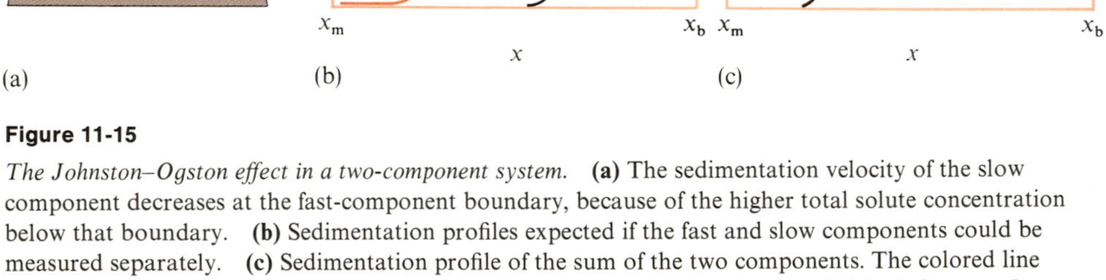

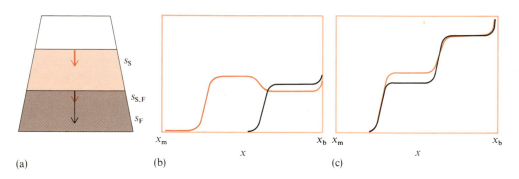

Figure 11-15

The Johnston–Ogston effect in a two-component system. **(a)** The sedimentation velocity of the slow component decreases at the fast-component boundary, because of the higher total solute concentration below that boundary. **(b)** Sedimentation profiles expected if the fast and slow components could be measured separately. **(c)** Sedimentation profile of the sum of the two components. The colored line shows the actual profile; the black line shows the profile that would be expected if the Johnston–Ogston effect could be neglected.

sitting on the *moving* fast boundary and watching the fluxes of slow component. Looking toward the top of the cell, we would see slow component moving away with a flux $J = \omega^2 x(s_F - s_S)c_S$. Looking toward the bottom, we would see slow component approaching with $J' = \omega^2 x(s_F - s_{S,F})c_{S,F}$. Therefore, J' is larger than J, implying that slow substance should accumulate continuously at the fast boundary. Such accumulation is impossible, because it would lead to a high local density. The fast component would slow down because of the $1 - \bar{V}_2\rho$ term in Equation 11-28, and a gravitation instability would result as well. The boundary would be denser than the solution below it, and it would fall to the bottom by convective mixing.

What happens instead, when the problem is properly solved, is that the concentration of the slow component readjusts to prevent this catastrophe. For there to be no flux change in the slow component at the fast boundary, $J = J'$, which means that

$$c_{S,F}(s_F - s_{S,F}) = c_S(s_F - s_S) \tag{11-46}$$

Because $s_{S,F} < s_S$, the concentration c_S of the slow component above the fast boundary is greater than the concentration $c_{S,F}$ below the boundary (Fig. 11-15b). The result is a slow boundary larger than naively expected and a fast boundary smaller than expected, because most optical monitoring techniques detect the sum of the concentrations of both components.

Zonal sedimentation of multicomponent systems

The Johnson–Ogston effect and other multicomponent effects are particularly troublesome in velocity sedimentation because all components, except the slowest one, are always present as mixtures. This also means that sedimentation, as described thus far, is rather ineffective at separating or purifying components, unless their sedimentation rates are very different.

The technique of zonal sedimentation circumvents both of these problems. A thin band of macromolecule sample is layered on top of a much larger volume of solvent. This can be done in the analytical ultracentrifuge by using a band-forming centerpiece (Fig. 11-16a), which is similar in principle to the synthetic boundary cell described earlier. A mixture of several components with different sedimentation constants will rapidly separate into bands of pure components. Each will travel at its characteristic sedimentation velocity. Diffusion will broaden each band into a Gaussian shape (Eqn. B of Box 10-3). In practice, the shapes are more complicated. Because the acceleration force increases as $\omega^2 x$, the leading edge of a band will travel faster than the trailing edge. The concentration dependence of sedimentation further distorts the band shape. The concentration is low at the leading and trailing edges, and molecules move rapidly. The concentration is high in the center of a band, and velocities are slower. The result is a band with a broad leading edge and a sharp trailing edge (Fig. 11-16b).

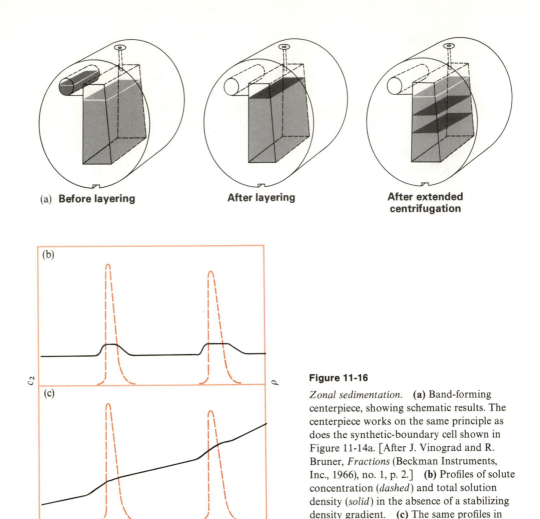

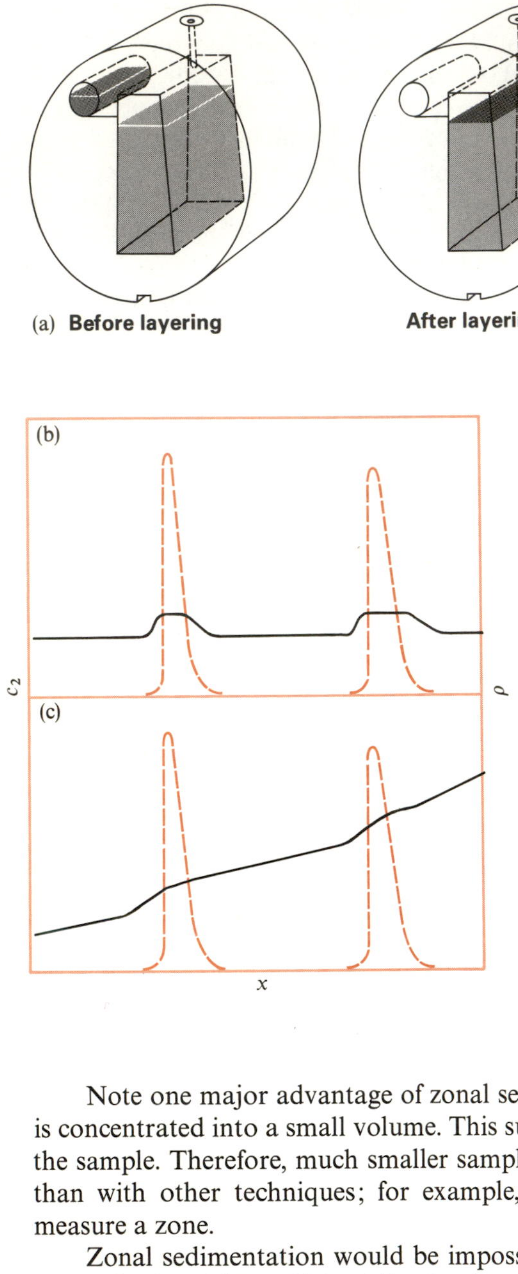

Figure 11-16

Zonal sedimentation. **(a)** Band-forming centerpiece, showing schematic results. The centerpiece works on the same principle as does the synthetic-boundary cell shown in Figure 11-14a. [After J. Vinograd and R. Bruner, *Fractions* (Beckman Instruments, Inc., 1966), no. 1, p. 2.] **(b)** Profiles of solute concentration (*dashed*) and total solution density (*solid*) in the absence of a stabilizing density gradient. **(c)** The same profiles in the presence of a stabilizing density gradient.

Note one major advantage of zonal sedimentation. All of the biological sample is concentrated into a small volume. This substantially enhances the ability to detect the sample. Therefore, much smaller samples can be used with zonal sedimentation than with other techniques; for example, less than 1 μg of DNA will suffice to measure a zone.

Zonal sedimentation would be impossible in a homogeneous solution because of gravitational instabilities. The density of the sample would be higher in a band than in the solution below it, and the bulk solution region containing the band would simply collapse. For example, a DNA solution with concentration 0.01 mg ml^{-1}

would have a density roughly 2×10^{-5} g cm^{-3} larger than that of the solvent. However, this density difference is magnified by the 4×10^5 g forces typical in the ultracentrifuge. The result is roughly equivalent to trying to float mercury on top of water. It doesn't work.

A density gradient must be created in the solvent to eliminate this problem. This can be done with the moderately dilute salt solutions used in the analytical ultracentrifuge by having the original sample in a more dilute salt. When the sample transfers in a zone on top of the bulk solvent, rapid small molecule diffusion creates a shallow density gradient (Fig. 11-16b). This gradient can differ as little as 1% from top to bottom of the cell. Thus, it does not perturb sedimentation velocities too seriously. The gradient will be stabilized (and even enhanced) in the ultracentrifuge, because salt molecules are denser than water. The high forces in the ultracentrifuge can be felt even by small ions. However, these dilute salt gradients are not very stable outside of the centrifuge. If one wanted to recover a pure, separated band, it would be necessary to pump it out while the rotor was still spinning. This is possible, but there is a simpler solution in most cases.

Preparative zonal sedimentation is mainly a purification technique. A preformed gradient of a somewhat viscous solute is used. A linear gradient of 5% to 20% sucrose is typical. Glycerol is another common solute. A linear gradient is easy to prepare using the mixing device shown in Figure 11-17 (see Box 11-2). Such a gradient is

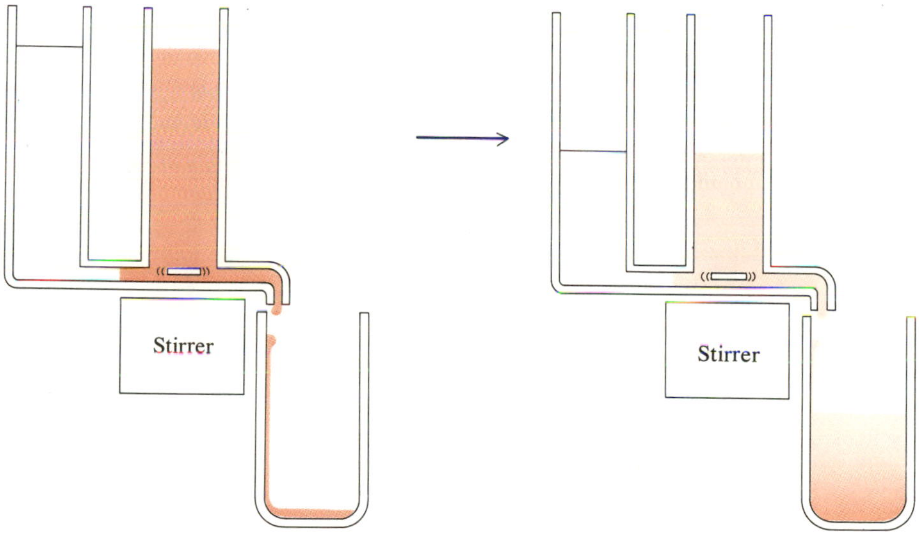

Figure 11-17

Apparatus for generation of a linear density gradient. The right-hand vessel is assumed to be stirred to homogeneity at all times.

more stable outside the centrifuge than a dilute salt gradient, and it will remain fairly constant in the ultracentrifuge because the viscosity of the concentrated sucrose impedes redistribution. Therefore, a sample can be layered on top of the preformed gradient in a centrifuge tube so long as its density is less than that at the top of the gradient. The solute molecules of the sample will sediment through the gradient because $(1 - \bar{V}_2\rho)$ is greater than zero throughout. After spinning for a time sufficient to move the solute molecules part of the way down the centrifuge tube, the centrifuge is stopped. The gradient is collected and gently removed from the tube, either by puncturing the bottom and allowing it to drop out, or by pumping in more dense sucrose at the bottom and displacing the gradient out the top. Individual fractions are collected and analyzed in any way one wishes.

It is very difficult to determine an absolute sedimentation constant by the analysis of a sucrose gradient. Only a single time point is available, so there are uncertainties in the meaning of the band position, because acceleration and decel-

Box 11-2 THE GRADIENT PRODUCED BY A MIXING CHAMBER

The mixing-chamber apparatus is illustrated in Figure 11-17. We wish to show that a linear concentration gradient will be delivered to the tube. Because the heights of the two mixing vessels must remain equal, the volume of solution in the right-hand container is $V_R = V_0 - \frac{1}{2}V_d$, where V_0 is the initial volume, and V_d is the volume that has dripped into the tube. The mass of solute in the right-hand container is $M_R = c_R V_R$, where c_R is the concentration in the right-hand container. The rate of mass change with each drop must be $dM_R/dV_d = -c_R + \frac{1}{2}c_L$, because the right-hand vessel loses one drop of its contents at c_R, but gains one-half drop from the left-hand vessel at concentration c_L. However, one also can evaluate this derivative from the chain rule:

$$\begin{aligned} dM_R/dV_d &= V_R(dc_R/dV_d) + c_R(dV_R/dV_d) \\ &= (V_0 - \tfrac{1}{2}V_d)(dc_R/dV_d) - \tfrac{1}{2}c_R \end{aligned}$$

Equating the two expressions for dM_R/dV_d, we obtain a simple differential equation:

$$-2dc_R/(c_R - c_L) = dV_d/(V_0 - \tfrac{1}{2}V_d)$$

Integration from $c_R(0)$ to $c_R(V_d)$, as V_d goes from 0 to V_d, yields

$$[c_R(V_d) - c_L]/[c_R(0) - c_L] = 1 - V_d/2V_0$$

The concentration decreases linearly in the right-hand vessel as solution drips into the tube, and therefore the concentration in the tube must decrease linearly with height.

eration times of the rotor may not be known with any accuracy. More serious is the direct effect of the sucrose itself. The density and viscosity of the solution are not constant, and therefore s is not constant. Thus Equation 11-18 cannot be used to compute s. In practice, standards usually are run along with an unknown sample. Convenient standards are enzymes or isotopically labeled substances that can be added in trace amounts and subsequently assayed after the gradient has been fractionated. It is a common practice to determine the sedimentation constant of an unknown by bracketing it between two standards of known $s_{20,w}$ and assuming that s is a linear function of the distance between the two standards. This assumption is not correct for most gradients, and the procedure should be used only when the standards are virtually coincident with the unknown.

If accurate sedimentation constants must be obtained through preparative gradient centrifugation, the best procedure is to construct an isokinetic gradient. This gradient is designed so that $[1 - \bar{V}_2\rho(x)]/[\eta(x)]$ cancels the x-dependence of the accelerating force. It can be approximated by a properly chosen exponential gradient of a solute that affects both ρ and η linearly with concentration. Then the sedimentation constant will be a linear function of the distance traveled. One still must worry about attempting to match the $\bar{V}_2$ values of samples and standards. One also must realize that there could be significant three-component thermodynamic effects in concentrated sucrose. However, when properly used, sucrose-gradient sedimentation is a powerful analytical tool.

11-3 EQUILIBRIUM ULTRACENTRIFUGATION

Suppose a macromolecule is subjected to ultracentrifugation at speeds insufficient to transport it to a packed band at the bottom of the cell. Some transport must occur, but eventually an equilibrium will be established (Fig. 11-18). Solute is removed from

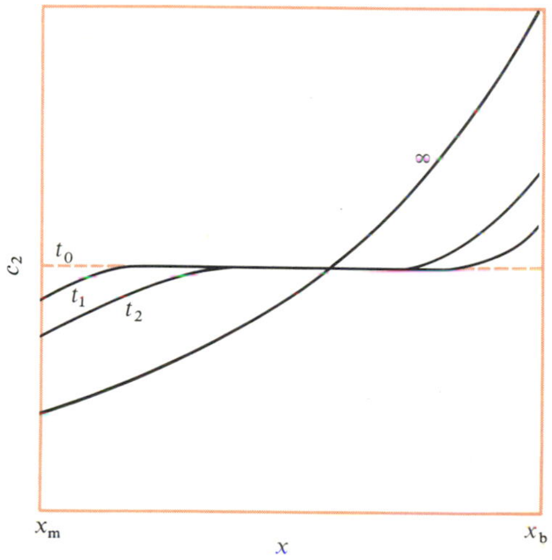

Figure 11-18

Sedimentation to an equilibrium solute distribution. [After C. Tanford, *Physical Chemistry of Macromolecules* (New York: Wiley, 1961), p. 385.]

the meniscus region as time progresses, and it tends to accumulate near the bottom of the cell. Diffusion forces are too strong to let a boundary be formed, and so a smooth concentration gradient eventually develops throughout the cell. The time required for this gradient to reach equilibrium can be shown to depend on $(x_m - x_b)^2$. For the standard sector-shaped centerpieces used in velocity sedimentation, this time is impractically long; however, the time can be brought within practical limits by the use of much shorter column heights.

Determining molecular weight with equilibrium ultracentrifugation

At equilibrium, the solute flux J_2 must be zero everywhere in the sample. Using either Equation 11-31 (corrected for nonideality) or Equation 11-4 and the appropriate molecular expressions for s and D, we can write

$$[c_2\omega^2 xM(1 - \bar{V}_2\rho)]/N_0 f = (kT/f)[1 + \partial(\ln \hat{\gamma}_2)/\partial(\ln c_2)](\partial c_2/\partial x) \quad (11\text{-}47)$$

Rearranging to place all measurable quantities on one side, and canceling f, we obtain

$$\begin{aligned} M[1 + \partial(\ln \hat{\gamma}_2)/\partial(\ln c_2)]^{-1} &= [RT/(1 - \bar{V}_2\rho)\omega^2](1/c_2 x)(\partial c_2/\partial x) \\ &= [2RT/(1 - \bar{V}_2\rho)\omega^2][\partial(\ln c_2)/\partial x^2] \end{aligned} \quad (11\text{-}48)$$

Note that only thermodynamic variables such as T and $\bar{V}_2$ remain. This is as it should be; the frictional coefficient can have no role at equilibrium. In ideal solutions, or at low concentrations, the activity-coefficient term vanishes, and Equation 11-48 can be integrated, using as a boundary condition the concentration c_2 at some reference point x_0:

$$c_2(x) = c_2(x_0) \exp\{[M(1 - \bar{V}_2\rho)\omega^2/2RT(x^2 - x_0^2)\} \quad (11\text{-}49)$$

Equations 11-48 and 11-49 imply that the slope of a plot of ln c_2 versus x^2 is all that is needed to measure the molecular weight.

Equation 11-49 can be used to estimate the rotation speed needed for an equilibrium measurement. With a typical optical system, accurate measurements can be made when the solute concentration falls by a factor of two across a 1 mm sample height. If $M = 50{,}000$ d, and $x_0 = 6$ cm, and $\bar{V}_2 = 0.75\ \text{cm}^3\ \text{g}^{-1}$, then ω is 10,400 rpm. This centrifuge speed is 5-fold slower (and the resulting forces are 25-fold weaker) than what is used for typical velocity sedimentation experiments.

Conservation of mass can be used to provide an alternative form of Equation 11-49 that is useful for determining molecular weight. The total solute in the cell at equilibrium must be equal to the total that was initially present. The initial weight of solute is

$$\int_{x_m}^{x_b} \phi a c_0 x\, dx = \frac{1}{2} \phi a c_0 (x_b^2 - x_m^2) \qquad (11\text{-}50)$$

where c_0 is the uniform initial concentration. The total mass present at equilibrium can be obtained by rearranging the first and second expressions of Equation 11-48 and integrating. For ideal solutions, we obtain

$$\begin{aligned} \phi a \int_{x_m}^{x_b} c_2(x) x\, dx &= [RT/M(1 - \bar{V}_2\rho)\omega^2]\phi a \int_{x_m}^{x_b} dc_2 \\ &= [RT/M(1 - \bar{V}_2\rho)\omega^2]\phi a (c_b - c_m) \qquad (11\text{-}51) \end{aligned}$$

where c_b and c_m are the solute concentrations at the bottom and at the meniscus, respectively. Setting Equations 11-50 and 11-51 equal, we obtain an expression for the molecular weight that depends only on concentration ratios:

$$(c_b - c_m)/c_0 = M(1 - \bar{V}_2\rho)\omega^2(x_b^2 - x_m^2)/2RT \qquad (11\text{-}52)$$

In practice, the use of Equation 11-49 or 11-52 is complicated by the difficulty of making actual concentration measurements in the ultracentrifuge. Some optical methods give only a derivative of c_2 or provide a measured quantity that is only indirectly related to c_2. Thus they do not provide the absolute solute concentration at any position in the cell. Although only concentration ratios are needed for Equations 11-49 and 11-52, these ratios cannot be calculated unless the absolute concentration is known at some position for calibration.

One way to circumvent the problem of concentration determination is to perform the equilibrium ultracentrifugation at sufficiently high rotor speeds so that no solute is left at the meniscus. This technique (the Yphantis meniscus-depletion method) allows use of the meniscus as a reference point for the optical system.

When appropriate precautions are taken, rather accurate molecular weights can be obtained from equilibrium centrifugation; Table 11-2 gives some representative results, which show that among various methods compared, equilibrium centrifugation is the most consistently accurate.

Equilibrium ultracentrifugation of mixtures of macromolecules

Suppose that the solution is not homogeneous, but instead is a mixture with i components. In such cases, curved lines may be observed when $\ln c_2$ is plotted against x^2. If there are c_i g cm^{-3} of material with molecular weight M_i, then Equation 11-49 becomes

$$c_T(x) = \sum_i c_i(x) = \sum_i c_i(x_0) \exp[M_i(1 - \bar{V}_{2i}\rho)\omega^2(x^2 - x_0^2)/2RT \qquad (11\text{-}53)$$

where $\bar{V}_{2i}$ is the partial specific volume of the ith component, and c_T represents the total weight concentration of solute. In principle, it is possible to fit such a function and derive values of $M(1 - \bar{V}_2\rho)$ for each of the individual components. Very accurate data are required, and it is rare to have sufficient accuracy in practice to justify a fit to more than two components. Substances with close molecular weights can never be resolved accurately by such a fitting approach.

One alternative is to settle for average molecular weight data. If the partial specific volume is the same for each component, differentiation of Equation 11-53 yields

$$\begin{aligned} d[\ln c_T(x)]/dx^2 &= [(1 - \bar{V}_2\rho)\omega^2/2RT]\left[\sum_i M_i c_i(x) \Big/ \sum_i c_i(x)\right] \\ &\equiv [(1 - \bar{V}_2\rho)\omega^2/2RT]\bar{M}_w(x) \end{aligned} \qquad (11\text{-}54)$$

The average molecular weight $\bar{M}_w(x)$ given by Equation 11-54 is a weight average, because c_T and c_i are weight concentrations. (See Box 11-3.) Equation 11-54 allows us to determine the weight-average molecular weight $\bar{M}_w(x)$ at each point x in the cell. The value of $\bar{M}_w(x)$ should increase with x, because heavy molecules will tend to be located farther from the rotation axis at equilibrium.

Often what we desire is the average molecular weight of the entire sample. We could obtain this value by integrating Equation 11-54 from the meniscus to the cell bottom, but it is simpler to obtain it another way. Starting from Equation 11-52, for a mixture of components, we can write

$$\sum_i c_{ib} - \sum_i c_{im} = \sum_i c_{io} M_i(1 - \bar{V}_2\rho)\omega^2(x_b^2 - x_m^2)/2RT \qquad (11\text{-}55)$$

where c_{ib} and c_{im} are the equilibrium concentrations of the ith component at the bottom and at the meniscus of the cell, respectively, and c_{io} is the initial concentration of the ith component. Dividing both sides of this equation by $\sum_i c_{io} = c_{T,0}$, and recognizing that the two sums in the left-hand side of the equation are just the total concentrations at the bottom and the meniscus at equilibrium, we obtain

$$[c_T(x_b) - c_T(x_m)]/c_{T,0} = [(1 - \bar{V}_2\rho)\omega^2/2RT](x_b^2 - x_m^2)\bar{M}_w \qquad (11\text{-}56)$$

where $\bar{M}_w$ clearly is the weight-average molecular weight of the entire mixture. (It often is called a cell-average $\bar{M}_w$.)

Other average molecular weights can be obtained if the data are treated differently. It is difficult to obtain the number-average molecular weight from centrifugation, but the z average at each cell position (see Box 11-3) can be calculated from observed data. It can be shown that

$$\bar{M}_z(x) = \sum_i M_i^2 c_i \Big/ \sum_i M_i c_i$$

$$= [2RT/(1 - \bar{V}_2\rho)\omega^2](d\{\ln[c_T(x)\bar{M}_w(x)]\}/dx^2) \qquad (11\text{-}57)$$

Then the cell-average $\bar{M}_z$ is obtained by integrating from the meniscus to the cell bottom. Equations 11-54 and 11-57 provide a sensitive test of the homogeneity of a substance. A pure single component with no tendency to self-associate should have identical values of $\bar{M}_z(x)$ and $\bar{M}_w(x)$ at each position in the cell.

Ultracentrifugation of a monomer–dimer equilibrium system

Suppose one is dealing with an associating macromolecular system. The various components must be in equilibrium at every point in the centrifuge cell. If there is no change in the partial specific volume during the association, then the equilibrium constant will be the same as that for a sample unperturbed by centrifugation. However, the equilibrium will shift with pressure if there is a net change in molecular volumes. The pressures generated in an ultracentrifuge can be considerable. The pressure gradient will be $dP/dx = \rho\omega^2 x$; therefore, $P(x) = (P/2)\omega^2(x^2 - x_m^2)$. The pressure can amount to several hundred atmospheres at the bottom of the cell.

To keep things simple, consider the monomer–dimer equilibrium, $2\,P \rightleftarrows P_2$, characterized by the equilibrium constant

$$\hat{k} = c_2/c_1^2 \qquad (11\text{-}58)$$

Pressure effects will be ignored, which means that $\bar{V}_2$ must be the same for all associating species. Because weight units are used, $\hat{k}$ differs from the usual equilibrium constant by a factor of $2/M_1$, where M_1 is the monomer molecular weight.

First, let us show that this equilibrium expression is obeyed at every position x in the cell. Suppose we have a mixture of a monomer and a dimer that are not equilibrating. The distribution of each separate species within the cell is given by Equation 11-49:

$$c_1(x) = c_1(x_0)\exp\{[M_1(1 - \bar{V}_2\rho)\omega^2/2RT](x^2 - x_0^2)\} \qquad (11\text{-}58a)$$

$$c_2(x) = c_2(x_0)\exp\{[2M_1(1 - \bar{V}_2\rho)\omega^2/2RT](x^2 - x_0^2)\} \qquad (11\text{-}58b)$$

Dividing these two equations yields

$$c_2(x)/c_1(x) = [c_2(x_0)/c_1(x_0)]\exp\{[M_1(1 - \bar{V}_2\rho)\omega^2/2RT](x^2 - x_0^2)\} \qquad (11\text{-}58c)$$

But the exponential term is simply $c_1(x)/c_1(x_0)$, so we have the following result:

$$c_2(x)/c_1^2(x) = c_2(x_0)/c_1^2(x_0) \tag{11-58d}$$

This relationship is independent of the rotor speed ω, and it is the same at all positions

Box 11-3 MOLECULAR WEIGHT AVERAGES

Consider a sample containing a distribution of species with different molecular weights. Let $n(M)\,dM$ be the number of moles of species with molecular weight between M and $M + dM$. The total number n_T of moles of species in the sample is

$$n_T = \int_0^\infty dM n(M)$$

The function $n(M)$ is called a molecular weight distribution function. If it were known explicitly, it would provide a complete description of all species in the sample. Often, however, it is desirable or necessary to deal with a less precise description. For example, we can define moments of the distribution function $n(M)$. The kth moment of the function is

$$m_k = \int_0^\infty dM n(M) M^k$$

Thus, n_T is the zeroth moment; it is just the area under the function.

Average molecular weights are defined as the ratios of various higher (kth) moments of $n(M)$ to the $(k-1)$th moment. The number-average molecular weight is defined as

$$\bar{M}_n \equiv m_1/m_0 = \int_0^\infty dM n(M) M \bigg/ \int_0^\infty dM n(M)$$

Consider a discrete distribution of species, containing n_i moles of components with molecular weight i. In this case, we can write the expression for $\bar{M}_n$ as

$$\bar{M}_n = \left(\sum_i n_i M_i\right) \bigg/ \sum_i n_i$$

We can write $\bar{M}_n$ in terms of molar concentrations, simply by dividing both numerator and denominator of this expression by the volume V of the sample:

$$\bar{M}_n = \left[\left(\sum_i n_i M_i\right) \bigg/ V\right] \bigg/ \left[\left(\sum_i n_i\right) \bigg/ V\right]$$

The weight concentration units used in hydrodynamics are g cm^{-3}, so $c_i = n_i M_i / V$. Thus, in the cell. Thus, if we load a centrifuge cell with a monomer–dimer mixture at equilibrium, so that Equation 11-58 is obeyed initially, the equilibrium distribution will be preserved at any speed and at all positions in the cell.

In the ultracentrifuge, we cannot observe the monomer and dimer separately, but only the total weight concentration c_T as described by Equation 11-53. The

the number-average molecular weight can be written as

$$\bar{M}_n = \left(\sum_i c_i\right) \Big/ \sum_i (c_i/M_i)$$

The weight-average molecular weight is defined as the ratio of the second moment to the first moment of the distribution function:

$$\bar{M}_w \equiv m_2/m_1 = \left(\sum_i n_i M_i^2\right) \Big/ \sum_i n_i M_i$$

Substituting weight concentrations just as before, we can write

$$\bar{M}_w = \left(\sum_i c_i M_i\right) \Big/ \sum_i c_i$$

Because $\sum_i c_i$ is the total weight concentration of solute (c_T), each term in $\bar{M}_w$ is just equivalent to the weight fraction w_i with molecular weight M_i in the sample.

A third molecular weight average commonly used is the z-average molecular weight:

$$\bar{M}_z \equiv m_3/m_2 = \left(\sum_i n_i M_i^3\right) \Big/ \sum_i n_i M_i^2 = \left(\sum_i c_i M_i^2\right) \Big/ \sum_i c_i M_i$$

For a mixture of components with very different molecular weights, the values of $\bar{M}_n$, $\bar{M}_w$, and $\bar{M}_z$ are quite disparate. Consider a 1:1 weight mixture of two macromolecules, one with molecular weight 10^3, and the other of 10^5 mol wt. The resulting averages are

$$\bar{M}_n = (1 + 1)/[(1/10^3) + (1/10^5)] = 1{,}950$$

$$\bar{M}_w = [(1 \times 10^3) + (1 \times 10^5)]/(1 + 1) = 50{,}500$$

$$\bar{M}_z = \{[1 \times (10^3)^2] + [1 \times (10^5)^2]\}/[(1 \times 10^3) + (1 \times 10^5)] = 99{,}020$$

Different experimental techniques provide different average molecular weights. Thus, if two or more different average values were measured, it would be very easy to see that the sample being studied must be a mixture. Ultracentrifugation provides values for $\bar{M}_w$ and $\bar{M}_z$. Light scattering measures $\bar{M}_w$ (see Chapter 14). To measure $\bar{M}_n$ directly, one must use techniques such as osmotic-pressure or vapor-pressure measurements.

following is one approach for obtaining the equilibrium constant. The total weight concentration of a monomer–dimer equilibrium mixture is

$$c_T = c_1 + c_2 = c_1 + \hat{k}c_1^2 \tag{11-59}$$

It is convenient to differentiate Equation 11-59:

$$dc_T/dc_1 = 1 + 2\hat{k}c_1 \tag{11-60}$$

Because of the equilibrium reaction, the weight-average molecular weight at each position in the cell now is a function of c_T as well:

$$\bar{M}_w(c_T) = (M_1c_1 + 2M_1c_1^2\hat{k})/c_T = [M_1c_1(1 + 2c_1\hat{k})]/c_T \tag{11-61}$$

Inverting Equation 11-60, and using Equation 11-61 to replace the term $1 + 2c_1\hat{k}$, we can write

$$dc_1/dc_T = 1/(1 + 2c_1\hat{k}) = (c_1/c_T)[M_1/\bar{M}_w(c_T)] = w_1M_1/\bar{M}_w(c_T) \tag{11-62}$$

where we have defined $w_1 = c_1/c_T$ as the weight fraction of monomer. If we differentiate this definition of w_1, we obtain an alternative expression for dc_1/dc_T:

$$dc_1/dc_T = w_1 + c_T(dw_1/dc_T) \tag{11-63}$$

Setting Equation 11-62 equal to Equation 11-63, and rearranging, we obtain

$$d(\ln w_1)/d(\ln c_T) = [M_1/\bar{M}_w(c_T)] - 1 \tag{11-64}$$

Equation 11-64 can be integrated between two positions in the cell, x_0 and x, to give

$$\ln [w_1(x)/w_1(x_0)] = \int_{c_T(x_0)}^{c_T(x)} \{[M_1/\bar{M}_w(c_T)] - 1\}(dc_T/c_T) \tag{11-65}$$

To use this equation, we must have data that extend to such a low concentration that $c_T(x_0) \cong 0$. Then $w_1(x_0) = 1$. We can obtain $c_T(x)$ directly in an equilibrium ultracentrifugation experiment. Using Equation 11-54, we can obtain $\bar{M}_w(x)$, and hence we can calculate $\bar{M}_w(c_T)$. Thus, as long as M_1 is known independently, we can obtain $\ln w_1$ at each position in the cell by integrating a plot of $\{[M_1/\bar{M}_w(c_T)] - 1\}c_T^{-1}$ versus c_T. We then can compute the equilibrium constant as

$$\hat{k} = (1 - w_1)/w_1^2c_T \tag{11-66}$$

More elaborate equilibria require more complex treatments, but the overall results are similar. Equilibrium ultracentrifugation is an excellent method for studying macromolecular associations. In real cases, however, one must use equations more sophisticated than Equation 11-65—equations that take nonideality into account.

Analysis of the approach to equilibrium

Equilibrium ultracentrifugation is a very powerful technique, but it is a very time-consuming measurement. Even with short columns, a day or two is required for 500,000 d samples to reach equilibrium, and a few hours for 50,000 d samples. A popular alternative is to analyze low-speed sedimentation data of samples on the way to equilibrium. Such data are typified by the intermediate time point in Figure 11-18. In 1947, W. J. Archibald (who had been working for years on solutions of the Lamm equation) realized that, at the meniscus or at the bottom of the cell, the flux equation can be solved trivially at all times in a sedimentation experiment. The flux J_2 must be zero at the meniscus and at the bottom of the cell, because no material can pass through these points. This fact is uninteresting at high rotor speeds, because there is no solute at the meniscus, and a pellet is formed at the bottom. However, at low speeds, Equation 11-4 set equal to zero immediately yields

$$\omega^2 s/D = (\partial c_2/\partial x)(1/xc_2) \qquad \text{for } x = x_b \text{ or } x_m \text{ only} \tag{11-67}$$

Substituting expressions for s and D, we can write

$$M_m = [RT/(1 - \bar{V}_2\rho)\omega^2](\partial c_2/\partial x)_{x_m}(1/x_m c_m) \tag{11-68a}$$

$$M_b = [RT/(1 - \bar{V}_2\rho)\omega^2](\partial c_2/\partial x)_{x_b}(1/x_b c_b) \tag{11-68b}$$

where c_m and c_b are the concentrations at the meniscus and at the bottom, respectively. With these expressions, we can obtain two estimates of the molecular weight at any time if we can measure the solute concentration and its gradient at top and at bottom. For a homogeneous sample, the two molecular weights should be equal. The Archibald method thus provides a good test of homogeneity.

The difficulty in using Equation 11-68 lies in measuring both the concentrations and their gradients. If absorption optics are used, the concentrations are obtained, and these must be numerically differentiated to obtain gradients. Such a procedure magnifies the noise in a concentration-versus-distance plot, resulting in imprecision. Schlieren optics help, because they give gradients directly. A plateau region remaining in the cell makes it easier to extract the concentrations at the bottom and at the meniscus from Schlieren data. One can always write

$$c_m = c_p - \int_{x_m}^{x_p} (\partial c/\partial x)\, dx \tag{11-69a}$$

$$c_b = c_p + \int_{x_p}^{x_b} (\partial c/\partial x)\, dx \tag{11-69b}$$

By considering the total material between x_m and x_p, or between x_p and x_b, and conserving mass, one can show that

$$c_m = c_0 - (1/x_m^2) \int_{x_m}^{x_p} x^2(\partial c/\partial x)\, dx \tag{11-70a}$$

$$c_b = c_0 + (1/x_b^2) \int_{x_p}^{x_b} x^2(\partial c/\partial x)\, dx \tag{11-70b}$$

where c_0 is the initial uniform concentration of solute. If no plateau exists, more complex approaches can be used (see Schachman, 1959). Any of these transient-state methods rapidly provide a value for the molecular weight.

Density-gradient ultracentrifugation: simple theory

Suppose a concentrated solution of a small molecule is spun at high speeds in the ultracentrifuge until equilibrium is reached. According to Equation 11-49, the small solute (component 3) will redistribute in the cell in just the same way as a large molecule might. If we use the meniscus of the cell as a reference point, and we expand the exponential in Equation 11-49 because M is small, we obtain

$$c_3(x)/c_3(x_m) = 1 + [M_3(1 - \bar{V}_3\rho)\omega^2/2RT](x^2 - x_m^2) \tag{11-71}$$

This expression is only approximate, because ρ is a function of x for a concentrated solution, and nonideality effects cannot be neglected. So Equation 11-49 is not exact to begin with. But, ignoring these complications for the moment, one can see that a parabolic concentration distribution is predicted. The density of the solution is roughly linear in the concentration of solute. This means that an equilibrium density gradient is established. From Equation 11-71, we have $\partial\rho/\partial x \propto \partial c_3/\partial x \propto \omega^2 x$. In practice, using salts of heavy metals such as CsCl, one can create a gradient of more than 10% in density across a typical centrifuge tube at the top speeds of modern ultracentrifuges.

Now consider the effect of adding a small amount of a macromolecule to the salt solution prior to centrifugation. If the macromolecule (component 2) is more dense than the highest density of CsCl at the bottom of the gradient, it will pellet to the bottom. If it is less dense than the lowest CsCl density, it will float to the top. However, if the macromolecular density falls within the range of the gradient, it will collect in a band at the density where $(1 - \bar{V}_2\rho)$ is zero (Fig. 11-19). If the density gradient is known, then the density (and thus $\bar{V}_2$) of the macromolecule can be determined from its position in the cell. At the center of the band, $\bar{V}_2^{-1}$ is just equal to the solution density ρ_0 at that point. This density is called the buoyant density. A mixture of macromolecules with different densities can be separated preparatively

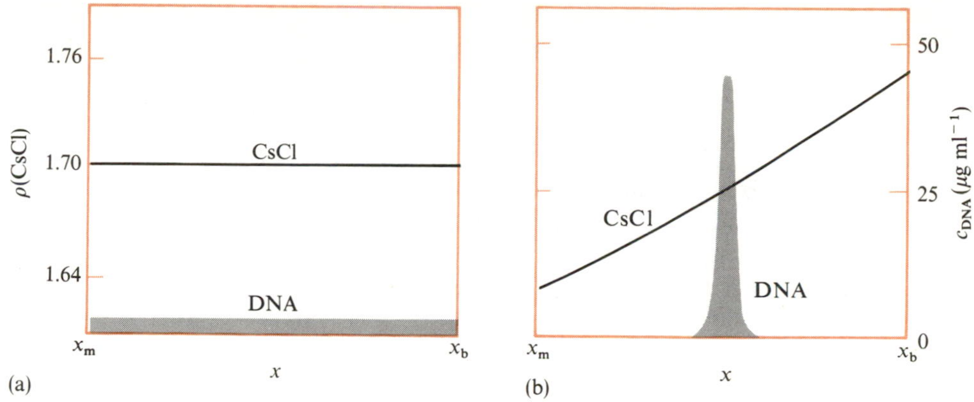

Figure 11-19

Equilibrium ultracentrifugation of a small amount of DNA in a CsCl gradient. **(a)** The density distribution at the beginning of the experiment. **(b)** The density distribution at equilibrium. [After W. Szybalski, *Fractions* (Beckman Instruments, Inc., 1968). no. 1, p. 1.]

by such centrifugation. The resolution is potentially very sharp, because (in principle) the density gradient can be made quite shallow. Thus, it is important to examine the factors that determine the width of the band of a macromolecule.

Still ignoring three-component thermodynamic effects, consider just a pure physical density gradient. In the region of the macromolecule, we can expand the density in a Taylor series:

$$\rho(x) = \rho_0 + (x - x_0)(d\rho/dx)_{x_0} \tag{11-72}$$

where x_0 is the center of the equilibrium band position of the macromolecule. Substitute this expression into Equation 11-48, the equilibrium distribution of the macromolecule. Neglecting nonideality, we obtain

$$M_2(1 - \bar{V}_2)\{[\rho_0 + (x - x_0)(d\rho/dx)_{x_0}]\}x = (RT/\omega^2)(1/c_2)(dc_2/dx) \tag{11-73}$$

This equation is simplified by recognizing that $\bar{V}_2\rho_0 = 1$, as we have mentioned earlier. Then letting $y = x - x_0$, and $dy = dx$, we can rewrite Equation 11-73 as

$$d(\ln c_2) = (-M_2\omega^2/RT)(d\rho/dx)_{x_0}\bar{V}_2(y^2 + x_0 y)\,dy \tag{11-74}$$

Because y is much smaller than x_0, the term in y^2 can be dropped. Integrating from $y = 0$ to $y = x$, and replacing y with $x - x_0$, we obtain

$$c_2(x) = c_2(x_0)\exp[-(x - x_0)^2/2\sigma^2] \tag{11-75}$$

Thus, a macromolecule in a density gradient distributes in a Gaussian band with a width characterized by standard deviation σ, which is given by

$$\sigma^2 = RT/M_2\bar{V}_2\omega^2 x_0(d\rho/dx)_{x_0} \tag{11-76}$$

Density-gradient centrifugation: three-component theory

To solve the density-gradient equilibrium problem correctly, one must explicitly consider the thermodynamics of all three components. The molar concentration of macromolecule within the band is high enough that it cannot be neglected. Preferential interaction between components 2 and 3 can alter the density gradient, the shape of the band, and the apparent buoyant density (ρ_0). It is quite a complicated problem, and we sketch here only a few features.

We start with the general phenomenological equation for the flux of the ith component in a three-component system (see Eqn. 10-38):

$$J_i = \sum_{j=1}^{3} L_{ij}X_j \tag{11-77}$$

where X_j is the generalized force on component j. In the ultracentrifuge, we know (by comparing Eqns. 11-77 and 11-29) that the force on the jth component is

$$X_j = \omega^2 x - \partial\hat{\mu}_j/\partial x \tag{11-78}$$

At mechanical and thermal equilibrium, this force must be zero. Hence,

$$\omega^2 x = \partial\hat{\mu}_j/\partial x \tag{11-79a}$$

The gradient of the chemical potential of component j can be evaluated in the same way as that shown earlier for a two-component system (Eqns. 11-30 and 11-31). The result allows Equation 11-79a to be rewritten as

$$(1 - \bar{V}_j\rho)\omega^2 x = \sum_{i=1}^{3} (\partial\hat{\mu}_j/\partial c_i)(\partial c_i/\partial x) \tag{11-79b}$$

In density-gradient centrifugation, the third component is the concentration of a heavy salt such as CsCl. We first seek a description of the density gradient generated by this component. This gradient can arise from concentration effects and pressure effects:

$$d\rho/dx = (\partial\rho_C/\partial x) + (\partial\rho_P/\partial x) \tag{11-80}$$

One starts by making the approximation that the chemical potential of component 3 is independent of the concentrations of components 1 and 2. Then,

$$(1 - \bar{V}_3\rho)\omega^2 x = (\partial\hat{\mu}_3/\partial c_3)_p(\partial c_3/\partial x) \tag{11-81}$$

Using Equation 11-81, we can write the density gradient due to the composition of component 3 (low-molecular-weight solute such as salt) as

$$\partial\rho_C/\partial x = (\partial\rho/\partial\hat{\mu}_3)(\partial\hat{\mu}_3/\partial c_3)(\partial c_3/\partial x) = g\omega^2 x \tag{11-82}$$

where the constant $g = (1 - \bar{V}_3\rho)(\partial\rho/\partial\hat{\mu}_3)$. The derivative $\partial\rho/\partial\hat{\mu}_3$ can be measured from the dependence of density on the activity of component 3. There is also a compression density gradient:

$$\partial\rho_P/\partial x = (\partial\rho/\partial P)(\partial P/\partial x) = k\omega^2 x \tag{11-83}$$

where $k = \partial\rho/\partial P$ is another constant. The overall density gradient is the sum of these two effects (from Eqn. 11-80):

$$d\rho/dx = (k + g)\omega^2 x \tag{11-84}$$

Next we calculate the position of the macromolecule band. In a three-component system, the third component need not be included explicitly, because there are only two independent chemical potential variables in a three-component system (see also Box 10-1). It is convenient to ignore the salt and to work only with solvent and macromolecule. Now, however, components 1 (water) and 2 (macromolecule) must be treated by two-component thermodynamics. Equation 11-79b becomes

$$(1 - \bar{V}_1\rho)\omega^2 x = (\partial\hat{\mu}_1/\partial c_1)_{c_2}(\partial c_1/\partial x) + (\partial\hat{\mu}_1/\partial c_2)_{c_1}(\partial c_2/\partial x) \tag{11-85a}$$

$$(1 - \bar{V}_2\rho)\omega^2 x = (\partial\hat{\mu}_2/\partial c_2)_{c_1}(\partial c_2/\partial x) + (\partial\hat{\mu}_2/\partial c_1)_{c_2}(\partial c_1/\partial x) \tag{11-85b}$$

where the concentration of component 2 is now explicitly allowed to affect the chemical potential of component 1, and vice versa.

The difficulty in solving Equation 11-85 arises from the two-component effects: $(\partial\hat{\mu}_1/\partial c_2)_{c_1}$ and $(\partial\hat{\mu}_2/\partial c_1)_{c_2}$. However, because the chemical potential of component 1 is a function of both components, a change in this potential resulting from a change in solution can be described as

$$d\hat{\mu}_1 = (\partial\hat{\mu}_1/\partial c_1)_{c_2}\, dc_1 + (\partial\hat{\mu}_1/\partial c_2)_{c_1}\, dc_2 \tag{11-86}$$

If the change $d\hat{\mu}_1$ is carried out by varying component 1 at constant $\hat{\mu}_1$, the result is

$$0 = (\partial\hat{\mu}_1/\partial c_1)_{c_2}(\partial c_1/\partial c_1)_{\hat{\mu}_1} + (\partial\hat{\mu}_1/\partial c_2)_{c_1}(\partial c_2/\partial c_1)_{\hat{\mu}_1} \tag{11-87}$$

By rearranging this, we can define a solvation parameter as

$$\Gamma' = (\partial c_1/\partial c_2)_{\mu_1} = -(\partial\hat{\mu}_1/\partial c_2)_{c_1}/(\partial\hat{\mu}_1/\partial c_1)_{c_2} \tag{11-88}$$

This parameter is essentially the weight of solvent that also must be added when adding a gram of macromolecule in dilute solution to keep the chemical potential of solvent constant. It is, therefore, a measure of the amount of bound solvent.

The parameter Γ' is the three-component equivalent of the gram/gram hydration parameter we used previously. Using Equation 11-88, we can solve Equation 11-85 as follows. First, note that $(\partial\hat{\mu}_2/\partial c_1) = (\partial\hat{\mu}_1/\partial c_2)$, because both are equal to $\partial^2 G/\partial c_1 \partial c_2$, where G is the Gibbs free energy. We can use $-\Gamma'\partial\hat{\mu}_1/\partial c_1$ to replace $\partial\hat{\mu}_1/\partial c_2$. Then Equations 11-85a and 11-85b can be solved simultaneously to eliminate the unknowns $(\partial\hat{\mu}_1/\partial c_1)_{c_2}$ and $\partial c_1/\partial x$. After considerable tedious manipulation, the resulting equation can be arranged in the form of Equation 11-81:

$$(1 + \Gamma')\{1 - [(\bar{V}_2 + \Gamma'\bar{V}_1)/(1 + \Gamma')]\rho\}\omega^2 x = (\partial\hat{\mu}_2/\partial c_2)_{c_1}(\partial c_2/\partial x) \tag{11-89}$$

In a density-gradient equilibrium, the buoyant density (ρ_0) is operationally the point at which the macromolecule concentration is at a maximum. Thus $\partial c_2/\partial x = 0$, and

$$\boxed{1/\rho_0 = (\bar{V}_2 + \Gamma'\bar{V}_1)/(1 + \Gamma') \equiv \bar{V}_s} \tag{11-90}$$

where we define a solvated partial specific volume $\bar{V}_s$ as the inverse of the buoyant density.

Equation 11-89 can be put in a more familiar form by recalling that

$$(\partial\hat{\mu}_2/\partial c_2)_{c_1} = (RT/M_2)[\partial(\ln a_2)/\partial c_2] = (RT/M_2 c_2)\{1 + [\partial(\ln \hat{y}_2)/\partial(\ln c_2)]\} \tag{11-91}$$

Thus we can write

$$M_2(1 + \Gamma')(1 - \bar{V}_s\rho)\omega^2 x/\{1 + [\partial(\ln \hat{\gamma}_2)/\partial(\ln c_2)]\} = (RT/c_2)(\partial c_2/\partial x) \tag{11-92}$$

Comparison of Equations 11-92 and 11-48 shows that the apparent molecular weight that enters into Equation 11-92 is a solvated molecular weight: $M_s = M_2(1 + \Gamma')$.

The solvated quantities $\bar{V}_s$ and M_s vary with density because the concentration of solvent (and thus the solvation) changes with position in the gradient. Thus the effective density gradient, $(\partial\rho/\partial x)^{\text{eff}}$, must be the physical gradient (Eqn. 11-84) corrected for this effect. It is reasonable that the macromolecule perturbs the gradient by interacting with solvent. When all of these factors are taken into account (including the nonideality of the macromolecular solution), an equation can be derived that has the same Gaussian form as Equation 11-75, with a width of

$$\sigma^2 = RT/M_s^{\text{app}}(x_0)\bar{V}_s(x_0)(\partial\rho/\partial x)_{x_0}^{\text{eff}}\omega^2 x_0 \tag{11-93}$$

where the apparent solvated molecular weight (M_s^{app}) is related to the true solvated weight (M_s) by a virial expansion:

$$M_s^{\text{app}} = M_s[1 + 2B_2c_2(x_0) + \cdots]^{-1} \tag{11-94}$$

where B_2 is a constant.

It must be apparent that the use of Equations 11-90 and 11-93 to obtain absolute molecular weights and densities is a formidable problem. The usual practice is to use Equation 11-84 to approximate $(\partial\rho/\partial x)^{\text{eff}}$, because all the quantities in this equation are measurable. A macromolecule with known or assumed ρ_0 is placed in the gradient, and unknowns are compared to the standard by their relative positions in the gradient. For more accurate work, $(\partial\rho/\partial x)^{\text{eff}}$ can be determined by using isotopically substituted samples, such as $[^{15}N]$-DNA versus $[^{14}N]$-DNA. In this case, it can be assumed that the isotopic substitution changes none of the thermodynamic parameters, but alters only the anhydrous molecular weight and density of the macromolecule. Once $(\partial\rho/\partial x)^{\text{eff}}$ is known, then Equation 11-93 will yield an accurate molecular weight from the shape of the band.

Note that the considerations leading to Equation 11-92 should apply to any three-component system, whether or not the third component has a substantial effect on the density of the solution. This sensitivity to other components might appear to introduce serious complications into any ultracentrifugation experiment. Fortunately, methods have been developed to circumvent such complications (see Box 11-4).

Equilibrium density-gradient sedimentation occasionally has been applied to proteins, but most work involves nucleic acids. Chapter 22 gives examples of results from this technique.

Box 11-4 ULTRACENTRIFUGATION IN THREE-COMPONENT SYSTEMS

E. F. Casassa and H. Eisenberg have developed a useful technique for determining the molecular weight of a solute in a three-component system, unaffected by three-component effects such as preferential solvation. Here we only sketch the results; for a detailed treatment, see H. Fujita (1975).

Casassa and Eisenberg showed that, in the limit of low macromolecule concentration ($c_2 \to 0$), it is possible to evaluate the terms in the left-hand side of Equation 11-89 as

$$(1 + \Gamma')\{1 - [(\bar{V}_2 + \Gamma'\bar{V}_1)/(1 + \Gamma')]\rho\} = (\partial\rho/\partial c_2)_{\hat{\mu}_1,\hat{\mu}_3}$$

Experimentally, this evaluation is accomplished by dialyzing a solution of macromolecule and salt against a large volume of salt solution, and measuring the density of the macromolecule solution after equilibrium is reached. Then the results of an equilibrium ultracentrifugation experiment can be evaluated from Equation 11-92 (with nonideality terms deleted because $c_2 \to 0$) as

$$d(\ln c_2)/d(x^2) = (M_2\omega^2/2RT)(\partial\rho/\partial c_2)_{\hat{\mu}_1,\hat{\mu}_3}$$

Similarly, velocity sedimentation experiments can be analyzed by replacing $1 - \bar{V}_2\rho$ in Equation 11-28 with the three-component equivalent, $(\partial\rho/\partial c_2)_{\hat{\mu}_1,\hat{\mu}_3}$.

Summary

Sedimentation velocity experiments measure the rate at which a macromolecule moves when subjected to radial acceleration in an ultracentrifuge. The rate per unit field, s, is proportional to the molecular weight and inversely proportional to the frictional coefficient. Thus, for spherical particles, $s \propto M^{2/3}$. In the ultracentrifuge, each pure component normally gives rise to a boundary. Above the boundary, the component is absent; beneath the boundary, in the plateau region, the concentration of the component is constant with distance, but varies with time. One can measure s by examining either the boundary or the plateau. The shape of the boundary is affected by diffusion but, in practice, the concentration dependence of s also has a strong influence on boundary shape. Very complex effects on boundary shape and height can result when several different macromolecular species are present. These effects can be circumvented by zonal centrifugation, in which one starts with a thin band of macromolecule-containing solution layered on top of a supporting solvent, rather than with a homogeneous sample. Stabilizing gradients allow recovery of physically separated samples after the solution is removed from the centrifuge.

At rotor speeds much smaller than those used for sedimentation velocity experiments, an equilibrium distribution of macromolecules is created in the ultracentrifuge. The shape of this distribution is proportional to the molecular weight, and it does not depend on frictional properties. When mixtures of components are present, equilibrium centrifugation yields a weight-average molecular weight. In some cases, the equilibrium distribution of associating macromolecules can be analyzed to yield the association constant. Density-gradient equilibrium ultracentrifugation is a powerful technique for separating macromolecules according to their buoyant densities. Here the macromolecule is allowed to equilibrate in a gradient of a heavy salt (such as CsCl) that spans its density. The macromolecule at equilibrium is found in a narrow band centered about its buoyant density. This buoyant density is not equal to the density of pure macromolecule because of complex three-component thermodynamic effects present in the concentrated salt solutions used to create the gradient. The width of the band of macromolecule is inversely proportional to the square root of the molecular weight.

Problems

11-1. Estimate the velocity (cm sec^{-1}) with which a 1 g copper sphere (a) will sink in a lake; (b) will move in aqueous solution in a centrifuge spinning at 1 rpm, 10 cm from the rotation axis.

11-2. Suppose you wish to measure the sedimentation constant of a substance present in small amounts contaminated by many impurities. You obviously cannot detect this substance in a meaningful way by optical methods in the ultracentrifuge. However, you can assay it in some biochemical way, such as "rats killed per cm^3 of solution." You make use of a separation cell (Fig. 11-20), initially filled with a solution at concentration c_0 as determined

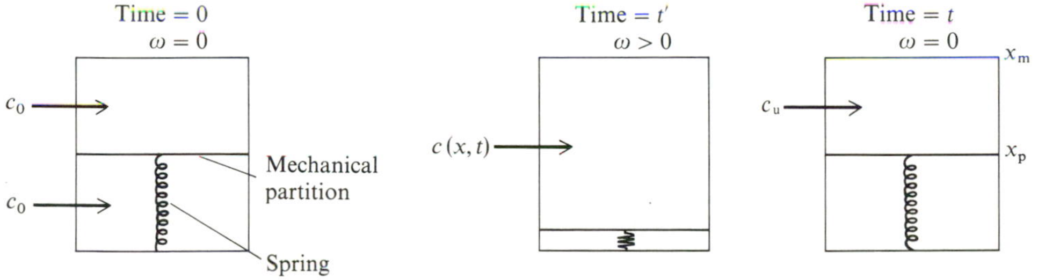

Figure 11-20
Separation-cell experiment for Problem 11-2.

by biochemical assay. At time t', during ultracentrifugation, the partition moves downward. (Assume that movements of the partition during ultracentrifugation do not mix the solution.) At time t, when the ultracentrifuge is stopped, the partition rises to the position x_p (assume that this position is in the plateau region). The solution above the partition is then mixed, and is found to have a concentration of c_u after mixing (again using the biochemical assay on the mixed solution). Derive an expression that will let you calculate s from the measured values of ω, t, x_m, x_p, c_u, and c_0. HINT: It's all a question of conserving mass. (A modern version of this approach is active enzyme centrifugation; see Cohen et al., 1967.)

11-3. Straightase is an (hypothetical) enzyme known, from electron microscopic studies, to be a prolate ellipsoid with an axial ratio of 20. When it is titrated with excess *p*-chloromercuribenzoate (PCMB), the sedimentation coefficient decreases by a factor of 1.58, although only a single sedimenting band is observed both before and after the titration. The diffusion coefficient increases by a factor of 2.52 after titration.
   a. Into how many fragments has straightase been broken by PCMB?
   b. Calculate the axial ratio of the fragments.
   c. Decide whether the fragments are more probably prolate or oblate, and draw a plausible schematic structure for straightase.

11-4. A spherical protein with an anhydrous radius R has a sedimentation constant $s_{20,w} = 8$ S. Suppose that a spherical core of the protein, extending from $r = 0$ to $r = R/2$, is removed and replaced with solvent. How will this solvent core change the measured $s_{20,w}$?

11-5. When a protein is placed in pure D_2O, proton exchange leads to an increase of 1.55% in the molecular weight. Using this fact, show that it is possible to measure $\bar{V}_2$ by comparing equilibrium ultracentrifugation measurements on a protein in D_2O and in H_2O. You can assume that the molecular volume in H_2O is the same as that in D_2O.

References

GENERAL

Freifelder, D. 1976. *Physical Biochemistry*. San Francisco: W. H. Freeman and Company.
Fujita, H. 1975. *Foundations of Ultracentrifugal Analysis*. New York: Wiley.
Schachman, H. K. 1959. *Ultracentrifugation in Biochemistry*. New York: Academic Press.
Tanford, C. 1961. *Physical Chemistry of Macromolecules*. New York: Wiley.
Van Holde, K. E. 1975. Sedimentation analysis of proteins. In *The Proteins*, 3d ed., ed. H. Neurath and R. Hill (New York: Academic Press), vol. 1, p. 228.

SPECIFIC

Adams, E. T., Jr. 1967. Analysis of self-association systems by sedimentation equilibrium experiments. *Fractions* (Beckman Instruments, Inc.), no. 3, p. 1.
Bloomfield, V., D. Crothers, and I. Tinoco, Jr. 1974. *Physical Chemistry of Nucleic Acids*. New York: Harper & Row.

Cann, J. R. 1970. *Interacting Macromolecules: The Theory and Practice of Their Electrophoresis, Ultracentrifugation and Chromatography*. New York: Academic Press.

Cohen, R., B. Giraud, and A. Messiah. 1967. Theory and practice of analytical centrifugation of an active substrate–enzyme complex. *Biopolymers* 5:203.

Dishon, M., G. H. Weiss, and D. A. Yphantis. 1967. Numerical solutions of the Lamm equation, III: Velocity centrifugation. *Biopolymers* 5:697.

Eisenberg, H. 1976. *Biological Macromolecules and Polyelectrolytes in Solution*. Oxford: Clarendon Press. [A useful advanced treatise.]

Hearst, J. E., and C. W. Schmid. 1973. Density gradient sedimentation equilibrium. In *Methods in Enzymology*, vol. 17, ed., C. H. W. Hirs and S. N. Timasheff (New York: Academic Press), p. 111.

Hinton, R., and M. Dobrota. 1976. *Density Gradient Centrifugation*. New York: North-Holland.

Kuntz, I. D., Jr., and W. Kauzmann. 1974. Hydration of proteins and polypeptides. In *Advances in Protein Chemistry*, vol. 28, ed. C. B. Anfinsen, J. T. Edsall, and F. M. Richards (New York: Academic Press), p. 239.

Van Holde, K. E. 1971. *Physical Biochemistry*. Englewood Cliffs, N.J.: Prentice-Hall.

12
Other hydrodynamic techniques

12-1 VISCOMETRY

In this chapter we discuss a number of additional methods for obtaining information on the sizes and shapes of biopolymers. We discuss these techniques in much less detail than our discussion of ultracentrifugation in Chapter 11. Some of these techniques are used less frequently than ultracentrifugation; the theories of some are complicated and are of less instructional value; and some are commonly used as preparative procedures, but are less often employed for obtaining quantitative structural information. We begin with the technique of viscometry.

Measurement of viscosity

The appealing feature of viscosity measurements is their experimental simplicity. In many cases, one inexpensive piece of glass apparatus, the capillary viscometer, suffices for an accurate measurement of the viscosity of a solution (Fig. 12-1a). We have defined viscosity η (Chapter 10) as the force needed to maintain a velocity gradient dv/dx in a fluid, normal to the applied force:

$$F = \eta A(dv/dx) \tag{12-1}$$

where A is the area of the fluid parallel to the flow. Hydrostatic pressure provides the force in a capillary viscometer. In a capillary of radius a, the force is the pressure times the cross-sectional area: $F = P\pi a^2$.

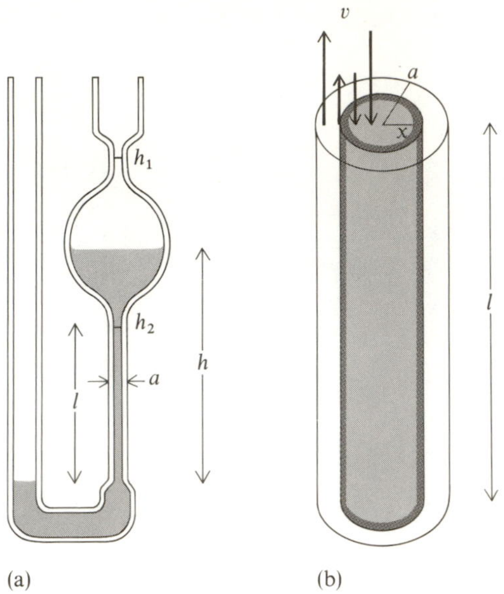

Figure 12-1
Capillary viscometry. **(a)** A typical Ostwald viscometer. The actual capillary is a length l below h_2. Solution is drawn up to h_1, and the time required for the liquid level to fall to h_2 is measured. **(b)** Enlarged view of the capillary, showing the cylindrical volume of fluid considered in the derivation of Poiseuille's law. At top, the fluid velocity profile, as viewed from the cylinder at radius x, is depicted schematically.

The differential force on a small cylindrical sheet of fluid at a radial distance x (Fig. 12-1b) is just

$$dF = 2\pi x P\, dx \tag{12-2}$$

because the cross-sectional area is $2\pi x\, dx$. If the fluid is flowing through the capillary at a steady rate, this force must be balanced at each position x by the frictional force. The lateral surface area of the cylindrical sheet is $2\pi x l$. The force exerted by the fluid on one surface of this sheet is given by Equation 12-1, with a negative sign because it is in the direction opposite to the applied force:

$$F = -\eta l 2\pi x(dv/dx) \tag{12-3}$$

The net force on the sheet due to fluid motion is the differential force felt by the two sides of the sheet:

$$dF = -\eta l 2\pi\left(\frac{d[x(dv/dx)]}{dx}\right)dx \tag{12-4}$$

Now we can equate the two differential forces of Equations 12-2 and 12-4, removing the common factor of $2\pi\, dx$:

$$Px = -\eta l\left(\frac{d[x(dv/dx)]}{dx}\right) \tag{12-5}$$

This expression is readily integrated to yield

$$Px^2/2 + C_1 = -\eta l x(dv/dx) \qquad (12\text{-}6)$$

where C_1 is an integration constant. Dividing through by x and integrating again, we obtain

$$Px^2/4 + C_1 \ln x + C_2 = -\eta l v \qquad (12\text{-}7)$$

We can evaluate the integration constants C_1 and C_2 from the boundary conditions. At the center of the capillary, $x = 0$, and the velocity of flow must be finite; therefore, $C_1 = 0$. At the outside edge, the velocity must be zero if we use stick boundary conditions (see Chapter 10); therefore, $C_2 = -Pa^2/4$. The result is well known. The velocity profile of the fluid in a capillary will have a parabolic shape:

$$\boxed{v = (P/4\eta l)(a^2 - x^2)} \qquad (12\text{-}8)$$

Equation 12-8 represents the flow velocity. However, a more readily measurable quantity is the volume rate of flow, which is the velocity of each cylindrical sheet multiplied by its area $2\pi x\, dx$ and integrated from the center of the capillary to the radius:

$$dV/dt = \int_0^a 2\pi x v\, dx = (\pi P/2\eta l) \int_0^a (a^2 - x^2)x\, dx = \pi P a^4/8\eta l \qquad (12\text{-}9)$$

Equation 12-9 is Poiseuille's law. In a typical measurement with a capillary viscometer, the fluid of density ρ is allowed to fall from height h_1 to h_2 (Fig. 12-1), and the time for this fall is determined. The only pressure is hydrostatic pressure, which is $\rho g h$, where h is the difference in height between the two ends of the column of fluid in the capillary, and g is the acceleration due to gravity. This pressure decreases as the height of the fluid in the capillary decreases. The time required for the total volume V to flow is obtained by integration of Equation 12-9:

$$\boxed{t = (8\eta l/\pi g \rho a^4) \int_{h_1}^{h_2} dV/h} \qquad (12\text{-}10)$$

The integral is a constant for a given apparatus. It usually is found by calibration with a fluid of known density—most commonly, the solvent that will be used for a subsequent study on a macromolecule.

Effect of shear on measured viscosity

Capillary viscosity measurements are quite straightforward and easy, as long as the capillary is kept clean. These measurements have two disadvantages. First, a rather large volume of solution is required. Second (and more serious), the shearing forces generated by the flow gradients in a typical capillary viscometer are quite large. The shear stress S is defined as the force per unit area resulting from solution flow. The shear is the velocity gradient perpendicular to the force (Fig. 10-8). Thus, rearrangement of Equation 12-1 reveals that shear stress and shear are related by

$$S = F/A = \eta(dv/dx) \tag{12-11}$$

Note that this equation is an oversimplification because, even in the case of parallel plates (Fig. 10-8b), S and dv/dt actually are vectors. In more general cases, both the shear and the shear stress are tensors. However, it will be sufficient here to ignore these complications.

Shear stress causes orientation of rigid long molecules such as small DNAs. It can cause distortion of the coil distribution of more flexible molecules. Thus, shear stress can alter the viscosity measured for a sample. The effect of shear stress on the observed viscosity of DNA solutions is very large. Figure 12-2 shows some typical results. Very high shear stress can actually lead to cleavage of macromolecules. Thus, it is imperative to make measurements at very low shear, and preferably to extrapolate to zero shear.

The average shear stress in a capillary viscometer can be calculated by starting with Equation 12-6. We have already evaluated C_1 as zero. Thus Equation 12-6 can be rearranged in the form of Equation 12-11 to give an expression for the shear

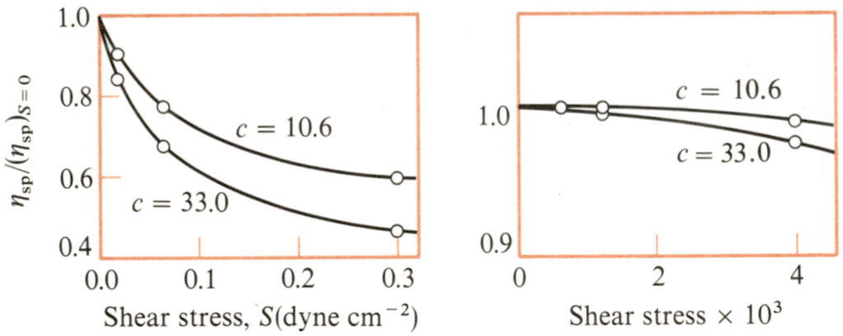

Figure 12-2

Shear-stress dependence of specific viscosity. The specific viscosity of phage T2 DNA (relative to the value at zero shear stress) is shown for two solutions at different concentrations (μg cm^{-3}) over a range of shear stress. Data for extremely low shear are given at the right. Note that the viscosity becomes essentially shear-independent at very low shear. It should be clear from the shapes of the curves that attempts to extract [η] by plotting η_{sp} as a function of concentration will be meaningful only at very low shear. An intrinsic viscosity of 31,600 cm^3 g^{-1} was determined for these DNA samples. [After D. M. Crothers and B. H. Zimm, *J. Mol. Biol.* 12:525 (1965).]

stress in a cylindrical sheet of fluid with radius x:

$$S_x = \eta(dv/dx) = -Px/2l \qquad (12\text{-}12)$$

The average shear stress is obtained by integrating this expression over all sheets, weighted by their area, and dividing by the integrated area:

$$S = \int_0^a 2\pi xl\,dxS_x \Big/ \int_0^a 2\pi xl\,dx = -Pa/3l \qquad (12\text{-}13)$$

The pressure depends on the difference in height between the two ends of the column of fluid in the capillary. Thus, the shear stress can be varied in practice by adjusting the height of the fluid in the capillary. Note that the pressure does not remain constant during an actual capillary viscosity measurement, because the height changes continuously. Thus, in practice, to compute the true average shear stress during an experiment, one must integrate Equation 12-13 over time and divide by Equation 12-10.

For a typical capillary viscometer with water as the sample, S is about 10 dyn cm^{-2}. Much lower shear stresses can be obtained by using a rotating cylinder viscometer; Figure 12-3 shows one version of this apparatus. A rotating magnetic

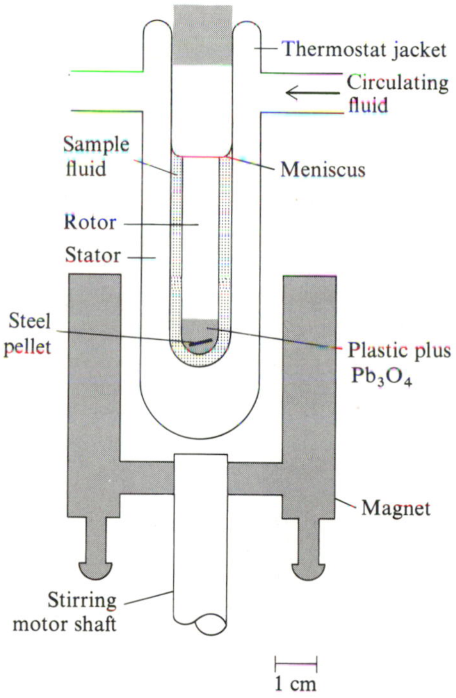

Figure 12-3

Cross-section of the Zimm–Crothers low-stress rotating cylinder viscometer. [After B. H. Zimm and D. M. Crothers, *Proc. Natl. Acad. Sci. USA* 48:905 (1962).]

field applies a constant torque on an inner cylinder, freely floating in the solution to be studied. In response, the inner cylinder will rotate with a constant angular velocity ω. This velocity is less than that of the rotating magnetic field. The relative viscosities of any two solutions can be shown to be simply $\eta_1/\eta_2 = \omega_2/\omega_1$. The shear can be altered, either by changing the dimensions of the inner cylinder, or (much more conveniently) by altering the strength of the applied magnetic field. The Zimm–Crothers viscometer (shown in Fig. 12-3) can work with water at shear stresses as low as 0.0006 dyn cm^{-2}.

Effect of solute molecules on viscosity

Although viscosity is easy to measure, derivation from first principles of the effect of suspended particles on the solution viscosity poses formidable problems. The physical problem is to compute how a particle distorts the flow lines of a solution containing a velocity gradient. For a two-dimensional solution (Fig. 10-8a), this problem is complicated enough. However, the true three-dimensional result is even worse. To calculate the flow-line distortion, one must take explicitly into account the fact that shear and shear stress are tensors. Then it is easiest to compute the energy required per unit of time to maintain the shear. Using the schematic parallel-plate apparatus of Figure 10-8b, this energy per time[§] is

$$Fv_b = \eta(dv/dz)^2 Ah \tag{12-14}$$

where the velocity of the moving bottom plate relative to the fixed top plate is $v_b = h(dv/dz)$. Equation 12-14 allows η to be defined as the rate of energy dissipation per unit volume ($Ah = 1$) at unit shear ($dv/dz = 1$). The reader skilled in tensor calculus is encouraged to look elsewhere to see how the viscosity of solution containing suspended spheres is calculated (Yamakawa, 1971). Here we can only sketch the results.

The viscosity of a solution of particles must be a complex function of the concentration because, at high concentration, the distortion of the velocity pattern of the liquid by one particle can affect the shear at a neighboring particle. Thus, in general, one can expect that the viscosity of a solution should be expressed as a power series relative to the viscosity η_0 of pure solvent:

$$\eta = \eta_0(1 + k_1c_2 + k_2c_2^2 + \cdots) \tag{12-15}$$

where c_2 is the solute concentration in g cm^{-3}. The ratio η/η_0 is defined as η_{rel}, the relative viscosity. The specific viscosity is defined as

$$\eta_{sp} = \eta_{rel} - 1 = k_1c_2 + k_2c_2^2 + \cdots \tag{12-16}$$

[§] The dimensions of the left-hand side of Equation 12-14 are force × length ÷ time = energy ÷ time.

This means that η_{sp} contains all the ways in which the solute can influence the viscosity of the solution. In such a virial expansion, the coefficient k_1 can be associated with the contribution of individual solute molecules, k_2 with pairwise effects of clusters of two molecules, and so on. For extracting properties of individual macromolecules, k_1 is the quantity of major interest. It is defined as $[\eta]$, the intrinsic viscosity, and it can be obtained experimentally as

$$[\eta] = \lim_{c_2 \to 0} (\eta_{sp}/c_2) = \lim_{c_2 \to 0} (k_1 + k_2 c_2 + \cdots) = k_1 \qquad (12\text{-}17)$$

Beware that the units of $[\eta]$, η_{sp}, and η are all different. The units used for η are poise; η_{sp} is dimensionless; and $[\eta]$ has units of $cm^3\ g^{-1}$.

It is $[\eta]$ that can be computed by considering the extra energy dissipation that a suspended spherical molecule of molecular weight M causes in a fluid under shear. Some idea of the properties of $[\eta]$ can be obtained by dimensional analysis. The most likely variables to enter are the radius r and the mass m of the sphere. We know that the frictional properties will depend on the radius, and that the energy that can be put into a tumbling sphere depends on the mass. Thus, we can expect

$$[\eta] \propto m^a r^b, \qquad \text{with units of } g^a\ cm^b \qquad (12\text{-}18)$$

Because the units of $[\eta]$ are $cm^3\ g^{-1}$, we choose $a = -1$ and $b = 3$, which leads to

$$[\eta] \propto r^3/m \propto V_h N_0/M \qquad (12\text{-}19)$$

where V_h is the hydrated volume of a solute molecule, and M/N_0 is its weight.

Albert Einstein was the first to obtain a complete expression for $[\eta]$. He expressed the result as $\eta_{sp} = 2.5\ \phi_2$, where ϕ_2 is the volume fraction occupied by the spherical solute molecules. Note the surprising implication that η_{sp} (and thus $[\eta]$) is independent of the size of the spheres. Because V_h is the hydrated volume of a molecule, $V_h N_0/M$ is the volume per gram of solute. Thus, $\phi_2 = V_h N_0 c_2/M$, because c_2 is the number of grams solute per cm^3 of solution. Einstein's equation can be rewritten as

$$\eta_{sp} = 2.5(V_h N_0/M)c_2 \qquad \text{or} \qquad [\eta] = 2.5(V_h N_0/M) \qquad (12\text{-}20)$$

Comparing Equations 12-19 and 12-20, we see that dimensional analysis led to the correct functional form, but of course we had no way of predicting the magnitude of the dimensionless constant 2.5.

We have already derived an expression for V_h (Chapter 10). From Equation 10-11, the intrinsic viscosity becomes

$$[\eta] = 2.5(\bar{V}_2 + \delta_1 \bar{V}_1) \qquad (12\text{-}21)$$

where $\bar{V}_2$ and $\bar{V}_1$ are the partial specific volumes of solute and water, respectively, and δ_1 is the hydration in grams per gram. The absence of any dependence on molecular weight is clear, and this independence is confirmed by the typical experimental results shown in Table 12-1.

Effect of molecular shape on viscosity

What good then is measurement of viscosity if there is no molecular-weight dependence? It turns out that $[\eta]$ is exquisitely sensitive to shape. For coiled molecules, it is a function of the dimensions of the coil and of the extent to which solvent can drain freely through the coil without becoming entrapped by hydrodynamic interactions (see Chapter 19). For molecules that can be modeled as rigid ellipsoids of revolution, Equation 12-21 becomes

$$[\eta] = v(\bar{V}_2 + \delta_1 \bar{V}_1) = v V_h N_0 / M \qquad (12\text{-}22)$$

The factor v is called a Simha factor; it contains all the shape dependence.[§] In all cases, $v \geqslant 2.5$, and Table 10-2 gives tabulated values for prolate and oblate ellipsoids. Figure 12-4 shows a plot of v versus axial ratio a/b for small axial ratios. For axial ratios greater than 10, the Simha factor is given by the asymptotic equations

$$v = \frac{(a/b)^2}{5[\ln(2a/b) - 1/2]} + \frac{(a/b)^2}{15[\ln(2a/b) - 3/2]} + \frac{14}{15} \qquad (12\text{-}23a)$$

for prolate ellipsoids, and

$$v = (16/15)(a/b)/\tan^{-1}(a/b) \qquad (12\text{-}23b)$$

for oblate ellipsoids.

Long rods (such as DNA pieces still short enough to be fairly stiff) can be modeled as prolate ellipsoids. Their intrinsic viscosity generally is fit to an equation such as $[\eta] = kM^\alpha$ (where k and α are constants) or slightly more complicated equations. For a homologous series of rods of constant diameter, the molecular weight is proportional to the length. Approximating the lengths as the axial ratio of a prolate ellipsoid, we can find the coefficient α as $d(\ln[\eta])/d[\ln(a/b)]$. From Equation

[§] Concentrations in viscosity experiments sometimes are measured in grams per 100 cm^3, instead of g cm^{-3}. With these units, Equation 12-22 becomes $[\eta] = (v/100)(\bar{V}_2 + \delta_1 \bar{V}_1)$, so the derived values of $[\eta]$ are reduced by a factor of 100. When using published viscosity data, be sure to check the units.

Table 12-1
Results of viscosity measurements

Sample	Molecular weight	$[\eta]$ ($cm^3\ g^{-1}$)	$\bar{V}_2$ ($cm^3\ g^{-1}$)	Maximum possible δ_1 (g/g)	Maximum possible prolate a/b	Scheraga–Mandelkern β ($\times 10^6$)
Near-spherical particles:						
Ribonuclease A (bovine)	13,683	3.3	0.728	0.59	3.9	2.01
Lysozyme (chicken)	14,211	2.7	0.688	0.39	3.2	—
Serum albumin (bovine)	66,296	3.7	0.734	0.75	4.4	2.04
Hemoglobin	68,000	3.6	0.749	0.69	4.1	—
Bushy stunt virus	10,700,000	3.4	0.74	0.62	4.0	—
Rodlike particles:						
Tropomyosin	93,000	52.	0.74	20.	29.	—
Fibrinogen	330,000	27.	0.71	10.	20.	2.15
Poly-γ-benzyl-L-glutamate α helix	340,000	720.	—	—	—	—
Myosin	493,000	217.	0.728	86	68	—
DNA	6,000,000	5,000	—	—	—	—
Tobacco mosaic virus	40,000,000	37	—	—	—	2.61
Coillike particles:						
Serum albumin (bovine-urea)	66,296	22	—	—	—	~2.05
Serum albumin (bovine-guanidinium)	66,296	52	—	—	—	—
Poly-γ-benzyl-L-glutamate coil	340,000	184	—	—	—	—

SOURCE: Results mostly from C. Tanford, *Physical Chemistry of Macromolecules* (New York: Wiley, 1961).

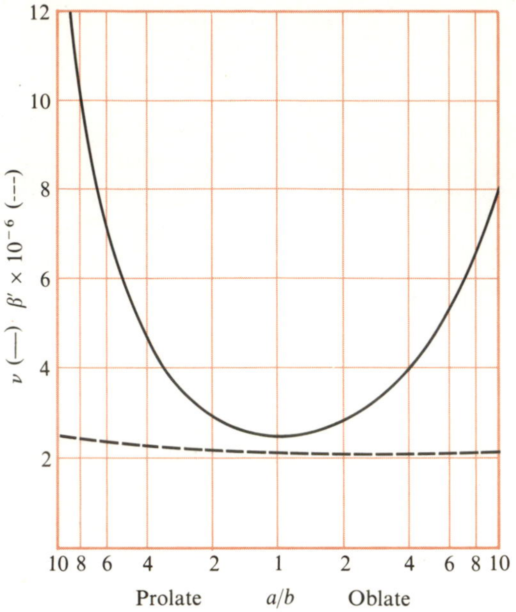

Figure 12-4

Simha factor and Scheraga–Mandelkern parameter. The Simha factor ν (*solid line*) and the the Scheraga–Mandelkern parameter β (*dashed line*) are plotted against the logarithm of the axial ratio for small axial ratios. Compare this graph with Figure 10-10 to judge the relative sensitivities of various hydrodynamic measurements to molecular shape.

12-23 (ignoring the terms −1/2, −3/2, and 14/15), we have

$$\nu = (20/75)(a/b)^2/\ln(2a/b) \tag{12-24}$$

and the constant α is

$$\alpha = d(\ln \nu)/d[\ln(a/b)] = 2 - 1/\ln(2a/b) \tag{12-25}$$

For axial ratios on the order of 10^3, α is about 1.8.

A more realistic model treats a rod as a linear array of n spherical monomers with diameter d and length $L = nd$. J. G. Kirkwood computed the viscosity of such arrays, using procedures similar to the frictional coefficient calculations described in Equations 10-24 through 10-33. For anhydrous monomers of molecular weight M_0 and volume V_h, the result is

$$\begin{aligned}[\eta] &= 2\pi N_0 L^2 d/45 M_0 \ln(L/d) = (N_0 V_h/M_0)[2\pi L^2 d/(4/3)\pi(d/2)^3]/45 \ln(L/d) \\ &= (V_h N_0/M_0)(20/75)(L/d)^2/\ln(L/d)\end{aligned} \tag{12-26}$$

Compare this with Equation 12-22; note that, for long lengths, the terms equivalent to the Simha factor become identical to those in Equation 12-24. Call L/d an axial ratio a/b, and recognize in Equation 12-24 that (for large a/b) $\ln(2a/b) = \ln 2 + \ln(a/b) \cong \ln(a/b)$.

It is interesting to compare the Simha and Perrin factors for prolate and oblate ellipsoids as a function of axial ratio. Note from Table 10-2 that Simha factors are much more sensitive to shape. It can be much easier to detect an elongated or flattened structure by viscosity measurements than by sedimentation or diffusion measurements. Furthermore, we need not even know the molecular weight in order to make a shape estimate. On the other hand, unless the molecule is a coil, we *cannot* learn the molecular weight from viscosity measurements alone. Viscosity is more sensitive than the frictional coefficient in discriminating between prolate and oblate shapes. In practice, an observed Simha factor of 15 or more virtually guarantees that one is working with a prolate or rodlike macromolecule. The equivalent oblate shape would have to be almost unrealistically thin.

Using viscosity to estimate molecular weight

In general, both the Perrin shape factor F and the Simha factor ν increase monotonically and smoothly with increasing axial ratio. This observation suggests that there should be a way to combine $[\eta]$ with sedimentation or diffusion measurements to eliminate most shape effects and so provide a good estimate of molecular weight.

Equation 10-68 (for the frictional coefficient of an ellipsoid) can be written in terms of the hydrated volume:

$$f = 6\pi\eta(3/4\pi)^{1/3}V_{\mathrm{h}}^{1/3}F \tag{12-27a}$$

If we combine this expression with the sedimentation constant $s = M(1 - \bar{V}_2\rho)/N_0 f$, we obtain

$$6\pi(3/4\pi)^{1/3}F = M(1 - \bar{V}_2\rho)/N_0 s V_{\mathrm{h}}^{1/3}\eta \tag{12-27b}$$

Equation 12-22 (for the intrinsic viscosity) also depends on the hydrated volume. When this equation is rearranged to place all constant and shape information on one side, and then the cube root is taken, we obtain

$$N_0^{1/3}\nu^{1/3} = [\eta]^{1/3}M^{1/3}/V_{\mathrm{h}}^{1/3} \tag{12-28}$$

If we divide Equation 12-28 by Equation 12-27b, we can eliminate the hydrated volume:

$$\beta' \equiv (N_0/162\pi^2)^{1/3}\nu^{1/3}/F = N_0 s[\eta]^{1/3}\eta/M^{2/3}(1 - \bar{V}_2\rho) \tag{12-29}$$

Equation 12-29 is called the Scheraga–Mandelkern equation. Its usefulness results from the properties of the parameter β'. It is traditional to use intrinsic viscosities measured in dl g^{-1} when working with this equation. Therefore, tabulated values of the Scheraga–Mandelkern parameter usually are presented as $\beta = 100^{1/3}\beta'$. Table 10-2 shows values of this Scheraga–Mandelkern parameter as a function of axial ratio, and Figure 12-4 plots it as a function of log axial ratios. An equation equivalent to Equation 12-29 can be derived by combining viscosity and diffusion relationships. It is rarely used because of the paucity of reliable diffusion data.

It is clear that β is much less sensitive to axial ratio than is either the Simha factor or the Perrin factor. In fact, for an oblate ellipsoid, β increases only from 2.12×10^6 to 2.15×10^6 as axial ratio increases from 1 to 200. Therefore, if viscosity and sedimentation measurements show $\beta > 2.15 \times 10^6$ for a globular sample of known molecular weight, the particle must be prolate (see Table 10-2 or Fig. 12-4). However, the insensitivity of β to shape suggests that the most efficient way to calculate the axial ratio of this prolate ellipsoid is to use either viscosity or sedimentation data alone. The major use of the Scheraga–Mandelkern equation is in the calculation of molecular weights. It is quite rare to come upon a protein with an axial ratio greater than 10. In these rare cases, there is usually some hint of the high ratio, such as a very high intrinsic viscosity. Therefore, except for these cases, β can be set equal to a value such as 2.20×10^6. The true β for an unknown protein is likely to be within 5% of this value. Because $M \propto \beta^{-3/2}$, the molecular weight derived from the Scheraga–Mandelkern equation should be accurate to within 10%. Table 11-2 compares examples of molecular weights determined in this way with other estimates.

Some applications of viscosity measurements

The insensitivity of β to shape is particularly frustrating in one respect: the hydration does not appear in the Scheraga–Mandelkern equation. Therefore, a combination of viscosity and sedimentation, in principle, allows the axial ratio to be determined, unaffected by any uncertainty in the knowledge of δ_1. Then, once the shape is known, δ_1 could be obtained by reusing either the sedimentation or viscosity data separately. Unfortunately, shapes computed from the Scheraga–Mandelkern equation are not that accurate, and neither are the derived hydrations. One can examine this in a different way by asking: given the axial ratios of proteins known from x-ray crystallography, what hydrations can be computed from individual hydrodynamic measurements, and how consistent are these? Table 12-2 shows a few examples of such self-consistency tests. The agreement among viscosity, diffusion, and sedimentation measurements is not bad. However, for small axial ratios, β, ν, and F are all so insensitive to shape that substantial variations in predicted axial ratio would result if one picked a given hydration value and used it for the analysis of all three sets of data.

A few generalizations about viscosity are useful to remember. Intrinsic viscosity

Table 12-2
Hydrations of biopolymers computed by using shapes known from x-ray diffraction or electron microscopy

Sample	Known axial ratio[§]	Hydration (δ_1 in g/g) based on		
		Viscosity	Diffusion	Sedimentation
Bushy stunt virus	1.0	0.65	0.71	0.71
Carboxypeptidase	1.25	—	0.30	0.69
Cytochrome *c*	1.48	—	0.18	0.24
Hemoglobin	1.3	0.62	0.52	0.75
Lysozyme	1.5	0.34	0.52	0.52
Myoglobin	1.76	0.44	0.50	0.42
Tobacco mosaic virus	18	0.32	0.1–0.7	0.26

[§]Axial ratios are for prolate ellipsoids, except for cytochrome *c*, which is oblate.
SOURCE: After I. D. Kuntz, Jr., and W. Kauzmann, in *Advances in Protein Chemistry*, vol. 28, ed. C. B. Anfinsen, J. T. Edsall, and F. M. Richards (New York: Academic Press, 1974), p. 239.

$[\eta]$ is quite sensitive to conformation. If a globular particle (such as a protein) is denatured, the molecular weight stays the same, but $[\eta]$ often increases dramatically, whereas $s_{20,w}$ shows a significant but smaller decrease. If a long rigid rod (such as DNA) becomes more flexible or coillike (as in DNA denaturation), $[\eta]$ decreases substantially, and $s_{20,w}$ increases. In contrast, if shape remains the same but molecular weight decreases (as in dissociation of an oligomeric structure into subunits), $s_{20,w}$ decreases, and $[\eta]$ either stays the same or shows a small decrease or increase. (Table 12-1 illustrates some of these types of viscosity changes.)

Viscosity long has been used to distinguish coils from rods. We showed earlier that, for a rod, $[\eta] \propto M^{1.8}$. In Chapter 19, we show that, for a coil, $[\eta] \propto M^{0.5}$ to $M^{1.0}$. Figure 12-5 shows a striking example of the sensitivity of viscosity measurements to molecular shape. A series of samples of poly-γ-benzyl-L-glutamate of different molecular weights showed marked solvent effects on the molecular-weight dependence of $[\eta]$. In $CHCl_3$, $[\eta] \propto M^{1.7}$, indicating rods and suggesting an α helix. In $CHCl_2COOH$, $[\eta] \propto M^{0.83}$, suggesting a somewhat expanded coil. The explanation is straightforward. A strong hydrogen-bonding solvent (such as $CHCl_2COOH$) will compete effectively for peptide hydrogen-bond donors and acceptors, and thus the solvent will denature the α helix.

Viscoelastic relaxation

The tendency of short DNA to orient (or of long DNA wormlike coils to stretch) in a shear gradient is a nuisance for many studies. However, Lynn Klotz and Bruno Zimm have turned this difficulty into an elegant and powerful technique for the study of the longest DNAs. Figure 12-6 shows schematically the fundamentals

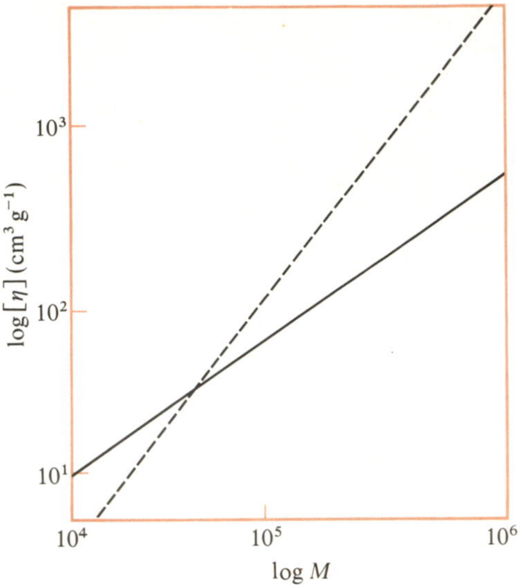

Figure 12-5

Intrinsic viscosity versus molecular weight. This log–log plot shows the intrinsic viscosity of poly-γ-benzyl-L-glutamate as a function of molecular weight. The solid line represents samples measured in dichloroacetic acid; the dashed line represents samples measured in chloroform–formamide (99.5:0.5) or dimethylformamide. [After P. Doty, J. H. Bradbury, and A. M. Holtzer, *J. Am. Chem. Soc.* 78:947 (1956).]

of their viscoelastic relaxation measurements. When a torque is applied to a DNA solution in a rotating cylinder viscometer, the average chain configuration of DNA is altered from a stiff coil to a more elongated structure. The energy for this change comes from the applied torque coupled through the shear gradient in the fluid. Eventually, a steady state is reached. The DNA molecules then have some stored energy in their chain configuration. When the torque is removed, the rotation would soon slow down and stop if there were no DNA. However, with DNA present, the rotor cannot stop until the DNA molecules have relaxed back to a random coil. The rotor reverses direction rapidly and then creeps to a stop very slowly. The forces that coupled the original torque to produce extended chains now exert a torque on the rotor as the chains relax. It turns out that this effect can last an impressively long time.

As shown originally by Zimm and by P. E. Rouse, the kinetics of the transition of a homogeneous polymer random coil from a distorted state to the final equilibrium statistical distribution is complex. It occurs as a sum of exponential terms e^{-t/τ_k}, each characterized by a relaxation time

$$\tau_k = (M\eta_0[\eta]/RT)(1/\lambda_k) \tag{12-30}$$

where η_0 is the solvent viscosity. The constants λ_k are functions of the eigenvalues obtained when the dynamics of the relaxation of the macromolecules are computed from equations of motion. The important thing is that λ_k^{-1} decreases rapidly with increasing k. For linear-coiled molecules, $\lambda_1^{-1} = 0.451$, and the terms $\lambda_1^{-1}:\lambda_2^{-1}:\lambda_3^{-1}$ decrease in the ratio 1:0.31:0.16. For circular-coiled molecules, each λ_i^{-1} is close to half the value for the corresponding linear-molecule constant.

For a viscoelastic relaxation measurement in the limit of dilute macromolecule

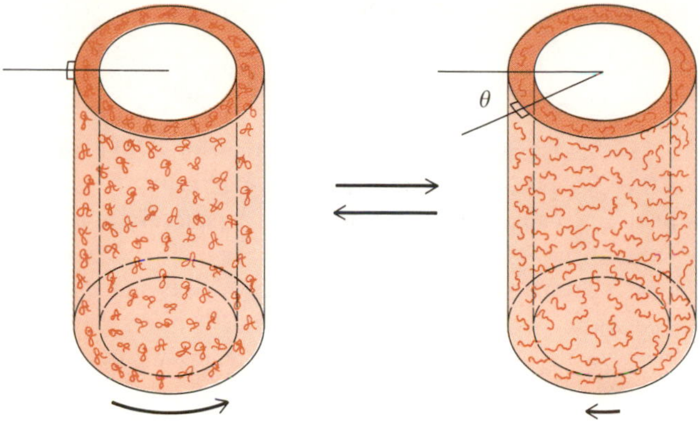

Figure 12-6

Schematic relaxation viscosity experiment. The actual apparatus used would be more similar to that shown in Figure 12-3. The outer cylinder is slowly rotated to an angle θ away from the initial position. Then external torque is removed, and the cylinder is allowed to move freely. It relaxes back toward its original position, and θ is measured as a function of time until a final value $\theta(\infty)$ is reached.

concentrations, it can be shown that the decay of the angle of the rotor, $\theta(t)$, to its infinite time value $\theta(\infty)$ is given by

$$\theta(t) - \theta(\infty) = \left(\omega_r \eta_{sp} \Big/ \sum_{k=1}^{n} \tau_k\right)\left(\sum_{k=1}^{n} \tau_k^2 e^{-t/\tau_k}\right) \tag{12-31}$$

where ω_r is the initial rotor velocity (when the driving torque is removed). The attractive feature of Equation 12-31 is that each exponential is weighted by τ_k^2, and the terms decrease rapidly with increasing k. Thus, at long times (to a good approximation), the decay should be a single exponential, and the relaxation time τ_1 can be found by the expression

$$d\{\ln[\theta(t) - \theta(\infty)]\}/dt = -t/\tau_1 \tag{12-32}$$

Because λ_1 is known from the previous mechanical calculations, Equation 12-30 will yield the molecular weight if τ_1, η_0, and $[\eta]$ have been measured. This approach does not completely eliminate all of the difficulties, because it is very hard to measure the intrinsic viscosities of long DNAs. An alternative approach is to use empirically derived expressions relating $[\eta]$ to molecular weight. For DNA, quite an accurate fit to measured data is obtained from

$$M = k([\eta] + b)^{3/2} \tag{12-33}$$

where $k = 19.8$ for linear DNA, and $b = 500 \text{ cm}^3 \text{ g}^{-1}$. Neglecting b for large DNAs (where a typical intrinsic viscosity might be $10^4 \text{ cm}^3 \text{ g}^{-1}$ or more), Equations 12-30 and 12-33 can be combined to yield (at 298 °K)

$$M = 1.45 \times 10^8 \tau_1^{3/5} \tag{12-34}$$

Equation 12-34 has several interesting implications. The molecular weight is proportional to the 3/5 power of the measured quantity. It is an ideal situation, because errors in the measurements are not very sensitively felt in the final derived result. Because the observed τ_1 is proportional to $M^{5/3}$, the weighting factor in Equation 12-31 is proportional to $M^{10/3}$. Suppose the sample is heterogeneous; almost any preparation of huge DNAs has some broken pieces. The relaxation behavior described by Equation 12-31 will be extremely insensitive to all but the largest pieces.

$$\theta(t) - \theta(\infty) \propto \sum_i f_i M_i^{10/3} \exp[-t(2.2 \times 10^8/M_i)^{5/3}] \tag{12-35}$$

where f_i is the number fraction of the molecules with molecular weight M_i. Thus, even half-molecules cannot contribute very much to the longest-time relaxation behavior, and small pieces will be "invisible." However, any aggregates could pose a serious problem.

Figure 12-7 shows some typical results from viscoelastic relaxation of DNA.

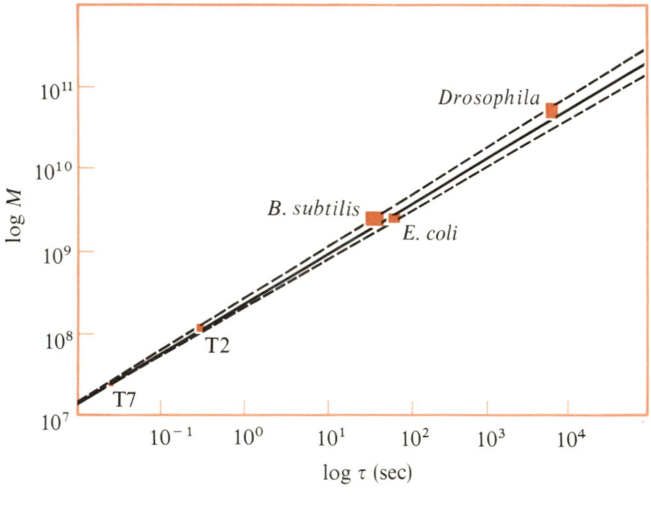

Figure 12-7

Viscoelastic relaxation time and molecular weight. This log–log plot shows the molecular weight of various DNA samples as a function of the longest relaxation time τ seen in viscoelastic relaxation. The molecular weights of DNAs from phages T7 and T2, *E. coli*, and *B. subtilis* were known independently. They were used to establish the experimental relationship shown, from which the molecular weight of chromosome-sized *Drosophila* DNA could be determined. [After R. Kavenoff, L. C. Klotz, and B. Zimm, *Cold Spring Harbor Symp. Quant. Biol.* 38:1 (1974).]

Note the extremely long relaxation times that occur with molecular weights of 10^9 d or more. The excellent fit for DNAs of known molecular weight shows that Equation 12-34 is valid. This technique has been used successfully to determine molecular weights for DNAs up to almost 10^{11} d. These huge DNAs, such as a whole unfolded bacterial or eukaryotic chromosome, are so fragile that they cannot even be moved from one container to another. For study, the cells containing the DNA must be lysed right in the rotating cylinder viscometer, and the DNA freed from associated proteins by detergent treatment and enzymatic digestion in situ. Convection alone must be used to mix these reagents with the cell lysate, because no mixing can be tolerated. Naturally, a very low rotor speed is applied to reduce the possibility of shear breakage. The rotor is not allowed to turn even one revolution before the torque is removed and relaxation begins.

12-2 TECHNIQUES BASED ON ROTATIONAL MOTION

As shown in Chapter 10, more than a single frictional coefficient is needed to describe rotational motion for nonspherical bodies. Therefore, the friction must be described as $\underset{\sim}{\mathbf{f}}$, a tensor. Rotational diffusion measurements will monitor the distribution $W(\theta, \phi, t)$ of orientation of molecules in the same way that translational diffusion monitors the spatial distribution; $W(\theta, \phi, t)$ is the fraction of molecules with orientation angles between θ and $\theta + d\theta$ and between ϕ and $\phi + d\phi$ at time t.

The equivalent of Fick's second law can be written as

$$\partial W(\theta, \phi, t)/\partial t = -\underset{\sim}{\mathbf{L}} \cdot \underset{\sim}{\mathbf{D}}_{\text{rot}} \cdot \underset{\sim}{\mathbf{L}} W(\theta, \phi, t) \tag{12-36}$$

where $\underset{\sim}{\mathbf{L}}$ is the angular momentum operator, and $\underset{\sim}{\mathbf{D}}_{\text{rot}}$ is the rotational diffusion tensor. In many limiting cases, a single component of $\underset{\sim}{\mathbf{D}}_{\text{rot}}$ dominates, and then Equation 12-36 simplifies to

$$\boxed{\partial W(\theta, \phi, t)/\partial t = D_{\text{rot}} \nabla^2 W(\theta, \phi, t)} \tag{12-37}$$

Because there are no concentration gradients in rotational diffusion measurements, one need not be so concerned about the concentration dependence of D_{rot} in solving Equation 12-37. By analogy to the Einstein–Sutherland equation, $D_{\text{rot}} = kT/f_{\text{rot}}$. However, one must decide which of the three possible frictional rotational coefficients to use: f_a, f_b, or f_h (see Chapter 10). The choice turns out to depend on the method used to observe $W(\theta, \phi, t)$.

Two prerequisites must be met in order to observe rotational diffusion. The isotropic equilibrium distribution of orientations in normal aqueous solution must be perturbed. Also, some property of the system must be found that can measure a preferential orientation. In most methods, a force is applied to the system to cause partial orientation. When the force is on, Equation 12-37 will have an additional term describing the torque that the force exerts on the molecules. It then becomes analogous to the Lamm equation, in which opposing effects of sedimentation and diffusion are considered. The analog of Equation 12-37 must consider the opposing effects of torque and rotational diffusion. In sedimentation or other linear transport, a sufficiently high force will cause all the solute to move to one edge of the experimental chamber; a useful equilibrium distribution is not reached. In contrast, a high torque can produce only some maximum orientation, with no translation.

The solution of Equation 12-37 in the presence of an external force is formidable, and it rarely has been accomplished in analytical form. It is customary to work in two limiting cases. The first case involves measuring $W(\theta, \phi, \infty)$, the extent of orientation produced by a given force in a state of mechanical equilibrium after all transient time-dependent effects have died off; this approach is analogous to equilibrium ultracentrifugation. The second case involves turning off the force and watching the distribution $W(\theta, \phi, t)$ decay to an isotropic solution; this decay is governed by Equation 12-37, which can be solved analytically. An equivalent ultracentrifugation experiment would be to start a sedimentation experiment, then suddenly turn the rotor speed down to a negligible low value and watch the boundaries broaden by diffusion.

Measuring flow orientation by linear dichroism

The various methods of examining rotation motions are named by the forces used to cause the orientation, and by the observable quantity used to monitor it. For example, consider flow dichroism. Molecules will orient if there is hydrodynamic shear. Suppose this shear is produced by placing the sample between two concentric cylinders, one stationary and one kept rotating at a constant angular velocity in an apparatus like that of Figure 12-3 or Figure 12-8. At equilibrium it can be shown that Equation 12-37 becomes

$$D_{\text{rot}} \nabla^2 W(\theta, \phi, \infty) = \nabla \cdot W(\theta, \phi, \infty) \boldsymbol{\omega}(\theta, \phi) \tag{12-38}$$

where $\boldsymbol{\omega}$ is the angular velocity of the particle produced by the velocity gradient.

This equation can be solved, but it is more useful to gain some physical intuition by examining the motion of a rod in a two-dimensional velocity gradient (Fig. 12-8). Assume that the diameter of the cylinders is sufficiently large that curvature can be neglected; essentially, then, the shear arises from the motion of two parallel plates, and therefore the velocity gradient in the fluid is linear. Let θ be the angle between the long axis of the rod and the laminar flow of the shear. The torque is smallest at $\theta = 0°$, and the rod tends to stay in this orientation for a long time. It is not a stable orientation because, for a rod of finite thickness, there is still a finite shear across it and therefore a finite torque. At $\theta = 90°$, the torque is maximal, and the rod will quickly move away from this orientation. Thus, shear produces an anisotropic distribution. Rotational

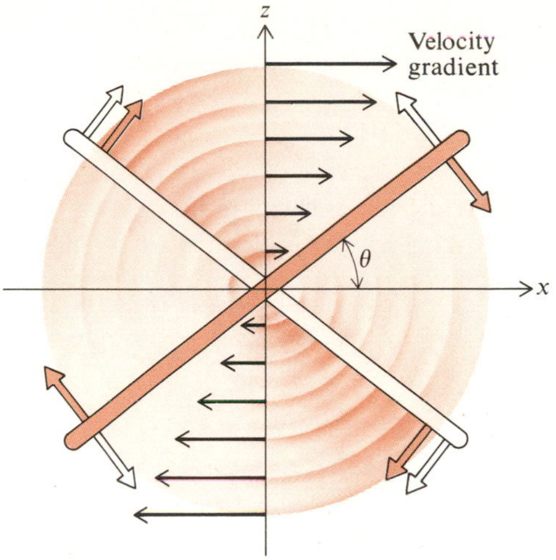

Figure 12-8

Forces acting on a rigid rod in a velocity gradient. As shown, shear forces exert a torque causing clockwise rotation of the rod. These forces tend to orient the rod for longer times at $\theta = 0°$ (where the shear is least) rather than at $\theta = 90°$ (where the shear is greatest). Rotational Brownian motion acts to oppose any preferential orientation and therefore applies an effective torque to rotate all molecules toward $\theta = 90°$. Only the tangential components of all forces are considered, because other components cause translation but no change in orientation.

Brownian-motion forces will tend to oppose this orientation (Fig. 12-8). At low shear, the most probable orientation is at $\theta = 45°$, where Brownian-motion forces and shear forces are equal and opposite. At high shear, all the rods tend to orient near $\theta = 0°$.

Orientation induced by flow affects only the position of the long axis of a prolate ellipsoid. If the shear force is suddenly turned off at $t = 0$, only motion about the short axis (b) can alter the position of the long axis (a). The average orientation $\langle W(\theta, \phi, t)\rangle$ decays as

$$\langle W(\theta, \phi, t)\rangle = \langle W(\theta, \phi, 0)\rangle e^{-t/\tau_a} \tag{12-39}$$

where $\tau_a = f_b/2kT$. As shown in Equation 10-23, f_b is the frictional coefficient for rotation about the short axis of an ellipsoid.

A serious complication is that $\langle W(\theta, \phi, t)\rangle$ cannot be measured directly. To detect anisotropic orientation of molecules, one must examine some spectroscopic or electrical property that depends on the average orientation. For example, one can measure the linear dichroism or the linear birefringence of the system. (Linear dichroism, which originates from the preferential absorption of polarized light by a molecule along certain molecular axes, is discussed in Chapter 7.) For ellipsoidal macromolecules, it often is sufficient to describe the anisotropy of light absorption by two extinction coefficients: $\varepsilon_{\parallel}$ for absorption of light polarized parallel to the long

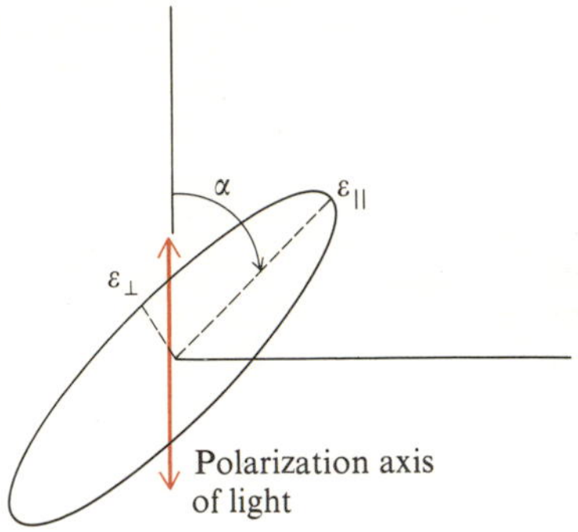

Figure 12-9

Coordinate system used to describe different absorbance of polarized light along the principal axes of an ellipsoidal molecule. The angle α is between the long axis of the prolate ellipsoid and the polarization of the light.

axis of a prolate ellipsoid, and $\varepsilon_{\perp}$ for absorption of light polarized perpendicular to this axis (Fig. 12-9). Linear dichroism describes the difference between $\varepsilon_{||}$ and $\varepsilon_{\perp}$. A convenient measure of this difference is simply $\Delta\varepsilon_L = \varepsilon_{||} - \varepsilon_{\perp}$.

The linear dichroism of a partially oriented macroscopic sample is measured by studying the absorbance or extinction of light polarized along a particular laboratory axis. The observed absorbance will be an average over the orientational distribution of molecules weighted by the contribution each molecule makes to the polarized absorption. Define α as the angle between the long axis of a particular molecule and the laboratory polarization direction (Fig. 12-9). The amount of light absorbed by each molecule will depend on the angle between the laboratory polarization direction and the two extinction coefficients $\varepsilon_{||}$ and $\varepsilon_{\perp}$. As shown in Chapter 7, absorption intensity for light polarized at an angle α relative to a preferred absorption direction depends on $\cos^2 \alpha$. Therefore, from Figure 12-9, it is clear that the extinction coefficient ε_α for a molecule oriented in the plane of the page at angle α will be

$$\varepsilon_\alpha = \varepsilon_{||} \cos^2 \alpha + \varepsilon_{\perp} \cos^2(90^\circ - \alpha) = \varepsilon_{||} \cos^2 \alpha + \varepsilon_{\perp} \sin^2 \alpha \qquad (12\text{-}40)$$

With no net orientation or polarization, the total extinction is simply $\varepsilon = (\varepsilon_{||} + 2\varepsilon_{\perp})/3$ (where the factor of two enters because there are two short axes and only one long axis, and each axis has the same probability of assuming any particular orientation). It turns out that it is convenient experimentally to measure the difference between ε_α and ε. For a solution, the quantity measured will be $\varepsilon_\alpha - \varepsilon$ averaged over the orientation of all molecules:

$$\begin{aligned}\langle \varepsilon_\alpha - \varepsilon \rangle &= \langle \varepsilon_{||} \cos^2 \alpha + \varepsilon_{\perp} \sin^2 \alpha - \varepsilon_{||}/3 - 2\varepsilon_{\perp}/3 \rangle \\ &= \langle (\varepsilon_{||} - \varepsilon_{\perp}) \cos^2 \alpha - \varepsilon_{||}/3 + \varepsilon_{\perp}/3 \rangle = \Delta\varepsilon_L(\langle \cos^2 \alpha \rangle - 1/3) \qquad (12\text{-}41)\end{aligned}$$

where we have used the identity $\sin^2 \alpha + \cos^2 \alpha = 1$. Thus, experimental determination of $\langle \varepsilon_\alpha - \varepsilon \rangle$ will yield an estimate of the average degree of orientation, $\langle \cos^2 \alpha \rangle$, providing that $\Delta\varepsilon_L$ can be calculated or estimated independently. This can be done in two ways—either by applying a force strong enough to fully orient the sample, or by attempting to extrapolate $\langle \varepsilon_\alpha - \varepsilon \rangle$ to infinite force. Having obtained $\langle \cos^2 \alpha \rangle$, it can then be related to $\langle W(\theta, \phi, t) \rangle$ and thus data are available to use in Equation 12-39.

Measuring flow orientation by linear birefringence

Linear birefringence is an effect exactly analogous to linear dichroism. It originates from the difference in refractive index along molecular axes, $\Delta n = n_{||} - n_{\perp}$. The birefringence observed for a solution of oriented molecules will consist of two contributions, intrinsic and form birefringence. Intrinsic birefringence is a property of the optical anisotropy of individual molecules. It can be related to $\Delta\varepsilon_L$ by the Kronig–Kramers transform (Chapter 8). Even if Δn is zero for an individual molecule, a solution of oriented nonspherical molecules will be birefringent if the solute refractive index is different from that of the solvent; this effect is called form birefringence. As with linear dichroism, actual measured birefringence is dependent on how the molecular axes are situated with respect to experimental axes.

The result of all these considerations is that, rather than measuring $W(\theta, \phi, t)$, we actually observe an orientation function, $f(\theta, \phi, t)$. This function contains $W(\theta, \phi, t)$, weighted by the effects of orientation on the optical parameter being studied. For birefringence, it turns out that

$$f(\theta, \phi, t) = f(\theta, \phi, 0)e^{-3t/\tau_a} \tag{12-42}$$

Starting with equations given in Chapter 10, for long prolate ellipsoids, it can be shown that

$$\tau_a = 8\pi\eta a^3/3kT[2\ln(2a/b) - 1] \tag{12-43}$$

Thus, the relaxation time increases essentially as the cube of the length of the long axis. This great sensitivity provides a strong inducement for attempts to handle some of the difficult optical and hydrodynamic problems that underlie flow birefringence or dichroism.

An alternative measure of τ_a can come from the degree of molecular orientation induced by a constant applied shear. The orientation can be measured by the birefringence of the sample. A weakly birefringent solution will appear to rotate the plane of polarized light except when the plane of polarization is parallel or perpendicular to an optical axis (Fig. 12-10). The optical axes are directions (within an individual molecule) parallel or perpendicular to the orientations showing maximal and minimal indices of refraction for an ellipsoid. The optical axes usually are parallel to molecular axes. Therefore, the angle χ, between the macroscopic optical axes measured for a solution of molecules and the direction of solvent flow, is a measure of the degree of orientation.

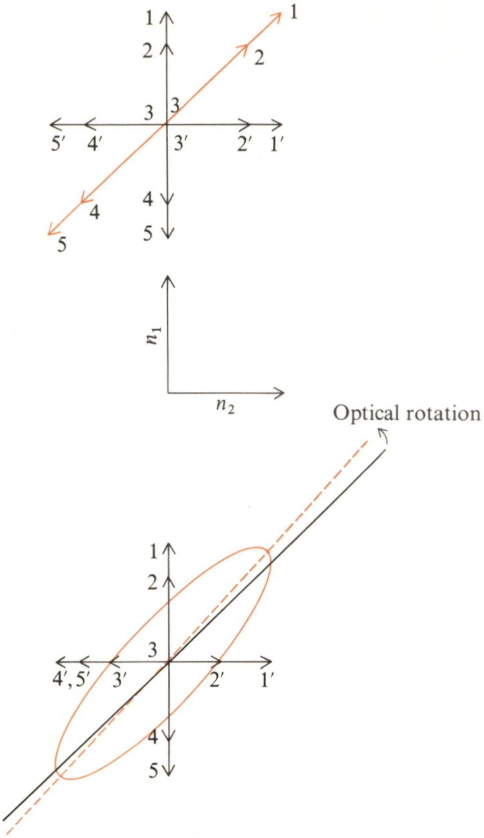

Figure 12-10

Effect of a birefringent medium on transmitted plane-polarized light. (*Top*) Plane-polarized light (*color*) is incident on a sample with two different refractive indices, ($n_2 > n_1$) oriented as shown in the center. One can resolve the incident light into two plane-polarized beams, each parallel to an optic axis (a direction with only a single refractive index). Arrows show the positions of the electric vector at equally spaced, increasing time intervals. (*Bottom*) Light emerging from the birefringent medium is elliptically polarized (*color*) because the component along n_2 is retarded in phase relative to that along n_1. For thin (or only slightly birefringent) samples, the ellipse is so elongated that it still resembles plane-polarized light. Therefore, one can say that the birefringent medium has produced optical rotation. Compare the orientation of the long axis of the ellipse (*colored dashed line*) with the original orientation of the plane-polarized light. (Note that light incident parallel to an optic axis is not rotated, because it "sees" only a single refractive index.)

The angle χ can be determined by viewing the sample in the shear field between crossed polarizers (Fig. 12-11). For an isotropic sample, these polarizers allow no transmission. However, because of the birefringence, light will be transmitted everywhere except for four regions where the optic axis is parallel to the polarization. Simple geometric considerations indicate that the angle between the zones of no light transmission and the polarizers is equal to χ.[§] For low shear, χ in radians is

$$\chi = (\pi/4) - (\tau_a/6)(dv/dx) + \cdots \tag{12-44}$$

where terms in $(dv/dx)^2$ and higher powers of the derivative have been omitted. Thus, a measurement χ as a function of the shear will yield τ_a. In the limit of low shear, you can see that $\chi \rightarrow 45°$; this is exactly the result predicted from the simple mechanical argument given earlier.

[§] When optical axes are parallel to molecular axes, the angle χ is equal to the angle θ in Figure 12-8.

Figure 12-11

Schematic illustration of a flow birefringence measurement. The sample is partially oriented by a shear gradient created by concentric rotating and stationary cylinders. It is placed between two perpendicular polarizers. Rodlike molecules will tend to orient as described in Figure 12-8. Optical axes will occur parallel and perpendicular to the long axes of the rods. Therefore, whenever rods have an average orientation parallel or perpendicular to the polarizer, no optical rotation will occur, and thus the analyzer will completely block all of the polarized light. The angle χ is thus a direct measure of the angle between the long axis of the rod and the solvent flow direction.

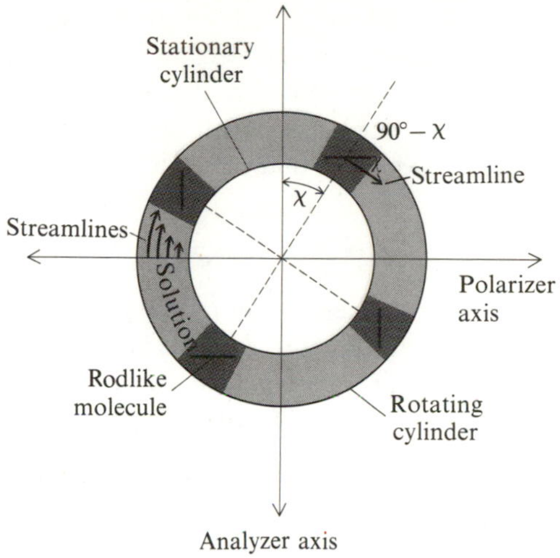

Orientation in electric fields

A strong electric field can be used instead of flow to orient the sample (Fig. 12-12). The electric field usually is applied as a square-wave pulse. It must be short in duration to minimize electrical heating of the solution and bulk translation of the solute due to electrophoresis in the applied field. The solution also must have a low ionic strength to prevent excessive heating. Electric birefringence and dichroism are simply optical measurements of the orientation caused by the electric field. As in flow orientation, analysis of the motions of the molecules in the presence of the orienting force is a

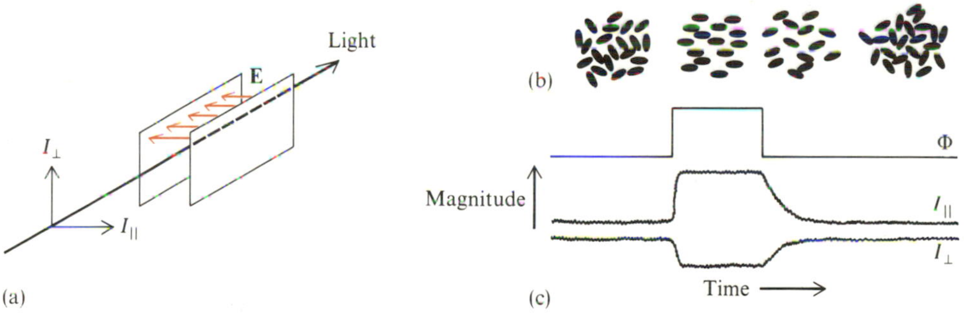

Figure 12-12

Schematic illustration of an electric dichroism experiment. **(a)** The experimental geometry. **(b)** The orientation of solute molecules as a function of time. **(c)** The time dependence of the applied voltage Φ, and the intensity I of light transmitted parallel and perpendicular to the applied electric field **E**.

complex problem. Thus, it is usual to measure the steady-state orientation or the decay of orientation once the electric field is removed.

The orientation of molecules induced by an electric field is easy to describe for a weak field. A net charge can cause only translational motion. The leading term in the interaction energy that can result in molecular orientation is

$$E_\Phi = -\boldsymbol{\mu} \cdot \mathbf{E} \tag{12-45}$$

where $\boldsymbol{\mu}$ is the permanent dipole moment of the ground state. At higher fields, terms such as $\mathbf{E} \cdot \underset{\sim}{\boldsymbol{\alpha}} \cdot \mathbf{E}$ enter, where $\underset{\sim}{\boldsymbol{\alpha}}$ is the polarizability; these terms will be ignored here. The principles of statistical mechanics allow us to write the relative probability (at equilibrium) of a state or orientation with energy E_Φ as $\exp(-E_\Phi/kT)$. Thus, the normalized distribution $W(\theta)$ induced by an electric field is

$$W(\theta) = e^{\mu E_z \cos\theta/kT} \Big/ \int_0^{2\pi} d\phi \int_0^{\pi} e^{\mu E_z \cos\theta/kT} \sin\theta \, d\theta$$

$$= \exp(\mu E_z \cos\theta/kT)/4\pi kT \sinh(\mu E_z/kT) \tag{12-46}$$

where E_z is the z component of $\mathbf{E}$, which is assumed to originate from parallel plates separated in the z direction, and θ is the angle between the molecular dipole moment and the field (Fig. 12-13), and $\sinh x$ is $\frac{1}{2}(e^x - e^{-x})$.

Various properties can be evaluated using $W(\theta)$. For example, consider the linear dichroism of a sample in which the only extinction (ε) is parallel to the permanent dipole moment. One can measure the extinction coefficient with light polarized

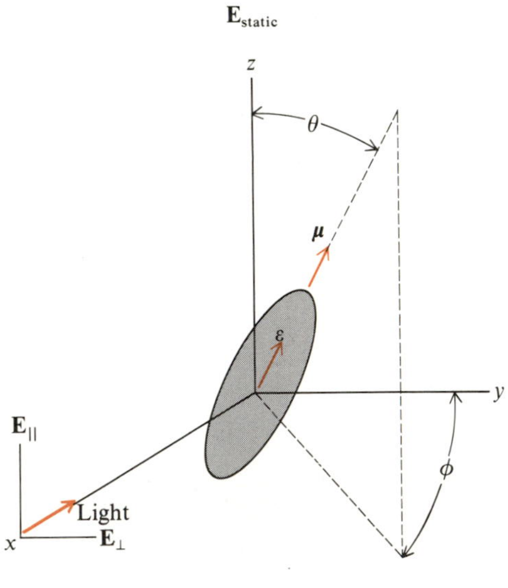

Figure 12-13

The coordinate system used to analyze electric dichroism. Light is incident along the x axis and is observed along that axis. The applied static electric field $\mathbf{E}$ is along the z axis. In the particular case shown, it is assumed that the extinction coefficient parallel to the permanent dipole is ε, and that perpendicular is zero.

parallel (E_z) or perpendicular (E_y) to the applied electric field. The difference in extinction will be $\varepsilon_z - \varepsilon_y$ (Fig. 12-13). For a given molecule, $\varepsilon_z = \varepsilon \cos^2 \theta$, and $\varepsilon_y = \varepsilon \sin^2 \theta \cos^2 \phi$, for the reasons discussed in connection with flow dichroism. The extinction difference ($\varepsilon_z - \varepsilon_y$) must be averaged over the orientations of all molecules, using Equation 12-46[§]

$$\langle \varepsilon_z - \varepsilon_y \rangle = \varepsilon \int_0^{2\pi} d\phi \int_0^{\pi} W(\theta)(\cos^2 \theta - \cos^2 \phi \sin^2 \theta) \sin \theta \, d\theta$$

$$= (\varepsilon/10)(\mu E_z/kT)^2 - (\text{terms in } E_z^4) \qquad (12\text{-}47)$$

This obviously is a highly artificial case and, in general, one must consider individual molecules with different extinction coefficients parallel and perpendicular to the dipole moment. However, Equation 12-47 shows the important general result that the equilibrium electric dichroism increases as the square of the applied field. Similar equations can be derived for electric birefringence.

You can see that the steady-state electric dichroism does not explicitly contain any useful information about shape. This situation is analogous to that with equlibrium ultracentrifugation, where all shape information was lost. However, there is implicit size and shape information in the dipole moment. In the case shown in Equation 12-47, μ can be determined from measured $\langle \varepsilon_z - \varepsilon_y \rangle$ if ε is known. In a real case, it will be hard to do this, because extinction can take place along axes not parallel to $\boldsymbol{\mu}$, so the form of Equation 12-47 is more complex. Birefringence involves similar difficulties. One difference between electric and flow orientation is instructive. Forces that lead to flow orientation are hydrodynamic, and the steady-state orientation is a function of the hydrodynamic properties of the molecule. Electrical orientation forces are not explicitly hydrodynamic, and information about size and shape is lost at equilibrium.

When the electric field is turned off, the oriented distribution of macromolecules decays back to an isotropic distribution. For the likely case in which the permanent dipole lies along the long axis of a prolate ellipsoid, the decay is the same as that given for flow birefringence or dichroism in Equation 12-42. This is because the symmetries of the oriented distribution prior to field turn-off are equivalent in the various cases. In a more general case, when $\boldsymbol{\mu}$ is not parallel to the long axis, a decay with one or two exponential terms may be observed weighting both rotational relaxation times, τ_α and τ_b.

It is also possible to compute the kinetics of orientation for certain limiting cases when the field is suddenly turned on. The results comes from solving Equation 12-37 with the addition of the applied electric force. For a molecule with $\boldsymbol{\mu}$ parallel to the long axis, the kinetics of appearance of birefringence will be

$$\Delta n/\Delta n_\infty = 1 - Ae^{-t/\tau_a} + Be^{-t/3\tau_a} \qquad (12\text{-}48)$$

[§] The reader is encouraged to try actually to work out the integral in Equation 12-47. It is an excellent test of one's facility with series expansions and trigonometric integrals.

where Δn is $\langle n_z - n_y \rangle$ (using the coordinate system of Fig. 12-13), and Δn_∞ is the final equilibrium birefringence. The constants A and B are functions of the permanent dipole moment (μ^2) and the anisotropy of the polarizability. The important thing is that the rate of orientation is governed by the same rotational relaxation times as are relaxation rates, but the complex form of the equation makes analysis more difficult.

Dielectric dispersion

Another quantity that can be computed using Equation 12-46 is the average dipole moment along the z axis of all of the solute molecules. This is

$$\langle \mu \rangle = \int_0^{2\pi} d\phi \int_0^{\pi} \sin d\theta\, \mu \cos\theta\, W(\theta) = \mu^2 E_z/3kT - (\text{terms in } E_z^3) \qquad (12\text{-}49)$$

The average dipole moment $\langle \mu \rangle$ will contribute to the overall dielectric constant of the solution. In general, the molecular polarizability will be $\alpha = \alpha_i + \alpha_d$, where α_i is due to field-induced distortion of the electron distribution of the molecule (and possibly also to field-induced migration of counterions), and α_d is due to field-induced molecular orientations. Because an induced dipole has the form $\boldsymbol{\mu} = \underset{\sim}{\boldsymbol{\alpha}} \cdot \mathbf{E}$, it is evident from Equation 12-49 that $\alpha_d = \mu^2/3kT$. Here we focus just on the contribution α_d makes to the molecular polarizability.

The polarizability of pure fluids can be related to the dielectric constant ε by

$$(\varepsilon - 1)/(\varepsilon + 2) = (4/3)\pi N'\alpha \qquad (12\text{-}50)$$

where N' is the number of molecules per cm^3. More complex expressions exist for polar liquids. Well-established techniques exist for measuring ε. The dielectric constant is a function of the frequency of the electric field. This effect, called dielectric dispersion, can yield information about molecular sizes and shapes. At the limit of low frequency, the orientation term $\mu^2/3kT$ contributes its full value to Equation 12-50 because molecules can move fast enough to achieve the equilibrium degree of orientation at all times as the field varies. In the limit of high frequency, $\mu^2/3kT$ cannot contribute at all because the field changes so fast (relative to hydrodynamic drag) that the molecules are effectively frozen. At intermediate frequencies, the rates of orientation and relaxation will determine the extent of the contribution of $\mu^2/3kT$. The detailed theory is quite involved and will not be treated here.

Other ways of measuring and interpreting rotational motions

Three other methods are in general use for studying the orientational motions of macromolecules: fluorescence polarization, EPR, and NMR. Each of these techniques is specialized and quite complex, and each is treated elsewhere in this book. Several other new techniques can be used to examine molecular rotations. Rotational motion affects the angular correlation of two successive γ rays emitted from a nucleus such as Ir during the short time between the emission of one γ ray and the other. This technique can even be applied to study the motion of a molecule in vivo if an appropriate Ir-containing derivative can be prepared. Inelastic scattering of depolarized light also contains information about molecular rotations; this technique has not been used very much yet, but it holds considerable promise for the future.

The results of any of the measurements of rotational motion just described can be put in the general form $\tau_{obs} = \tau_0 F_s(V_h/V_0)$, where τ_0 is the rotational relaxation time of an anhydrous sphere, V_h/V_0 is the ratio of the volumes of hydrated and anhydrous spheres, and F_s is a function of shape (see Eqns. 10-20 and 10-23, and Fig. 10-10). Depending on the measurement, τ_{obs} may be a single rotational relaxation time (τ_a or τ_b) or the harmonic mean of τ_a and τ_b. Thus the rotational results can be put in the following form for comparison with translation hydrodynamic measurements:

$$\text{rotation} \qquad \tau_{obs} = K_1 M(\bar{V}_2 + \delta_1 \bar{V}_1)F_s \qquad (12\text{-}51)$$

$$\text{viscosity} \qquad [\eta] = K_2(\bar{V}_2 + \delta_1 \bar{V}_1)\nu \qquad (12\text{-}52)$$

$$\text{translation} \qquad f = K_3 M^{1/3}(\bar{V}_2 + \delta_1 \bar{V}_1)^{1/3} F \qquad (12\text{-}53)$$

where K_1, K_2, and K_3 are constants. Earlier, we derived the Scheraga–Mandelkern equation by combining viscosity and sedimentation data and eliminating hydration. This approach allows unequivocal determination of shape and hydration (except for prolate–oblate ambiguity), if the molecular weight is known. However, it is insensitive. Rotation data can be combined similarly with either viscosity or translation data to eliminate hydration by solving any pair of the preceding equations simultaneously; again, both shape and hydration can be determined. Table 12-3 shows the results of such calculations for several proteins. The consistency among several different techniques appears to be fairly satisfactory.

Table 12-3
Protein hydration and axial ratios determined simultaneously by combining hydrodynamic data

Data combined	Bovine serum albumin a/b	δ_1	β-Lactoglobulin monomer a/b	δ_1	Ovalbumin a/b	δ_1	Lactic dehydrogenase a/b	δ_1
ν, F	2.5	0.6	2.5	0.3	4.0	0.2	4.0	0.1
ν, f_a	2.8	0.4	—	—	—	—	—	—
ν, f_b	3.5	0.2	—	—	4.5	0.1	—	—
ν, f_h	2.5	0.5	1.5	0.4	4.5	0.1	3.8	0.2
F, f_a	3.5	0.4	—	—	—	—	—	—
F, f_b	2.5	0.6	—	—	4.0	0.2	—	—
F, f_h	4.0	0.3	2.0	0.4	5.0	0.1	4.0	0.2

NOTE: All axial ratios shown are for prolate ellipsoids.
SOURCE: After I. D. Kuntz, Jr., and W. Kauzmann, in *Advances in Protein Chemistry*, vol. 28, ed. C. B. Anfinsen, J. T. Edsall, and F. M. Richards (New York: Academic Press, 1974), p. 239.

In a favorable case, rotation measurements can yield both f_a and f_b. Because these values are very different for oblate and prolate ellipsoids, the type of ellipsoid as well as its axial ratio can then be determined. The general conclusion from all such treatments is that the power of individual hydrodynamic measurements is greatly enhanced by parallel experiments using other techniques on the same system.

12-3 MOLECULAR-SIEVE CHROMATOGRAPHY

It is hard to imagine a biochemist who has not used gel exclusion chromatography to fractionate samples according to molecular size. Figure 12-14 is a schematic diagram of one particle of a molecular-sieve resin. Large-pore molecular sieves such as Sephadex (cross-linked polydextrans) and Biogel (cross-linked polyacrylamide) are easy to use and are available from commercial sources. (Porous glass beads and agarose gels also are used for biochemical preparative work.) Use of such resins is a powerful technique for separation of a mixture into rough size classes.

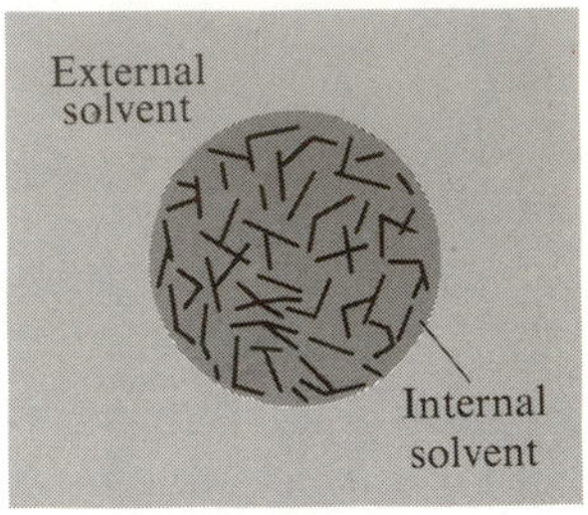

Figure 12-14

Schematic diagram of one bead of a gel partition resin. The external solvent occupies the volume V_0 around the bead. The internal solvent occupies the volume V_i within the bead. The gel matrix (shown by dark lines) occupies the volume V_g, from which solvent is excluded.

Principles of gel filtration

A qualitative explanation of the mode of action of molecular sieves is simple. A column containing the loosely packed resin particles has a total volume (V_T) made up of three components: the volume of exterior solvent (the void volume), V_0; the solid volume of gel particles, V_g; and the internal volume of the pores of the particles that are accessible to solvent, V_i.

$$V_T = V_0 + V_g + V_i \tag{12-54}$$

If a solute is added to such a gel column, it will partition between the external and internal solvent regions; the result can be described by a partition coefficient σ. The mass m_i of solute found in internal regions will be

$$m_i = \sigma V_i c \tag{12-55}$$

where c is the solute concentration in the external regions. The significance of σ is the following.

$\sigma = 0$ No solute molecules can enter the pores of the gel matrix.

$\sigma < 1$ Solute is less likely to be found in the pores than in the bulk solution. The simplest explanation is that, in the distribution of pore sizes, some pores are too small to permit entry of solute.

$\sigma = 1$ Solute partitions equally between pores and external solution.

$\sigma > 1$ Solute is preferentially attracted to the pores on the resin surface; some kind of adsorption phenomenon is occurring.

The amount of solute material inside the gel can be written as $m_i = c_p V_p$, where V_p is the penetrable volume (that part of V_i that is accessible to a particular solute), and c_p is the concentration inside this volume. If thermodynamic ideality is assured, the concentration of solute should be everywhere equal ($c_p = c$), so we can write σ as

$$\sigma = V_p/V_i \tag{12-56}$$

Thus σ is the fraction of internal volume available to solute. (An alternative way of looking at this is to assume that all internal volume could, in principle, have been accessible to solute. In this case, $V_p \equiv V_i$, and $\sigma = c_p/c = c_i/c$, the ratio of the internal concentration to the external concentration.)

The most common application of molecular sieves is elution chromatography. A thin band of solution containing one or more solutes is run onto the top of a column of resin, and solvent is then allowed to flow through the column at a rate slow enough to allow solutes to equilibrate fully between external and internal volume at each level. The general procedure is to measure the elution volume, the volume of solvent that must flow through the column before a species emerges. If diffusion is ignored, we can consider the thin band of solution as a separate phase that must displace solvent ahead of it as it passes through the column. A solute that is completely excluded by the resin ($\sigma = 0$) will have to displace a volume corresponding to the entire void volume V_0. But a solute that can enter some of the pores will have to displace the accessible internal volume $V_p = \sigma V_i$ in addition to the void volume (Fig. 12-15). At a constant rate of flow, it will travel down the column more slowly and will emerge after eluting a volume

$$\boxed{V_e = V_0 + \sigma V_i} \tag{12-57}$$

Equation 12-57 shows that σ can be calculated from a measurement of the elution volume, because $V_i + V_0$ can be determined by the weight of water uptaken by dry

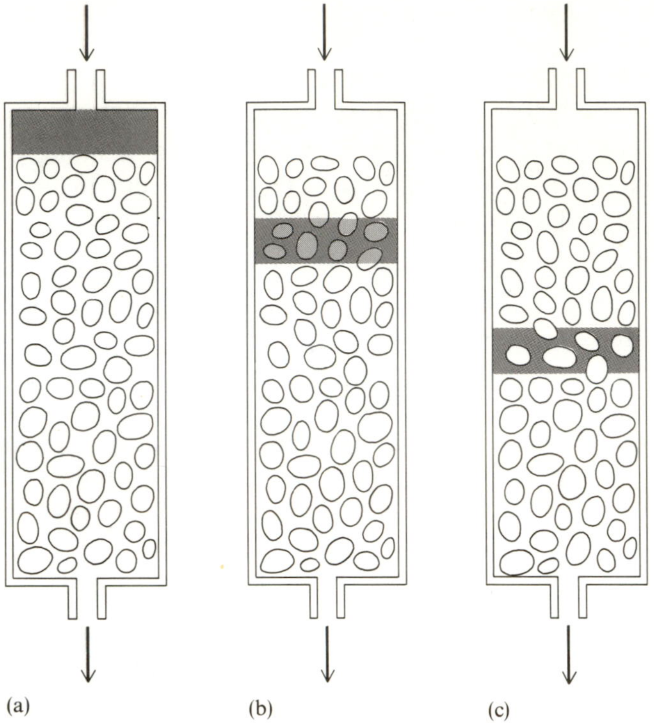

Figure 12-15

Solute transport in gel partition chromatography. **(a)** A zone of solute prior to contact with any resin beads. **(b)** The zone after a total volume V has passed through the column, for a solute that can enter most of the internal volume of the gel. **(c)** The zone after the same volume V has passed for a solute that cannot enter any of the internal volume of the gel.

resin, and V_0 can be determined by measuring the elution volume of a particle much larger than any gel pores.

Analysis of the shape of the eluting bands

Equation 12-57 is the starting point for a description of the actual shape of an eluting band of solute. Define ξ as the fraction of a column cross-sectional area that is available to solute:

$$\xi = (\alpha + \beta\sigma)/A \tag{12-58}$$

where A is the total cross-sectional area, α is the cross-sectional void area, and β is the cross-sectional internal area. [Note that $(\alpha + \beta) < A$.] Suppose the column is eluted at a constant total volume rate of flow, F_V. The mean linear transport rate v_x along the column is volume flow divided by accessible area:

$$v_x = F_V/\xi A \tag{12-59}$$

The units of v_x are $\text{cm}^3\ \text{sec}^{-1}/\text{cm}^2 = \text{cm sec}^{-1}$. The time it takes a band to flow down the entire column length l is $t = l/v_x$. During that time, the total volume that has passed through the column (the elution volume) is $V_e = F_V t$. Substituting for t and then for v_x, we obtain

$$V_e = A\xi l \tag{12-60}$$

Equations 12-60 and 12-57 are identical, as you can see by substituting Equation 12-58 for ξ in Equation 12-60.

The flux J of solute in the x direction (along the column) can be written as

$$J = v_x c - L(\partial c/\partial x) = F_V c/\xi A - L(\partial c/\partial x) \tag{12-61}$$

where the first term describes the effect of flowing solvent, and the second term will contain any effects of spreading of the zone of solute. The factor L is called the coefficient of axial dispersion. If diffusion were the only cause of spreading, L would simply be the diffusion constant corrected for the fraction of cross-sectional area applicable. However, L is more complex because flow times are not long enough for complete diffusional equilibrium to occur. Furthermore, because of friction, the flow rate of the solvent will be slower near the surface of the resin than in resin-free regions. Mass must be conserved (as always) by Equation 10-49: $\partial c/\partial t = -\partial J/\partial x$. Differentiating with respect to x, we can convert Equation 12-61 to the following form:

$$(\partial c/\partial t) + (F_V/\xi A)(\partial c/\partial x) = L(\partial^2 c/\partial x^2) \tag{12-62}$$

Equation 12-62 is simply Fick's second law for a moving frame of reference. From the mean flow rate in Equation 12-59, the distance traveled by the zone center is $x = v_x t = F_V t/\xi A$. If we change variables with $y = x - F_V t/\xi A$, we obtain a frame of reference in which the zone center appears motionless, and Equation 12-62 becomes

$$\partial c/\partial t = L(\partial^2 c/\partial y^2) \tag{12-63}$$

The initial boundary condition is an infinitely sharp band, which is located at the top of the column ($x = 0$) when $t = 0$. Thus $y = 0$ when $t = 0$, and the initial band can be represented by a delta function, $c(y, 0) = w_0\,\delta(0)$, where w_0 is the weight of solute in the band. We already have solved this equation with precisely these boundary conditions; using the result shown in Box 10-2, we obtain

$$c(y, t) = (w_0/\sqrt{4\pi Lt}) \exp(-y^2/4Lt) \tag{12-64}$$

Now, to compare with experiment, we change the variables to remove y $(= x - F_V t/\xi A)$, replace t by V/F_V (where V is the total volume that has flowed through the column), and remove x by the fact that one usually makes measurements at the bottom of the column where $x = l = V_e/\xi A$ (Eqn. 12-60). We now can write the concentration of solute eluting from the column as a function of the flow rate and the volume of flow:

$$c(F_V, V) = (w_0/\xi A\sqrt{4\pi LV/F_V}) \exp[-F_V(V_e - V)^2/4\xi^2 A^2 LV] \tag{12-65}$$

Thus, at a constant flow rate F_V, Equation 12-65 shows that the elution profile $c(F_V, V)$ from a gel exclusion column is not a simple Gaussian in the volume of flow, V.

Note that it is a common practice to state that an eluted band is a single Gaussian, and that therefore the sample is a homogeneous species. Obviously this practice is incorrect. However, on the column at any fixed time t, the profile of solute is a simple Gaussian in y and thus in x. Gary Ackers (1975) has developed elegant procedures for direct scanning of solute distributions on gel columns. These procedures allow easy determination of the coefficient L; more importantly, they simplify quantitative analysis of mixtures of solutes or equilibrating systems.

Molecular-sieve behavior and macromolecular size and shape

The problem that remains is to relate σ to properties of macromolecular solutes. [From the equations derived above, we see that this is equivalent to evaluating ξ (Eqn. 12-58) or V_e (Eqn. 12-60).] If the pores of the gel were all of uniform size, then σ should be a step function, equaling one for molecules smaller than the pore size, and equaling zero for molecules bigger than the pore size. For a distribution of pore sizes, σ will vary gradually with molecular size. However, the critical feature is that the size, rather than the molecular weight, of a macromolecule should determine its partition coefficient. Of course, in molecules with similar shapes, size is related to molecular weight. Many workers have shown that σ can be correlated with the molecular weight. For chromatography of various proteins on any particular resin,

$$\sigma = -A \log M + B \tag{12-66a}$$

where A and B are constants. Figure 12-16 shows an example of such a correlation. Many molecules fit well, but some definitely fall off the predicted curve.

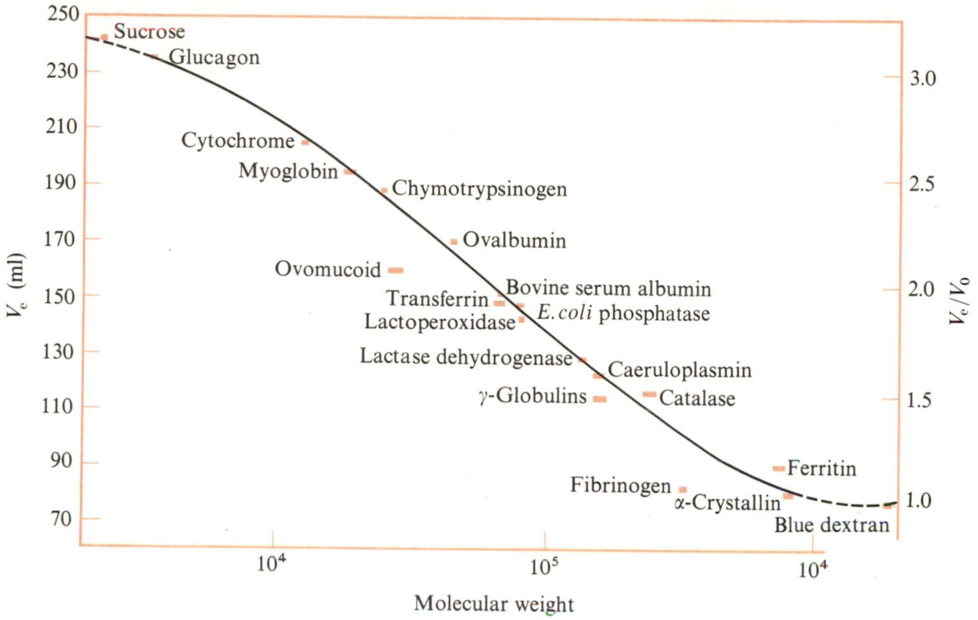

Figure 12-16

Elution volume of various proteins on a column of the gel partition resin Sephadex G-200 as a function of molecular weight. The right-hand vertical axis shows the ratio of the elution volumes to that of blue dextran, a high-molecular-weight polysaccharide that is believed to be completely excluded from the internal volume. [After P. Andrews, *Biochem. J.* 96:595 (1965).]

More detailed treatments predict equations of the form

$$\sigma = -A' \log r_h + B' \tag{12-66b}$$

where r_h is the effective hydrated radius. This is the quantity that would be measured in a diffusion experiment, ignoring shape and just using Stokes' law, $kT/D = f = 6\pi\eta r_h$. Equation 12-66b agrees with experiment better than does Equation 12-66a, indicating that gel exclusion chromatography can be used to measure the frictional coefficient once a suitable calibration procedure has been established to determine the constants A' and B'. Because gel exclusion chromatography yields r_h, it often is assumed to be equivalent to diffusion measurements. The results of gel exclusion chromatography can be combined with measured sedimentation constants to yield molecular weights.

For mixtures, it can be shown that the measured σ will be a weight average of all solutes in bulk solution. Gel chromatography is an especially useful technique for examining interacting systems. For a detailed discussion of this approach, see Ackers (1975).

12-4 ELECTROPHORESIS

If a macromolecule has a net charge q, then application of an electric field $\mathbf{E}$ will result in an applied force $\mathbf{F} = q\mathbf{E}$. This force will cause acceleration of the particle in a fluid until a steady-state velocity $\mathbf{v}$ is reached. At this velocity, frictional forces are equal and opposite to the applied force, so

$$\mathbf{v} = q\mathbf{E}/f \tag{12-67}$$

For a spherical macromolecule of radius a with a charge equal to ze (where e is the charge on the electron), we have

$$\mathbf{v} = ze\mathbf{E}/6\pi\eta a \tag{12-68}$$

If the electrical field originates from parallel plates or the equivalent, the molecule travels in a straight line. By analogy with the definition of the sedimentation coefficient, the mobility u can be defined as the velocity per unit field, $u = \mathbf{v}/\mathbf{E}$.

Unfortunately, this description of electrophoresis (the transport of charged particles in the presence of an electric field) is completely inadequate. An immediate difficulty arises over what to call the net charge on a macromolecule.

Predicting electrophoretic mobility

In any aqueous solution, there are counterions. Some may be associated rather tightly with the polymer, others more loosely. Consider what will happen if an electric field is applied to a positively charged macromolecule with only enough anions in the solution to neutralize this charge. The macromolecule will tend to migrate in one direction, the counterions in the opposite direction. The electrostatic energy involved in any net charge separation is enormous, and therefore little net overall motion will actually occur. If there is any net mobility at all, it will be an average of the mobilities of macromolecule and counterions, just as we discussed in Chapter 10 for diffusion of electrolytes. As in that case, a large amount of supporting electrolyte can be introduced to weaken the effective character of the ion pairing. This procedure resolved the complication for diffusion, but in electrophoresis it introduces more difficulties. Now the electrolyte forms an ion atmosphere around the polymer; there are two ways to view its effect. The ion atmosphere will effectively reduce the net charge on the polymer, because oppositely charged ions will tend to be attracted closely to the polymer (Fig. 12-17). Alternatively, one can say that the ion atmosphere sets up an electric field that is felt by the polymer and that must be considered along with any applied field in computing its mobility.

It is customary to treat this atmosphere as a continuous distribution and to ignore individual charges. The electrostatic potential Φ at a distance r away from a

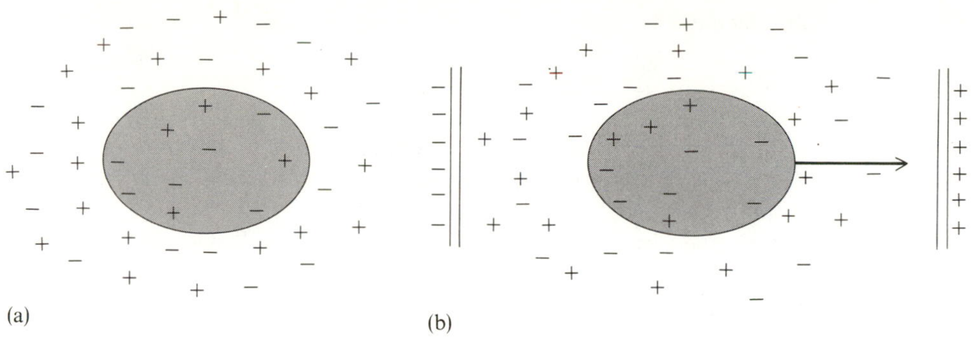

Figure 12-17

A protein with a net negative charge, and its counterion atmosphere. **(a)** In the absence of an applied electric field. **(b)** While the molecule is being transported by the applied electric field shown. Note the distortion of the ion atmosphere. A change in the charge distribution of the protein itself also is shown; this is not a necessary consequence of an applied field, but it certainly is a realistic possibility.

uniformly charged sphere of radius a in a medium of dielectric constant ε is

$$\Phi = ze/\varepsilon r \qquad \text{for } r > a \tag{12-69}$$

For the same sphere in an electrolyte solution, it can be shown, by solving an approximate form of the Poisson–Boltzmann equation (Rice and Nagasawa, 1961), that

$$\Phi = (ze/\varepsilon r)\exp[\kappa(a - r)]/(1 + \kappa a) \tag{12-70}$$

where κ (a screening parameter with the dimensions of inverse length) is given by

$$\kappa = (8\pi N_0 e^2/1{,}000\varepsilon kT)^{1/2} I^{1/2} \tag{12-71}$$

The ionic strength, I, is a sum over all of the molar concentrations C_i and charges of species in the solution:

$$I = \frac{1}{2}\sum_i C_i z_i^2 \tag{12-72}$$

The parameter κ regulates the extent to which the potential of an ion falls off more rapidly in solution than it would if there were no counterions. It is evident that, the higher the ionic strength, the more effectively the potential of an ion is screened.

If the screened potential stays constant in an applied electric field, a difficult (but moderately straightforward) analysis leads to a predicted electrophoretic mobility for a spherical molecule with radius a of

$$u = (ze/6\pi\eta a)X_1(\kappa a)/(1 + \kappa a) \tag{12-73}$$

where $X_1(\kappa a)$ is a function that has been tabulated by D. C. Henry (see Rice and Nagasawa, 1961). However, the applied field—and the resulting motions of the macromolecule and small ions—distorts the ion atmosphere (Fig. 12-17). This distortion alters the potential given by Equation 12-70. To calculate all of these effects is a formidable problem. The result is an equation with the same leading term as Equation 12-73, but with many additional terms. The most satisfactory treatment of this problem was worked out by F. Booth (see Rice and Nagasawa, 1961, for a detailed discussion). The simple theory of Equation 12-73 predicts that mobilities will increase linearly with increasing charge on the macromolecule. The more complete theories show that, at sufficiently high charges, the mobility is less than that expected from Equation 12-73 and can even start to decrease as z increases.

As complicated as all this is, the Booth theory does not adequately explain observed macromolecular electrophoresis results. It holds only for spheres of uniform charge density, and this model is a very poor representation of a protein or nucleic acid. Just imagine an ellipsoid with an asymmetric charge distribution. The electric field will apply torques as well as net displacement. There will be preferred orientation, and the motion of the macromolecule no longer can be described by a rotationally averaged frictional coefficient. Furthermore, all of these effects will couple into the distortion of the ion atmosphere.

Applications of electrophoresis

Despite all these complications, electrophoresis is a powerful and practical tool in the analysis and separation of proteins and nucleic acids, so long as one does not demand quantitative structural data from it. All of the theories predict, in common, that the mobility is proportional to the ratio of net charge to frictional coefficient. This relationship allows one to use electrophoresis to obtain information about relative charge (for molecules of the same size and shape) or about relative size (for molecules of the same charge). The most common use of electrophoresis is the separation or qualitative analysis of mixtures, based on sizes and shapes of individual components.

In moving-boundary (or free) electrophoresis, a broad zone of macromolecules in solution is subjected to an electric field, and the motion of the zone is measured. Electrophoresis will cause a net displacement; diffusion will cause broadening. This is the form of electrophoresis most amenable to mathematical treatment. Separations can be followed by direct optical monitoring, as in diffusion or ultracentrifugation. However, moving-boundary electrophoresis is not especially convenient as a preparative technique because, as in velocity sedimentation, actual separation takes place only at the edges of the zone. Various artifacts, convection, experimental difficulties, and multicomponent effects weaken its usefulness as a general analytical procedure.

These problems can be minimized in zonal electrophoresis, where (in analogy to zonal sedimentation) there is initially only a very thin band of macromolecule. Convection can be suppressed by a density gradient, but it is much more common in electrophoresis to use a solid support permeated with buffer. Some supports interact only weakly or nonspecifically with the macromolecular solutes; these include paper, thin-layer cellulose, and cellulose acetate. Other materials retard the motion of certain molecules relative to one another; these include polyacrylamide and agarose gels (which can discriminate by size) and ion-exchange papers (which selectively retard charged molecules.

The use of a solid support makes a quantitative analysis of mobility nearly impossible. In addition to all of the problems cited thus far, the molecule is forced to trace a tortuous path through the support medium. This obscures the relationship between the observed net mobility and the actual molecular mobility. Any specificity of interaction with surface, such as equilibration of bound and free solute, is an added complication.

However, there are two variations of zone electrophoresis that can yield some quantitative molecular information. One is based on the fact that the charge on a protein or other macromolecule depends on the pH. A study of electrophoretic mobility as a function of pH will allow the isoelectric point to be found. This is the pH at which the average net charge on the macromolecule is zero. The mobility at the isoelectric pH will also be zero. Furthermore, the particle will move toward the cathode at a pH below the isoelectric point, and will move in the opposite direction at a higher pH.

It is tedious to perform a set of separate electrophoreses at closely spaced pH values. A sensible alternative is to carry out electrophoresis in a pH gradient. This technique, called isolectric focusing is a direct analog of equilibrium density-gradient ultracentrifugation. Ampholytes are molecules with positive and negative charges—for example, polymers containing numerous amino and carboxyl groups. A mixture of ampholytes with a wide range of isoelectric points is allowed to distribute in a column under the influence of an electric field. This procedure establishes a pH gradient in which each particular ampholyte comes to rest at a position near its isoelectric point. Loss of biological material to the cathode or anode is prevented by having such extremes of pH there that no biological materials have isoelectric points in this range. The pH gradient is created on a support medium to block convection. A small amount of protein is added to the system. It migrates until it reaches the pH of its isoelectric point. It will remain there as a sharp band. To analyze the results of the experiment, we can scan the pH gradient for protein absorbance, can stain it, or can cut it up into slices and analyze each of these for pH and protein content, enzymatic activity, or radioactivity.

Electrophoresis in sodium dodecylsulphate to obtain molecular weights

A second very popular technique is zonal electrophoresis of proteins in sodium dodecylsulphate (SDS). Here the basic idea is to try to convert all proteins into similar structures that differ only in molecular weight. SDS is an effective protein denaturant. It binds to all proteins qualitatively the same, at about 1.4 g per gram of amino acid. There is one negative charge on each SDS molecule. The resulting charge density due to all the SDS in the SDS–protein complex more-or-less overwhelms variations in the charges of different protein molecules. Reducing agents, such as β-mercaptoethanol, are added to break any intrachain or interchain disulfide bonds and to allow the denatured forms to take whatever equilibrium form the protein–SDS interactions dictate. Charles Tanford and others have studied the structure of SDS–protein complexes. Hydrodynamically, these appear to be prolate ellipsoids or rods with a constant diameter of about 18 Å and a length that is a linear function of the molecular weight.

Polyacrylamide gels generally are used as the support medium in SDS electrophoresis. These gels can be prepared at various ratios of solid gel to fluid solution. It is a general observation that, the higher the gel concentration, the lower the apparent electrophoretic mobility. When careful quantitative measurements are made, a surprising result emerges (Fig. 12-18a). The mobilities of a set of similar proteins, extrapolated to zero gel concentration, are the same. The data can be fit to an equation of the form

$$\ln u(C) = -k_x C + \ln u(0) \tag{12-74}$$

where $u(C)$ is the apparent mobility at a gel concentration of C (in percent), k_x is a constant that depends on the extent of cross-linking of the gel and on the shapes and molecular weights of the particular proteins, and $u(0)$ is a constant for a set of similar kinds of molecules.

The mobility at zero gel concentration should reflect the true electrophoretic mobility of the SDS–protein complexes, unencumbered by any interactions with the gel. The simple theories of electrophoresis outlined above can account for the fact that $u(0)$ is a constant. If the protein–SDS complexes have a constant weight percentage of SDS, and if protein charge can be neglected, the net charge z will be proportional to the molecular weight, and thus proportional to the length l because the complexes are rods. The frictional coefficient also is roughly linear in the molecular weight; this can be seen by writing f as $6\pi\eta V_{\mathrm{h}}^{1/3}F$. The hydrated volume of a rod will be directly proportional to its length. The shape factor F can be approximated from Equation 10-19a for prolate ellipsoids. As a/b becomes very large, $F = (a/b)^{2/3} \ln(a/b)$. But the axial ratio of a rod is just proportional to its length, and so $f \propto l^{1/3}l^{2/3}/\ln l \cong l$ for large l. Thus, $u(0) \propto z/f$ will be independent of the molecular weight. So it appears that, in SDS electrophoresis, the electric field is simply causing a constant drift to all the molecules, and any separation based on molecular weight is originating from the specifics of interaction of different-sized SDS–protein rods with the supporting gel matrix.

Equation 12-74 indicates that the mobility decreases as the amount of *solvent* in the gel support decreases. The gel itself should have a structure like the beads shown in Figure 12-14. However, all of the volume of the system is occupied by gel; there is no external volume as there is between the beads in a partition column. When

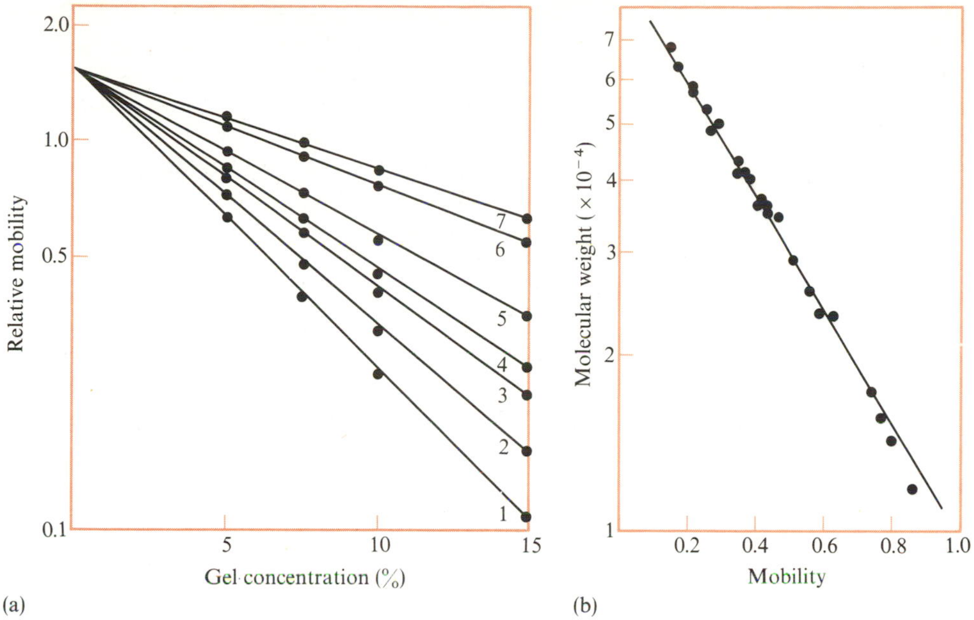

Figure 12-18

SDS polyacrylamide-gel electrophoresis of proteins. **(a)** Relative mobility of seven proteins as a function of the concentration of the acrylamide gel. The molecular weights of the proteins range from 14,000 d for the coat protein of phage R17 (curve 7) to 60,000 d for catalase (curve 1). **(b)** Relative mobility of various proteins in 10% acrylamide gels as a function of molecular weight. [After K. Weber and M. Osborn, in *The Proteins*, 3d ed., vol. 1 (New York: Academic Press, 1975), p. 179.]

macromolecules are forced by an electric field to move through the solid gel, the smallest ones should find many pores through which they can travel. Larger molecules will have to travel longer distances, because only a subset of the available pores are large enough to let them pass. The result is that, in the gel, larger molecules will have a lower net mobility. From Equation 12-58, we see that the partition coefficient σ is directly proportional to the cross-sectional area available to solute because, for a solid gel, the void area α is zero. Therefore, the net mobility should be proportional to σ. From Equation 12-66a, this lets us predict

$$u = b - a \log M \tag{12-75}$$

The constant b should contain the gel-free mobility $u(0)$, as well as properties of the gel pore-size distribution. The parameter a must be a function of the gel concentration C in order for Equations 12-74 and 12-75 to be consistent.

In practice, Equation 12-75 fits the observed mobilities of proteins in SDS–gel electrophoresis extremely well. Figure 12-18b shows some typical results. The same arguments we have used for proteins in SDS apply to nucleic acids in normal aqueous buffers. However, for very long molecules, it is no longer accurate to treat them as rigid rods. The analog of Equation 12-75 must be used, in which the effective hydrated radius of a coiled molecule replaces the molecular weight. Once a gel system has been calibrated, V_h can be obtained, and then it can be combined with sedimentation data to yield the true molecular weight, just as described earlier for gel partition chromatography.

The major advantage of gel electrophoresis or chromatography is that very sharp bands are maintained, because there is no external volume to allow for rapid diffusion. The separation power of SDS electrophoresis is based totally on molecular weight, and that of isoelectric focusing is based totally on charge. It is logical that, if the two could be combined into a single separation scheme, its resolution should be quite impressive. This can be done by two-dimensional electrophoresis, in which one dimension is SDS and the other is isoelectric focusing. Figure 12-19 shows an example of this from the work of Patrick O'Farrell. More than 1,000 different proteins from *E. coli* can be distinguished in a single analysis.

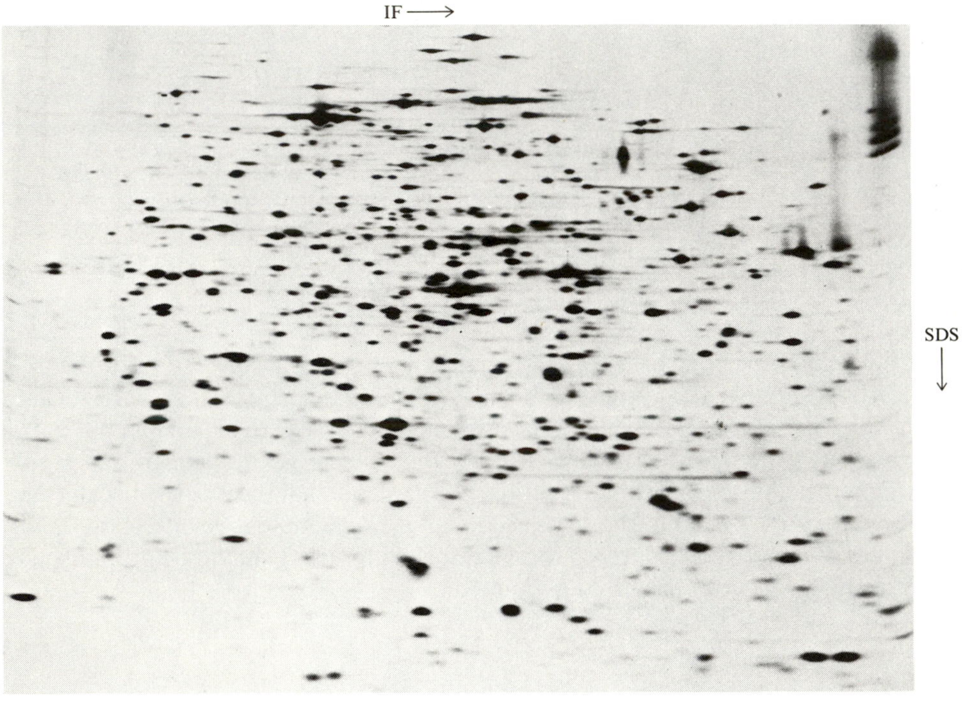

Figure 12-19

Two-dimensional SDS–isoelectric focusing gel electrophoresis. First the sample is run in a one-dimensional pH gradient gel (isoelectric focusing). Then this gel is placed alongside a square SDS gel, into which the sample moves at right angles to its motion in the first gel. The sample shown is total *E. coli* protein; it is visualized by autoradiography. For details of the technique, see P. O'Farrell, *J. Biol. Chem.* 250:4007 (1975). [Autoradiograph courtesy of Patrick O'Farrell.]

Summary

Intrinsic viscosity measures the increase in solution viscosity that results from the presence of dissolved macromolecules. For spherical particles, the intrinsic viscosity is independent of molecular weight, but it is very sensitive to the shape of macromolecules. Prolate ellipsoids have much higher intrinsic viscosities than do corresponding oblate ellipsoids. Molecular weights can be estimated fairly reliably by combining viscosity and sedimentation data. The viscosity measured for solutions of very long molecules will depend markedly on the velocity gradient (shear) in the apparatus used for the measurements. Shear forces can actually distort the average chain configuration of large DNAs. This effect is exploited to measure molecular weights in the technique of viscoelastic relaxation.

Size and shape information are available also from measurements of rotational motion. One approach is to measure the anisotropic equilibrium distribution of molecular orientations generated by an applied force. Shear forces or electric fields typically are used to generate the anisotropic distribution. A second approach is to watch the kinetics of decay of such a distribution back to an isotropic solution after the force is removed. Linear dichroism and birefringence are two of the techniques used to monitor the presence of an anisotropic distribution.

Gel exclusion chromatography is a simple, yet powerful technique to estimate molecular sizes and to fractionate molecules on the basis of size. A molecule that cannot enter the pores of a molecular sieve will move faster through a column of gel exclusion resin than will a smaller molecule. Electrophoresis is another extremely useful tool for the purification or qualitative analysis of mixtures of macromolecules. It is rather difficult quantitatively to relate observed electrophoretic mobilities to molecular charge and shape. However, certain specialized applications of electrophoresis are amenable to quantitative analysis. Isoelectric focusing yields the isoelectric points of proteins. Electrophoresis of proteins, in the presence of SDS, or of nucleic acids on a molecular-sieve supporting medium allows estimates of the molecular weight to be obtained very easily.

Problems

12-1. The same quantity of a molecular-sieve resin is placed in each of three containers. Sufficient solvent is added to cover the resin and provide some excess supernatant. N_P moles of a protein are added to one sample; N_L moles of a small molecule are added to the second sample; both N_P and N_L are added to the third sample. After equilibration with the resin, the ligand and protein *concentrations* are measured in the supernatant. Show that is is possible to compute the average number of ligands bound per protein. You may assume that protein and all protein–ligand complexes are completely excluded from the internal volume of the resin, whereas free ligand can gain access to all the internal volume of the resin.

12-2. Estimate the hydration of tobacco mosaic virus from the measured intrinsic viscosity $[\eta] = 37\ \text{cm}^3\ \text{g}^{-1}$ and the known axial ratio of 18:1 (prolate). You can assume that the partial specific volume is in the low range of a typical protein because the virus is more than 90% protein by weight.

12-3. A protein exists in two forms, a monomer and a dimer. The monomer is an oblate ellipsoid, but no accurate measurements of its axial ratio are available. The diffusion constant of the monomer is 1.96 times the diffusion constant of the dimer. The intrinsic viscosity of the dimer is 6.45 times that of the monomer. What is the axial ratio of the dimer? You can assume (a) the same partial specific volume for monomer and dimer, (b) the same degree of hydration for monomer and dimer, and (c) that the experiments were all done in water at 25°C.

12-4. Calculate the extinction angle expected for myosin in a flow birefringence experiment performed in a shear gradient of 4,000 sec^{-1}. (See Table 12-1 for necessary information about myosin.)

12-5. Explain why proteins with molecular weights smaller than 15,000 d behave anomalously in SDS gel electrophoresis. Calibrations established between mobility and *M* using larger proteins are not valid for these small proteins.

References

GENERAL

Ackers, G. K. 1975. Molecular sieve methods of analysis. In *The Proteins*, 3d ed., vol. 1, ed. H. Neurath and R. L. Hill (New York: Academic Press), p. 1.

Tanford, C. 1961 *Physical Chemistry of Macromolecules*. New York: Wiley.

SPECIFIC

Bloomfield, V., D. Crothers, and I. Tinoco, Jr. 1974. *Physical Chemistry of Nucleic Acids*. New York: Harper & Row.

Charney, E. 1971. Linear dichroism with special emphasis on electric field-induced linear dichroism. In *Procedures in Nucleic Acid Research*, vol. 2, ed. G. C. Cantoni and D. R. Davies (New York: Harper & Row), p. 176.

Eisenberg, H. 1976. *Biological Macromolecules and Polyelectrolytes in Solution*. Oxford: Clarendon Press. [A useful advanced treatist.]

Kasai, M., and F. Oosawa. 1972. Flow birefringence. In *Methods in Enzymology*, vol. 26, ed. C. H. W. Hirs and S. W. Timasheff (New York: Academic Press), p. 289.

Klotz, L. C., and B. H. Zimm. 1972. Size of DNA determined by viscoelastic measurements: Results on bacteriophages, *Bacillus subtilis* and *Escherichia coli*. *J. Mol. Biol.* 72:779.

Rice, S. A., and M. Nagasawa. 1961. *Polyelectrolyte Solutions*. New York: Academic Press.

Righetti, P. G., and J. W. Drysdale. 1976. *Isoelectric Focusing*. New York: North-Holland.

Van Holde, K. E. 1971. *Physical Biochemistry*. Englewood Cliffs, N.J.: Prentice-Hall.

Weber, K., and M. Osborn. 1975. Proteins and sodium dodecyl sulfate: Molecular weight determination on polyacrylamide gels and related procedures. In *The Proteins*, 3d ed., vol. 1, ed. H. Neurath and R. L. Hill (New York: Academic Press), p. 179.

Yamakawa, H. 1971. *Modern Theory of Polymer Solutions*. New York: Harper & Row.

Zimm, B. H. 1971. Measurement of viscosity of nucleic acid solutions. In *Procedures in Nucleic Acid Research*, vol. 2, ed. G. C. Cantoni and D. R. Davies (New York: Harper & Row), p. 245.

13
X-ray crystallography

13-1 X-RAY SCATTERING BY ATOMS AND MOLECULES

X-ray diffraction is the most powerful technique currently available for studying the structure of large molecules. In many cases, x-ray diffraction studies on protein or nucleic acid crystals have yielded the complete tertiary structure at a level of resolution of 3 Å or better. If only a less-well-ordered sample (such as an oriented fiber) is available, x-ray diffraction still provides a wealth of structural information. Though insufficient to determine the structure uniquely, this information in many cases can provide decisive tests of structural models. Here we develop the theory of x-ray diffraction and describe some of the steps involved in obtaining structures from diffraction data.

Outline and limitations of our treatment

As one might expect, a technique that can provide so many structural details is intrinsically rather complex. We omit as many of the complications as possible and try to focus on the essential features of the method. Thus, atoms are treated as motionless, even though in crystals there is appreciable motion at finite temperatures. Crystals are treated as perfectly ordered arrays, even though they may actually be ordered only in local domains, X-ray radiation is treated as monochromatic, even though a distribution of wavelengths is always used in practice. Finally, diffraction data are considered to be very precise, even though experimental errors often are a significant problem in practice.

To understand x-ray diffraction, one must know how x rays interact with atoms and the manner in which atoms can be organized into crystals. Most traditional descriptions of the technique start with a discussion of the symmetry and structure of crystals. Diffraction of x rays is described in terms of reflections from crystal planes. The structure of molecules within the crystal is introduced into the discussion only later. The reader probably has seen this approach before in more elementary texts. Here we use a different approach, elaborated by H. Lipson and C. A. Taylor (1958). The x-ray scattering of single atoms is explained. Then we build in complexity to describe the x-ray scattering of sets of atoms (one-dimensional arrays) and, finally, of the three-dimensional arrays found in crystals. Although this treatment requires somewhat more sophisticated mathematics, there seems to be a consensus among practicing crystallographers that it ultimately affords much greater insight and understanding.

X rays: short-wavelength electromagnetic radiation

X rays are photons with wavelengths in the range of 0.1 Å to 100 Å. They usually are generated by bombarding a target with electrons of energies of 10,000 electron volts (eV) or more. Upon collision, these high-energy electrons can knock electrons out of the target atoms, leaving vacancies in atomic shells. If, for example, a vacancy is produced in the innermost (K) shell of an atom, it rapidly will be filled by an electron descending from the next (L) shell, or one from the one after that (M). The photons emitted as a result of these transitions are called, respectively, K_α and K_β x rays. Their wavelengths are

$$\lambda_{K_\alpha} = hc/(E_L - E_K) \quad \text{and} \quad \lambda_{K_\beta} = hc/(E_M - E_K) \tag{13-1}$$

where h is Planck's constant, c is the speed of light, and E refers to the energy of a particular state (K, L, or M). Typical x rays used in structure determination are Cu K_α ($\lambda = 1.54$ Å) and Mo K_α ($\lambda = 0.71$ Å).

Parameters that describe an electromagnetic wave

X rays, like any other photons, are electromagnetic waves. A general expression for the propagation of one such wave in the **k** direction through space and time is

$$\begin{aligned} E(\mathbf{r}, t) &= E_0 e^{2\pi i(\hat{\mathbf{k}} \cdot \mathbf{r}/\lambda - \nu t + \delta')} \\ &= E_0\{\cos[2\pi(\hat{\mathbf{k}} \cdot \mathbf{r}/\lambda - \nu t + \delta')] + i \sin[2\pi(\hat{\mathbf{k}} \cdot \mathbf{r}/\lambda - \nu t + \delta')]\} \end{aligned} \tag{13-2}$$

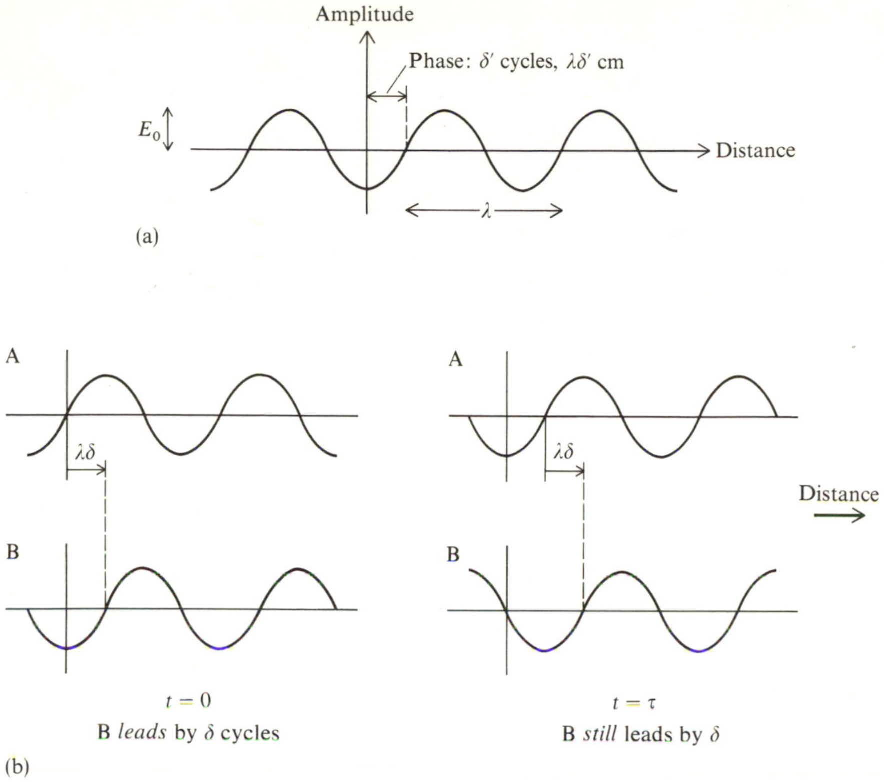

Figure 13-1

Characteristics of electromagnetic waves. **(a)** The electric field amplitude as a function of distance at time zero. **(b)** The relative phase of two waves remains constant with time. The phase difference is δ cycles ($\lambda\delta$ cm) both at $t = 0$ and at t.

where $E(\mathbf{r}, t)$ is the electric field at point $\mathbf{r}$ and time t; $\hat{\mathbf{k}}$ is a unit vector in the $\mathbf{k}$ direction; λ is the wavelength in cm cycle^{-1}; ν is the frequency in cycles sec^{-1} past a fixed point; δ' is the phase of the wave (in cycles) that defines its amplitude at $\mathbf{r} = 0$ and $t = 0$; and E_0 is the maximal amplitude (Fig. 13-1). Such a transverse wave oscillates periodically in both time and space.

It would be equally accurate to describe the wave by a real function, such as $\sin[2\pi(\hat{\mathbf{k}} \cdot \mathbf{r}/\lambda - \nu t + \delta')]$, rather than by a complex function (Box 13-1). However, the measured radiation intensity of a wave depends on the square of the amplitude, and this always will be a real quantity. We choose to describe x rays by complex exponentials because of the great mathematical convenience of working with such functions. For example, $e^{a+b} = e^a e^b$, whereas $\sin(a + b) = \sin a \cos b + \cos a \sin b$.

Two waves propagating in the same direction with the same amplitude, wavelength, and frequency can differ only in phase. We can describe them as

$$E_1(\mathbf{r}, t) = E_0 e^{2\pi i(\hat{\mathbf{k}} \cdot \mathbf{r}/\lambda - \nu t + \delta'_1)}$$

$$E_2(\mathbf{r}, t) = E_0 e^{2\pi i(\hat{\mathbf{k}} \cdot \mathbf{r}/\lambda - \nu t + \delta'_2)} = E_1(\mathbf{r}, t) e^{2\pi i \delta}$$

where $\delta = \delta'_2 - \delta'_1$ is the phase shift. Note that δ is constant for all space and time. If two such phase-shifted waves are combined, the net amplitude is $E_1(\mathbf{r}, t)(1 + e^{2\pi i \delta})$. When δ is zero, this net amplitude is just twice the individual amplitude but, when δ is one-half cycle, the net amplitude is zero because $e^{i\pi}$ is -1. Clearly, in situations where an observable is a superposition of many waves, their relative phases are quite critical.

Geometry of an x-ray scattering experiment

Consider the geometry of the typical x-ray scattering experiment shown in Figure 13-2a. A collimated beam of x rays is allowed to impinge on a sample consisting of a

Box 13-1 RELATIONSHIP BETWEEN SINES, COSINES, AND EXPONENTIALS

It is possible to express periodically varying functions either in terms of sines and cosines or as complex exponentials. The basic relationship between these two representations is

$$e^{ix} = \cos x + i \sin x$$

One easy way to justify this relationship is to expand each of the functions in an infinite series:

$$e^{ix} = 1 + ix - x^2/2! - ix^3/3! + x^4/4! + ix^5/5! - \cdots$$

$$\cos x = 1 - x^2/2! + x^4/4! - x^6/6! + \cdots$$

$$i \sin x = ix - ix^3/3! + ix^5/5! - ix^7/7! + \cdots$$

Because $\cos(-x) = \cos x$, and $\sin(-x) = -\sin x$, it is obvious that

$$e^{-ix} = \cos x - i \sin x$$

Therefore, we can always represent trigonometric functions in terms of complex exponentials as follows:

$$\cos x = (1/2)(e^{ix} + e^{-ix})$$

$$\sin x = (1/2i)(e^{ix} - e^{-ix})$$

single electron located at the origin of the coordinate system. A unit vector, $\hat{\mathbf{s}}_0$, describes the direction of the incoming radiation. Scattering will deflect a certain fraction of the incident x rays, and will lead to radiation propagating away from the sample in all directions. Suppose we could place an x-ray detector at some location in space and measure the amplitude and phase of radiation scattered in that direction. The position of the detector is denoted by another unit vector, $\hat{\mathbf{s}}$. The scattering angle θ is defined as one-half the angle of deflection of $\hat{\mathbf{s}}$ relative to $\hat{\mathbf{s}}_0$ (Fig. 13-2a). We are

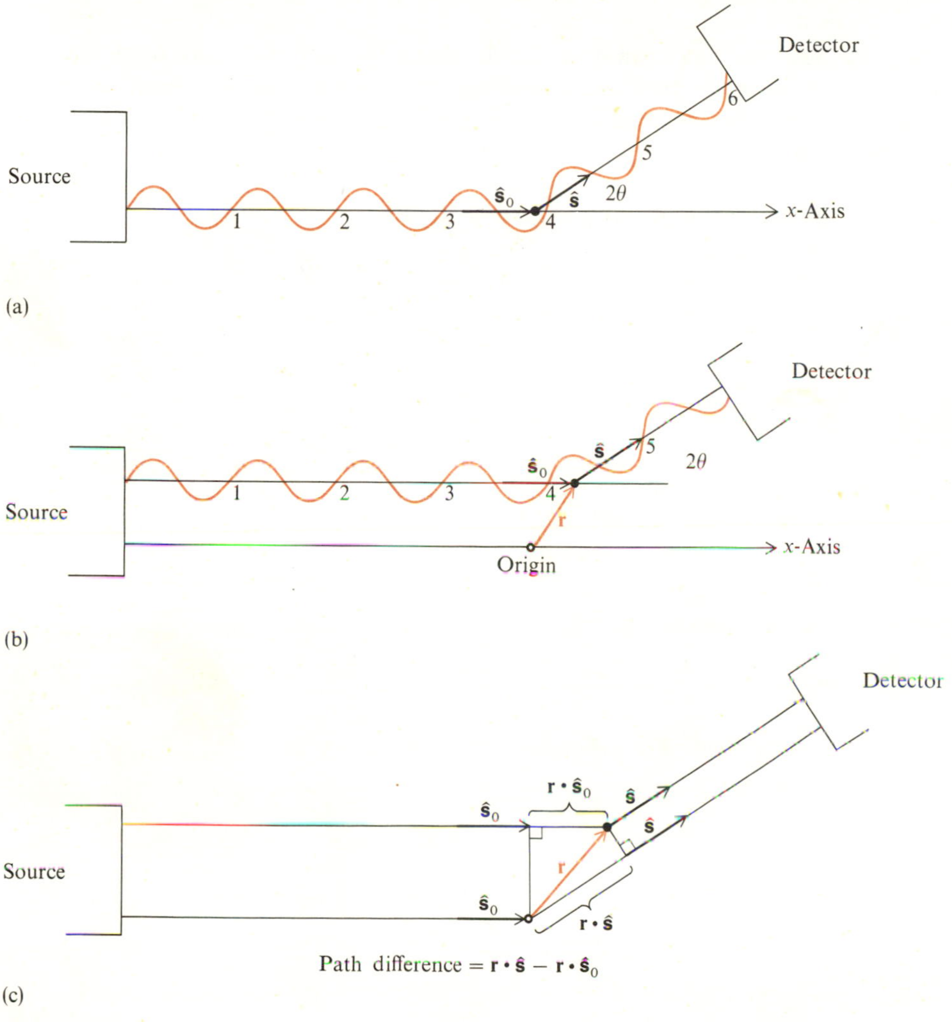

Figure 13-2

X-ray scattering by a single electron. The angle of deflection (2θ) between the source and the detector is the same in all three cases. **(a)** An electron at the origin. **(b)** An electron located at position **r** relative to the origin. **(c)** An expanded view near the origin, showing the path difference between radiation scattered by an electron at **r** and that scattered by an electron at the origin. The numbers shown in parts a and b measure the x-ray path in units of wavelength. The vectors $\hat{\mathbf{s}}_0$ and $\hat{\mathbf{s}}$ are unit vectors describing the direction of incident rays and that of scattered rays seen by the detector, respectively.

concerned here only with the elastic scattering of x rays. This means that the wavelength of the incident and scattered radiation is the same.

The intensity of x-ray scattering will depend on the orientation of the sample relative to the incident and scattered rays. It is convenient mathematically and, as you will see shortly, very convenient conceptually to define a new single variable **S**, called the scattering vector:

$$\mathbf{S} = (\hat{\mathbf{s}}/\lambda) - (\hat{\mathbf{s}}_0/\lambda) \tag{13-3}$$

Figure 13-3a shows the meaning of **S**. The direction of **S** bisects the angle between incident and scattered radiation. The dimensions of **S** are inverse length, so that **S** measures the number of cycles of radiation per cm. The length of **S** is a function of the total scattering angle (Fig. 13-3b).

$$\begin{aligned}\mathbf{S} \cdot \mathbf{S} &= (\hat{\mathbf{s}}^2 + \hat{\mathbf{s}}_0^2 - 2\hat{\mathbf{s}} \cdot \hat{\mathbf{s}}_0)/\lambda^2 \\ &= 2(1 - \cos 2\theta)/\lambda^2 = (4 \sin^2 \theta)/\lambda^2 \end{aligned} \tag{13-4}$$

Therefore the length of **S** is

$$|\mathbf{S}| = 2|\sin \theta|/\lambda \tag{13-5}$$

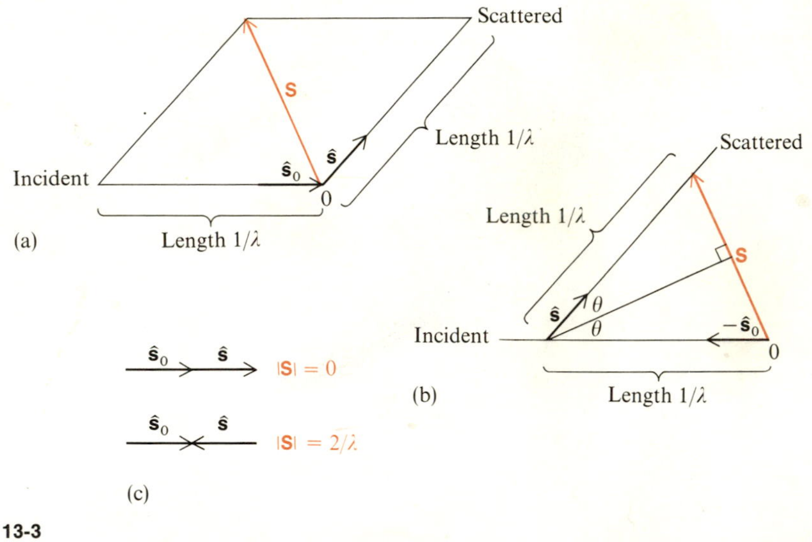

Figure 13-3

Basic geometry of an x-ray scattering experiment. The unit vectors $\hat{\mathbf{s}}_0$ and $\hat{\mathbf{s}}$ are defined in Figure 13-2. **(a)** The scattering vector **S** is defined by Equation 13-3. **(b)** When the unit vector $\hat{\mathbf{s}}$ describing the direction of scattered radiation is translated a distance $1/\lambda$ along the $\hat{\mathbf{s}}_0$ axis, it points directly toward the tip of the scattering vector **S**. **(c)** Arrangements of $\hat{\mathbf{s}}_0$ and $\hat{\mathbf{s}}$ that lead to maximal and minimal values of $|\mathbf{S}|$.

The value of $|\mathbf{S}|$ can vary from 0 to $2/\lambda$ (Fig. 13-3c). Thus, the vector **S** is described in a finite coordinate system in which each axis has the dimensions of a reciprocal distance. This coordinate system is called reciprocal space. Like any other coordinate system, the space containing **S** can be expressed by many different possible axes. We derive later a particularly convenient representation that allows **S** to be related to the axes of a crystalline sample.

Scattering as a function of electron position

The radiation $E(\mathbf{S})$ seen by the detector (Fig. 13-2a) that results from the scattering of a single electron at the origin can be computed by a proper consideration of the quantum mechanics of photons interacting with matter.[§] If we had more than one electron located at the origin, the scattered radiation at any angle should simply increase in amplitude in direct proportion to the number of electrons.

In crystallography, one is interested not so much in the scattering properties of individual electrons as in the effect of relative electron position on the pattern of scattering. Therefore, we can simply ask how the scattering changes as an electron is moved away from the origin. The structure factor, $F(\mathbf{S})$, is defined as the ratio of the radiation scattered by any real sample to that scattered by a single electron at the origin.

Suppose a sample contains a single electron located at position **r**, instead of at the origin (Fig. 13-2b). The source and detector are very far away from the sample, and are large compared to **r**. Therefore, to a very good approximation, the scattering angle, $\theta = (1/2)\cos^{-1}(\hat{\mathbf{s}} \cdot \hat{\mathbf{s}}_0)$, is the same for this sample as it is for a sample with an electron at the origin. The only difference in the two samples is the path length that x rays must travel from source to sample to detector. This path length is simply $(\hat{\mathbf{s}} - \hat{\mathbf{s}}_0) \cdot \mathbf{r}$ (Fig. 13-2c). Such a path length is equal to $(\hat{\mathbf{s}} - \hat{\mathbf{s}}_0) \cdot \mathbf{r}/\lambda = \mathbf{S} \cdot \mathbf{r}$ cycles, for x rays of wavelength λ. Therefore, if the radiation scattered by an electron at the origin is $E(\mathbf{S})$, moving the electron from the origin to a position **r** simply causes a phase shift of $\mathbf{S} \cdot \mathbf{r}$ cycles. The scattered radiation is $E(\mathbf{S})e^{2\pi i \mathbf{S} \cdot \mathbf{r}}$, and the structure factor $F(\mathbf{S})$ is $e^{2\pi i \mathbf{S} \cdot \mathbf{r}}$.

In general, because electrons are not localized, it is better to describe an electron density $\rho(\mathbf{r})$ in a volume element $d\mathbf{r}$, located at **r**; the scattering then is proportional to $\rho(\mathbf{r})\,d\mathbf{r}$. For continuous electron density at position **r**, the structure factor is

$$F(\mathbf{S}) = \rho(\mathbf{r})e^{2\pi i \mathbf{S} \cdot \mathbf{r}}\,d\mathbf{r} \tag{13-6}$$

where $\rho(\mathbf{r})\,d\mathbf{r}$ is the number of electrons in the volume element $d\mathbf{r}$.

A sample with many discrete scattering sites has a structure factor that is simply a sum over many terms corresponding to Equation 13-6. For a continuous electron

[§] The result shown in Figure 13-2a contains one serious oversimplification. In actuality, all scattered radiation experiences a phase shift of one-half cycle relative to the phase of the exciting radiation. We can ignore this.

distribution, the sum is replaced by an integral:

$$F(\mathbf{S}) = \int d\mathbf{r}\rho(\mathbf{r})e^{2\pi i\mathbf{S}\cdot\mathbf{r}} \tag{13-7}$$

The integral is taken over the entire sample. Equation 13-7 is the single fundamental equation that governs all x-ray scattering and diffraction. If the electron density distribution $\rho(\mathbf{r})$ of a sample is known, one can compute the structure factor, and from this one can compute the expected x-ray scattering for all scattering geometries.

X-ray scattering in terms of Fourier transforms

The mathematical form of Equation 13-7 is equivalent to a Fourier transform. This is an integral with very convenient properties (Box 13-2). Note that, outside the sample, $\rho(\mathbf{r})$ is zero. Therefore, the integral in Equation 13-7 can be extended over all space without changing its value. Thus, the physical meaning of Equation 13-7 is that the structure factor is a Fourier transform of the object.

Because $F(\mathbf{S})$ is the Fourier transform of $\rho(\mathbf{r})$, a second Fourier integral must exist that relates these two quantities. This is the inverse Fourier transform:

$$\rho(\mathbf{r}) = (1/V)\int d\mathbf{S}e^{-2\pi i\mathbf{S}\cdot\mathbf{r}}F(\mathbf{S}) \tag{13-8}$$

The integral is taken over all reciprocal space. V is a constant that contains $(2\pi)^3$ and other constants that compensate for the difference in the unit of volume of sample space $\mathbf{r}$ and reciprocal space $\mathbf{S}$. In what follows, we sometimes ignore the constant V.

Equation 13-8 means that, if one had measured or calculated values of $F(\mathbf{S})$ extending over all reciprocal space, one could readily compute the electron density distribution of the object. Thus, Equations 13-7 and 13-8 form a relationship that lets one interconvert structure factors and electron densities freely, providing each is known over all space. It is similar in spirit to the relationship given by the Kronig–Kramers transforms in Chapter 8, which let CD and ORD data be interconverted.

An example of the properties of Fourier transforms

To illustrate the properties of Equations 13-7 and 13-8, we shall derive the latter from the former. Set up the integral $I(\mathbf{r}') = \int F(\mathbf{S})\,d\mathbf{S}e^{-2\pi i\mathbf{S}\cdot\mathbf{r}'}$ in some new coordinate system $\mathbf{r}'$. Substitute for $F(\mathbf{S})$ from Equation 13-7:

$$I(\mathbf{r}') = \int d\mathbf{S} e^{-2\pi i \mathbf{S}\cdot\mathbf{r}'} \int d\mathbf{r}\rho(\mathbf{r}) e^{2\pi i \mathbf{S}\cdot\mathbf{r}} \tag{13-9}$$

We can exchange the order of the two integrals to write

$$I(\mathbf{r}') = \int d\mathbf{r}\rho(\mathbf{r}) \int d\mathbf{S} e^{-2\pi i \mathbf{S}\cdot\mathbf{r}'} e^{2\pi i \mathbf{S}\cdot\mathbf{r}} = \int d\mathbf{r}\rho(\mathbf{r}) \int d\mathbf{S} e^{2\pi i \mathbf{S}\cdot(\mathbf{r}-\mathbf{r}')} \tag{13-10}$$

The integral over $d\mathbf{S}$ in the right-hand expression of Equation 13-10 has a very unusual property. As shown in Box 13-3, it is the Dirac delta function:

$$\delta(\mathbf{r}-\mathbf{r}') = \int d\mathbf{S} e^{2\pi i \mathbf{S}\cdot(\mathbf{r}-\mathbf{r}')} \tag{13-11}$$

This function has the following characteristics. If $\mathbf{r} \neq \mathbf{r}'$, then $\delta(\mathbf{r}-\mathbf{r}') = 0$. If $\mathbf{r} = \mathbf{r}'$, then $\delta(\mathbf{r}-\mathbf{r}') = \infty$. However, $\int d\mathbf{r}\,\delta(\mathbf{r}-\mathbf{r}') = 1$ and [for some arbitrary function $g(\mathbf{r})$] $\int d\mathbf{r} g(\mathbf{r})\delta(\mathbf{r}-\mathbf{r}') = g(\mathbf{r}')$ if the integrals include the point $\mathbf{r} = \mathbf{r}'$. Thus, $\delta(\mathbf{r}-\mathbf{r}')$ will simply sample a function at $\mathbf{r} = \mathbf{r}'$. Equation 13-10 becomes $I(\mathbf{r}') = \rho(\mathbf{r}')$. Recognizing that $\mathbf{r}$ and $\mathbf{r}'$ are equivalent variables, this result is identical to Equation 13-8 except for the constant V.

Measuring the structure factor

Unfortunately for x-ray scattering studies, no way is known to measure $F(\mathbf{S})$ directly. F is a complex number that can be written as the product of two terms,

$$\boxed{F = |F|e^{i\phi}} \tag{13-12}$$

or as the sum of real and imaginary parts,

$$F = F_r + iF_i \tag{13-13}$$

The term $|F|$ is called the amplitude of the structure factor, and $e^{i\phi}$ is the phase. Figure 13-4 shows the relationship between the two representations of $F(\mathbf{S})$.

$$F_r = |F| \cos\phi; \qquad F_i = |F| \sin\phi \tag{13-14}$$

$$|F| = (F_r^2 + F_i^2)^{1/2}; \qquad \phi = \tan^{-1}(F_i/F_r) \tag{13-15}$$

Box 13-2 PROPERTIES OF FOURIER TRANSFORMS

Representing a Function by a Fourier Series

Consider a completely arbitrary function $f(\theta)$, defined in the interval $\theta = -\pi$ to $\theta = \pi$. It is possible to represent this function as an expansion in a series of functions with known properties. Only certain sets of functions are suitable for such an expansion and, in the interval $-\pi$ to π, sines and cosines together constitute such a set:

$$f(\theta) = \sum_{n=0}^{\infty} a_n \cos(n\theta) + a'_n \sin(n\theta)$$

where the index n runs through all positive integers. This expansion is called a Fourier series. The coefficients a_n and a'_n are numbers determined by the properties of $f(\theta)$.

As shown in Box 13-1, sines and cosines can be expressed in terms of complex exponentials. Therefore, the Fourier series just given can instead be written as

$$f(\theta) = \sum_{n=-\infty}^{\infty} b_n e^{in\theta}$$

where the index n now runs through both positive and negative values because these are necessary to describe sines and cosines. The coefficients b_n can be found in a simple way by making use of the following result.

For any two integers n and m,

$$\begin{aligned}\int_{-\pi}^{\pi} e^{in\theta} e^{-im\theta}\, d\theta = \int_{-\pi}^{\pi} e^{i(n-m)\theta} &= [1/i(n-m)](e^{i(n-m)\pi} - e^{-i(n-m)\pi}) \\ &= [2/(n-m)] \sin(n-m)\pi = 0 \quad \text{if } n \neq m \\ &= 2\pi \quad \text{if } n = m\end{aligned}$$

where the result for $n = m$ can be proven by expanding the sine expression in a power series. Therefore, to find a particular b_m, one performs the integral

$$(1/2\pi) \int_{-\pi}^{\pi} f(\theta) e^{-im\theta}\, d\theta = (1/2\pi) \int_{-\pi}^{\pi} d\theta \sum_{n=-\infty}^{\infty} b_n e^{in\theta} e^{-im\theta} = b_m$$

Note that the integral is carried out over the entire range of θ over which $f(\theta)$ is defined. It often is convenient to be able to work with an arbitary range $-L/2$ to $L/2$ rather than with $-\pi$ to π. This is accomplished by defining a new variable, $x = L\theta/2\pi$, such that when $\theta = \pi$, then $x = L/2$, and when $\theta = -\pi$, then $x = -L/2$. Incorporating this variable into the above equations, and using the fact that $dx = (L/2\pi)d\theta$, we obtain

$$f(x) = \sum_{n=-\infty}^{\infty} b_n e^{2\pi inx/L}$$

$$b_n = (1/L) \int_{-L/2}^{L/2} e^{-2\pi inx/L} f(x)\, dx$$

Fourier Transforms in One Dimension

The function $f(x)$ is defined at all x, whereas the set of coefficients b_n represents an infinite array of numbers, which must be tabulated. Therefore, it is convenient to find an analog of the Fourier series in which the coefficients b_n are replaced by a function, and the summation is replaced by an integral. This representation is called a Fourier transform when the interval over which the function is defined extends from $-\infty$ to $+\infty$.

We define a new continuous variable, $S = 2\pi n/L$, and a new continuous function $g(S) = Lb_n$. Using these, the equation for b_n is transformed to

$$g(S) = \int_{-\infty}^{\infty} e^{-2\pi iSx} f(x)\,dx \tag{A}$$

in the limit as $L \to \infty$. The series expansion for $f(x)$ becomes

$$f(x) = \sum_{n=-\infty}^{\infty} [g(S)/L]e^{2\pi iSx}$$

To replace the sum by an integral, note that the interval ΔS corresponds to $(2\pi/L)\,\Delta n$ from the definition of S. But $\Delta n = 1$ in the summation, and therefore each increment dS in an integral is equivalent to $2\pi/L$ in the sum. Thus,

$$f(x) = (L/2\pi) \int_{-\infty}^{\infty} [g(S)/L]e^{2\pi iSx}\,dS = (1/2\pi) \int_{-\infty}^{\infty} g(S)e^{2\pi iSx}\,dS \tag{B}$$

Equations A and B constitute a pair of Fourier transforms that allow $f(x)$ to be calculated if $g(S)$ is known, and vice versa. They are particularly interesting because the variables x and S have opposite dimensions. For example, if x is distance, then S is reciprocal distance. The factor of $(1/2\pi)$ in equation B often is written instead as $(1/\sqrt{2\pi})$ in front of the integrals in both equations A and B.

Fourier Transforms in Three Dimensions

Suppose the function f is now defined in a Cartesian coordinate system with axes x, y, z. For fixed y and z, the function $f(x, y, z)$ can be expanded in a Fourier series in $e^{2\pi iS_x x}$, and the Fourier transform becomes (by analogy to Equation A)

$$g_{yz}(S_x) = \int_{-\infty}^{\infty} e^{-2\pi iS_x x} f(x, y, z)\,dx$$

This expression, in turn, can be expanded in the function $e^{2\pi iS_y y}$ for fixed z, and finally as a function of $e^{2\pi iS_z z}$. The resulting three-dimensional Fourier transform is

$$g(S_x, S_y, S_z) = \int_{-\infty}^{\infty} dz e^{-2\pi iS_z z} \int_{-\infty}^{\infty} dy e^{-2\pi iS_y y} \int_{-\infty}^{\infty} dx e^{-2\pi iS_x x} f(x, y, z)$$

If we use the vector $\mathbf{S}$ to represent the three variables S_x, S_y, and S_z, and we use $\mathbf{r}$ to represent x, y, and z, then the three-dimensional transform can be written very compactly as

$$g(\mathbf{S}) = \int_{-\infty}^{\infty} d\mathbf{r}\, e^{-2\pi i\mathbf{S}\cdot\mathbf{r}} f(\mathbf{r})$$

Similarly, the analog of Equation B becomes

$$f(\mathbf{r}) = (1/2\pi)^3 \int_{-\infty}^{\infty} d\mathbf{S}\, e^{2\pi i\mathbf{S}\cdot\mathbf{r}} g(\mathbf{S})$$

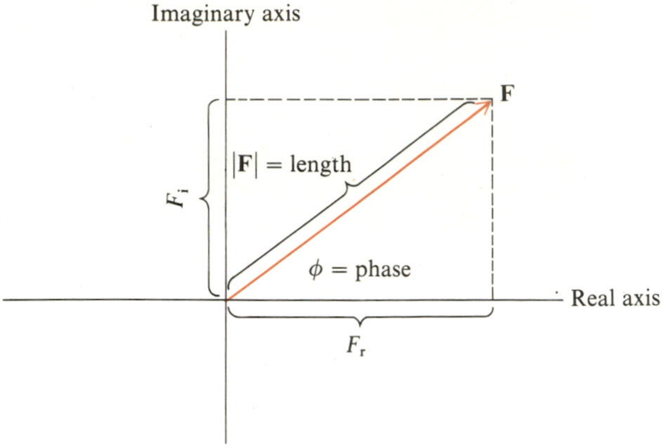

Figure 13-4

The structure factor plotted as a vector in the complex plane.

Box 13-3 THE DIRAC DELTA FUNCTION

We wish to demonstrate that the following integral is a representation of the one-dimensional Dirac delta function:

$$\delta(x - x') = \int_{-\infty}^{\infty} e^{2\pi i(x-x')S}\, dS$$

The results can easily be generalized to three dimensions. If this is the delta function, it must obey three properties.

First, if $x' = x$, then $\delta(x - x') = \infty$. It is obvious that, with $x = x'$, the exponential in the above integral is just unity; therefore, the integral is infinite.

Second, if $x' \neq x$, then $\delta(x - x') = 0$. It is not so obvious that the integral meets this requirement. The way to realize that it does is to note that the complex exponential is a periodic function that continually oscillates from -1 to 1 throughout all space. For each positive lobe there exists an adjacent (absolutely equivalent) negative lobe. The areas underneath these lobes cancel identically.

Third, if x' lies between a and b, then

$$\int_b^a dx\, \delta(x - x') = 1$$

Let $a = x' + \varepsilon$, and $b = x' - \varepsilon$. Then the area under the delta function is

$$\int_{x'-\varepsilon}^{x'+\varepsilon} dx \int_{-\infty}^{\infty} e^{2\pi i(x-x')S}\, dS = \int_{-\infty}^{\infty} dS \int_{x'-\varepsilon}^{x'+\varepsilon} dx\, e^{2\pi i(x-x')S}$$

$$= \int_{-\infty}^{\infty} e^{-2\pi i x' S}\, dS \int_{x'-\varepsilon}^{x'+\varepsilon} e^{2\pi i x S}\, dx$$

$$= \int_{-\infty}^{\infty} e^{-2\pi i x' S}[(1/2\pi i S)(e^{2\pi i(x'+\varepsilon)S} - e^{2\pi i(x'-\varepsilon)S})]\, dS$$

Experimentally, all one can observe is the intensity of radiation scattered at an angle 2θ. If we express this intensity relative to the intensity scattered by a single electron at the origin, it is

$$I(\mathbf{S}) = F(\mathbf{S})F^*(\mathbf{S}) = |F|^2 \qquad (13\text{-}16)$$

We must multiply by the complex conjugate, rather than simply by $F(\mathbf{S})$, because F is a complex number. The intensity is a pure observable and must be real. It is given by the square of the amplitude of the structure factor. Thus $|F|$ can be measured experimentally. The phase term ($e^{i\phi}$) of $F(\mathbf{S})$ is not directly measurable; this is the major obstacle in x-ray scattering and diffraction studies. In order to use Equation 13-8 to calculate $\rho(\mathbf{r})$, one first must guess, calculate, or indirectly estimate $e^{i\phi}$.

$$= \int_{-\infty}^{\infty} e^{-2\pi i x' S}[(e^{2\pi i x' S}/2\pi i S)2i \sin 2\pi\varepsilon S]\, dS$$

$$= (1/\pi) \int_{-\infty}^{\infty} [(\sin 2\pi\varepsilon S)/S]\, dS = 1$$

because

$$\int_0^{\infty} [(\sin x)/x]\, dx = \int_{-\infty}^{0} [(\sin x)/x]\, dx = \pi/2$$

If x' is not between a and b, then the integral $\int_a^b dx\, \delta(x - x')$ is zero, because the function is everywhere zero. Thus we see that the integral originally given meets all the requirements, and is in fact the Dirac delta function.

A most important property of the delta function is the ability to shift the location of another function:

$$\int_{-\infty}^{\infty} dx\, f(x)\, \delta(x - x') = f(x')$$

We can demonstrate this by choosing a narrow interval $x' - \varepsilon$ to $x' + \varepsilon$ near x' and breaking up the integral into three parts:

$$\int_{-\infty}^{x'-\varepsilon} dx\, f(x)\, \delta(x - x') + \int_{x'-\varepsilon}^{x'+\varepsilon} dx\, f(x)\, \delta(x - x') + \int_{x'+\varepsilon}^{\infty} dx\, f(x)\, \delta(x - x')$$

The first and third integrals are zero for any finite-valued function $f(x)$, because everywhere within them $\delta(x - x') = 0$. The second integral can be evaluated if we choose ε small enough so that $f(x) = f(x')$; then it becomes

$$f(x') \int_{x'-\varepsilon}^{x'+\varepsilon} dx\, \delta(x - x') = f(x')$$

The electron density in Equations 13-7 and 13-8 is, in principle, measurable directly, and therefore it must be real. As probed by x-ray scattering experiments, $\rho(\mathbf{r})$ behaves as a real quantity so long as there is no anomalous scattering (vide infra). The reality of $\rho(\mathbf{r})$ allows a constraint on $F(\mathbf{S})$ to be developed. Because $\rho(\mathbf{r})$ is real, $\rho(\mathbf{r}) = \rho^*(\mathbf{r})$. Substituting Equation 13-13 into Equation 13-8, and taking the complex conjugate, we obtain

$$\int [F_r(\mathbf{S}) + iF_i(\mathbf{S})]e^{-2\pi i\mathbf{S}\cdot\mathbf{r}}\,d\mathbf{S} = \int [F_r(\mathbf{S}) - iF_i(\mathbf{S})]e^{+2\pi i\mathbf{S}\cdot\mathbf{r}}\,d\mathbf{S} \qquad (13\text{-}17a)$$

Note that $\mathbf{r}$ can take on any value. For Equation 13-17a to hold for any arbitrary value of $\mathbf{r}$, it is necessary (for every value of $\mathbf{S}$) that

$$F_r(\mathbf{S}) = F_r(-\mathbf{S}) \qquad \text{and} \qquad F_i(\mathbf{S}) = -F_i(-\mathbf{S}) \qquad (13\text{-}17b)$$

In other words, the real part of the scattering function must be symmetrical about the origin of $\mathbf{S}$ space, whereas the imaginary part is antisymmetric. When such a relationship holds, the function $F(\mathbf{S})$ is called a conjugate function.

When the results of Equation 13-17b are substituted into the definition of the intensity, an interesting result emerges:

$$I(\mathbf{S}) = |F(\mathbf{S})|^2 = F_r^2 + F_i^2 = |F(-\mathbf{S})|^2 = I(-\mathbf{S}) \qquad (13\text{-}18)$$

Equation 13-18 reveals that the observed pattern of scattered intensity is symmetric about the origin of reciprocal space at $\mathbf{S} = 0$. The result, that $I(\mathbf{S})$ has a center of symmetry, is called Friedel's law. It means that one has to measure only half the scattering to obtain all the information it contains.

A requirement for heterogeneities in electron density

Suppose that an experimental sample consists of a uniform distribution of electron density, $\rho(\mathbf{r}) = \rho$. Then the expected structure factor is

$$F(\mathbf{S}) = \rho \int d\mathbf{r} e^{2\pi i\mathbf{S}\cdot\mathbf{r}} \qquad (13\text{-}19)$$

But this is just the Dirac delta function $\delta(\mathbf{S} - 0)$. The only x rays that emerge from the sample are $F(0)$. From Figure 13-3, $\mathbf{S} = 0$ corresponds to scattered radiation parallel to the incident beam. In other words, a uniform sample cannot deflect x rays at all, just as a medium of constant refractive index cannot bend or focus collimated light.

A fundamental principle of scattering is the requirement for spatial (or temporal) heterogeneities. Scattering is caused by the contrast between a given region and its neighbors. We now must calculate the scattering that results from the presence of discrete atoms, and then that resulting from arrangements of atoms found in molecules or crystal lattices.

Scattering from a single atom at the origin

Suppose the sample consists solely of a single atom located at the origin. The detailed pattern of electron density around an individual atom depends on the bonding it is involved in. However, in almost all x-ray diffraction experiments, the resolution is not high enough to detect this detailed pattern. Thus, it is a good approximation to model the electron distribution of an atom as spherically symmetric. Then $\rho(\mathbf{r})$ becomes $\rho(r)$. If we express Equation 13-7 in spherical polar coordinates (Fig. 13-5a),

$$F(\mathbf{S}) = \int_0^{2\pi} d\phi \int_0^{\pi} \sin\theta \, d\theta \int_0^{\infty} dr \rho(r) r^2 e^{2\pi i \mathbf{S}\cdot\mathbf{r}}$$

$$= 2\pi \int_0^{\infty} dr \rho(r) r^2 \int_0^{\pi} d\theta \sin\theta \, e^{2\pi i S r \cos\theta} \qquad (13\text{-}20)$$

where S and r are the lengths of the vectors $\mathbf{S}$ and $\mathbf{r}$, respectively. By making the substitution $x = \cos\theta$, we can easily evaluate the θ integral, obtaining

$$\boxed{F(\mathbf{S}) = 4\pi \int_0^{\infty} dr \rho(r) r^2 [(\sin 2\pi S r)/2\pi S r] \equiv f(S)} \qquad (13\text{-}21)$$

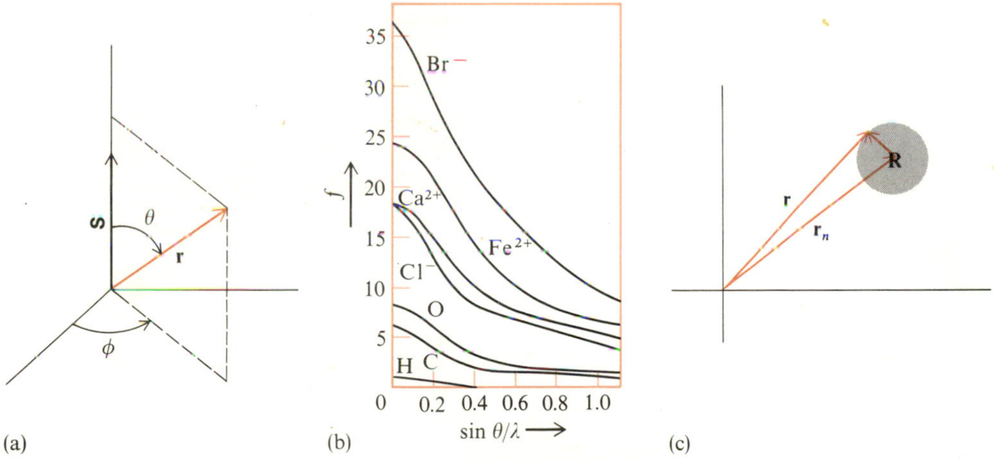

Figure 13-5

X-ray scattering from atoms. **(a)** Coordinate system used to evaluate Equation 13-20. **(b)** Atomic scattering factor for various atoms as a function of the scattering angle (2θ). [After J. P. Glusker and K. N. Trueblood, *Crystal Structure Analysis: A Primer* (London: Oxford Univ. Press, 1972).] **(c)** Coordinate system used to describe an atom not at the origin.

The function $f(S)$ is defined as the atomic scattering factor. It depends only on $|\mathbf{S}|$ and thus, from Equation 13-5, it depends only on the angle between $\hat{\mathbf{s}}_0$ and $\hat{\mathbf{s}}$, and not on the orientation of the sample. Note that, because $\rho(r) = \rho(-r)$, the function $f(S)$ is real. Therefore, the intensity measured in a scattering experiment on a single atom, $I(S)$, can be used to compute $f(S)$ directly by Equation 13-16:

$$f(S) = \pm[I(S)]^{1/2} \tag{13-22}$$

The only ambiguity is the choice of sign; we can arbitrarily define this sign to be positive.

For real atoms, $\rho(r)$ can be crudely approximated by a Gaussian distribution of electron density: $\rho(r) = zNe^{-kr^2}$, where z is the number of electrons, N is a normalization constant, and k is related to the width of the Gaussian. Then Equation 13-21 can be integrated to yield.

$$f(S) = ze^{-(\pi^2/k)S^2} \tag{13-23}$$

This relationship shows that the atomic scattering factor has the same sign everywhere in space. The atomic scattering factor for forward-scattered radiation ($S = 0$) is simply the number of electrons. Equation 13-23 shows that spherical atoms scatter x rays most efficiently in the forward directions. The scattering factor drops fairly rapidly with increasing scattering angle (Fig. 13-5b).

Scattering from atoms not located at the origin

Next, suppose the sample is still a single spherical atom, but it now is centered at the position $\mathbf{r}_n$. As before, $\mathbf{r}$ is a vector from the coordinate system origin to a point within the electron density distribution of the atom. $\mathbf{R}$ is a vector from the atom center to the point $\mathbf{r}$. It is defined by $\mathbf{r} = \mathbf{R} + \mathbf{r}_n$ (Fig. 13-5c). Thus, from Equation 13-7, the x-ray scattering is

$$F(\mathbf{S}) = \int d(\mathbf{R} + \mathbf{r}_n)\rho(\mathbf{R} + \mathbf{r}_n)e^{2\pi i\mathbf{S}\cdot(\mathbf{R}+\mathbf{r}_n)} \tag{13-24}$$

Because $\mathbf{r}_n$ is constant, $d(\mathbf{R} + \mathbf{r}_n) = d\mathbf{R}$, and the term $e^{2\pi i\mathbf{r}_n\cdot\mathbf{S}}$ can be removed from the integral:

$$F(\mathbf{S}) = e^{2\pi i\mathbf{S}\cdot\mathbf{r}_n}\int d\mathbf{R}\rho(\mathbf{R} + \mathbf{r}_n)e^{2\pi i\mathbf{S}\cdot\mathbf{R}} \tag{13-25}$$

The integral in Equation 13-25 is taken over all space. Because ρ is the electron density distribution around the atom, the constant vector $\mathbf{r}_n$ specifying the original coordinate system is irrelevant. Thus, this integral is identical to Equation 13-20. It is just the atomic scattering factor, and so Equation 13-25 becomes

$$F(\mathbf{S}) = f(S)e^{2\pi i\mathbf{S}\cdot\mathbf{r}_n} \qquad (13\text{-}26)$$

For a set of N atoms, each located at position $\mathbf{r}_n$ with atomic scattering factor f_n, the total structure factor expected is

$$F(\mathbf{S}) = \sum_{n=1}^{N} f_n(S)e^{2\pi i\mathbf{S}\cdot\mathbf{r}_n} \qquad (13\text{-}27)$$

where f_n is the scattering factor of the nth atom. If the N atoms happen to belong to a single molecule, then Equation 13-27 is called the molecular structure factor, $F_{\mathrm{m}}(S)$.

Consider the case of a sample with a center of symmetry. If that center is placed at the origin, then for each atom at $\mathbf{r}_n$ contributing $f_n(S)e^{2\pi i\mathbf{S}\cdot\mathbf{r}_n}$ in Equation 13-27, there must be an equivalent atom at $-\mathbf{r}_n$ contributing $f_n(S)e^{-2\pi i\mathbf{S}\cdot\mathbf{r}_n}$. Because $e^{\pm ix} = \cos x \pm i \sin x$ (Box 13-1), the structure factor of the sample can be written as the centrosymmetric function:

$$F_{\mathrm{cs}}(\mathbf{S}) = \sum_{n=1}^{N/2} 2f_n(S)\cos(2\pi\mathbf{S}\cdot\mathbf{r}_n) \qquad (13\text{-}28)$$

which is a sum over $N/2$ symmetry-related pairs of atoms. This is a real function, and therefore the problem of determining the phase of $F_{\mathrm{cs}}(\mathbf{S})$ is dramatically simplified. From Figure 13-4, note that ϕ must be either 0 or π. Thus the term $e^{i\phi}$ must be simply $+1$ or -1 at each point $\mathbf{S}$, corresponding respectively to $\phi = 0$ or $\phi = \pi$.

13-2 X-RAY DIFFRACTION

Interference fringes from sets of atoms

In general, moving an atom to position $\mathbf{r}_n$ (away from the origin) introduces a phase shift, $e^{2\pi i\mathbf{S}\cdot\mathbf{r}_n}$, in the x-ray scattering. Note that, for a single atom, this will lead to no observable change in scattering because the intensity is still $I(S) = f^2(S)$. Suppose, however, a sample contains one atom at the origin and an identical atom at position $\mathbf{r}_n$. The total structure factor will be

$$F(\mathbf{S}) = f(S)(1 + e^{2\pi i\mathbf{S}\cdot\mathbf{r}_n}) \qquad (13\text{-}29a)$$

The scattering intensity will be

$$I(\mathbf{S}) = f^2(S)(1 + e^{2\pi i\mathbf{S}\cdot\mathbf{r}_n})(1 + e^{-2\pi i\mathbf{S}\cdot\mathbf{r}_n}) = 2f^2(S)[1 + \cos(2\pi\mathbf{S}\cdot\mathbf{r}_n)] \quad (13\text{-}29b)$$

Thus—in addition to the scattering seen from each of the atoms separately, $f^2(S)$—there is an interference pattern generated by the cos $2\pi\mathbf{S}\cdot\mathbf{r}_n$ term (see Fig. 13-6b). This is exactly comparable to the interference fringes seen in a two-slit experiment in optical diffraction (see Box 13-4). The term $e^{2\pi i\mathbf{S}\cdot\mathbf{r}_n}$ in Equation 13-29 often is called a fringe function.

If it were possible to measure the scattering from a sample containing just a few atoms, the pattern of fringes would yield information on the spatial arrangement. Such measurements are impossible because the intensity of radiation scattered from just a few atoms is too small. The number of terms in the structure factor increases as the total number of atoms (N_T) increases, and therefore the observed intensities increase as N_T^2. To increase the number of atoms without loss of information, it is necessary to work with periodic arrays of atoms, such as atoms in crystals. Here we shall demonstrate the pattern of fringes introduced by such arrays.

Calculation of x-ray diffraction from a one-dimensional array

Start with a one-dimensional row of $2N + 1$ identical atoms. Locate the central atom at the origin of the coordinate system. As shown in Figure 13-6a, the position of each atom in the array is generated from that of its neighbor by translation along a vector **a**. The position of the nth atom in the array is $n\mathbf{a}$. The structure factor resulting from this atom is given by Equation 13-26:

$$F_n(\mathbf{S}) = e^{2\pi in\mathbf{S}\cdot\mathbf{a}}f(S) \quad (13\text{-}30)$$

Thus the scattering from any of the atoms can be written in terms of the atomic scattering factor $f(S)$ for an atom at the origin. The structure factor for the whole array can be written as

$$F_{\text{Tot}}(\mathbf{S}) = f(S) \sum_{n=-N}^{N} e^{2\pi in\mathbf{S}\cdot\mathbf{a}} \quad (13\text{-}31)$$

The sum is the fringe function resulting from the array.

We could treat a linear array of molecules in a similar way. If this array is generated by translation, the resulting scattering will be given by an equation identical to Equation 13-31, except that the molecular scattering factor $F_m(\mathbf{S})$ will replace the atomic scattering factor $f(\mathbf{S})$. However, with molecules, more complex arrays

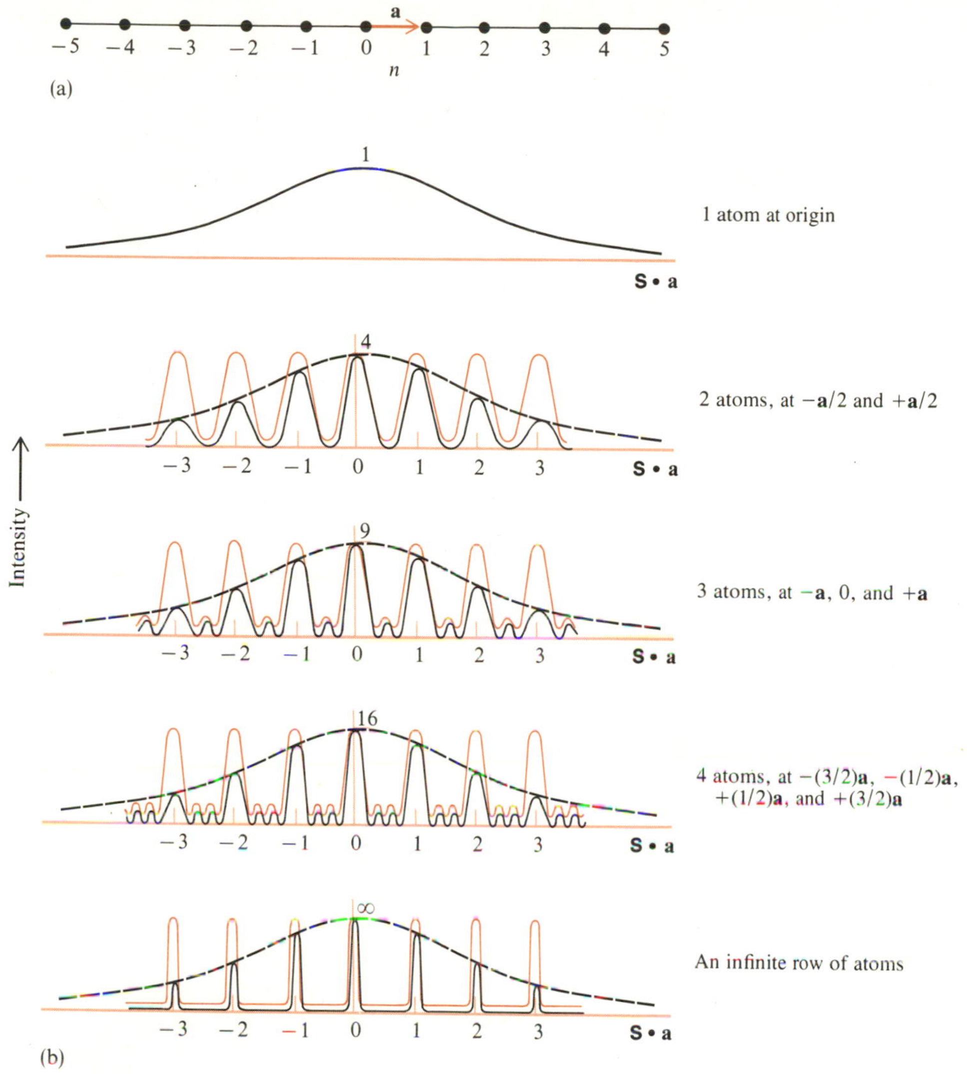

Figure 13-6

X-ray scattering from a one-dimensional array of atoms. **(a)** The array, as defined by the vector translation **a**. **(b)** X-ray scattering intensity as a function of the number of atoms in the array. Shown are the actual observed scattering (*black line*), the scattering expected for a single atom (*dashed line*), and the fringe function produced by the array (*colored line*). The observed scattering is the product of the fringe function and the single-atom scattering. Note the changes in vertical scale as the number of atoms increases. The horizontal scale is in units of **S** · **a** and is the same for all cases.

can be generated if there are rotations as well as translations relating adjacent molecules. These cases can be handled by simple extension of the methods used here; some examples are given in Chapter 14, where we discuss scattering from helices.

Equation 13-31 is simply a geometric series with an initial term $e^{-2\pi i N\mathbf{S}\cdot\mathbf{a}}$, a constant ratio $e^{2\pi i\mathbf{S}\cdot\mathbf{a}}$, and a final term $e^{2\pi i N\mathbf{S}\cdot\mathbf{a}}$. The sum of a geometric series is $t(1 - r^m)/(1 - r)$, where r is the ratio between terms, m is the number of terms, and t is the first term. Using this expression, Equation 13-31 becomes

$$F_{\text{Tot}}(\mathbf{S}) = f(S)\frac{e^{-2\pi i N\mathbf{S}\cdot\mathbf{a}}(1 - e^{2\pi i(2N+1)\mathbf{S}\cdot\mathbf{a}})}{1 - e^{2\pi i\mathbf{S}\cdot\mathbf{a}}} \tag{13-32}$$

Equation 13-32 can be simplified by multiplying both numerator and denominator by $e^{-\pi i\mathbf{S}\cdot\mathbf{a}}$. Then

$$F_{\text{Tot}}(\mathbf{S}) = f(S)\frac{e^{-\pi i(2N+1)\mathbf{S}\cdot\mathbf{a}} - e^{\pi i(2N+1)\mathbf{S}\cdot\mathbf{a}}}{e^{-\pi \mathbf{S}\cdot\mathbf{a}} - e^{\pi i\mathbf{S}\cdot\mathbf{a}}}$$

$$= f(S)\frac{\sin[(2N+1)\pi\mathbf{S}\cdot\mathbf{a}]}{\sin(\pi\mathbf{S}\cdot\mathbf{a})} \tag{13-33}$$

We have used the fact that $e^{\pm ix} = \cos x \pm i \sin x$ to reach the final form of Equation 13-33.

The intensity of scattering from the array will be

$$I_{\text{Tot}}(\mathbf{S}) = |F_{\text{Tot}}(\mathbf{S})|^2 = [f(S)]^2\left(\frac{\sin[(2N+1)\pi\mathbf{S}\cdot\mathbf{a}]}{\sin(\pi\mathbf{S}\cdot\mathbf{a})}\right)^2 \tag{13-34}$$

This is shown schematically as a function of the size of the array in Figure 13-6b. You can see that, as N becomes large, the intensity tends to zero everywhere except where $\mathbf{S}\cdot\mathbf{a}$ is integral. The term $\mathbf{S}\cdot\mathbf{a}$ that appears in Equation 13-34 measures the relative orientation of the sample and the detector. Note that the scattering intensity is a maximum when $\mathbf{S}\cdot\mathbf{a} = 0$. This occurs when $\mathbf{S}$ is in a plane perpendicular to the long axis of the array.

Discontinuous diffraction pattern from a one-dimensional array

It is helpful to demonstrate explicitly the behavior of Equation 13-33 as N becomes large. For most values of $\mathbf{S}\cdot\mathbf{a}$, the value of $\sin(\pi\mathbf{S}\cdot\mathbf{a})$ lies between 0.1 and 1.0 or between -0.1 and -1.0. Around these values of $\mathbf{S}\cdot\mathbf{a}$, the value of $\sin[(2N+1)\pi\mathbf{S}\cdot\mathbf{a}]$ oscillates wildly between 0 and 1. Therefore, the quotient in Equation 13-33 falls in the range of about -10 to 10, regardless of the value of N. However, what happens when $\sin(\pi\mathbf{S}\cdot\mathbf{a})$ approaches zero? It is easiest to examine this behavior in the limit where $\mathbf{S}\cdot\mathbf{a} \to 0$. If we use the series expansion for $\sin x = x - x^3/3! + \cdots$, and if we keep only the first term as $x \to 0$, the quotient in Equation 13-33 becomes $(2N + 1)(\pi\mathbf{S} \cdot \mathbf{a})/(\pi\mathbf{S} \cdot \mathbf{a}) = 2N + 1$.

In a crystalline array of molecules, N can be 10^6 or more. Therefore, the structure factor becomes enormous each time $\sin(\pi\mathbf{S} \cdot \mathbf{a})$ goes to zero. This occurs every time $\mathbf{S} \cdot \mathbf{a}$ approaches an integer. Compared with the sharp peak in scattering for integral $\mathbf{S} \cdot \mathbf{a}$, all other values are negligible. Therefore, the fringe function of a linear array leads to a discontinuous scattering pattern (Box 13-4 illustrates the analogous effect in optical diffraction). Only for certain orientations of sample and x-ray detector will any scattering be observed at all. This result is called a von Laue condition:

$$\mathbf{S} \cdot \mathbf{a} = h, \qquad \text{where } h = 0, \pm 1, \pm 2, \ldots \tag{13-35}$$

The vector $\mathbf{a}$ is a property of the particular one-dimensional crystalline sample and its orientation in space. The vector $\mathbf{S}$ depends on the geometry of the scattering experiment. The observed scattering depends only on $\mathbf{S} \cdot \mathbf{a}$ and is intense only when Equation 13-35 is satisfied. Figure 13-7 shows the geometrical significance of this. $\mathbf{S} \cdot \mathbf{a}$ is the projection of $\mathbf{S}$ onto $\mathbf{a}$. Suppose $\mathbf{a}$ is fixed. Then $\mathbf{S} \cdot \mathbf{a} = 0$ means that $\mathbf{S}$ can be any vector in a plane perpendicular to $\mathbf{a}$ and passing through the origin (Fig. 13-7a). $\mathbf{S} \cdot \mathbf{a} = 1$ means that $\mathbf{S}$ can be any vector from the origin to a plane perpendicular to $\mathbf{a}$ and spaced a distance $1/a$ away from the origin (Fig. 13-7b). For example, if $\mathbf{S}$ is parallel to $\mathbf{a}$, then $\mathbf{S} \cdot \mathbf{a} = 1$ implies that $|\mathbf{S}| = 1/|\mathbf{a}|$.

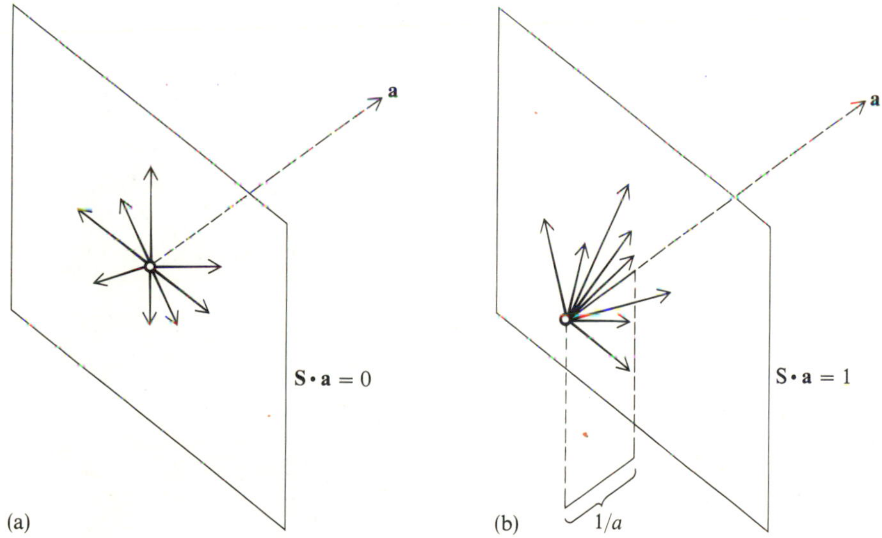

Figure 13-7

The von Laue scattering conditions for a one-dimensional array. Scattering vectors are shown as solid arrows. **(a)** For $\mathbf{S} \cdot \mathbf{a} = 0$. **(b)** For $\mathbf{S} \cdot \mathbf{a} = 1$.

Box 13-4 OPTICAL DIFFRACTION PATTERNS FROM ARRAYS

The same mathematical formalism developed in the text to calculate x-ray diffraction from molecular arrays also applies to optical diffraction from arrays of slits or pinholes. The figures show the optical diffraction pattern from a series of opaque masks containing increasingly more elaborate arrays of pinholes. Such diffraction patterns can be created by the apparatus shown in Figure 10-4a by using the mask as a sample. The figures on the left show the sample masks used; the corresponding figures on the right indicate the diffraction patterns produced by the masks.

(a)

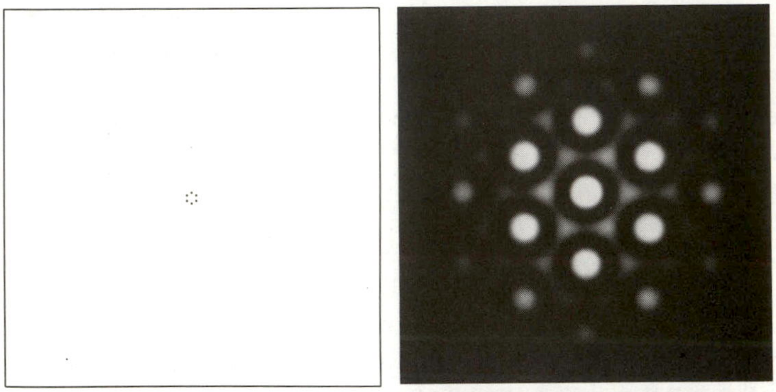

(b)

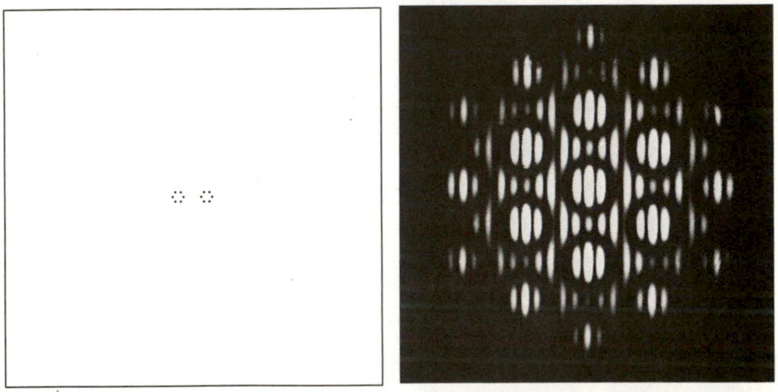

(c)

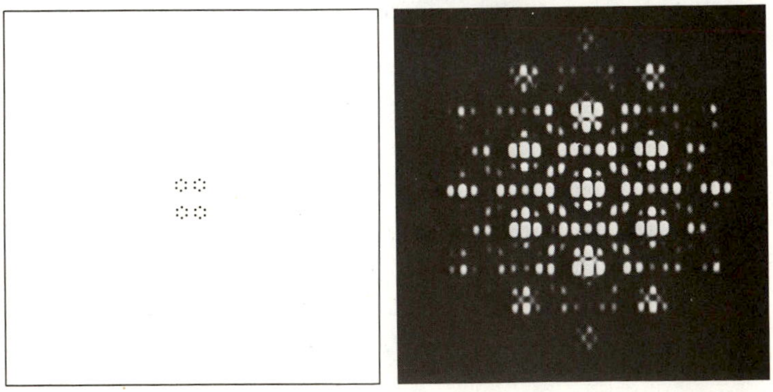

(a) A six-atom molecule, modeled by six pinholes. **(b)** Two six-atom molecules in a row. Note how the presence of two atoms introduces additional vertical fringes. **(c)** Four six-atom molecules. The horizontal repeat in structure leads to additional horizontal fringes. **(d)** A vertical row of many pairs of six-atom molecules. Note how the diffraction pattern sharpens in the vertical direction but remains broad in the horizontal direction. **(e)** A two-dimensional crystalline array of six-atom molecules. Note that the diffraction pattern is now a set of sharp spots. **(f)** A different crystalline array of the same molecules. The smaller reciprocal lattice results from the larger crystal lattice. [From G. Harburn, C. A. Taylor, and T. R. Welberry, *Atlas of Optical Transforms* (Ithaca, N.Y.: Cornell Univ. Press, 1975).]

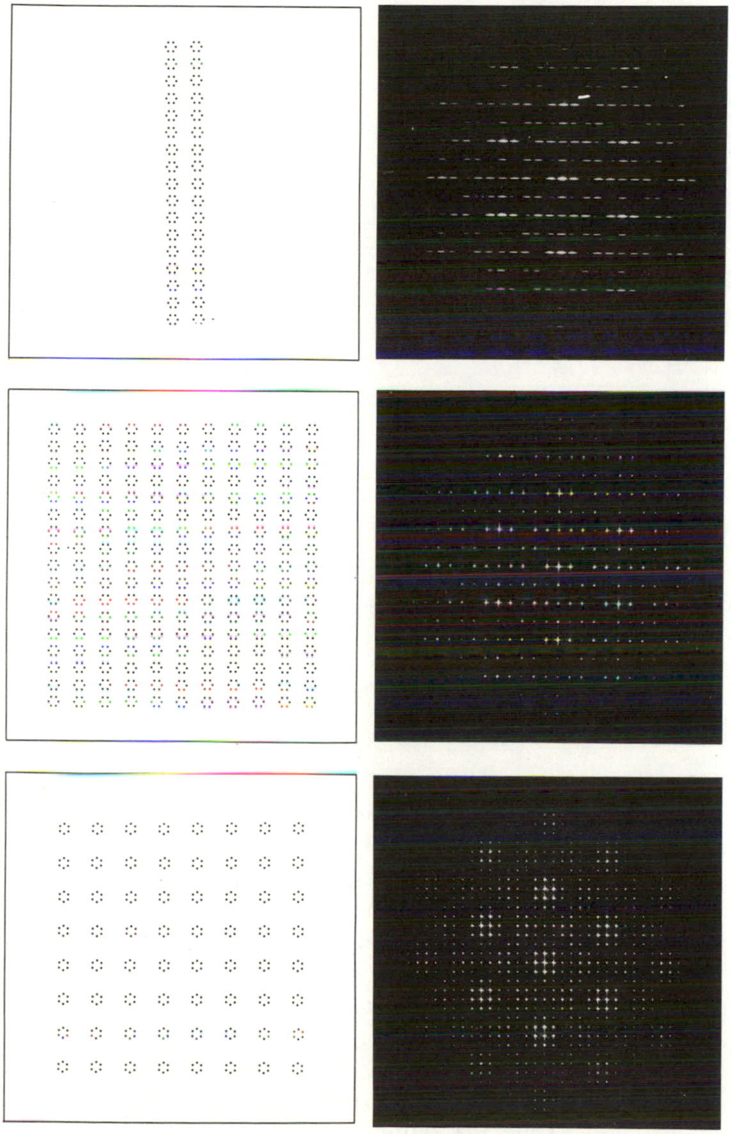

(d)

(e)

(f)

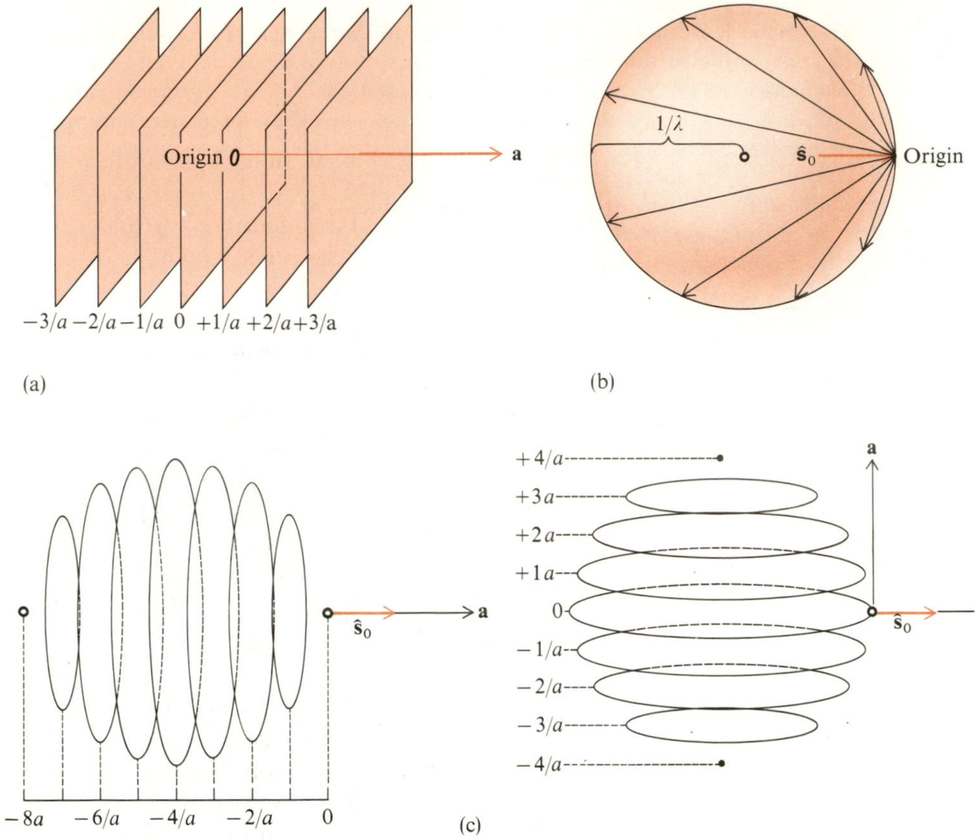

Figure 13-8

Experimental conditions for observation of scattering from the linear array of atoms shown in Figure 13-6. **(a)** A set of parallel planes, representing the von Laue condition imposed by the array of atoms. **(b)** For a fixed direction $\hat{\mathbf{s}}_0$ of incident x rays, the possible scattering vectors (*black*) must lie on the surface of the sphere. (See Fig. 13-3a for further information.) **(c)** The intersection of the two sets of conditions outlined in parts a and b is shown for two different relative geometries of **a** and $\hat{\mathbf{s}}_0$.

By extending this argument, it is clear that integral values of $\mathbf{S} \cdot \mathbf{a}$ define a set of parallel planes. The spacing between these planes is $1/a$ (Fig. 13-8a). The set of parallel planes in reciprocal space defines all those values of the scattering vector that produce measurable intensity. Further constraints are introduced if the wavelength of the incident x rays is held constant, and if their direction is fixed at $\hat{\mathbf{s}}_0$.

Once a particular $\hat{\mathbf{s}}_0$ is selected, the various possible observation directions $\hat{\mathbf{s}}$ lead to a restricted set of possible scattering vectors **S**. Note from Figure 13-3b that the tip of the vector **S** always extends from the origin to a point at a distance $1/\lambda$ along $\hat{\mathbf{s}}$. The locus of all points located $1/\lambda$ from $\hat{\mathbf{s}}$ will define a sphere of radius $1/\lambda$ centered at the tail of $\hat{\mathbf{s}}$. Thus all possible scattering vectors **S** must extend from the origin to the surface of a sphere of radius $1/\lambda$ (Fig. 13-8b). This sphere is called the sphere of reflection. It is always tangent to the plane drawn through the origin perpendicular to the direction of incident x rays, $\hat{\mathbf{s}}_0$.

The scattering of x rays will be observed only when both the von Laue conditions and the conditions of the sphere of reflection are satisfied. This means that the scattering vector must lie on points formed by the intersection of the surface of a sphere and a set of parallel planes. As shown in Figure 13-8c, this intersection is a set of parallel circles. The orientation of the circles, and the identity of the particular planes from which they originate, depend on the angle between $\hat{\mathbf{s}}_0$ and **a**.

Sampling the scattering from any atom or molecule in a periodic array

For a sample consisting of a single atom, the atomic scattering factor $f(S)$ would be measurable with a single sample orientation and $\hat{\mathbf{s}}_0$, at geometries **S**, anywhere on the surface of a sphere of radius $1/\lambda$. With a linear array of atoms oriented along **a**, this atomic transform now can be measured only where this sphere is intersected by a set of parallel planes with a spacing of $1/a$ (Fig. 13-8c). One describes this by saying that the originally broad atomic or molecular structure factor now is sampled at discrete places. Figure 13-6 shows an additional example of this sampling (see also Box 13-4). The orientation and spacing of the scattering planes contain all the information about the array and no information about the atom or molecule. The actual value (amplitude and phase) of the structure factor at these sampling positions retains information about the structure of the atom or molecule.

Note that all of the conditions restricting the observation of scattering have been plotted in Figures 13-7 and 13-8 in terms of the vector **S**. The dimensions of **S** are reciprocal distance, and so the coordinate system shown in these figures is reciprocal space. Increasing the distance between atoms of an array (in real space) will result in decreasing the spacing imposed by the von Laue conditions on the parallel planes (in reciprocal space).

A fixed orientation of **a** and $\hat{\mathbf{s}}_0$ allows only a restricted region of reciprocal space to be sampled in an x-ray scattering experiment. This region can be enlarged by changing the angle between **a** and $\hat{\mathbf{s}}_0$, by rotating either the sample or the angle of incidence of the x rays. The largest possible value of $|S|$ for any geometry is $2/\lambda$ (Fig. 13-3c). Therefore, the maximal region of reciprocal space that can be sampled, after all orientations of **a** and $\hat{\mathbf{s}}_0$ have been tried, is a sphere of radius $2/\lambda$ centered at the origin of reciprocal space. This sphere is called the limiting sphere (see Fig. 13-23b).

X-ray scattering actually observed in the laboratory frame

The scattering vector **S** is very convenient for mathematics, but it tends to obscure what is happening in an experiment. Therefore, Figure 13-9 shows the diffraction from a linear array of identical scatterers in the laboratory frame. Assume that the sample is placed at the center of a cylinder of x-ray film (Fig. 13-9a). X rays are incident along a fixed direction, and all of the scattered intensity is detected by the film. Dif-

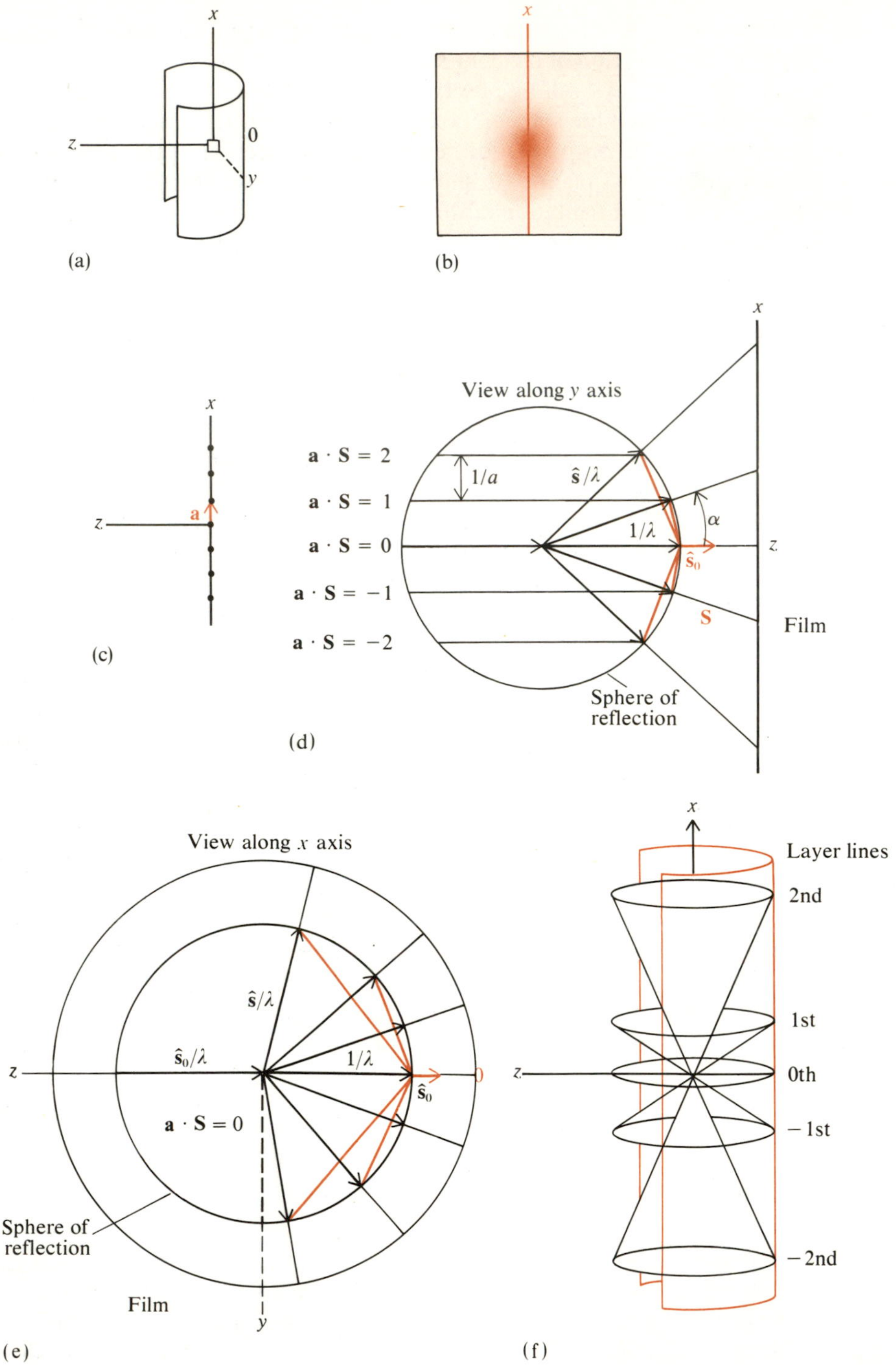
x
0
z
y
(a)
x
(b)
x
z
a
(c)
x
View along y axis
a · S = 2
a · S = 1
a · S = 0
a · S = −1
a · S = −2
1/a
ŝ/λ
1/λ
α
z
ŝ₀
S
Film
Sphere of reflection
(d)
View along x axis
ŝ/λ
ŝ₀/λ
1/λ
z
0
ŝ₀
a · S = 0
Sphere of reflection
Film
y
(e)
x
Layer lines
2nd
1st
z
0th
−1st
−2nd
(f)

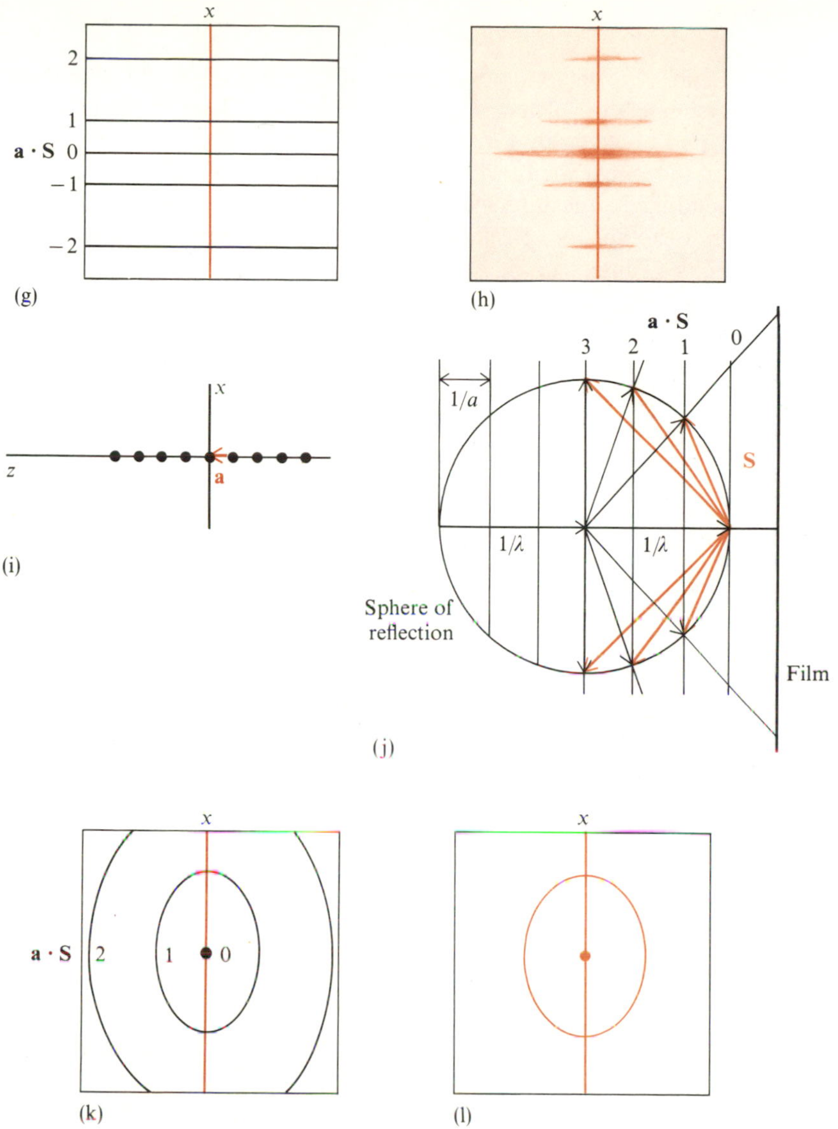

Figure 13-9

X-ray scattering from a one-dimensional array as seen in the laboratory. **(a)** X rays incident along the z axis strike a sample at the origin, and scattered rays are detected by a cylindrical film. **(b)** Scattering pattern produced by a single atom. The pattern is elliptical because of the cylindrical film; the pattern would be circular if flat film were used. **(c)** The linear array. **(d)** The scattered radiation allowed by the von Laue conditions, viewed in the x–z plane. **(e)** The scattered radiation allowed by the von Laue conditions, viewed in the y–z plane. **(f)** Cones of scattered radiation produced by the von Laue conditions, for the particular geometry shown in part a. All scattered rays will extend along the surface of one of the cones. **(g)** Diffraction pattern of the array, resulting from the intersection of the scattering cones with the cylindrical film. **(h)** The actual scattering seen is the product of the atomic scattering shown in part b with the diffraction pattern of part g. **(i)** An alternative scattering geometry with the array parallel to the direction of incident radiation. **(j)** The scattered rays allowed by the von Laue conditions in the geometry of part i. **(k)** The array diffraction pattern resulting from the geometry of part i. **(l)** The product of the diffraction pattern of part k with the atomic scattering of part b.

ferent vectors **S** now correspond to different scattering angles 2θ. Consider each scatterer just as a single atom. Then, if the sample had only a single atom at the origin, the scattering would be given by Equation 13-23 (Fig. 13-9b).

The effect of the linear array is to allow finite intensity only at scattering geometries corresponding to the intersection of the set of planes ($\mathbf{a} \cdot \mathbf{S} = h$, for $h = 0, \pm 1, \pm 2, \ldots$) with the sphere of reflection. For each scattering vector **S** drawn to one of these points of intersection, there corresponds a ray of scattered radiation. From Equation 13-3, this ray propagates in the direction $\hat{\mathbf{s}} = \lambda \mathbf{S} + \hat{\mathbf{s}}_0$ (Fig. 13-3).

To compute the pattern of scattered radiation, it is easiest to use the description of scattering shown in Figure 13-3b rather than the equivalent one shown in Figure 13-3a. The vector $\hat{\mathbf{s}}$ is placed along the $\hat{\mathbf{s}}_0$ axis a distance $1/\lambda$ from the origin. In other words, $\hat{\mathbf{s}}$ can be shown as emanating from the center of the sphere of reflection. When drawn in this manner, $\hat{\mathbf{s}}$ points toward the tip of the corresponding scattering vector **S** (Figs. 13-3b and 13-9e).

The vectors **a** and $\hat{\mathbf{s}}_0$ are fixed by the choice of orientation of the sample and the incident beam. Because of the von Laue constraint, each scattering vector **S** extends from the origin to one of the planes spaced $1/a$ apart. Figure 13-9d,e shows two cross sections through the origin. In the plane defined by $\mathbf{S} \cdot \mathbf{a} = 0$, a continuous set of scattering vectors **S** is allowed. This leads to a continuous distribution of scattered radiation, which emanates from the sample in a circle parallel to the $\mathbf{S} \cdot \mathbf{a} = 0$ plane (Fig. 13-9f).

In the plane parallel to **a**, only certain values of **S** are allowed. Thus, vectors describing scattered radiation appear only at certain deflection angles α (Figs. 13-9d). Elementary geometrical considerations indicate that $\sin \alpha = h\lambda/a$, where h is any integer such that $|h| \leqslant a/\lambda$. Each value of h leads to a cone of scattering (Fig. 13-9f). Where this cone intersects the cylindrical film, a ring of scattered intensity results. When the film is unrolled, this ring becomes a line, called a layer line. The various layer lines are all parallel, and their spacing increases progressively as $|h|$ increases. The lines are perpendicular to the linear array (Fig. 13-9g).

The film records scattering from the individual atom of Figure 13-9b only in the lines allowed by the array (Fig. 13-9g). This effect leads to the pattern of scattering shown in Figure 13-9h. Note that only a finite number of layer lines are seen because of the conditions imposed by the sphere of reflection. There is a maximal value of 1 for $\sin \alpha$. For the geometry shown in Figure 13-9c, this value will occur when $\hat{\mathbf{s}}$ is parallel to **a**. Here the largest value of h that can be included within the sphere of reflection is $\pm|\mathbf{a}|/\lambda$. For example, suppose $|\mathbf{a}|$ is 5 Å, and λ is 1 Å. Then the integer h can have values only in the range $-5 \leqslant h \leqslant +5$; the diffraction pattern will have 11 layer lines.

Figure 13-9h shows that the intensities drop off as h increases. This occurs because the atomic scattering decreases as e^{-S^2}. As demonstrated in Figure 13-9d,e, large values of h tend to correspond to large values of $|\mathbf{S}|$. The scattered intensity also drops off rapidly on each layer line as one moves from the center to the edges. This is because, along each line of scattering, larger angles inevitably correspond to longer lengths of the scattering vector, $|\mathbf{S}|$ (see Fig. 13-9e).

The appearance of the scattering pattern depends markedly on the relative orientation of the incident beam ($\hat{\mathbf{s}}_0$) and the repeating array (**a**). For example, if the array is rotated so that **a** and $\hat{\mathbf{s}}_0$ are parallel (Fig. 13-9i), the scattering pattern changes from a series of lines to a series of concentric curves (Fig. 13-9k). These curves are elliptical because the intersection of a circle (the cone of scattered radiation) and a cylindrical surface with its long axis parallel to the plane of the circle (the film) is an ellipse.

X-ray scattering from a two-dimensional array of atoms

The results shown in Figures 13-8 and 13-9 are the basic ideas behind all x-ray crystallographic measurements. However, if they are to be useful in practice, one must extend them from a one-dimensional array to a three-dimensional crystal. First, consider a two-dimensional net of molecules (Fig. 13-10a). The periodicity of this net is defined by two vectors, **a** and **b**. In general, **a** and **b** are not perpendicular to one another, nor are they of the same length. The periodicity along **a** will cause a set of scattering fringes at $\mathbf{S} \cdot \mathbf{a} = h$, exactly as for the one-dimensional array in Figure 13-8. The additional periodicity along **b** causes a comparable set of fringes defined by $\mathbf{S} \cdot \mathbf{b} = k$, where k is any integer. By the arguments used earlier, you can see that these latter fringes are parallel planes perpendicular to the vector **b** and spaced by equal increments $1/|\mathbf{b}|$ (Fig. 13-10b).

The x-ray scattering will have finite intensity only where both von Laue conditions ($\mathbf{S} \cdot \mathbf{a} = h$, and $\mathbf{S} \cdot \mathbf{b} = k$) are satisfied. This condition is met at the intersection of the two sets of planar fringes. That intersection is an array of parallel lines (Fig. 13-10c). The lines all are perpendicular to the plane defined by **a** and **b**. Experimentally,

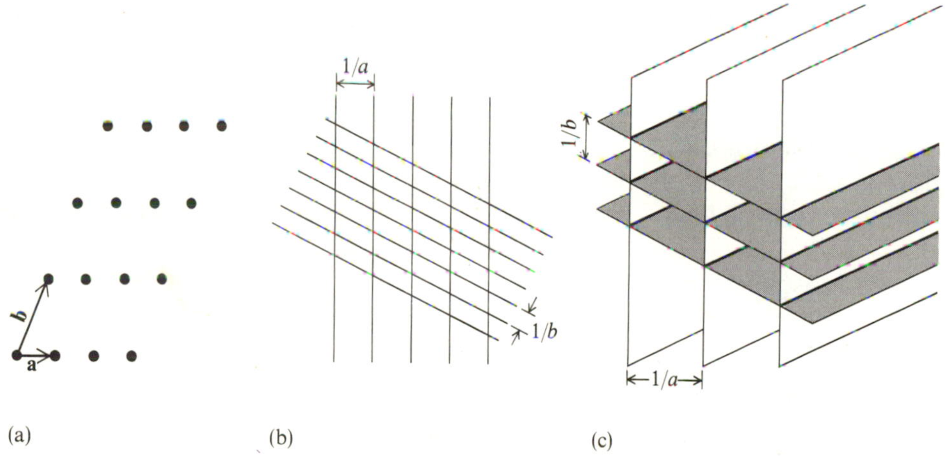

Figure 13-10

X-ray scattering from a two-dimensional array. **(a)** The array defined by **a** and **b**. **(b)** Parallel planes demonstrate the allowed positions of the scattering vector **S**. Shown in cross section are planes spaced by $1/a$ perpendicular to **a**. and planes spaced by $1/b$ perpendicular to **b**. **(c)** The array of lines resulting from the intersection of the parallel planes shown in part b. Scattering vectors extending from the origin to any point along one of these lines will yield observable x-ray scattering intensity.

scattering intensity will be observed whenever the geometry of incident and diffracted radiation leads to a scattering vector that extends from the origin to a point along one of the arrays of parallel lines.

The greater restriction of scattering caused by a two-dimensional array means that more of the intensity will be concentrated at a smaller number of sets of scattering angles, 2θ. The effect of a two-dimensional array on the actual experimental scattering pattern is shown schematically in Figure 13-11. The plane of the array is perpendicular to the direction of incident x rays. If the array is considered as the effect of two perpendicular one-dimensional arrays, each alone would produce a set of scattering fringes. The x-ray diffraction pattern is the product of the two sets of fringes. In this case, the scattered radiation detected by the x-ray film consists of a series of spots, each occurring at the intersection of two fringes. (See Box 13-4 for examples of the optical diffraction of two-dimensional arrays.)

X-ray diffraction from a three-dimensional array of atoms

It is easy mathematically to generalize x-ray scattering to three dimensions, but it is not so easy to visualize the results. For an array such as that found in a real three-dimensional crystal, there is now a third periodicity defined by the vector **c** (Fig. 13-12a). This leads to a third set of planar scattering fringes given by $\mathbf{S} \cdot \mathbf{c} = l$, where $l = 0$, ± 1, $\pm 2, \ldots$. This set of fringes intersects the parallel lines generated by $\mathbf{S} \cdot \mathbf{a} = h$ and $\mathbf{S} \cdot \mathbf{b} = k$. The result is a three-dimensional lattice of points, spaced evenly by $1/|\mathbf{a}|$ in the direction perpendicular to **a**, by $1/|\mathbf{b}|$ perpendicular to **b**, and by $1/|\mathbf{c}|$ perpendicular to **c** (Fig. 13-12b). Diffracted radiation will be observed only when the scattering vector **S** intersects one of the lattice points. It is not easy to illustrate the actual diffraction pattern from a three-dimensional crystal because it is a three-dimensional pattern.

Note that the lattice that describes allowed scattering geometries is not the same as the lattice of points that represents the positions of the atoms in the array. The position lattice has spacings **a**, **b**, and **c**, whereas the spacings in the diffraction lattice, **a***, **b***, and **c***, are related to the inverse of these. This scattering lattice is called the reciprocal lattice. The vector space it occupies is called the reciprocal space (Fig. 13-12c).

The array (crystal) we have described thus far contains only a single atom per repeating unit. The positions of the atoms define a set of cells bounded by **a**, **b**, and **c**

Figure 13-11

X-ray scattering observed in the laboratory for a two-dimensional array perpendicular to the direction of incident radiation. Each part shows patterns for both cylindrical and spherical film. **(a)** Sample geometries. **(b)** Layer lines resulting from the **a** periodicity of the array. **(c)** Layer lines resulting from the **b** periodicity of the array. **(d)** Actual scattering observed is the product of the functions shown in parts b and c with the atomic scattering pattern shown in Figure 13-9b.

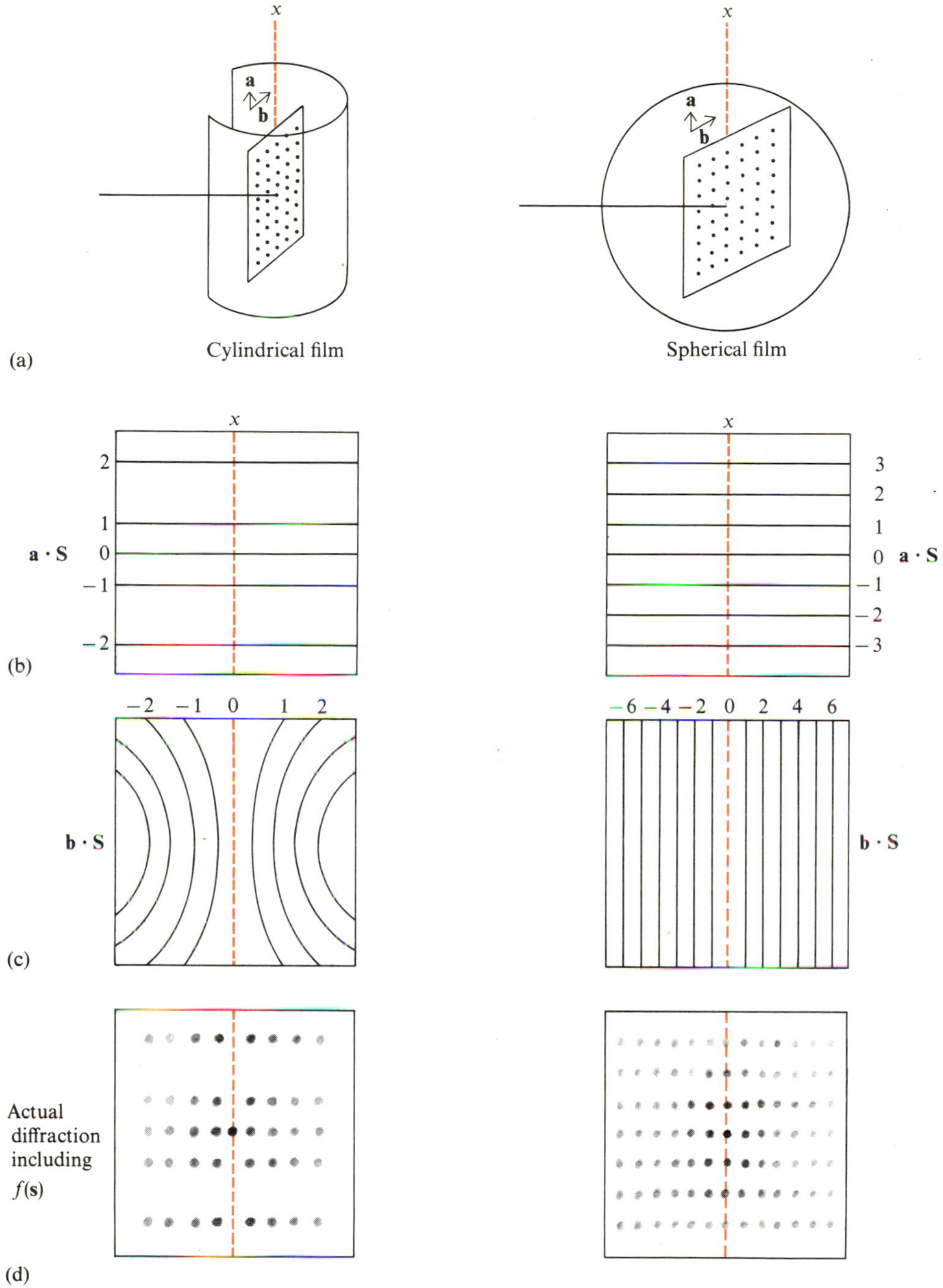
x
a
b
(a)
Cylindrical film
Spherical film
(b)
2
1
a · S
0
−1
−2
3
2
1
0
−1
−2
−3
(c)
−2 −1 0 1 2
b · S
−6 −4 −2 0 2 4 6
Actual diffraction including f(s)
(d)

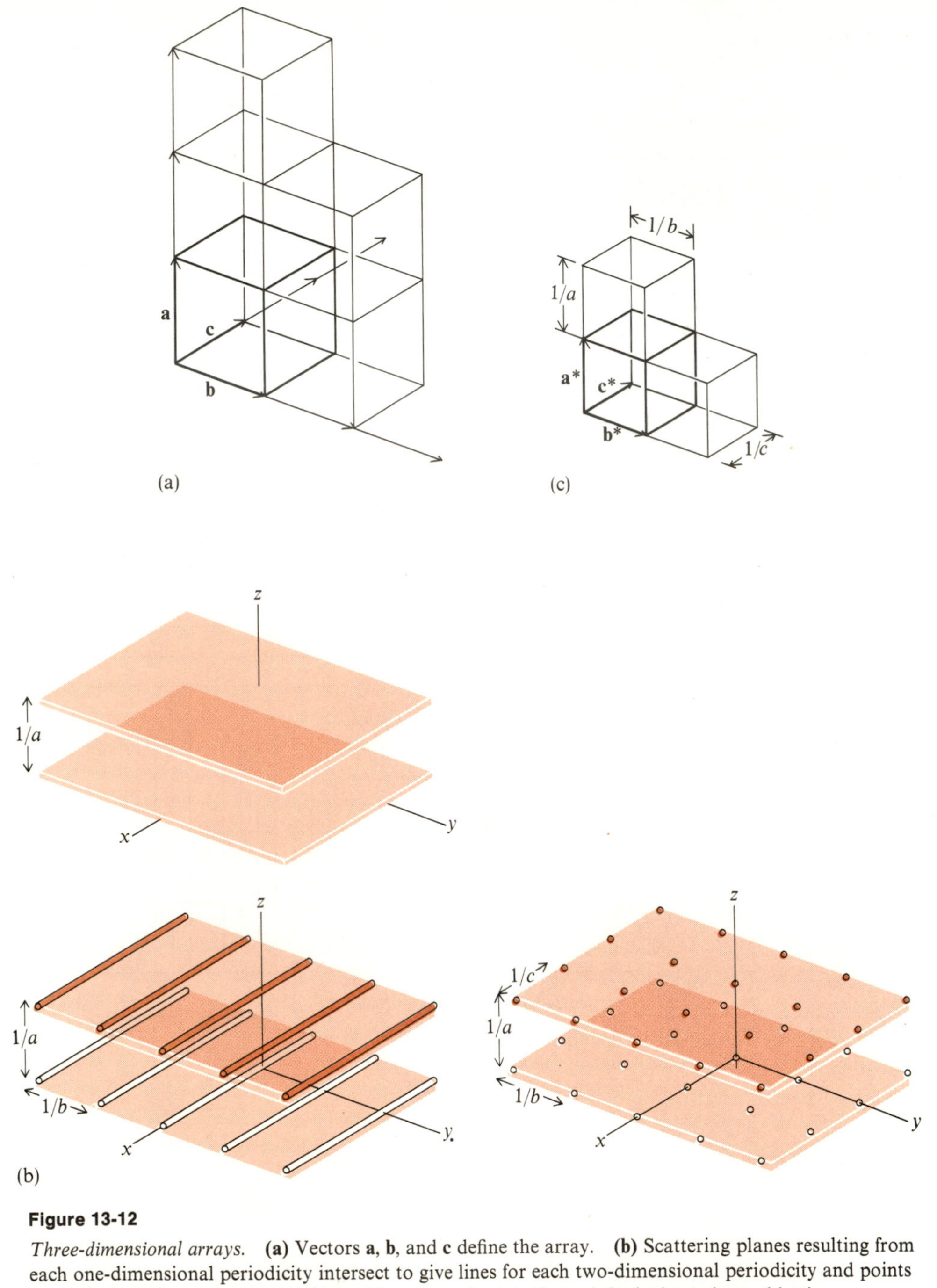

Figure 13-12

Three-dimensional arrays. **(a)** Vectors **a**, **b**, and **c** define the array. **(b)** Scattering planes resulting from each one-dimensional periodicity intersect to give lines for each two-dimensional periodicity and points for each three-dimensional periodicity. The array of points that results is the reciprocal lattice.
(c) The reciprocal lattice is generated by the vectors **a***, **b***, and **c***. Several cells of the reciprocal lattice are shown. Scattering vectors extending from the origin to these points result in observed intensity.

(Fig. 13-12a). It is convenient to express the position of the atoms in terms of a coordinate system defined by these vectors. A vector drawn from the origin to the jth atom position is $\mathbf{r} = x\mathbf{a} + y\mathbf{b} + z\mathbf{c}$. Because the atoms lie at the corners of the cells, x, y, and z must be integers.

The vector $\mathbf{r}$ can be used to calculate the x-ray scattering from the array. From Equation 13-27, the structure factor is a sum over all atom positions:

$$F_{\mathrm{Tot}}(\mathbf{S}) = \sum_x \sum_y \sum_z f(S) e^{2\pi i \mathbf{S} \cdot (x\mathbf{a} + y\mathbf{b} + z\mathbf{c})} \tag{13-36}$$

Inserting the von Laue conditions ($\mathbf{S} \cdot \mathbf{a} = h$, etc.), we can rewrite this as

$$F_{\mathrm{Tot}}(h, k, l) = \sum_x \sum_y \sum_z f(S) e^{2\pi i (hx + ky + lz)} \tag{13-37}$$

where h, k, and l are any integers. Every diffracted ray can be computed by choosing the appropriate integral values of h, k, and l and summing over the array. Note that, for an array of identical atoms, each exponential term in Equation 13-37 is simply unity, because h, k, l, x, y, and z are all integers. Therefore, Equation 13-37 becomes

$$F_{\mathrm{Tot}}(h, k, l) = Nf(S) \tag{13-38}$$

where N is the number of atoms in the array, and $f(S)$ is the atomic scattering factor, now evaluated only at the particular values of S allowed by integral choices of h, k, and l. Thus the x-ray scattering is just the single-atom scattering sampled at all points in reciprocal space allowed by the von Laue conditions imposed by the lattice.

Equation 13-37 is a Fourier series rather than a Fourier transform because of the discrete nature of the diffraction pattern. However, it can be inverted in exactly the same way as a transform. By analogy to Equation 13-8, the electron density distribution of the array will be given by

$$\rho(x, y, z) = (1/NV) \sum_{h=-\infty}^{\infty} \sum_{k=-\infty}^{\infty} \sum_{l=-\infty}^{\infty} F_{\mathrm{Tot}}(h, k, l) e^{-2\pi i (hx + ky + lz)} \tag{13-39}$$

where V is the volume of one unit cell, $V = \mathbf{a} \cdot \mathbf{b} \times \mathbf{c}$, so that NV is the volume of the entire array. The presence of the volume factor V in Equation 13-39 is easy to rationalize. $F(h, k, l)$ is proportional to the number of electrons, whereas ρ (an electron density) has units of electrons per unit volume.

X-ray diffraction from a three-dimensional molecular crystal

Although Equations 13-37, 13-38, and 13-39 were derived for an array with only one atom at the corner of each cell, it can be shown that similar equations hold for any real crystal. The repeating element of a crystal is defined as a unit cell. The crystal is a lattice of unit cells, each defined by the vectors **a**, **b**, and **c** (Fig. 13-12a). Whatever is inside each cell—whether a single atom, a molecule, or many molecules—the pattern of electron density repeats periodically throughout the crystal by translation along the vectors **a**, **b**, and **c**. Therefore, the von Laue conditions still apply and restrict the sampling of the structure factor to only those points defined by the reciprocal lattice. Now, however, what is sampled is not the atomic scattering factor. It is instead a molecular structure factor, or unit cell structure factor, given by Equation 13-27. In order to show this, it is useful to employ a mathematical device known as a convolution.

A repeating structure as a convolution

X-ray diffraction examines both the properties of the crystal lattice and the structure of individual molecules. The distribution of electron density within each unit cell is identical. The lattice describes how this distribution is replicated into a three-dimensional pattern. Such a repeating structure can be very conveniently described as a convolution.

Consider two arbitrary one-dimensional functions, $f(x)$ and $g(x)$. The functions f and g both are defined along the x axis. Their convolution product is defined as

$$\widehat{fg}(u) = \int_{-\infty}^{\infty} dx f(x) g(u - x) \tag{13-40}$$

where the variable u can take on any value that x can. It is equivalent to x, except that it is held constant in the integral.

We will try to develop a physical picture of the convolution product. See the example shown in Figure 13-13a. The function $g(u)$ plotted along the u axis is identical to a plot of $g(x)$ along the x axis. The function $g(u - x)$ plotted along the u axis is just the function $g(u)$ shifted in space by an amount x. Therefore, in the convolution product, the function g is placed successively at all points along the u axis, but it is multiplied by a weighting factor, $f(x)$, for each shift in position x. All the weighted values of g are added or integrated to produce the final convolution.

Physically, the convolution $\widehat{fg}(u)$ means that one is laying down successive images of g weighted by f.[§] It turns out to be equivalent to say that one is laying down images of f weighted by g. To see this, let $x' = u - x$ in Equation 13-40. Then $dx' = -dx$, and

[§] If this is not absolutely clear, stop right here. Study Figure 13-13a; try calculating a convolution product for the *simple* functions of your choice.

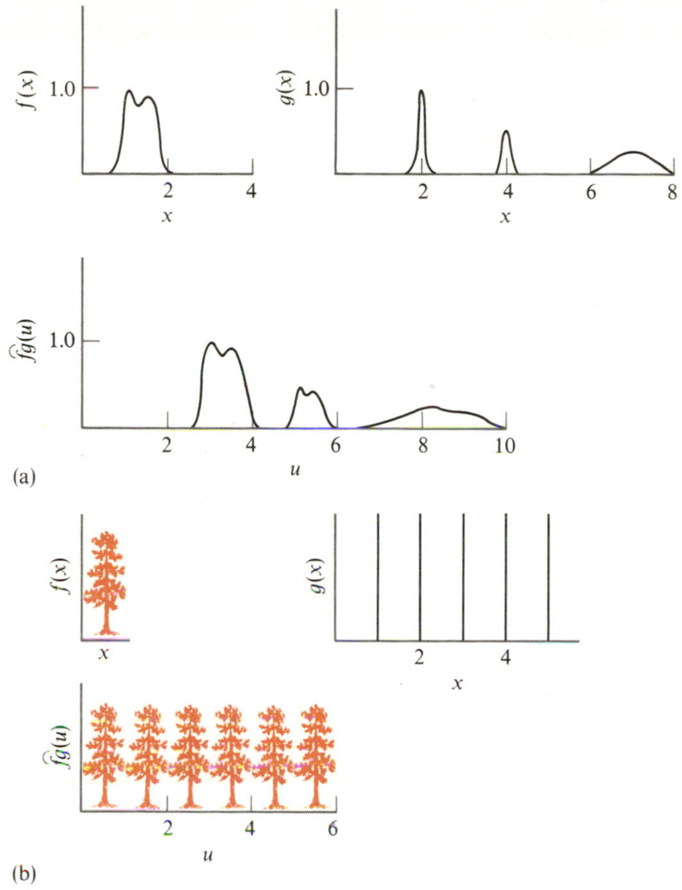

Figure 13-13

Convolution integrals. **(a)** Two functions f and g and their convolution calculated by Equation 13-40. **(b)** Two functions—a tree, f, and a lattice (set of delta functions), g—and their convolution.

(taking note of the reversal of the limits of integration)

$$\widehat{fg}(u) = -\int_{+\infty}^{-\infty} dx' f(u - x')g(x') = \widehat{gf}(u) \tag{13-41}$$

because x and x' are just dummy variables.

Suppose the function g is the Dirac delta function, $\delta(x - a)$. Then the convolution

$$\widehat{f\delta}(u) = \int_{-\infty}^{\infty} dx f(x)\,\delta[u - (x - a)] = f(u + a) \tag{13-42}$$

just shifts the function the distance a along the u coordinate system. In three dimensions, all the same results hold. The convolution product is

$$\widehat{fg}(\mathbf{u}) = \int d\mathbf{r} f(\mathbf{r}) g(\mathbf{u} - \mathbf{r}) \tag{13-43}$$

If g is the three-dimensional delta function, $\delta(\mathbf{r} - \boldsymbol{\rho})$, the convolution integral defined in Equation 13-43 merely shifts the function $f(\mathbf{r})$ along the vector $\boldsymbol{\rho}$ in the $\mathbf{u}$ coordinate system.

Delta functions can be used to describe a lattice. It is sufficient to define the origin of each unit cell. In one dimension, the origin can be any integral multiple of the vector $\mathbf{a}$. This restricts values of x that define the origin to $n\mathbf{a}$. A function that describes one of these values is $\delta(x - na)$. Thus, an infinite one-dimensional lattice is given by the function

$$L(x) = \sum_{n=-\infty}^{\infty} \delta(x - na) \tag{13-44}$$

This function really is just a list of all the lattice points.

Suppose that the electron density distribution within one cell of the lattice is $\rho(x)$. Then to describe the crystal, we want to replicate this electron density in each unit cell. From the properties of the convolution integral described above, this is done by

$$\text{Crystal} = \widehat{L\rho}(u) = \int_{-\infty}^{\infty} dx \rho(x) \sum_{n=-\infty}^{\infty} \delta[u - (x - na)] = \sum_{n=-\infty}^{\infty} \rho(u + na) \tag{13-45}$$

The convolution $\widehat{L\rho}$ simply lays down an image of the structure in each unit cell (Fig. 13-13b).

In three dimensions, an infinite lattice is described by

$$L(\mathbf{r}) = \sum_{n=-\infty}^{\infty} \sum_{m=-\infty}^{\infty} \sum_{p=-\infty}^{\infty} \delta(\mathbf{r} - n\mathbf{a} - m\mathbf{b} - p\mathbf{c}) \tag{13-46}$$

where n, m, and p are any integers. The electron density within one cell is $\rho(\mathbf{r})$, and the crystal is described by

$$\text{Crystal} = \widehat{L\rho}(\mathbf{u}) = \sum_{n=-\infty}^{\infty} \sum_{m=-\infty}^{\infty} \sum_{p=-\infty}^{\infty} \rho(\mathbf{u} + n\mathbf{a} + m\mathbf{b} + p\mathbf{c}) \tag{13-47}$$

Thus any crystal can be described as the convolution of the contents of one unit cell with the lattice.

The Fourier transform of a convolution

There is a property of convolutions that makes them especially useful for describing x-ray scattering. Suppose that $f(\mathbf{r})$ and $g(\mathbf{r})$ are functions that can be expressed as the Fourier transforms of the functions $F(\mathbf{S})$ and $G(\mathbf{S})$:

$$f(\mathbf{r}) = \int_{-\infty}^{\infty} d\mathbf{S}F(\mathbf{S})e^{-2\pi i\mathbf{S}\cdot\mathbf{r}} \tag{13-48a}$$

$$g(\mathbf{r}) = \int_{-\infty}^{\infty} d\mathbf{S}G(\mathbf{S})e^{-2\pi i\mathbf{S}\cdot\mathbf{r}} \tag{13-48b}$$

and, consequently, F and G are inverse transforms of f and g:

$$F(\mathbf{S}) = \int_{-\infty}^{\infty} d\mathbf{r}f(\mathbf{r})e^{2\pi i\mathbf{S}\cdot\mathbf{r}} \tag{13-49a}$$

$$G(\mathbf{S}) = \int_{-\infty}^{\infty} d\mathbf{r}g(\mathbf{r})e^{2\pi i\mathbf{S}\cdot\mathbf{r}} \tag{13-49b}$$

Then the convolution product of the two functions can be written as

$$\begin{aligned}\widehat{fg}(\mathbf{u}) &= \int_{-\infty}^{\infty} d\mathbf{r}f(\mathbf{r})g(\mathbf{u}-\mathbf{r}) \\ &= \int_{-\infty}^{\infty} d\mathbf{r}\int_{-\infty}^{\infty} d\mathbf{S}F(\mathbf{S})e^{-2\pi i\mathbf{S}\cdot\mathbf{r}}\int_{-\infty}^{\infty} d\mathbf{S}'G(\mathbf{S}')e^{-2\pi i\mathbf{S}'\cdot(\mathbf{u}-\mathbf{r})}\end{aligned} \tag{13-50}$$

Rearranging the order of the integrals, we obtain

$$\widehat{fg}(\mathbf{u}) = \int_{-\infty}^{\infty} d\mathbf{S}F(\mathbf{S})\int_{-\infty}^{\infty} d\mathbf{S}'G(\mathbf{S}')e^{-2\pi i\mathbf{S}'\cdot\mathbf{u}}\int_{-\infty}^{\infty} d\mathbf{r}\,e^{2\pi i(\mathbf{S}'-\mathbf{S})\cdot\mathbf{r}} \tag{13-51}$$

The third integral is just the Dirac delta function, $\delta(\mathbf{S}'-\mathbf{S})$ (see Box 13-3). Therefore, the result of the second integral is just to set $\mathbf{S}'=\mathbf{S}$, and Equation 13-51 becomes

$$\widehat{fg}(\mathbf{u}) = \int_{-\infty}^{\infty} d\mathbf{S}F(\mathbf{S})G(\mathbf{S})e^{-2\pi i\mathbf{S}\cdot\mathbf{u}} \tag{13-52}$$

Note that Equation 13-52 is just the Fourier transform of the product of the two functions $F(\mathbf{S})$ and $G(\mathbf{S})$.

This is an important conclusion. The Fourier transform of the product of two functions, F and G, is the convolution product of their two Fourier transforms, f and g. We can derive a second important result by Fourier-transforming Equation 13-52:

$$\int_{-\infty}^{\infty} d\mathbf{u}\,e^{2\pi i\mathbf{S}'\cdot\mathbf{u}}\widehat{fg}(\mathbf{u}) = \int_{-\infty}^{\infty} d\mathbf{u}\int_{-\infty}^{\infty} d\mathbf{S}F(\mathbf{S})G(\mathbf{S})e^{-2\pi i\mathbf{S}\cdot\mathbf{u}}e^{2\pi i\mathbf{S}'\cdot\mathbf{u}}$$

$$= \int_{-\infty}^{\infty} d\mathbf{S} F(\mathbf{S}) G(\mathbf{S}) \int_{-\infty}^{\infty} d\mathbf{u}\, e^{2\pi i(\mathbf{S}' - \mathbf{S}) \cdot \mathbf{u}} \tag{13-53}$$

However, the second integral is just the Dirac delta function, $\delta(\mathbf{S}' - \mathbf{S})$. Thus,

$$\int_{-\infty}^{\infty} d\mathbf{u}\, e^{2\pi i \mathbf{S}' \cdot \mathbf{u}} \widehat{fg}(\mathbf{u}) = F(\mathbf{S}')G(\mathbf{S}') \tag{13-54}$$

The Fourier transform of a convolution product is just the product of the Fourier transforms of the two convoluted functions.

Convolutions in the computation of x-ray scattering

Consider a one-dimensional molecular crystal. Within each unit cell, the electron density distribution is $\rho_m(\mathbf{r})$. The x-ray scattering from the contents of a single cell located at the origin is given by Equation 13-7:

$$F_m(\mathbf{S}) = \int_{-\infty}^{\infty} d\mathbf{r} \rho_m(\mathbf{r}) e^{2\pi i \mathbf{S} \cdot \mathbf{r}} \tag{13-55}$$

The lattice is generated by the vector **a**, and the lattice can be described by Equation 13-44 expressed in three dimensions:[§]

$$L(\mathbf{r}) = \sum_{n=-\infty}^{\infty} \delta(\mathbf{r} - n\mathbf{a}) \tag{13-56}$$

The crystal is generated by the convolution

$$\widehat{\rho_m L}(\mathbf{u}) = \int_{-\infty}^{\infty} d\mathbf{r} \rho_m(\mathbf{r}) \sum_{n=-\infty}^{\infty} \delta[\mathbf{u} - (\mathbf{r} - n\mathbf{a})] \tag{13-57}$$

The x-ray scattering from the crystal, using Equation 13-7, is

$$F_{\text{Tot}}(\mathbf{S}) = \int_{-\infty}^{\infty} d\mathbf{r} e^{2\pi i \mathbf{S} \cdot \mathbf{u}} \widehat{\rho_m L}(\mathbf{u}) \tag{13-58}$$

This expression can be evaluated by using Equations 13-54 and 13-49, and changing from the variable **u** to the equivalent variable **r**:

$$F_{\text{Tot}}(\mathbf{S}) = \left(\int_{-\infty}^{\infty} d\mathbf{r} \rho_m(\mathbf{r}) e^{2\pi i \mathbf{S} \cdot \mathbf{r}}\right)\left(\int_{-\infty}^{\infty} dr \sum_{n=-\infty}^{\infty} \delta(\mathbf{r} - n\mathbf{a}) e^{2\pi i \mathbf{S} \cdot \mathbf{r}}\right) \tag{13-59}$$

[§] Even with a one-dimensional molecular crystal, one must formally consider a three-dimensional diffraction pattern, because the relative orientation of the molecule within the lattice will affect the scattering.

The first term is just the structure factor of a single unit cell, $F_{\mathrm{m}}(\mathbf{S})$ (Eqn. 13-55). The second term is the sampling function generated by the lattice, $F_{\mathrm{L}}(\mathbf{S})$. It is evaluated simply by using the properties of the delta function. The result is

$$F_{\mathrm{Tot}}(\mathbf{S}) = F_{\mathrm{m}}(\mathbf{S})F_{\mathrm{L}}(\mathbf{S}) = F_{\mathrm{m}}(\mathbf{S}) \sum_{n=-\infty}^{\infty} e^{2\pi i n \mathbf{S} \cdot \mathbf{a}} \tag{13-60}$$

This result is identical in form with Equation 13-31. However, it is more general because it holds for molecular crystals. This example shows the correctness and simplicity of the convolution approach. However, its real advantage is the physical insight that can be gained once one is used to it.

Equations 13-59 and 13-60 mean that, to calculate x-ray scattering, one can simply multiply the scattering expected from one unit cell by the sampling function generated by the lattice.

Another example of this approach will demonstrate its usefulness. Consider an infinite one-dimensional crystal with a cell length $2\mathbf{a}$ and two identical atoms per cell, one at the vertex and one halfway between adjacent vertices:

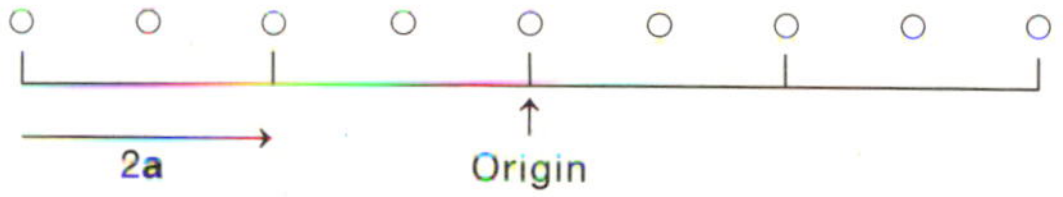

Let us calculate the x-ray scattering expected for such a crystal. From Equation 13-59, we evaluate the unit-cell structure factor as

$$F_{\mathrm{m}}(\mathbf{S}) = \int_{-\infty}^{\infty} d\mathbf{r}\rho_{\mathrm{m}}(\mathbf{r})e^{2\pi i \mathbf{S} \cdot \mathbf{r}} = f(S)(1 + e^{2\pi i \mathbf{S} \cdot \mathbf{a}}) \tag{13-61}$$

where we have used the definition of the atomic scattering factor. One must consider only two atoms—one at a vertex (say, the origin), and one at the center of the cell—because the atom spaced $2\mathbf{a}$ away from the origin will be counted as part of the next unit cell, and the atom $-\mathbf{a}$ from the origin is counted as part of the preceding unit cell.

The lattice shown leads to a sampling function

$$F_{\mathrm{L}}(\mathbf{S}) = \int_{-\infty}^{\infty} d\mathbf{r} \sum_{n=-\infty}^{\infty} \delta(\mathbf{r} - 2n\mathbf{a})e^{2\pi i \mathbf{S} \cdot \mathbf{r}} = \sum_{n=-\infty}^{\infty} e^{4\pi i n \mathbf{S} \cdot \mathbf{a}} \tag{13-62}$$

Thus, the x-ray scattering expected for the crystal is

$$F_{\mathrm{Tot}}(\mathbf{S}) = f(S)(1 + e^{2\pi i \mathbf{S} \cdot \mathbf{a}}) \sum_{n=-\infty}^{\infty} e^{4\pi i n \mathbf{S} \cdot \mathbf{a}}$$

$$= f(S) \sum_{n=-\infty}^{\infty} (e^{4\pi i n \mathbf{S} \cdot \mathbf{a}} + e^{4\pi i (n+1/2)\mathbf{S} \cdot \mathbf{a}}) \tag{13-63}$$

By writing out the sum in the final expression term by term, we can easily show it to be equal to

$$F_{\mathrm{Tot}}(\mathbf{S}) = f(S) \sum_{n=-\infty}^{\infty} e^{2\pi i n \mathbf{S} \cdot \mathbf{a}} \tag{13-64}$$

This is identical to Equation 13-31, illustrating the important result that the scattering calculated for a crystal does not depend on how we choose to define the unit cell.

Calculation of x-ray scattering from a molecular crystal using convolutions

To treat a real crystal, one must extend Equations 13-55 and 13-56 to three-dimensional arrays. The crystal is generated by the convolution $\rho_{\mathrm{m}} \frown L$, where $L(\mathbf{r})$ is given by Equation 13-46. Evaluating this (exactly in the way it was done in Equations 13-59 and 13-60) yields the structure factor of the crystal:

$$F_{\mathrm{Tot}}(\mathbf{S}) = F_m(\mathbf{S}) \sum_{n=-\infty}^{\infty} \sum_{m=-\infty}^{\infty} \sum_{p=-\infty}^{\infty} e^{2\pi i (n\mathbf{S} \cdot \mathbf{a} + m\mathbf{S} \cdot \mathbf{b} + p\mathbf{S} \cdot \mathbf{c})} \tag{13-65}$$

The triple sum in Equation 13-65 is the three-dimensional sampling function generated by the lattice. It limits the detection of scattered intensity to geometries allowed by the von Laue conditions. Applying these conditions, we can evaluate $\mathbf{S} \cdot \mathbf{a} = h$, and $\mathbf{S} \cdot \mathbf{b} = k$, and $\mathbf{S} \cdot \mathbf{c} = l$ to obtain

$$F_{\mathrm{Tot}}(h, k, l) = F_{\mathrm{m}}(\mathbf{S}) \sum_{n=-\infty}^{\infty} \sum_{m=-\infty}^{\infty} \sum_{p=-\infty}^{\infty} e^{2\pi i (nh + mk + pl)} \tag{13-66}$$

Now every exponential term is simply unity, and the triple sum simplifies to

$$\boxed{F_{\mathrm{Tot}}(h, k, l) = N F_{\mathrm{m}}(\mathbf{S})} \tag{13-67}$$

where N is the number of unit cells in the crystal.

It is convenient to write out the unit-cell structure factor, $F_{\mathrm{m}}(\mathbf{S})$, explicitly in terms of the positions of each atom in the unit cell, and of the corresponding atomic scattering factors. Using Equation 13-27 for the molecular structure factor, we choose a coordinate system based on the unit-cell vectors **a**, **b**, and **c**. The position of the jth atom in the unit cell is then

$$\mathbf{r}_j = x_j \mathbf{a} + y_j \mathbf{b} + z_j \mathbf{c} \tag{13-68}$$

where x_j, y_j, and z_j are now *fractions* of the corresponding unit-cell dimensions. Then Equation 13-27 becomes

$$F_m(\mathbf{S}) = \sum_j f_j(S) e^{2\pi i(x_j \mathbf{S}\cdot\mathbf{a} + y_j\mathbf{S}\cdot\mathbf{b} + z_j\mathbf{S}\cdot\mathbf{c})} \qquad (13\text{-}69)$$

where the sum is taken over all the atoms in one unit cell. However, $F_m(\mathbf{S})$ can be sampled only at geometries allowed by the von Laue conditions. When we apply these, equation 13-69 simplifies to

$$F_m(h, k, l) = \sum_j f_j(S) e^{2\pi i(hx_j + ky_j + lz_j)} \qquad (13\text{-}70)$$

This equation is called the structure factor equation. It represents the unit-cell x-ray scattering sampled at the reciprocal lattice points, h, k, and l.

Equation 13-70 is one of the key results in x-ray crystallography. It provides a direct way to calculate the x-ray diffraction of a crystal, provided that the structure of one unit cell is known. Alternatively, if the structure factor $F_m(h, k, l)$ is known, the electron density distribution of the crystal can be calculated. The equation used is identical to Equation 13-39. However, instead of using Equation 13-38 to describe the unit-cell contribution, one must use Equations 13-67 and 13-70.

Bragg's law of diffraction

Most elementary treatments of x-ray diffraction discuss the process as the reflection of x rays from certain planes in the crystal lattice. Because this is probably the formalism many readers have seen previously, it is worthwhile to show how our present treatment is equivalent. Lattice points are defined as the corners (vertices) of the unit cells. Lattice planes are a set of equidistant parallel planes constructed so that all lattice points lie on some member of the set. Clearly, the planes passing through the faces of the unit cells of the lattice are one such set of planes. Planes passing through opposite vertices of the unit cells also are a set of lattice planes (Fig. 13-14a). These cut the axes (**a**, **b**, and **c**) precisely at the values corresponding to one unit translation of the lattice. However, increasingly finer-spaced lattice planes can be drawn that cut the **a** axis at any $a/2$, $a/3$, . . . , a/h of the unit translation (Fig. 13-14b,c).

In three dimensions, lattice planes exist that cut one axis every $\mathbf{a}/h$ while cutting another at $\mathbf{b}/k$ and the third at $\mathbf{c}/l$. The planes can be described by specifying the Miller indices, h, k, and l. Note that the spacing (d) between adjacent planes is inversely related to the size of the indices of the planes. Therefore, it is reasonable that the planes could bear some relationship to the reciprocal lattice.

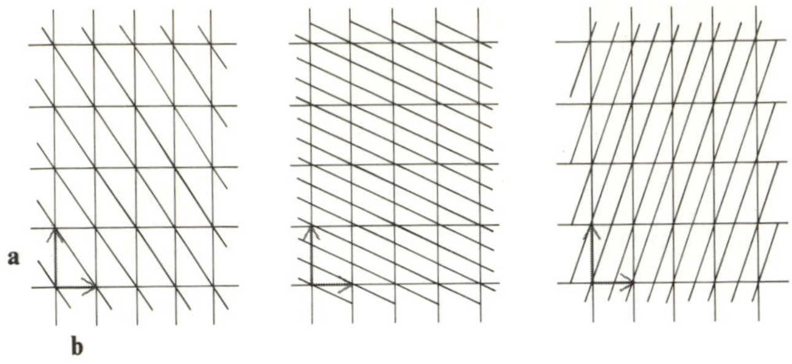

(a) (1, 1) (b) (3, 1) (c) (1, 2)

Figure 13-14

Three sets of lattice planes. Shown below each set are the Miller indices (h, k) that describe it.

In the Bragg's-law description of diffraction, an x-ray beam that impinges on a lattice plane at an angle θ is described as being reflected from that plane at an equal angle (Fig. 13-15a). This corresponds to a scattering angle 2θ. The Bragg conditions for observing diffraction require that the path difference between reflected beams from adjacent lattice planes be an integral number of wavelengths. From Figure 13-15a, we see that this condition clearly is met wherever

$$2d \sin \theta = n\lambda \tag{13-71}$$

where n is any integer, and d is the distance between two adjacent lattice planes.

To compare the Bragg treatment with our previous description, it is necessary to show how the scattering vector **S** is related to a lattice plane. Consider a lattice plane that intersects the three axes of a unit cell at $\mathbf{a}/h$, $\mathbf{b}/k$, and $\mathbf{c}/l$ (Fig. 13-15b). Let $\mathbf{r}$ be a vector drawn from the origin to any point in this plane. Consider the properties of a scattering vector **S** that happens to satisfy the equation $\mathbf{S} \cdot \mathbf{r} = 1$. For a fixed direction of **S**, the relation $\mathbf{S} \cdot \mathbf{r} = 1$ defines a plane perpendicular to **S**, because it simply means that the projection of **r** on **S** is a constant.

As we showed previously, not all values of **S** lead to detectable scattering. Only those values that satisfy the von Laue conditions are acceptable. These conditions are $\mathbf{S} \cdot \mathbf{a} = h$, with equivalent equations for **b** and **c**. We can put the von Laue conditions in the form

$$\mathbf{S} \cdot \mathbf{a}/h = \mathbf{S} \cdot \mathbf{b}/k = \mathbf{S} \cdot \mathbf{c}/l = 1 \tag{13-72}$$

In other words, $\mathbf{a}/h$, $\mathbf{b}/k$, and $\mathbf{c}/l$ are all values of **r** that satisfy the condition $\mathbf{S} \cdot \mathbf{r} = 1$. These three values uniquely define a plane (Fig. 13-15b). This plane is a lattice plane (as described in Fig. 13-14), but it is also a plane containing values of **S** that lead to x-ray diffraction.

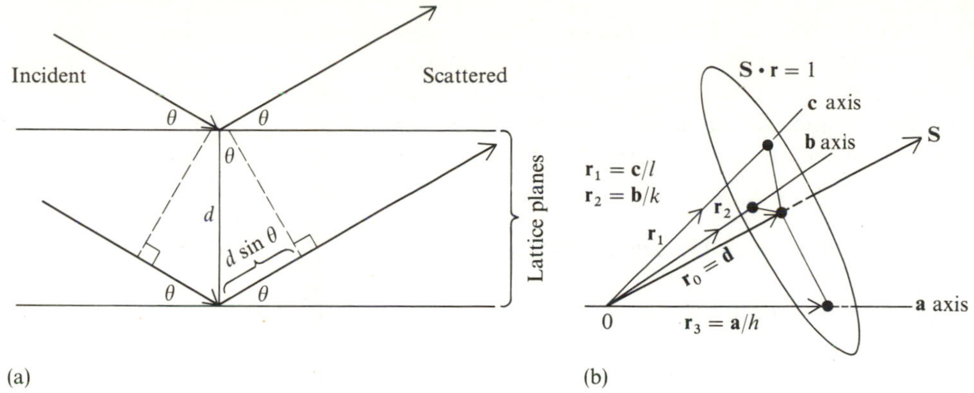

Figure 13-15

Derivation of Bragg's law. **(a)** Diffraction viewed as reflection of x rays from adjacent lattice planes. **(b)** The lattice plane (h, k, l) is a plane containing vectors that satisfy the von Laue scattering conditions.

The lattice plane adjacent to the plane defined by $\mathbf{S} \cdot \mathbf{r} = 1$ will pass through the origin. The spacing d between these two planes is the length of a vector $\mathbf{r}_0$ parallel to $\mathbf{S}$ (Fig. 13-15b). The condition $\mathbf{S} \cdot \mathbf{r}_0 = 1$ means that $d = |\mathbf{r}_0| = 1/|\mathbf{S}|$. However, we showed earlier that $|\mathbf{S}| = 2|\sin\theta|/\lambda$. Therefore, the scattering angle (2θ) produced by crystal planes separated by a spacing d, is given by

$$\sin\theta = \lambda/2d \qquad (13\text{-}73)$$

This is identical to Equation 13-71 with $n = 1$. Thus, the equivalence of the Bragg treatment and the von Laue conditions has been illustrated. (To derive the full Bragg equation, consider the properties of the plane defined by $\mathbf{S} \cdot \mathbf{r} = n$ where n is an integer.)

13-3 PROPERTIES OF CRYSTALS

A crystal is a three-dimensional ordered array of molecules. From the discussion of x-ray scattering in the previous section, it is clear that crystals are not a requirement for x-ray diffraction measurements. Any ordered (or partially ordered) array of molecules can, in principle, produce useful x-ray data. However, it is evident that crystals are the most favorable samples. A large ordered array leads to sharp diffraction spots, which concentrate the scattered intensity in small discrete regions of scattering angle (2θ). This greatly facilitates the acquisition of reliable intensity data.

If the sample is not a perfectly ordered crystal, intensity can reach observable levels over wider ranges of scattering angles. The diffraction pattern can smear into rings or streaks, and therefore considerable imprecision is introduced in assigning values of θ observed. In general, only if the sample has three-dimensional order will the diffraction pattern contain all the information needed to reconstruct the three-dimensional structure. Disorder corresponds to averaging over orientations of both the lattice and the molecules it contains. The resulting data then contains only information about the averaged structure.

Restrictions on possible crystal lattices

A crystal is essentially a three-dimensional mosaic. The unit cell defined by the vectors **a**, **b**, and **c** contains the fundamental repeating unit. The crystal is generated by successive translations of the unit cell along the axes **a**, **b**, and **c**; in just the same way, a mosaic is built up by placing down multiple copies of the same unit structure. It is a fundamental consequence of geometry that three-dimensional space can be filled only by mosaics of cells of certain shapes. There are, in fact, only seven fundamental types of unit cells. Each defines a crystal system (Fig. 13-16; Table 13-1).

Each unit cell consists of a *motif* that is the actual unit repeated throughout the crystal by the lattice translations. The crystal is a convolution of the motif and the lattice (Fig. 13-17a). A motif can be a single atom or molecule, or it can contain more than one molecule.

The simplest possible crystals would have one motif positioned with the same orientation at each corner of the unit cells. There are eight corners and each is shared by eight unit cells. Therefore, there is one motif per unit cell. Such lattices are called primitive, and they are denoted by the letter *P*. It is always possible to choose a primitive triclinic cell for any lattice. This is the least-symmetric unit cell. Each dimension and each angle are different. So it takes six parameters to specify such a cell.

There are many cases in which the symmetry of the lattice can be increased if a larger unit cell containing additional lattice points located on the faces or at the center is chosen. These nonprimitive lattices have more than one copy of the motif per unit cell. By choosing a nonprimitive lattice, one often can describe the crystal with fewer parameters. There are a total of seven nonprimitive lattices (Fig. 13-16). They are designated *I* for cells having an extra lattice point at the center, *C* for cells with two extra lattice points on one pair of opposite faces, and *F* for cells with extra lattice points on all faces. You should be able to convince yourself that *C* and *I* lattices have two motifs per unit cell, whereas *F* lattices have four.

It is important to recognize that the choice of lattice is not always unique. Figure 13-18 shows a few examples of alternative choices. There are certain conventions that help to decide which lattice to use, but we need not be concerned with them here and, in any event, they are not always rigidly adhered to. For each type of lattice, there are only certain arrangements of molecules or motifs that can be inserted without reducing the lattice to one of greater symmetry or one with a smaller unit cell.

The choice of lattice can simplify the analysis of x-ray scattering data. However, it is important to reiterate the point made earlier that the x-ray scattering calculated for a crystal of known structure is independent of the choice of lattice.

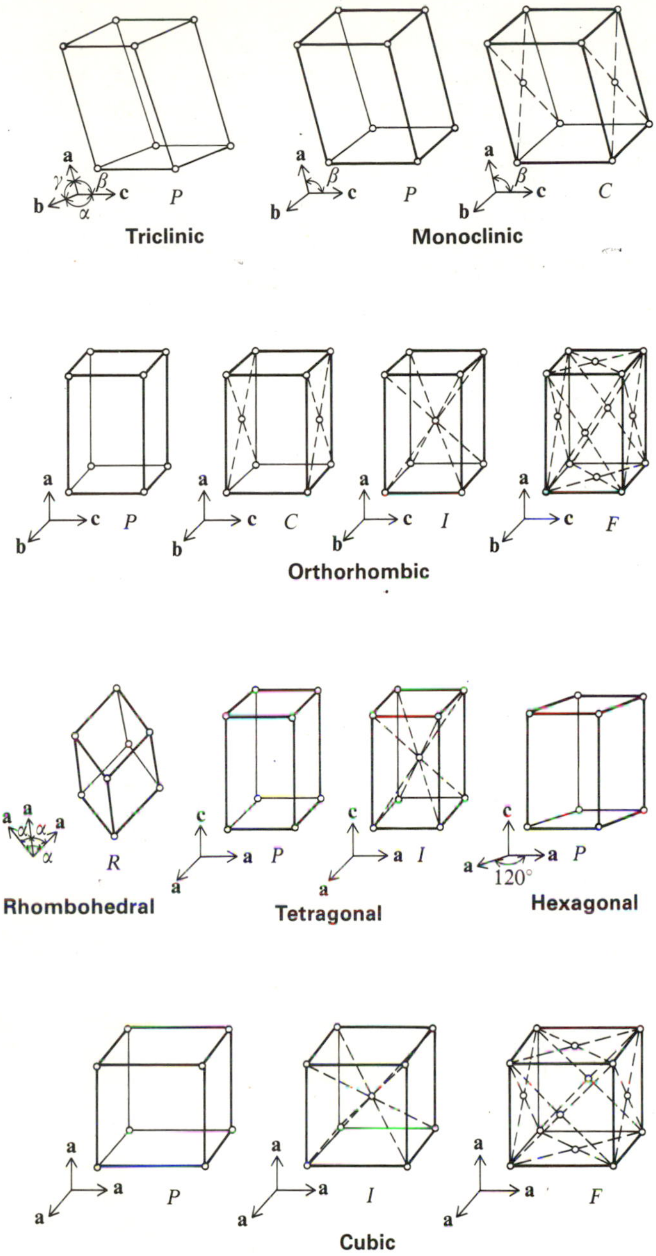

Figure 13-16

The fourteen Bravais lattices. For a list of their properties, see Table 13-1. [After G. H. Stout and L. M. Jensen, *X-Ray Structure Determination* (New York: Macmillan, 1968).]

Table 13-1

The 65 space groups allowed for molecules with no mirror or inversion symmetry

Crystal system	Number of independent parameters	Lattice	Minimum symmetry of unit cell	Unit cell edges and angles§	Diffraction pattern symmetry#	Space groups¶
Triclinic	6	P	None	$a \neq b \neq c$ $\alpha \neq \beta \neq \gamma$	$\bar{1}$	$P1$
Monoclinic	4	P	Twofold axis parallel to **b**	$a \neq b \neq c$ $\alpha = \gamma = 90°$ $\beta \neq 90°$	$2/m$	$P2$, $P2_1$ $C2$
Orthorhombic	3	P C I F	Three mutually perpendicular twofold axes	$a \neq b \neq c$ $\alpha = \beta = \gamma = 90°$	mmm	$P222$, $P2_12_12_1$, $P222_1$, $P2_12_12$ $C222$, $C222_1$ [$I222$, $I2_12_12_1$] $F222$
Tetragonal	2	P I	Fourfold axis parallel to **c**	$a = b \neq c$ $\alpha = \beta = \gamma = 90°$	$4/m$ $4/mmm$	$P4$, ($P4_1$, $P4_3$), $P4_2$ $I4$, $I4_1$ $P422$, ($P4_122$, $P4_322$), $P4_222$, $P42_12$, ($P4_12_12$, $P4_32_12$), $P4_22_12$ $I422$, $I4_122$
Trigonal/ rhombohedral	2	R§§ P§§	Threefold axis parallel to **c**	$a = b = c$ $\alpha = \beta = \gamma \neq 90°$	$\bar{3}$ $\bar{3}m$	$R3$ $P3$, ($P3_1$, $P3_2$) $R32$ [$P321$, $P312$], [($P3_121$, $P3_221$), ($P3_112$, $P3_212$)]
Hexagonal	2	P	Sixfold axis parallel to **c**	$a = b \neq c$ $\alpha = \beta = 90°$ $\gamma = 120°$	$6/m$ $6/mmm$	$P6$, ($P6_1$, $P6_5$), $P6_3$, ($P6_2$, $P6_4$) $P622$, ($P6_122$, $P6_522$), $P6_322$, ($P6_222$, $P6_422$)
Cubic	1	P I F	Threefold axes along cube diagonals	$a = b = c$ $\alpha = \beta = \gamma = 90°$	$m3$ $m3m$	$P23$, $P2_13$ [$I23$, $I2_13$] $F23$ $P432$, ($P4_132$, $P4_332$), $P4_222$ $I432$, $I4_143$ $F432$, $F4_132$

§ See Figure 13-16 for definitions of edge and angle symbols.

A number with an overbar indicates a rotary inversion axis; m = a mirror plane; $2/m$ = a mirror plane perpendicular to a twofold axis; $6/m$ indicates a mirror plane perpendicular to a sixfold axis.

¶ Pairs of space groups in parentheses differ from each other only in that they are enantiomorphs. Space groups enclosed in brackets (and also those in parentheses) cannot be distinguished from one another by systematic extinctions of reflections in the diffraction pattern. All other space groups can be assigned on the basis of the diffraction pattern.

§§ The rhombohedral system often is regarded as a subdivision of the hexagonal system, and unit cells in this system may be chosen on either hexagonal or rhombohedral axes.

SOURCE: After D. Eisenberg, in *The Enzymes*, 3d ed., vol. 1, P. D. Boyer (New York: Academic Press, 1970).

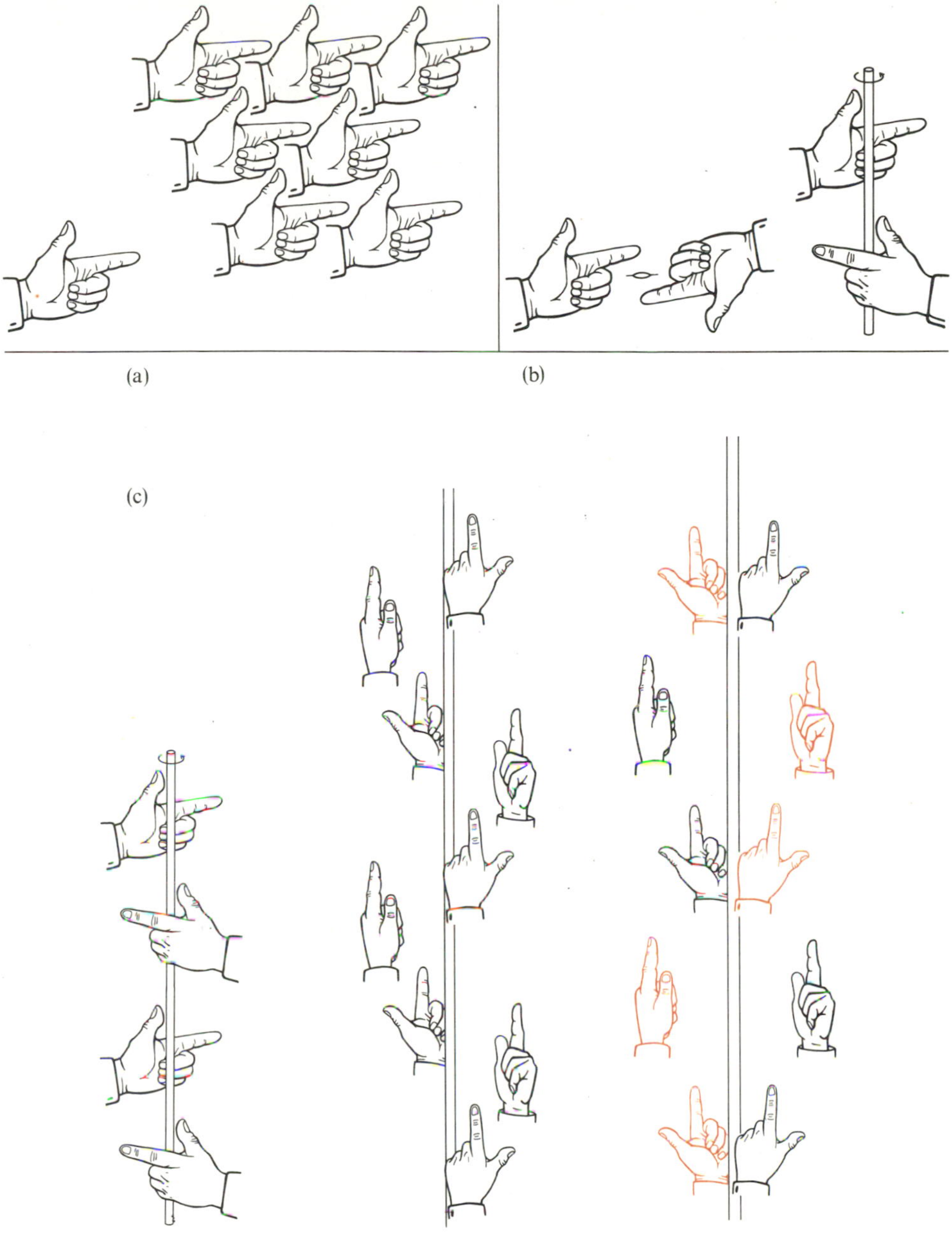

Figure 13-17

Motifs, lattices, and symmetry operations. **(a)** A lattice and simple motif (a single hand). The crystal is the convolution of the motif and the lattice. **(b)** Two motifs containing symmetry-related structures. The two hands on the left are related by a twofold rotation (C_2) axis. The two hands on the right are related by a 2_1 screw axis. In each structure, the motif consists of two hands. The asymmetric unit is just a single hand. **(c)** Arrays generated by screw axes. From left to right, the axes are 2_1, 4_1, and 4_2; the resulting unit cells have 2, 4, and 4 asymmetric units, respectively. [Drawings by Irving Geis.]

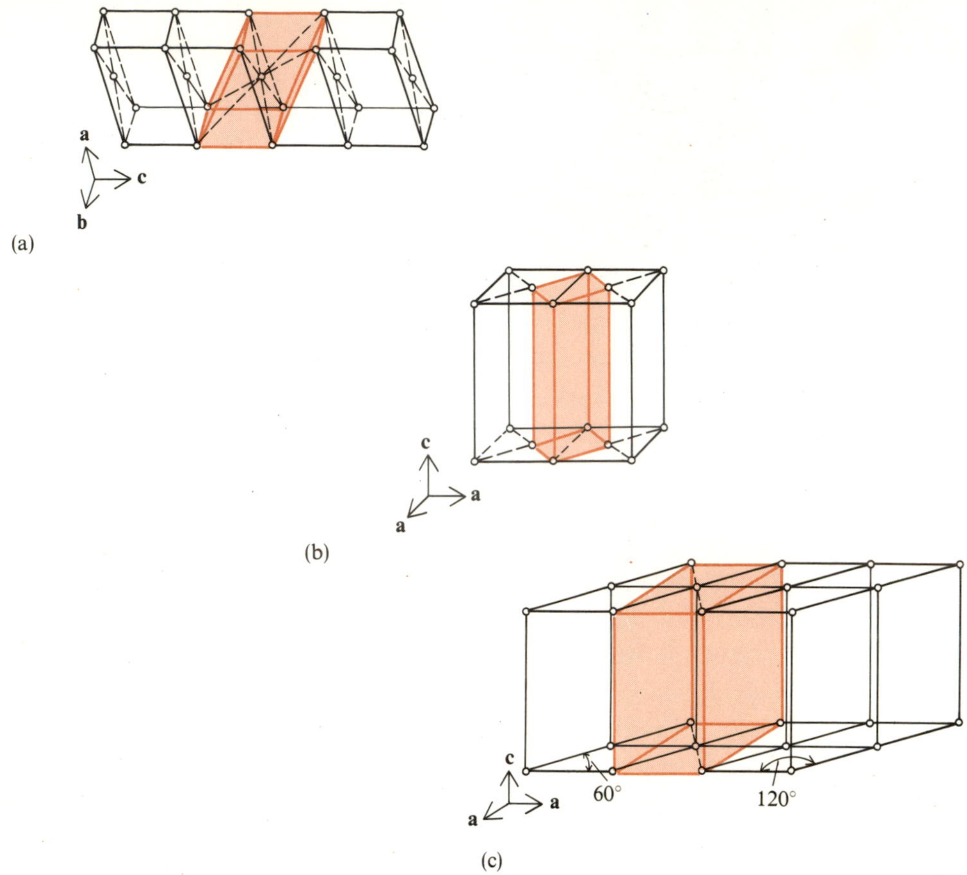

Figure 13-18

Choosing different unit cells for the same lattice. **(a)** Choice of *C* or *I* unit cells in a monoclinic lattice. **(b)** Choice of *C* or *P* unit cells in a tetragonal lattice. **(c)** Choice of an orthorhombic *C* lattice or a hexagonal *P* lattice. [After G. H. Stout and L. M. Jensen, *X-Ray Structure Determination* (New York: Macmillan, 1968).]

Symmetry properties of molecules and crystals

The overall symmetry of the crystal is called the space group. It is described by naming the type of unit cell and any symmetry relationships *within* the molecules that make up the motif. For arbitrary structures, it turns out that there are precisely 230 possible space groups. These contain two types of symmetry: point symmetry and space symmetry.

Point-symmetry operations consists of manipulations of an isolated object that leave at least one point in space unchanged (Box 2-3). These can include

1. *rotation axes*, named by a number (2 for twofold axis, 3 for threefold axis, and so on);
2. *mirror planes*, designated by m;
3. *rotation coupled with reflection* (for example, combining a twofold rotation axis with a mirror plane perpendicular to it, resulting in inversion of an object through an origin located at the intersection of the rotation axis and the mirror plane; this operation is designated $2/m$);
4. *rotation–inversion axes*, designated by a number with an overbar (for example, $\bar{4}$ indicates that each rotation of 90° is accompanied by inversion through the origin).

A point group is a list of all of the point-symmetry relationships possessed by an object. The object can be a molecule, a set of molecules, or an entire crystal. Several of the point groups possible for molecules consisting of multiple copies of identical subunits are illustrated in Chapter 2.

Space-symmetry operations involve translation of the object. These include *screw axes* (which are a rotation accompanied by translation) and *glide planes* (which are translations accompanied by reflection). Screw axes are called n_m, where n is the rotation axis, and m/n is the fraction of a unit cell along which the translation occurs. For example, 3_1 indicates a rotation of 120° accompanied by a translation 1/3 of the unit-cell length. The description of glide planes is complicated, because it depends on which face or diagonal the glide is along, as well as on how far the glide occurs. Figure 13-17c shows a few examples of motif-symmetry operations.

The presence of particular symmetry elements in the motif restricts the possible type of unit cell. For example, if a twofold rotation axis is present in the space group, this axis must be perpendicular to two unit-cell vectors. Otherwise, this symmetry operation would leave the motif unaltered internally, but would change its location within the unit cell. The presence of threefold or higher rotation axes requires that the two unit-cell vectors perpendicular to the axis must be equal in length.

Space groups available to biological molecules

The allowed combinations of the point and space symmetry possessed by the motif generate the 230 space groups. It is convenient to introduce the concept of an *asymmetric unit*. This is the smallest unit from which the crystal structure can be generated by making use of the symmetry operations of the space group. The asymmetric unit can be several molecules, one molecule, or a subunit of an oligomeric molecule. The crystal is generated, first by creating the motif by the space-group symmetry operations on the asymmetric unit, and then by translation of the motif through the lattice.

The number of asymmetric units per unit cell, n', is determined by the space group.

For biological molecules, the motifs inevitably contain asymmetric carbon atoms. Therefore, the symmetry arrangement of the molecules can never contain mirror planes, glide planes, centers of symmetry, or rotation inversion axes. Only 65 of the 230 space groups can apply to biological molecules. The biologically relevant space groups can contain 1, 2, 3, 4, 6, 8, 12, 24, 48, or 96 asymmetric units per unit cell (Table 13-1).

A practicing crystallographer presumably will learn to picture many of these space groups. However, molecules seem to prefer to crystallize in only a limited number of space groups. For example, 80% of 1,200 organic compounds surveyed fell into the triclinic, monoclinic, or orthorhombic crystal classes, and half of these occurred in just three space groups. Figure 13-19 shows a few of the most commonly found of the space groups allowable for biological molecules. Note that these different groups imply different numbers of molecules per unit cell.

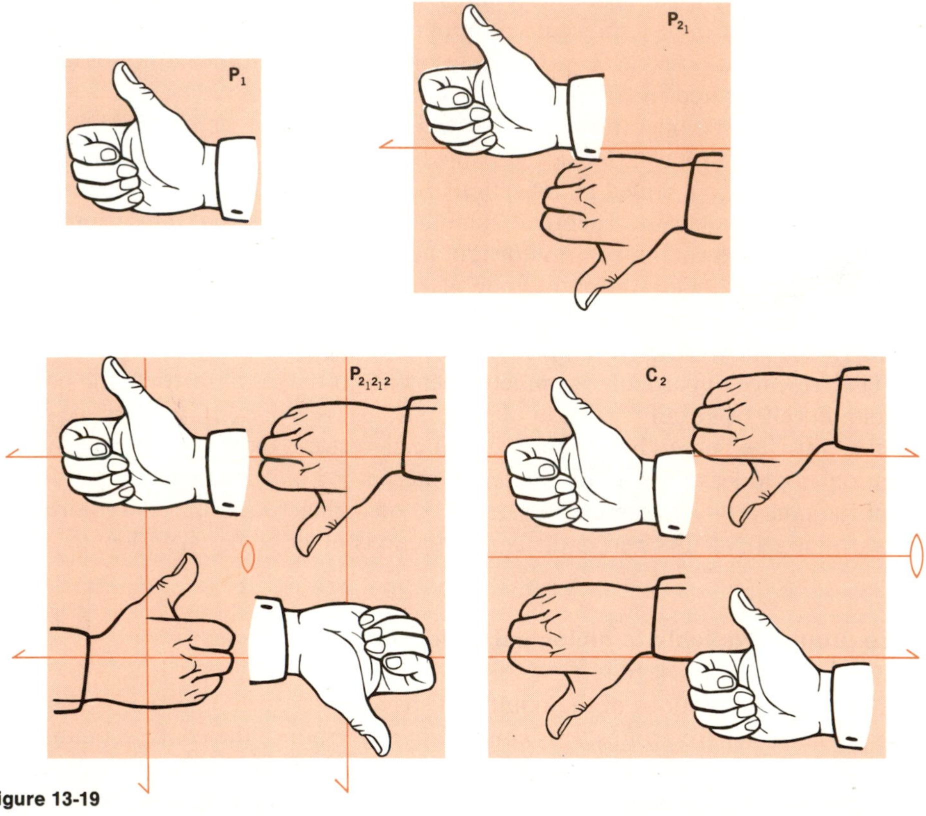

Figure 13-19

A familiar asymmetric unit as it might appear in four different space groups. $P1$ has no motif symmetry; $P2_1$ has a single 2_1 screw axis shown as a half-arrow; $P2_12_12$ has four screw axes (each 2_1) and a C_2 axis perpendicular to the plane of the page; $C2$ has two 2_1 screw axes and a C_2 rotational axis shown as the full arrow. [Drawings by Irving Geis.]

Determination of the dimensions of the crystal lattice

X-ray diffraction occurs whenever the scattering vector coincides with a reciprocal lattice point, as we have shown. That means that the resulting diffraction pattern can be used to construct an image of the reciprocal lattice. From the spacing between diffraction spots as actually observed in the laboratory and a knowledge of the geometry of the diffraction experiment, one can compute the spacing between points on the reciprocal lattice. This in turn allows the geometry of the unit cell to be calculated.

Here we shall demonstrate the determination of the spacing of a one-dimensional crystal. The crystal is placed in the center of a cylindrical film (Fig. 13-20). The von Laue conditions for the **a** crystal axis require that $\mathbf{S} \cdot \mathbf{a} = h$. Consider the first two diffraction planes, which will occur at $h = 0$ and $h = 1$. If we make measurements with the crystal oriented so that **S** is parallel to **a**, then the length of the scattering vector, S, is 0 for $h = 0$ and is $1/a$ for $h = 1$. The scattering angle is computed from Equation 13-5; $|\mathbf{S}| = 2|\sin\theta|/\lambda$. For $h = 0$, we have $\sin\theta = 0$ and $\theta = 0$. For $h = 1$, we have $\sin\theta = \lambda|\mathbf{S}|/2 = \lambda/2a$. Therefore, the angle between the two scattering planes is $2\theta = 2\sin^{-1}\lambda/2a$. Thus, if θ is measured experimentally, and if λ is known, the distance a can be computed.

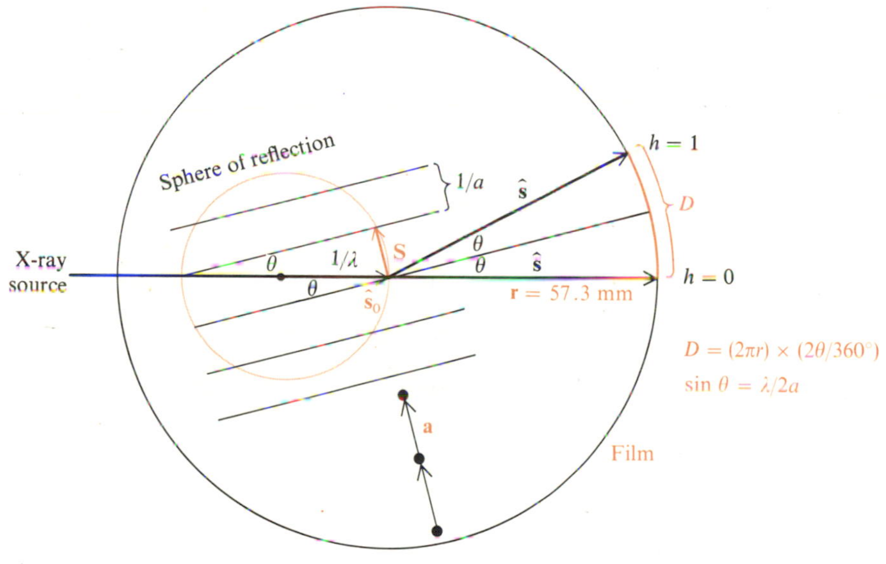

Figure 13-20

Experimental scattering geometry. Shown are a sample at the origin, a section of the reciprocal lattice, one scattering vector **S** (and the scattered radiation associated with it), and two layer lines as they intersect the film. Below the reciprocal lattice, three atoms in the actual crystal lattice are illustrated to show the orientation of the sample. In this example, for clarity, we show values of $\lambda/2a$ and θ much larger than the values typically encountered in actual experiments.

In a typical case, λ might be 1 Å, and a might be 10 Å. Therefore, $\sin \theta$ is 1/20, or θ is about 3°. A common x-ray camera would have film arranged in a cylinder 28.65 mm in radius. Its circumference is $2\pi \times 28.65$ mm. The angle between the $h = 0$ and $h = 1$ scattering planes is 2θ. This is 6°, or 6/360 the circumference of the film. Therefore, the distance between the two scattering planes as they intersect the film is $2\pi \times 28.65 \times 6/360 \cong 3$ mm.

Note that, although the actual crystal spacings are very small, the film is placed far away from the sample. This magnifies the diffraction pattern until planes are physically separated by a distance convenient for measuring. In an unknown case, all one has to do is work the calculation backwards to obtain a. The actual equation is $a = \lambda/2 \sin(360D/4\pi r)$, where D is the physically measured spacing on the x-ray film, and r is the radius of the camera.

The relationship between the crystal lattice and the reciprocal lattice

Real crystals are three-dimensional. The reciprocal lattice that one sees in an x-ray diffraction pattern also is three-dimensional. It is related in a simple way to the actual crystal lattice. By measuring the spatial pattern of diffracted spots, it is possible to compute the cell dimensions and shape of the reciprocal lattice. From this, the corresponding dimensions and shape of the unit cell of the actual crystal lattice can be derived.

Figure 13-21a shows that each of the vectors $\mathbf{a}^*$, $\mathbf{b}^*$, and $\mathbf{c}^*$ defining the reciprocal cell is located along lines formed by the intersection of two planes. For example, $\mathbf{c}^*$ is formed by the intersection of planes generated by successive values of h for the von Laue condition $\mathbf{a} \cdot \mathbf{S} = h$ (and therefore these planes are perpendicular to $\mathbf{a}$) and planes generated by $\mathbf{b} \cdot \mathbf{S} = k$ (and thus perpendicular to $\mathbf{b}$). This means that $\mathbf{c}^*$ must be perpendicular to both $\mathbf{a}$ and $\mathbf{b}$, and we can write, in general,

$$\mathbf{c}^* = r\mathbf{a} \times \mathbf{b} \tag{13-74a}$$

$$\mathbf{b}^* = q\mathbf{c} \times \mathbf{a} \tag{13-74b}$$

$$\mathbf{a}^* = p\mathbf{b} \times \mathbf{c} \tag{13-74c}$$

These constants control the magnitudes of the reciprocal-cell vectors. To determine the constants, we must use the von Laue conditions that generate the reciprocal lattice. For example, the condition $\mathbf{S} \cdot \mathbf{c} = l$ generates a set of planes spaced by $1/c$. The vector $\mathbf{c}^*$ extends between two such planes, although it is not necessary normal to them (Fig. 13-21b). However, the projection of $\mathbf{c}^*$ on $\mathbf{c}$ must be $1/c$ (Fig. 13-21b). Thus we can write

$$\mathbf{c} \cdot \mathbf{c}^* = |\mathbf{c}|(1/c) = 1 \tag{13-75}$$

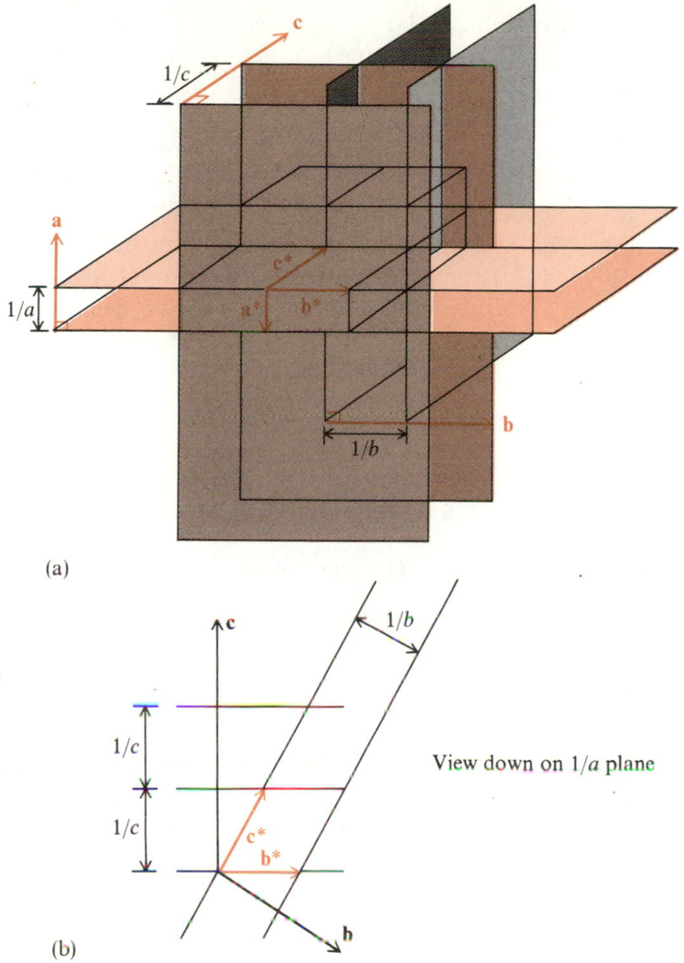

Figure 13-21

Geometrical properties of the reciprocal lattice. **(a)** Reciprocal-lattice vectors lie at the intersections of sets of parallel planes. For example, $\mathbf{c}^*$ *extends* between two planes spaced $1/|\mathbf{c}|$ apart, but it is *formed* by the intersections of planes perpendicular to $\mathbf{a}$ (spaced $1/|\mathbf{a}|$ apart) and planes perpendicular to $\mathbf{b}$ (spaced $1/|\mathbf{b}|$ apart). **(b)** Demonstration that $\mathbf{c}^*$ and $\mathbf{c}$ are not necessarily parallel.

Similarly, one can show that $\mathbf{a} \cdot \mathbf{a}^* = 1$ and $\mathbf{b} \cdot \mathbf{b}^* = 1$.

When Equation 13-74 is inserted into Equation 13-75, the result is $1 = \mathbf{c} \cdot \mathbf{c}^* = r\mathbf{c} \cdot \mathbf{a} \times \mathbf{b}$, from which we can evaluate r. Carrying out equivalent manipulations for $\mathbf{a}^*$ and $\mathbf{b}^*$, we obtain

$$r = 1/(\mathbf{c} \cdot \mathbf{a} \times \mathbf{b}) \qquad q = 1/(\mathbf{b} \cdot \mathbf{c} \times \mathbf{a}) \qquad p = 1/(\mathbf{a} \cdot \mathbf{b} \times \mathbf{c}) \tag{13-76}$$

Using the properties of the triple scalar product (see Box 8-2), each of these quantities is equal to the volume of the parallelopiped formed by the three vectors **a**, **b**, and **c**. Thus, $r = q = p = 1/V$, where V is the volume of the unit cell of the crystal. Using Equations 13-74, 13-75, and 13-76, we can construct the reciprocal cell if the actual unit cell of the crystal is known. Figure 13-22 shows two examples.

In practice, observations of the geometric pattern of diffraction spots allow measurement of the reciprocal lattice vectors $\mathbf{a}^*$, $\mathbf{b}^*$, and $\mathbf{c}^*$. Then one must compute the unit-cell vectors. The procedures are quite similar to that just outlined. Note, for example, that $\mathbf{b}^*$ and $\mathbf{c}^*$ lie in a plane perpendicular to **a** (Fig. 13-21). Therefore, $\mathbf{a} = r'\mathbf{b}^* \times \mathbf{c}^*$, and similar equations exist for **b** and **c**. To determine r', one uses the constraint $\mathbf{a} \cdot \mathbf{a}^* = 1$. Then $r' = 1/(\mathbf{a}^* \cdot \mathbf{b}^* \times \mathbf{c}^*) = 1/V^*$, where V^* is the volume of the reciprocal cell. Thus, the unit cell is constructed from the measured diffraction data by

$$\mathbf{a} = (1/V^*)(\mathbf{b}^* \times \mathbf{c}^*) \tag{13-77a}$$

$$\mathbf{b} = (1/V^*)(\mathbf{c}^* \times \mathbf{a}^*) \tag{13-77b}$$

$$\mathbf{c} = (1/V^*)(\mathbf{a}^* \times \mathbf{b}^*) \tag{13-77c}$$

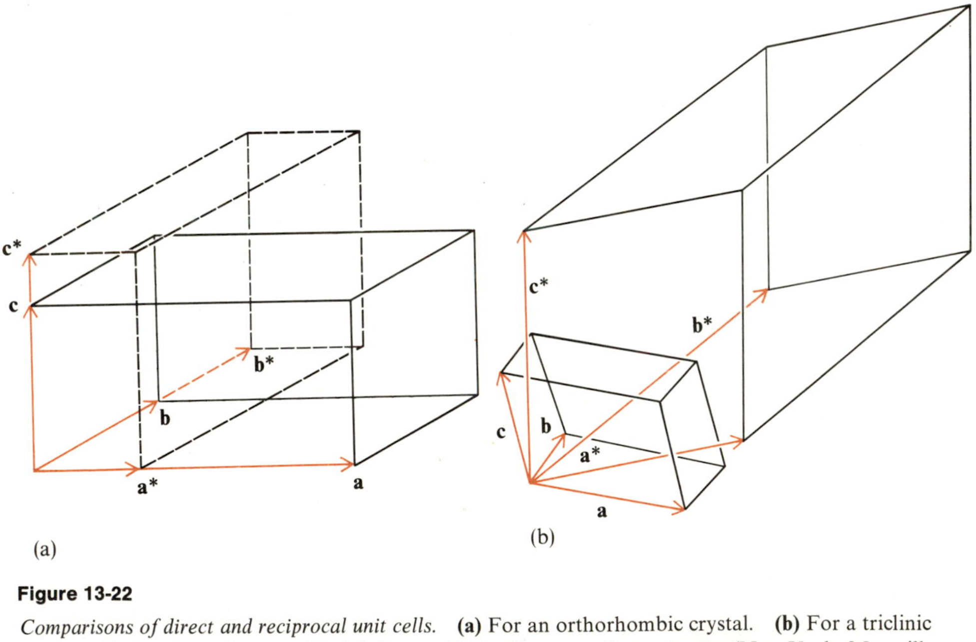

Figure 13-22

Comparisons of direct and reciprocal unit cells. **(a)** For an orthorhombic crystal. **(b)** For a triclinic crystal. [After G. H. Stout and L. M. Jensen, *X-Ray Structure Determination* (New York: Macmillan, 1968).]

A necessary consequence of Equations 13-74 and 13-77 is that the volumes of unit cells and reciprocal cells are inverse (see Guinier, 1963, p. 88):

$$V = 1/V^* \tag{13-78}$$

Determination of the space group

In addition to determining the unit cell, it is useful to establish the space group of the crystal. This shows if there is any internal symmetry, and whether one can use this symmetry to reduce the portion of the structure that must be solved. There is no general way to find the space group but, for many biological molecules, one can sharply narrow the possibilities by simple examination of the intensity of diffraction spots in the reciprocal lattice. Particular crystal classes show symmetries in the diffraction pattern that correspond to symmetries in the space group. For example, a twofold rotation axis in the crystal leads to a mirror plane in the diffraction pattern intensities.

Even more informative are systematic absences of intensity at certain points of the reciprocal lattice for many space groups. Consider a space group with a twofold screw axis along **c**. This axis rotates x to $-x$, rotates y to $-y$, and translates half of the unit-cell distance along **c**. Then, for each atom at $\mathbf{r} = x_j\mathbf{a} + y_j\mathbf{b} + z_j\mathbf{c}$, there must be an identical atom at $\mathbf{r}' = -x_j\mathbf{a} - y_j\mathbf{b} + (\mathbf{z}_j + 1/2)\mathbf{c}$. In calculating the structure factor, one can group the identical atoms by pairs. From Equation 13-70,

$$F_{\mathrm{m}}(h, k, l) = \sum_{j=1}^{N/2} f_j(S)(e^{2\pi i(hx_j + ky_j + z_j)} + e^{2\pi i(-hx_j - ky_j + lz_j + l/2)}) \tag{13-79}$$

When $h = k = 0$, the structure factor becomes

$$F_{\mathrm{m}}(0, 0, l) = \sum_{j=1}^{N/2} f_j(S)[e^{2\pi i l z_j}(1 + e^{2\pi i l/2})] \tag{13-80}$$

Whenever l is odd, the exponential in the last term becomes equal to -1, and so the scattering amplitude vanishes. Thus, there will be no diffraction in the special case $h = 0$, $k = 0$, $l =$ odd. If such absences are not sufficient to uniquely determine the space group, sometimes a statistical analysis of the pattern of intensities can complete the assignment.

Crystallographic estimation of molecular weight

Once the lattice and space group are known, it frequently is possible to determine the molecular weight of the molecules that compose the crystals. The density of the crystal, ρ_c, can be measured experimentally. Then the weight W of one unit cell can be computed as

$$W = \rho_c V \tag{13-81}$$

where V is the volume of the unit cell determined from the diffraction pattern.

In general, protein and other macromolecule crystals can be viewed as containing three components: anhydrous macromolecule (m), free solvent (s), and bound water (w). The weight of one unit cell will be the sum of the three components:

$$\rho_c V = \rho_m V_m + \rho_w V_w + \rho_s V_s \tag{13-82}$$

where V refers to the volume of each component, and ρ is its density. Usually, ρ_s is known experimentally, and ρ_w can be taken as the density of pure water.

We want to compute the weight of macromolecule per unit cell, $\rho_m V_m$. Thus, we must eliminate the unknown quantities V_w and V_s from Equation 13-82. The volume of bound water, V_w, can be written in terms of the hydration δ_1, in grams water per gram macromolecule.

$$V_w = \delta_1 \rho_m V_m / \rho_w \tag{13-83}$$

Hydration values are known approximately for proteins and nucleic acids (Chapter 10). The total volume of the unit cell is

$$V = V_m + V_w + V_s \tag{13-84}$$

Using Equation 13-83, the volume of free solvent can be written as

$$V_s = V - V_m(1 + \delta_1 \rho_m/\rho_w) \tag{13-85}$$

Inserting Equations 13-83 and 13-85 into Equation 13-82, and solving the resulting expression for $\rho_m V_m$, we obtain

$$W_m = \rho_m V_m = \frac{V(\rho_c - \rho_s)}{1 - \rho_s/\rho_m + \delta_1(1 - \rho_s/\rho_w)} \tag{13-86}$$

Thus the weight of macromolecule per unit cell (W_m) can be calculated if the anhydrous density (ρ_m) is known. Usually it is a good approximation to equate ρ_m^{-1} with the partial specific volume ($\bar{V}_m$) measured for the macromolecule in solution.

If there is a single macromolecule per unit cell, then the molecular weight is just $M = N_0 W_m$ where N_0 is Avogadro's number. If the space group indicates n'

asymmetric units per unit cell, the molecular weight of an asymmetric unit is

$$M = N_0 W_m / n' \tag{13-87}$$

Alternative methods for determining the molecular weight of molecules in a crystal are discussed by B. W. Matthews (1975).

Often an estimate of the molecular weight is already available from hydrodynamic measurements or primary structure data. Then n' can be computed from a measurement of W_m. This value must always be an integer. Therefore, once an estimate of n' is available, it can be used to refine the value of the molecular weight.

Using the space group for information on macromolecule symmetry

In most cases, the number of molecules per unit cell (n) is equal to the number of asymmetric units (n'). Here we consider the special case where the macromolecule is an oligomer of identical subunits. For example, a molecule with five subunits might have C_5 symmetry. But this symmetry can never correspond to a symmetry element of the space group, because there is no space group with a C_5 rotation axis. Therefore, the asymmetric unit must contain all five subunits.

On the other hand, in many cases, a molecule with C_2 or C_3 rotational symmetry crystallizes so that its axis is also a symmetry axis of the motif. Then it is possible that the number of asymmetric units per cell will be an integral multiple of the number of subunits per unit cell, rather than a multiple of the number of molecules. This relationship permits one to infer the presence of rotational axes of symmetry in the macromolecule. Note, however, that it is not necessary for all rotation axes of a molecule simultaneously to be rotation axes of the crystal. Therefore, the estimate of symmetry is a minimal estimate.

An example is aspartate transcarbamoylase, which was treated in Chapter 2. In one crystal form, this enzyme crystallizes in a space group with eight asymmetric units per cell, but there are only four molecules per cell. This indicates the presence of a twofold rotation axis. In a second crystal form, the space group has six asymmetric units per cell, but the cell dimensions allow for only two molecules per cell. Thus, a threefold rotation axis exists in the molecule. Because each subunit must be an asymmetric object, the only way that both of these axes can exist simultaneously is for the molecule to have six (or some integral multiple of six) subunits of each type. For aspartate transcarbamoylase, the subunit structure is c_6r_6, and a model of the symmetric arrangement is shown in Figure 2-49.

Varying scattering geometry to measure diffraction pattern

Reciprocal lattice points are precisely those locations in space that satisfy the von Laue conditions for scattering. Whenever the crystal and incident beam ($\hat{\mathbf{s}}_0$) are oriented so that the scattering vector **S** contacts a reciprocal lattice point, diffracted intensity is observed along the vector $\hat{\mathbf{s}} = \lambda\mathbf{S} + \hat{\mathbf{s}}_0$. In making measurements, one has the choice of varying the orientations of the crystal, the detector, or the incident beam. Usually it is the crystal that is allowed to move. The reciprocal lattice is fixed in space for a fixed crystal orientation. If the crystal is rotated through an angle about an axis, the reciprocal lattice will rotate the same angle about the same laboratory axis.

For given x-ray wavelength, crystal orientation, and incident x-ray beam, all possible scattering vectors extend from the origin to the surface of the sphere of reflection (Fig. 13-23a). The sphere can be generated from Figure 13-3a by allowing all possible orientations of **S**. It has a diameter of $2/\lambda$ because this is the largest possible value of $|\mathbf{S}|$, occurring at $\theta = 90°$. The surface of the sphere passes through the origin of the reciprocal lattice. Here, $h = k = l = 0$, the scattering vector **S** has length zero, and $\theta = 0°$; all radiation is forward scattered.

The sphere of reflection will enclose a set of reciprocal lattice points. However, diffraction will be observed only when these points intersect the surface of the sphere. Clearly, if the reciprocal lattice is literally composed of points, the probability of this occurring is infinitesimal. Fortunately, the actual radiation used in diffraction experiments is a distribution of wavelengths. This means that a spherical zone of **S** applies, rather than just a surface. Furthermore, in an actual crystal, zones of unit cells differ very slightly in orientation. This effect, called mosaicism, means that the reciprocal lattice points will be of a finite size. Nevertheless, not many lattice points will intersect the surface of the scattering sphere simultaneously (Fig. 13-23a).

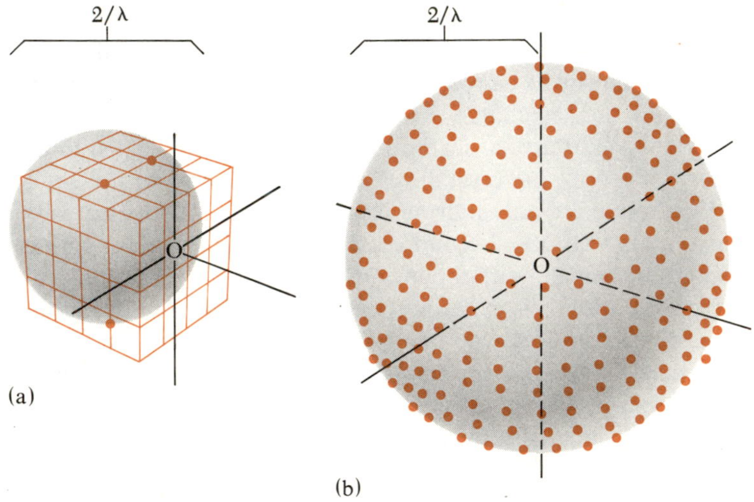

Figure 13-23

Experimental restrictions on the observation of x-ray diffraction. **(a)** For a fixed geometry and x-ray wavelength, scattering will be observed only when the surface of the sphere of reflection intersects reciprocal-lattice points. **(b)** Even if all possible geometries are sampled, only that portion of the reciprocal lattice that lies within a sphere of radius $2/\lambda$ (the limiting sphere) can be examined.

To collect sufficient diffraction data to solve a crystal structure, one must measure as many diffracted rays as possible. Therefore, what is usually done is to rotate or oscillate the crystal in a systematic way. This causes successive lattice points to intersect the scattering sphere and permits the resulting diffraction to be measured (Fig. 13-24). Note that it is the discontinuous nature of the reciprocal lattice that makes it difficult to collect diffraction data for a three-dimensional crystal. In a two-dimensional sample, the reciprocal lattice is a set of lines (Fig. 13-12). Most of these will intersect the sphere of reflection somewhere, and so a single sample geometry can yield many diffraction spots (Fig. 13-11).

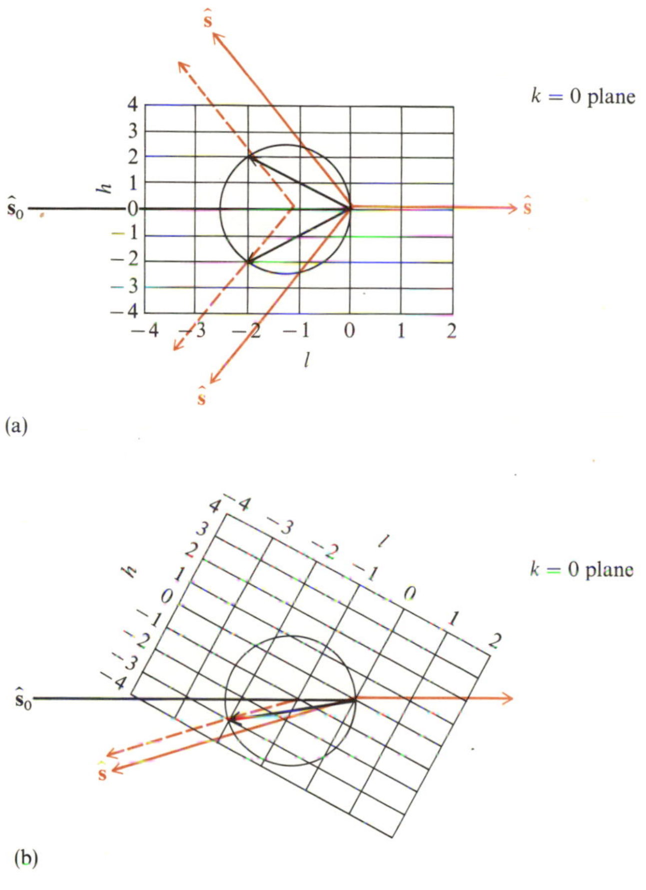

Figure 13-24

The effect on observed diffraction of rotating the sample. **(a)** One sample orientation where a pair of scattering vectors (*black arrows*) intersect the reciprocal lattice. Note that, if the scattered radiation (*colored arrows*) is viewed as originating from the center of the sphere of reflection, ŝ intersects the same reciprocal-lattice point as does **S**. **(b)** An alternative geometry, in which only a single reciprocal-lattice point is sampled.

Several methods for collecting scattering data

The spatial pattern of diffracted rays that emerges as one rotates a crystal is not necessarily a simple one. However, proper choices of rotation axes can lead to fairly regular patterns of diffracted spots. For example, suppose that the incident beam is perpendicular to axis **b**, and the crystal is rotated about this axis. In a rotation camera, a cylinder of film surrounds the sample (Fig. 13-25a). Diffracted rays passing through a given k level of the reciprocal lattice (say, $h, 0, l$) will all fall on the same line of the film. However, the order of spots, as a function of h and l values, is not regular, and the overall pattern is quite compressed. A rotation photograph projects a whole layer of the reciprocal lattice onto a single line. A typical example is shown in Figure 13-25b.

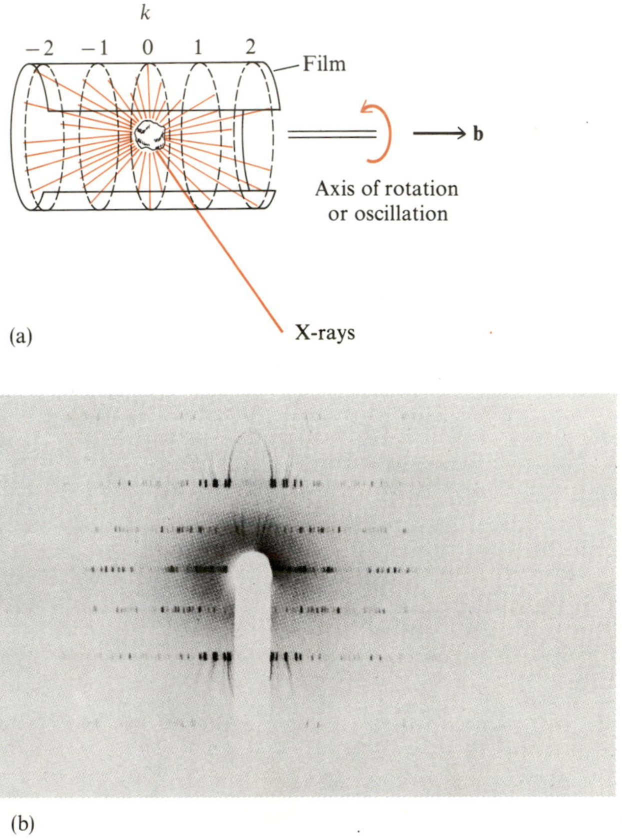

Figure 13-25

The rotation camera. This camera projects a *plane* of the reciprocal lattice onto a *line* of the film. **(a)** Schematic diagram of a rotation camera. **(b)** Example of a rotation photograph. [From G. H. Stout and L. M. Jensen, *X-Ray Structure Determination* (New York: Macmillan, 1968).]

Clearly, what one would like to have is a way of collecting diffraction data organized just like the reciprocal lattice. One way to do this is the precession camera. In essence, this camera rotates the sample and the film in such a coupled way that diffraction spots from all individual lines of the reciprocal lattice appear as properly spaced lines on the photographic film. The details of operation of such a camera are complex, and the interested reader can find them elsewhere. The results are photographs that each show one whole plane of reciprocal space. A set of such photographs permits one to reconstruct all accessible data about the diffraction pattern. Figure 13-26 shows an example. Two-dimensional scanning film densitometers are used to convert the x-ray photograph into a series of indexed integrated scattering intensities.

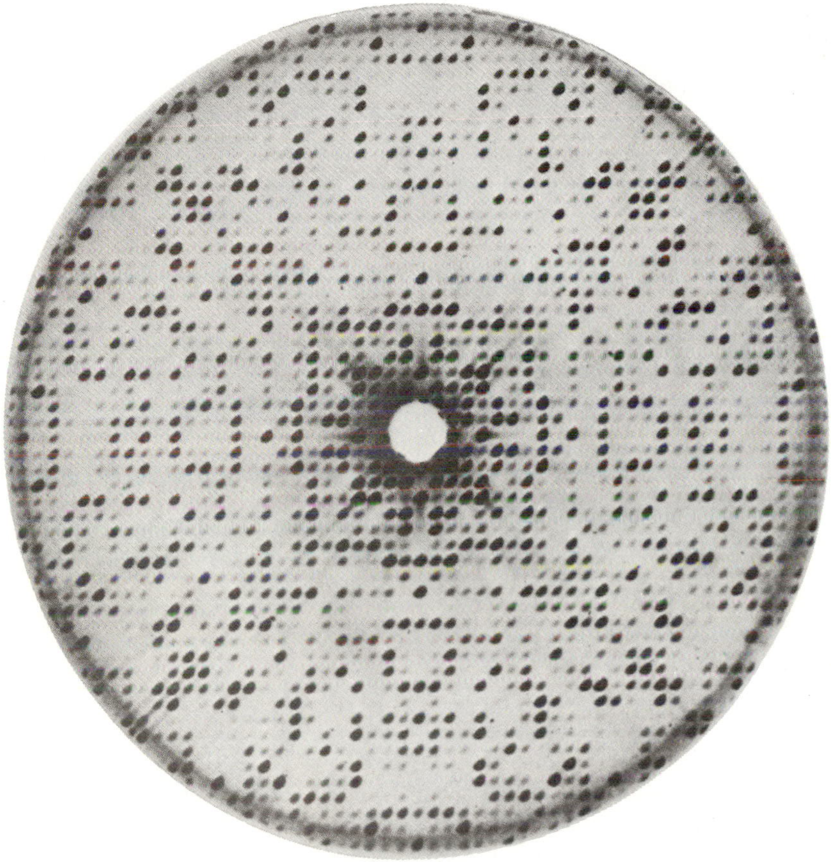

Figure 13-26

A precession photograph. An entire plane of the reciprocal lattice is displayed without distortion. The sample is a tetragonal crystal of lysozyme. Note the presence of a fourfold rotation axis and various mirror-symmetry planes in the diffraction pattern. [Courtesy of C. C. F. Blake.]

An alternative to the precession camera, now in much more common use, is the automated diffractometer. Once the crystal class and unit cell of the sample are known, its absolute orientation in space can be determined. Then, it is possible to predict the sample and detector geometry needed to produce a spot with particular indices *h*, *k*, *l*. This information is given to a computer, which finds the spot, measures the intensity, rotates the sample and detector to where the next spot should be, and so on. The x-ray intensity can be measured directly by solid scintillation detectors.

The limiting sphere of the reciprocal lattice

The reciprocal lattice, in principle, is infinite. Each of the indices *h*, *k*, *l* varies from $-\infty$ to $+\infty$. The Fourier inversion needed to calculate the electron density distribution from x-ray structure factors is an infinite sum over all three indices (Eqn. 13-39). It would not be practical to collect data over an infinite reciprocal lattice. More significantly, it is not even possible, because the finite wavelength of the x rays used limits the largest values of the indices *h*, *k*, *l* that yield diffraction intensity.

Return to Figure 13-23a and note the position of the sphere of reflection. Rotation of the crystal about any of the three laboratory axes will bring reciprocal lattice points into contact with the spherical surface, but cannot possibly reach any reciprocal lattice points that lie a distance farther from the origin than the sphere surface. The longest possible scattering vector has a length $2/\lambda$. Thus, even if all possible geometries are tried, no reciprocal lattice points farther from the origin than $2/\lambda$ can be sampled.

This limitation defines a sphere of radius $2/\lambda$, centered at the origin (Fig. 13-23b). The sphere is precisely twice the diameter of the sphere of reflection. It is called the limiting sphere. All reciprocal lattice points contained within the limiting sphere are measurable by a proper choice of experimental geometry. But no points outside of the limiting sphere can be detected. The only recourse would be to decrease the wavelength of the x rays and thus increase the diameter of the limiting sphere.

Limitations on the resolution of structures calculated from x-ray diffraction data

What is the result of our inability to measure scattering throughout the whole reciprocal space? Larger distances in reciprocal space correspond to smaller distances within the real crystal lattice. Therefore, with only a finite set of diffraction data, the ability to discriminate fine details of the electron density distribution is lost. In short, the resolution of the structure determination is decreased. It is useful to examine this statement more quantitatively.

What fraction of the data within the limiting sphere must be collected in order to produce a final structure determined to a given resolution? A vector **S** in reciprocal space is $h\mathbf{a}^* + k\mathbf{b}^* + l\mathbf{c}^*$. Its length is $|\mathbf{S}|$; the dimensions are Å^{-1}. Therefore, $|\mathbf{S}|$

corresponds to a distance $d = 1/|\mathbf{S}|$. One can estimate[§] that a collection of all diffraction data up to a value of $|\mathbf{S}|$ ought to contain the information needed to determine a structure with a resolution of around $1/|\mathbf{S}|$ Å.

The implications of limited resolution are best seen by a purely theoretical example. Figure 13-27a shows a section of β sheet. For simplicity, we shall view this as projected into the **a**–**b** plane. (See Box 13-5 for a discussion of how a projection is carried out mathematically.) The unit cell illustrated in Figure 13-27a repeats to form an infinite two-dimensional lattice. The structure factor produced by x-ray scattering from this array can be calculated exactly using the two-dimensional analog of Equation 13-70:

$$F_{\mathrm{m}}(h, k) = \sum_j f_j(S) e^{2\pi i(hx_j + ky_j)} \tag{13-88}$$

The indices h and k can be evaluated for any integral values we want, from $-\infty$ to $+\infty$.

Figure 13-27b shows part of the resulting set of structure factor data. Note that this plot illustrates both the phase and the amplitude of the structure factor. Because the two-stranded β sheet projected into two dimensions has a center of symmetry, the structure factor is real rather than complex, and the phase term reduces to just a sign ($+$ or $-$) as described earlier in the chapter.

Given a set of x-ray scattering data such as that shown in Figure 13-27b, we can calculate the structure that produced it by using Equation 13-39. In two dimensions, the result is

$$\rho(x, y) = (1/A) \sum_{h=-\infty}^{\infty} \sum_{k=-\infty}^{\infty} F_{\mathrm{m}}(h, k) e^{-2\pi i(hx + ky)} \tag{13-89}$$

where A is the area of one unit cell. However, in practice, we cannot measure values of $F_{\mathrm{m}}(h, k)$ with h and k extending all the way out to $\pm\infty$. Suppose that $|\mathbf{S}|$ could be measured only out to 1/4 Å^{-1}. This restricts h and k to values that fall within a circle of radius $|\mathbf{S}| = |h\mathbf{a}^* + k\mathbf{b}^*|$ drawn about the origin. (This is the innermost circle drawn in Fig. 13-27b.) It restricts h and k to values of -1, 0, and 1. If Equation 13-89 is used with just these terms, it produces an image of the structure at about 4 Å resolution. The result (Fig. 13-27c) suggests two strands of peptide, but obscures all molecular details.

Extending the data set used to larger values of $|\mathbf{S}|$ produces higher-resolution images (Fig. 13-27d,e). Note that in these images some regions of negative electron density (dashed contours) are included. These occur because the data set used is still finite. A perfect image of the structure can be obtained only when an infinite data set is available. Because such a set cannot be obtained in practice, corrections are used to compensate for the truncation of the series in Equation 13-89.

[§] From the theory of image formation, if all scattered waves are measured with wavelengths of d Å or more, one should be able to resolve structural features separated by $\geq 0.6d$ Å. In reality, x-ray data are not perfect, and a more realistic estimate is d Å.

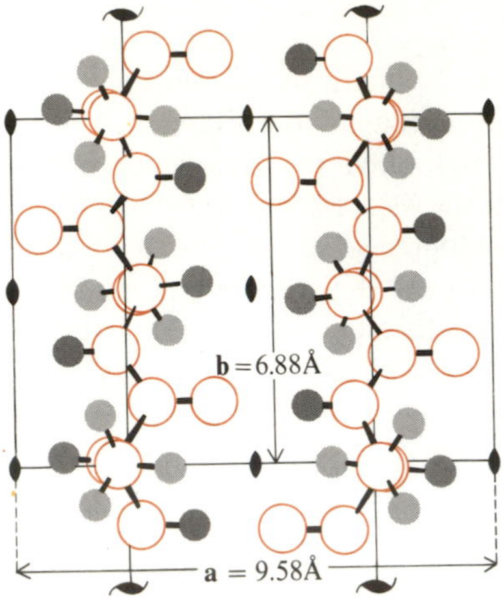

(a)

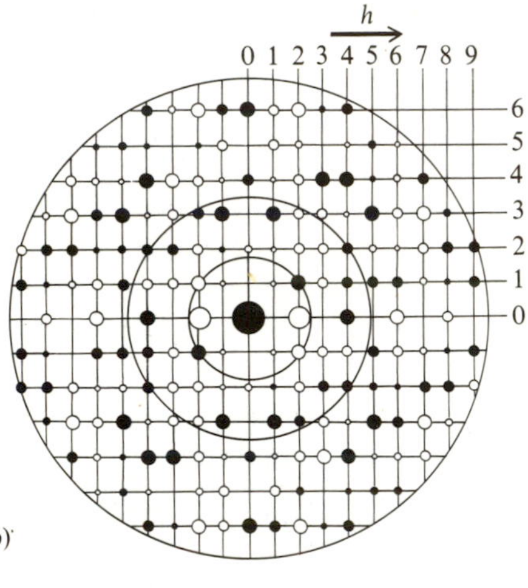

(b)

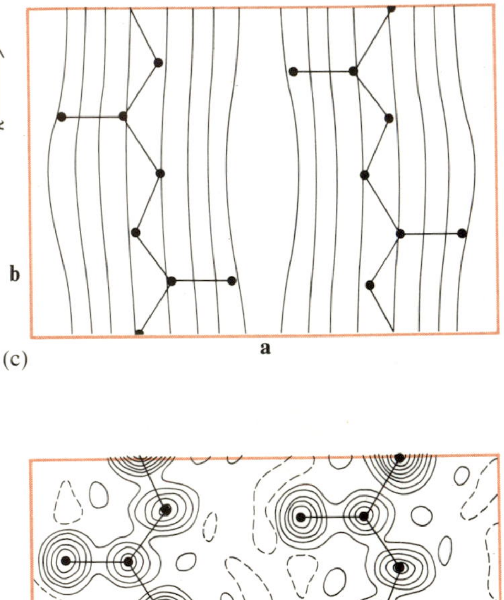

(c)

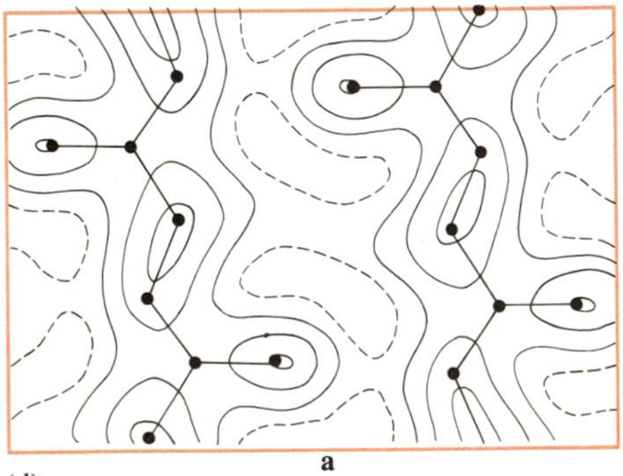

(d)

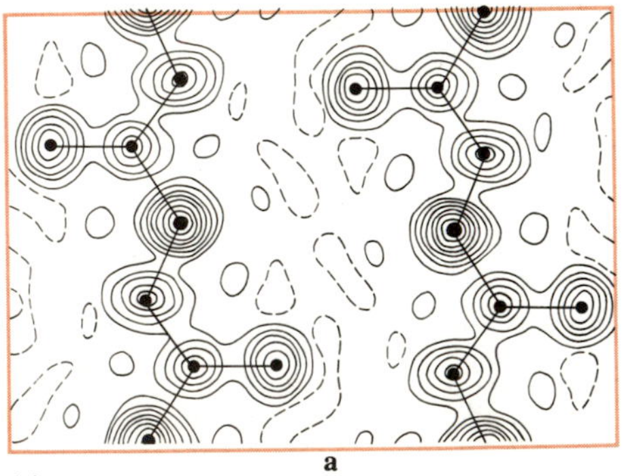

(e)

Figure 13-27

Electron density maps as a function of resolution. **(a)** Two strands of poly-L-alanine antiparallel β sheet within a two-dimensional unit cell. C_2 and α_1 symmetry axes of a planar projection of the structure are indicated. **(b)** Calculated structure factor data for a two-dimensional crystal, formed by projecting the structure shown in part a onto the **a**–**b** plane. The circles show the data that would be sampled for analysis at resolutions of 4 Å, 2 Å, and 1 Å (indicated by increasingly large circles). The filled dots indicate $F(h, k) > 0$; the open dots indicate $F(h, k) < 0$. The size of each dot is proportional to $|F(h, k)|$. **(c)** An electron density map at 4 Å resolution, calculated from part b. **(d)** An electron density map at 2 Å resolution, calculated from part b. **(e)** An electron density map at 1 Å resolution, calculated from part b. [After R. D. B. Fraser and T. P. McRae, in *Physical Principles and Techniques of Protein Chemistry*, part A, ed. S. J. Leach (New York: Academic Press, 1969).]

Experimental limitations on resolution

The example just discussed illustrates the limitations of reconstructing a structure even if perfect data were available. Experimentally, the wavelength of radiation used limits sampling of the reciprocal lattice to values of $|\mathbf{S}| \leqslant 2/\lambda$.

Why not always collect data sufficient for structure determination at the highest resolution allowed by this limiting sphere? There are three practical considerations. A given crystal will always have some disorder; thus, the x-ray data corresponding to short distances may be nonexistent. The amount of computation needed to compute the structure rises sharply with the number of data points. And the number of diffraction spots that is equal to the number of reciprocal lattice points contained within a sphere of radius $|\mathbf{S}|$ grows as the volume of the sphere.

The number of reciprocal lattice points within a sphere of radius $|\mathbf{S}|$ is approximately equal to the number n of reciprocal cells contained within the sphere. If V^* is the volume of one reciprocal lattice cell, and $(4/3)\pi|\mathbf{S}|^3$ is the volume of the sphere, then

$$n = (4/3)\pi|\mathbf{S}|^3/V^* = V(4/3)\pi/d^3 \tag{13-90}$$

where V is the volume of the real unit cell, and $d = |\mathbf{S}|^{-1}$ is the resolution. Therefore, the number of diffraction spots that must be measured increases as the cube of the desired resolution.

Two factors decrease the minimal number of diffraction spots or reciprocal lattice points needed to contain all structural information for a certain resolution. As shown in Equation 13-18, the fact that the electron density is real results in a center of symmetry for the diffraction pattern: $F(h, k, l) = F^*(-h, -k, -l)$, where the asterisk indicates the complex conjugate. Thus only one hemisphere of the limiting sphere must be measured. Furthermore, in most crystal classes, there is additional symmetry in the diffraction pattern when plotted in reciprocal space (Table 13-1).

A tetragonal crystal will have a fourfold rotation axis. The diffraction pattern of such a crystal is completely defined by only one octant of reciprocal space. Consider a crystal of cytochrome c in the tetragonal class. The unit-cell dimensions are

$a = b = 58.5$ Å and $c = 42.3$ Å. The unit-cell volume is $abc = 144{,}700$ Å^3. Equation 13-90 indicates that, for a limiting sphere sufficient to resolve structure to d Å, the number of reflections contained is $n = 606{,}400/d^3$. Because of the tetragonal class, the number of unique diffraction spots is only one-eighth of this: $75{,}800/d^3$. In

Box 13-5 PROJECTIONS OF ELECTRON DENSITY DISTRIBUTION

Many times it is convenient to work with a projection of the electron density in a plane rather than with the entire three-dimensional electron density distribution. Suppose we choose a plane perpendicular to an arbitrary direction $\mathbf{q}$. Any vector $\mathbf{r}$ to a given position in the crystal can be expressed as the sum of a component along $\mathbf{q}$ and a component $\mathbf{d}$ perpendicular to $\mathbf{q}$:

$$\mathbf{r} = \mathbf{d} + q\hat{\mathbf{q}}$$

where $\hat{\mathbf{q}}$ is a unit vector, and q is the magnitude of the projection along $\mathbf{q}$.

The electron density distribution of the crystal is

$$\rho(\mathbf{r}) = \int_{-\infty}^{\infty} d\mathbf{S}F(\mathbf{S})e^{-2\pi i\mathbf{S}\cdot\mathbf{r}} = \int_{-\infty}^{\infty} d\mathbf{S}F(\mathbf{S})e^{-2\pi i\mathbf{S}\cdot\mathbf{d}}e^{-2\pi iq\mathbf{S}\cdot\hat{\mathbf{q}}}$$

Its projection onto a plane perpendicular to $\mathbf{q}$ is simply the integral of $\rho(\mathbf{r})$ over all q:

$$\rho_q(\mathbf{d}) = \int_{-\infty}^{\infty} dq \int_{-\infty}^{\infty} d\mathbf{S}F(\mathbf{S})e^{-2\pi i\mathbf{S}\cdot\mathbf{d}}e^{-2\pi iq\mathbf{S}\cdot\hat{\mathbf{q}}} = \int_{-\infty}^{\infty} d\mathbf{S}F(\mathbf{S})e^{-2\pi i\mathbf{S}\cdot\mathbf{d}} \int_{-\infty}^{\infty} dqe^{-2\pi iq\mathbf{S}\cdot\hat{\mathbf{q}}}$$

The second integral is just the Dirac delta function, $\delta(\mathbf{S}\cdot\hat{\mathbf{q}})$. Therefore, it vanishes unless $\mathbf{S}\cdot\hat{\mathbf{q}} = 0$ or, in other words, unless $\mathbf{S}$ is in a plane perpendicular to $\hat{\mathbf{q}}$, whereupon the second integral is unity. So, if $\mathbf{S}_q$ represents all scattering vectors perpendicular to $\hat{\mathbf{q}}$, then

$$\rho_q(\mathbf{d}) = \int_{-\infty}^{\infty} d\mathbf{S}_qF(\mathbf{S}_q)e^{-2\pi i\mathbf{S}_q\cdot\mathbf{d}} \tag{A}$$

The projection integral is carried out only in a plane of reciprocal space perpendicular to the projection axis. If an inverse Fourier transform is performed,

$$F(\mathbf{S}_q) = \int_{-\infty}^{\infty} d\mathbf{d}\rho_q(\mathbf{d})e^{2\pi i\mathbf{S}_q\cdot\mathbf{d}} \tag{B}$$

Equations A and B are quite useful. They imply that, if one measures the x-ray scattering in a plane of reciprocal space, $F(\mathbf{S}_q)$, the electron density of a projection of the structure onto that plane can be computed. Alternatively, any plane in reciprocal space will contain information only about the electron density of the molecule projected onto a plane. It is common to choose a projection along crystal axes. For example, suppose we choose $\mathbf{q}$ to be the $\mathbf{c}$ axis. Then, after inserting the von Laue conditions, Equation A becomes

$$\rho(x, y) = (1/A) \sum_{n=-\infty}^{\infty} \sum_{k=-\infty}^{\infty} F(h, k, 0)e^{-2\pi i(hx+ky)}$$

practice, this implies that, for 4 Å resolution, about 1,200 diffraction spots must be measured. This number increases to 9,500 for 2 Å resolution, and to 75,800 for 1 Å resolution. Clearly, for large unit cells such as those found in macromolecular crystals, the amount of work needed to improve resolution can be quite formidable.

Note that this is the projection of the electron density onto a plane perpendicular to the **c** axis. That plane is not necessarily the **a**–**b** plane unless the crystal symmetry is such that **a** and **b** are perpendicular to **c**. A is the area of the projection of the **a**–**b** face of the unit cell perpendicular to the **c** axis, as shown in the figure.

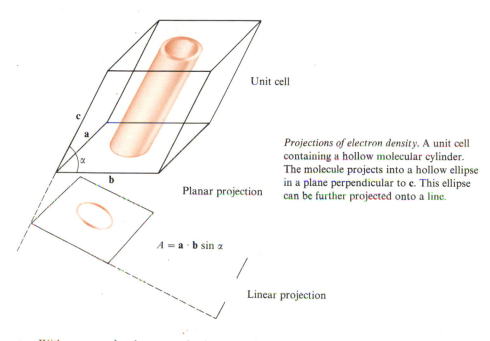

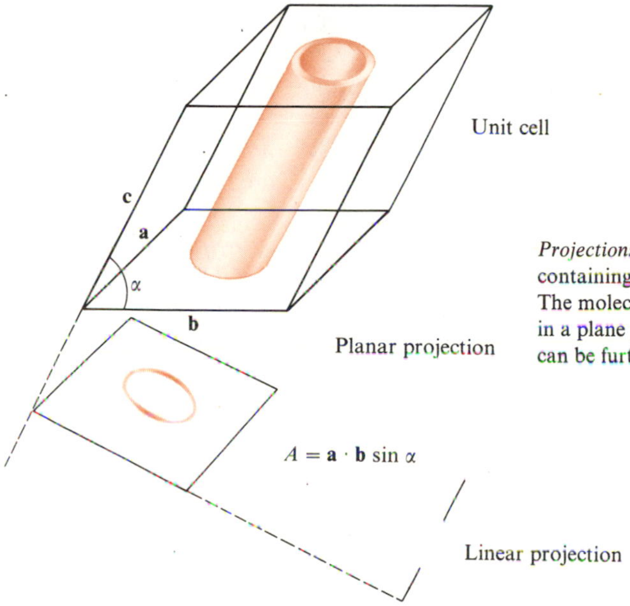

Projections of electron density. A unit cell containing a hollow molecular cylinder. The molecule projects into a hollow ellipse in a plane perpendicular to **c**. This ellipse can be further projected onto a line.

With many molecules, properly chosen projections may have an apparent symmetry not visible in the structure of the whole. Examination of the x-ray diffraction corresponding to such projections often can help simplify the determination of the structure. When crystallographers show diffraction patterns, they virtually always display data for one plane through the reciprocal lattice. Usually, this is a plane in which the index of one reciprocal-cell direction is zero—that is, $h, k, 0$ or $h, 0, l$ or $0, k, l$.

In a comparable way, a row of points of the reciprocal lattice will contain the data needed to calculate the projection of the electron density onto a line. That line will be the intersection of the planes perpendicular to the two projection directions, as shown in the figure. For example, where **a**, **b**, and **c** are all mutually perpendicular, the zero layer line $(h, 0, 0)$ describes the projection of the electron density along the **a** axis.

13-4 DETERMINATION OF MOLECULAR STRUCTURE BY X-RAY CRYSTALLOGRAPHY

The phase problem

We have seen that it is relatively easy to determine the properties of a crystal lattice from its measured diffraction pattern. However, the central problem in x-ray crystallography is to determine $\rho(\mathbf{r})$, the electron density distribution within the unit cell. In principle, we solved this problem with Equation 13-39:

$$\rho(x, y, z) = (1/NV) \sum_{h=-\infty}^{\infty} \sum_{k=-\infty}^{\infty} \sum_{l=-\infty}^{\infty} F(h, k, l)e^{-2\pi i(hx+ky+lz)} \tag{13-39}$$

However, an enormous practical deterrent exists. As mentioned earlier (Eqn. 13-12), each $F(h, k, l)$ structure factor is a complex number $|F|e^{i\phi}$, consisting of an amplitude $|F|$ and a phase term $e^{i\phi}$. Only the square of the amplitude, $|F|^2$, can be observed experimentally. The phase angle ϕ can have any value between 0 and 2π.

In the special case of a crystal with a center of symmetry, the phase is much more restricted. For such a crystal $\rho(\mathbf{r}) = \rho(-\mathbf{r})$. As shown earlier, $F(h, k, l)$ is real; the phase can be either 0 or π, and $e^{i\phi}$ is ± 1. This means that only the sign of each term in the Fourier synthesis of the electron density is unknown. However, even in these cases, there are 2^n possible choices of phase for a set of n diffraction spots.

For biological samples, the inability to measure phases experimentally poses a truly serious problem. Not only is the number of diffraction spots large, but also such crystals cannot have centers of symmetry because they contain asymmetric carbon atoms. Certain techniques we shall describe can help to infer partial phase information. Sometimes chemical insight or other previously known information about the molecule must be included in order to assist the structure determination. The methods usually are sufficient to permit phase estimates accurate enough to compute a three-dimensional structure. It must be borne in mind that occasionally the methods described below can converge on an incorrect structure.

Phases are more important than amplitudes

Because one has measured amplitudes but not phases, it is of interest to ask which of these two factors is most crucial in establishing the correct structure. This can be determined by taking a known structure and calculating the correct structure factors $|F|^{i\phi}$. If these are then inserted back into Equation 13-90, naturally an accurate image of the electron density distribution appears. This is shown for a two-dimensional projection of a portion of β sheet in Figure 13-27e.

Suppose instead that all the correct amplitudes are used, but each phase ϕ is arbitrarily assigned at the same value, 0°. The resulting Fourier synthesis (Fig. 13-28a) bears no resemblance to a β sheet. Next, suppose all amplitudes are set at the same

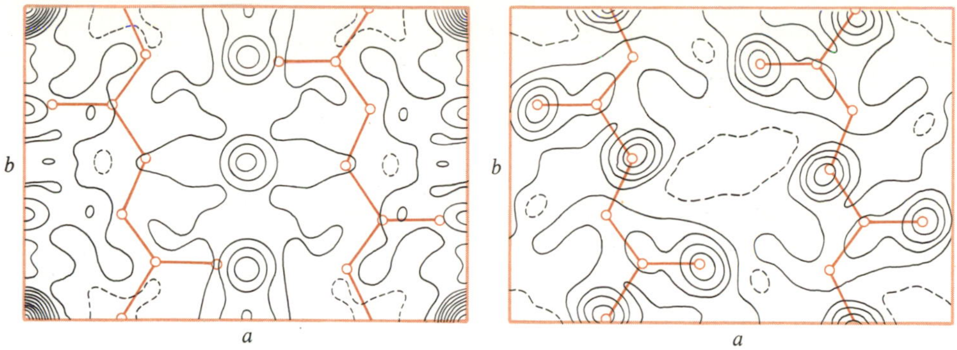

(a) (b)

Figure 13-28

Relative importance of intensities and phases in computing an electron density map from diffraction data. The sample and data are the same as those shown in Figure 13-27. **(a)** Correct amplitudes were used in this Fourier synthesis, but all phases were set equal to zero. **(b)** Correct phases were used in this Fourier synthesis, but all amplitudes were set equal to the same average value. [After R. D. B. Fraser and T. P. MacRae, in *Physical Principles and Techniques of Protein Chemistry*, part A, ed. S. J. Leach (New York: Academic Press, 1969).]

equal value, $|F| = (\sum_k |F_k|^2)^{1/2}$, where the sum is taken over all the square of all of the amplitudes of the diffraction pattern. This corresponds to an average over all measured intensities. If these are combined with the correct phases in a Fourier synthesis, the result (Fig. 13-28b) clearly has substantial resemblance to a β sheet. Thus, we have the unfortunate situation that the unmeasurable quantities are actually more useful than those that can be obtained experimentally.

General considerations in solving a crystal structure

The problem of solving a structure starts with an enormous number of unknowns: the location of each atom in the unit cell, the type of the atom and therefore the expected atomic scattering factor, and the phase associated with each diffraction spot. There is also a considerable amount of available data: the diffraction intensities, the space group and unit cell of the crystal, and usually a significant amount of information about the molecule being examined (for example, partial or complete chemical structure, and perhaps even some conformational data).

The most general goal is to find a structure for the molecule that represents a best fit to the available diffraction data and does not violate, without due cause, our chemical intuition and the set of available structural data. Putting it this way makes it clear that x-ray structure determination is not in practice an absolute technique. In most macromolecular structure cases, one must rely on other information besides diffraction data. That is, there are not enough pure x-ray data available to establish a unique location and identity for each atom in the structure. Even if all the phases of the scattering factors were experimentally measurable, there might not be enough information. One must marvel, then, at the courage of the first scientists to tackle macromolecular crystal structures.

Steps in determining the structure of a small molecule

Here we illustrate some of the procedures used to determine the structure of a small molecule. These are not necessarily the most powerful methods currently available, but they provide a useful comparison with our later listing of the techniques used on large molecules.

1. One attempts to prepare suitable crystals, and then determines the space group and the unit-cell dimensions, and collects a set of amplitude data $|F_0(h, k, l)|$.
2. One attempts to find the locations of a few of the atoms. This can be done by direct methods (see Blundell and Johnson, 1976), or by a search for a few heavy atoms using the Patterson function discussed later.
3. Once the position of a few atoms is known, the contribution F_H that these atoms make to the total scattering can be calculated by using Equation 13-70:

$$F_H(h, k, l) = \sum_j f_j(s) e^{2\pi i(hx_j + ky_j + lz_j)} \tag{13-91}$$

   where the index j runs over all atoms of known position. Note, however, that the structure factor observed experimentally (F_0) is the sum of contributions from the known atoms and from all atoms yet to be found (F_u):

$$F_0(h, k, l) = F_H(h, k, l) + F_u(h, k, l) \tag{13-92}$$

   What is crucial, however, is that, because we calculated it, $F_H(h, k, l)$ contains phase as well as amplitude data.
4. The phase of $F_H(h, k, l)$ can be used in several ways to estimate the phase associated with $F_0(h, k, l)$. Then a Fourier synthesis can be performed to compute an estimate of the electron density distribution from the known heavy atom positions:

$$\rho(x, y, z) = \sum_{h=-\infty}^{\infty} \sum_{k=-\infty}^{\infty} \sum_{l=-\infty}^{\infty} |F_0(h, k, l)| e^{i\phi_{hkl}} e^{-2\pi i(hx + ky + lz)} \tag{13-93}$$

Here, the structure factor has been explicitly divided into amplitude and phase terms. Note that it is essential to use *measured* amplitudes. If both calculated phases and amplitudes are used in Equation 13-93 (that is, if Eqn. 13-91 is simply inserted into Eqn. 13-93), all that can come out for $\rho(x, y, z)$ is precisely the known atom positions that were originally put into Equation 13-91 to compute $F_H(h, k, l)$.

5. The electron density distribution calculated by Equation 13-93 with even partial phase information will show some definite maxima corresponding to the locations of new atoms or groups of atoms. These in turn can be used in Equation 13-91 to compute more accurate phases, which then are used in Equation 13-93 to compute a new electron density map. This process, called a Fourier refinement, continues alternately until all atoms consistent with one's original or revised expectations have been found.

6. The structure at this point is still a very approximate one. The original subset of atoms used to start the bootstrap process rolling probably are not placed all that precisely. The electron density distribution that finally results is not always all that sharp. It usually is impossible to assign precise coordinates to all atoms. Furthermore, experimental errors in observed $F(h, k, l)$ affect the data, and these must be dealt with in a systematic way. Thus, the sixth and final phase of x-ray structure determination is to allow the molecular structure to vary somewhat in an attempt to maximize the agreement between the computed structure and the observed data. One way of doing this, called a least-squares refinement, is illustrated later.

In computing $\rho(x, y, z)$ from Equation 13-93 in practice, only a finite set of values of x, y, and z can be used. It is common to compute ρ for planar sections through the crystal (that is, to vary x and y, but leave z constant). Even then, only discrete values of x and y are used. From the resulting pattern of density for each plane, smooth contours are drawn representing areas of equal density. Usually, one interpolates available data to bring this about. Individual two-dimensional sections are produced by a computer. These can be drawn on transparent sheets and stacked up to give a three-dimensional image of the structure. Figure 13-29 shows an example. Alternatively, computer-generated displays can be viewed on a cathode ray tube from any desired perspective.

Calculating the Patterson function from measured scattering

The scattering intensities actually measured in an x-ray diffraction experiment are given by

$$I(\mathbf{S}) = |F(\mathbf{S})|^2 = F(\mathbf{S})F^*(\mathbf{S}) \tag{13-94}$$

where the asterisk indicates the complex conjugate. If we knew $F(\mathbf{S})$, we could

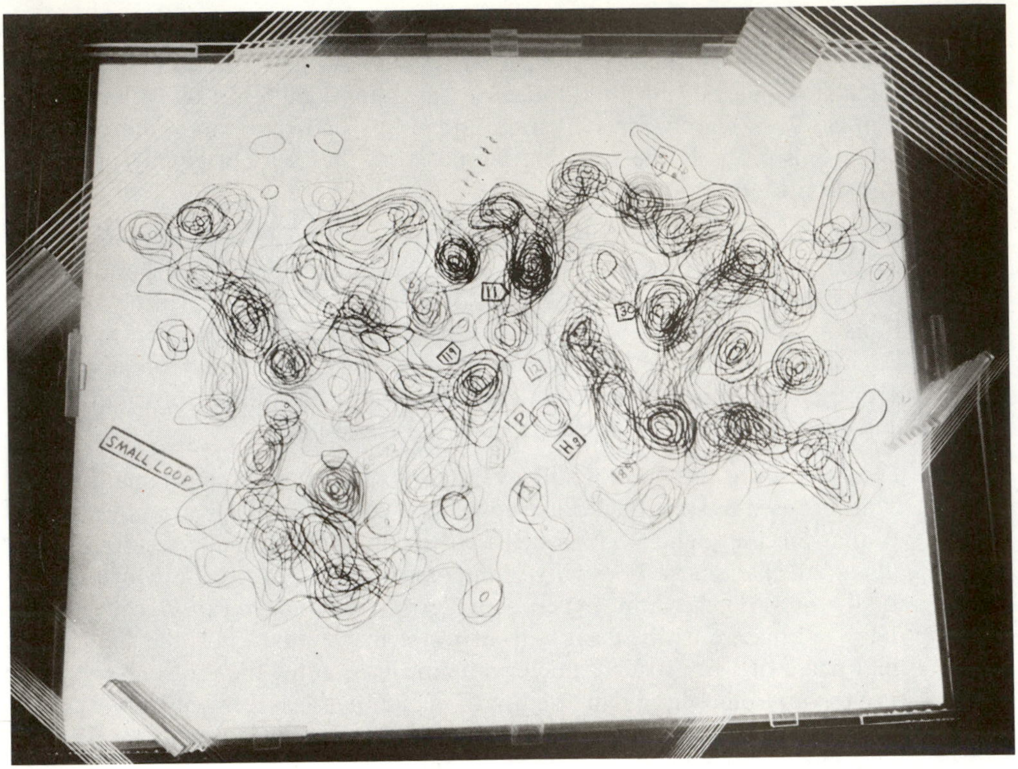

Figure 13-29

A three-dimensional electron density map of ribonuclease S, constructed by plotting density contours on lucite sheets and then stacking the sheets. [Provided by Frederick Richards.]

Fourier-transform it to obtain the electron density distribution of the entire crystal. If instead we Fourier-transform $I(\mathbf{S})$ directly, the result is called the Patterson function:

$$P = \int_{-\infty}^{\infty} I(\mathbf{S})e^{-2\pi i \mathbf{S}\cdot\mathbf{r}}\, d\mathbf{S} = \int_{-\infty}^{\infty} d\mathbf{S}F(\mathbf{S})F^*(\mathbf{S})e^{-2\pi i \mathbf{S}\cdot\mathbf{r}} \tag{13-95}$$

This equation is the Fourier transform of the product of two functions, $F(\mathbf{S})$ and $F^*(\mathbf{S})$. Therefore (by Eqn. 13-52) it is equal to the convolution of the Fourier transforms of $F(\mathbf{S})$ and $F^*(\mathbf{S})$.

The Fourier transform of $F(\mathbf{S})$ is

$$\int_{-\infty}^{\infty} d\mathbf{S}e^{-2\pi i \mathbf{S}\cdot\mathbf{r}}F(\mathbf{S}) = \rho(\mathbf{r}) \tag{13-96}$$

(as in Eqn. 13-8). What is the Fourier transform of $F^*(\mathbf{S})$? Because $\rho(\mathbf{r})$ is real, $F^*(\mathbf{S})$ is just (from Eqn. 13-7)

$$F^*(\mathbf{S}) = \left(\int_{-\infty}^{\infty} d\mathbf{r}e^{2\pi i\mathbf{S}\cdot\mathbf{r}}\rho(\mathbf{r})\right)^* = \int_{-\infty}^{\infty} d\mathbf{r}e^{-2\pi i\mathbf{S}\cdot\mathbf{r}}\rho(\mathbf{r}) = \int_{-\infty}^{\infty} d\mathbf{r}e^{2\pi i\mathbf{S}\cdot\mathbf{r}}\rho(-\mathbf{r}) \quad (13\text{-}97)$$

Therefore, the Fourier transform of $F^*(\mathbf{S})$, by analogy to what is shown in Equations 13-9 through 13-11, is just $\rho(-\mathbf{r})$. This is the electron density inverted through the origin. Thus

$$P = \rho(\mathbf{r})\frown\rho(-\mathbf{r}) = \int_{-\infty}^{\infty} d\mathbf{r}\rho(\mathbf{r})\rho(\mathbf{u}+\mathbf{r}) \quad (13\text{-}98)$$

The physical idea behind the convolution integral helps generate some feeling for the properties of the Patterson function. Figure 13-30 shows a simple case. It is a

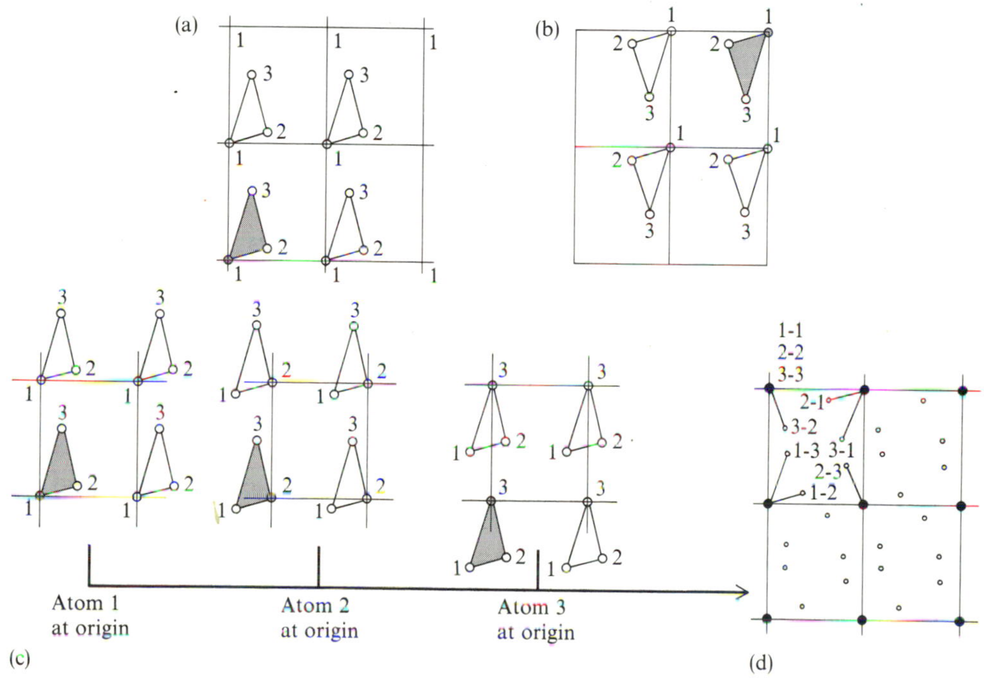

Figure 13-30

The Patterson function for a three-atom structure. **(a)** Four unit cells of the lattice; the molecule at the origin is shaded. **(b)** The same four unit cells, inverted through the origin. **(c)** Construction of the convolution of the arrays shown in parts a and b. The position of each atom in part b is used as the origin to lay down an image of the structure in part a. For example, note that, when atom 2 in part b is the origin, atom 1 in part a is displaced from the origin, but atom 2 in part a winds up at the origin. Therefore, an equivalent description of the convolution is simply to lay down successive images of part a with each atom in turn at the origin. **(d)** The overall convolution, adding the various contributions shown separately in part c. The numbers by each point indicate the way in which that point arose. For example, 1-2 means that an image of atom 2 resulted when the structure was laid down with atom 1 at the origin. [After J. P. Glusker and K. N. Trueblood, *Crystal Structure Analysis: A Primer* (London: Oxford Univ. Press, 1972).]

two-dimensional crystal containing one three-atom molecule per unit cell. Figure 13-30a shows the structure of the crystal, and Figure 13-30b shows the crystal inverted through the origin. We want the convolution of these two structures. First, consider just a single unit cell of the crystal and a single unit cell of the inverted structure. The convolution is formed by choosing each atom in the inverted structure one at a time. Lay down an image of the structure of the original unit cell, by superimposing the origin (lower left corner) of the cell on top of this atom, and weight this image by the electron density of the chosen atom.

Note that, when the origin of the original cell is superimposed on an atom in the inverted structure at $-\mathbf{r}$, the corresponding atom at $\mathbf{r}$ in the original structure now is located at the origin of the inverted cell. Therefore, the convolution can be constructed by ignoring the inverted structure, shifting the contents of a cell to place each atom in turn at the origin, and adding up weighted images of the structure (Fig. 13-30c).

There are three atoms in the structure. The image with each atom at the origin has three atoms. So, a total of nine atom images will appear in each unit cell of the convolution; three fall right at the origin (Fig. 13-30d). In general, for a molecule with N atoms in the unit cell, the Patterson function will have N^2 peaks in its unit cell. N of these peaks will occur at the origin, and the remaining $N(N-1)$ somewhere within the unit cell. It is clear that the Patterson function will be increasingly cumbersome to use or interpret as N grows.

Periodic repetition of Patterson functions

The convolution described by Equation 13-98 actually is operated over the whole crystal and not over just one unit cell. This fact has a simple consequence. Consider a lattice with only a single atom at each unit cell vertex. If one particular atom is used to superimpose an image of the whole crystal, the result is to place an atom at every vertex of every unit cell. Choosing any other atom results in exactly the same image.

The same argument applies in a molecular crystal. Choose one atom in a particular cell to lay down an image. Choose the corresponding atom in a different unit cell to lay down an image. The resulting images are coincident, except that the crystal is displaced in space by an integral number of unit cells. Thus, like the electron density, the Patterson function repeats periodically throughout the crystal. All the information of interest can be obtained by concentrating on a single unit cell. With actual data, the intensity $I(\mathbf{S})$ is not a continuous function, but is sampled only at the reciprocal lattice points. Thus, by analogy with Equation 13-39, the integral in Equation 13-95 is replaced by a sum:

$$P(x, y, z) = \sum_{h=-\infty}^{\infty} \sum_{k=-\infty}^{\infty} \sum_{l=-\infty}^{\infty} |F(h, k, l)|^2 e^{-2\pi i(hx+ky+lz)} \tag{13-99}$$

Correspondence of peaks in Patterson function and vectors between atoms

An alternative description of the Patterson function helps give a feeling for the structural information it contains. Note first that, if we change the origins used for the unit cells, the resulting Patterson function remains unaltered. That function still is constructed by placing each atom in turn at the origin of a cell. Hence, all of the peaks in the Patterson function must represent internal structural aspects of the unit cell.

Suppose the unit cell has three atoms, located at $\mathbf{r}_1$, $\mathbf{r}_2$, and $\mathbf{r}_3$. When an image is laid down by shifting the atom at $\mathbf{r}_1$ to the origin, the peaks in the image are placed at $\mathbf{r}_1 - \mathbf{r}_1$, at $\mathbf{r}_2 - \mathbf{r}_1$, and at $\mathbf{r}_3 - \mathbf{r}_1$. Therefore, these peaks in the Patterson function are simply the vectors from each atom in the structure to atom $\mathbf{r}_1$. When $\mathbf{r}_2$ is used as the origin, we get peaks for vectors from all atoms to $\mathbf{r}_2$, and so on. So the Patterson function is simply the set of all vectors between pairs of atoms in the structure. It is clear why this set does not depend on the choice of an origin. Using this physical description, the Patterson function can be rewritten as

$$P = \sum_j \sum_k \rho_j \rho_k (\mathbf{r}_j - \mathbf{r}_k) \qquad (13\text{-}100)$$

where each index j and k is summed over all atoms in the unit cell with electron density ρ.

The Patterson map contains more than enough information to determine the structure. The problem is that there is no efficient or easy strategy to use this information. Unfortunately, the peaks in the Patterson function are not labeled. There is no simple way of deciding which pair of atoms a given vector in the map represents. One must find a way to deconvolute the Patterson function to extract the structure. If just a few atoms are involved, this can easily be done by brute force. Alternatively, if the positions of a few atoms are already known, superposition techniques can be used to deconvolute the Patterson function (see Blundell and Johnson, 1976; Stout and Jensen, 1968).

Using Patterson maps to locate heavy atoms in small molecules

For complex molecules, the difficulty in interpreting the Patterson map is a direct consequence of the large number of interatom vectors. Suppose, however, that the structure contains two or more heavy atoms per unit cell. The atomic scattering factor is proportional to the number z of electrons (Eqn. 13-23). Thus observed intensities, and the resulting Patterson peaks, from atoms i and j are proportional to $z_i z_j$ (corresponding to the $\rho_i \rho_j$ terms in Eqn. 13-100). This means that vectors between pairs of heavy atoms are the most dominant feature of a Patterson map. With a limited number of heavy atoms, it usually is possible to find out enough about their locations to proceed with further refinement and structural analysis.

Most space groups contain molecules related by symmetry operations within the unit cell. If each molecule has a heavy atom, the Patterson vector between two symmetry-related atoms will fall in an easily identifiable region of the Patterson map. Consider the example shown in Figure 13-31. This is a monoclinic crystal in space

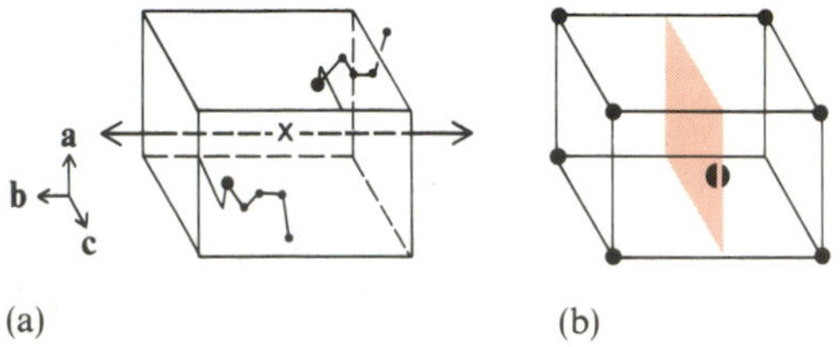

Figure 13-31

Locating heavy atoms. **(a)** A unit cell of a crystal in space group $P2_1$. There are two molecules per unit cell, with each molecule containing one heavy atom. **(b)** Patterson function calculated for the sample shown in part a. Only the heavy atom–heavy atom vectors are shown. The Harker section is colored.

group $P2_1$ with two molecules per unit cell. There is a twofold screw axis parallel to direction **b**. If there is a heavy atom at position $x'\mathbf{a} + y'\mathbf{b} + z'\mathbf{c}$, then the other heavy atom must be located at $-x'\mathbf{a} + (y' + 1/2)\mathbf{b} - z'\mathbf{c}$. The heavy atom–heavy atom vector must be located at the position $2x'\mathbf{a} + (1/2)\mathbf{b} + 2z'\mathbf{c}$ in the Patterson map. Thus x' and z' can be determined by looking for a peak in the plane of the Patterson map at $(1/2)\mathbf{b}$.

This procedure still leaves y' undetermined. However, for the $P2_1$ space group, the problem is not serious. There is no unique origin along the **b** axis, and therefore y' can be chosen to have any arbitrary value.

Planes or lines where symmetry-related Patterson vectors appear are called Harker sections. If one or more Patterson vectors can be assigned by examining these, it sometimes is possible to use superposition methods to find others. Each Harker section will contain not just the heavy atom–heavy atom vectors, but also all the other vectors related by the same symmetry operation. It will not contain any light atom–heavy atom vectors except for coincidences. The contrast afforded by the heavy-atom pair will be z_h^2 versus z_l^2 for each light-atom pair. This raises the question of how heavy an atom is needed. A general rule of thumb is that $z_h^2 \gtrsim \sum_l z_l^2$, where the sum is taken over all light atoms. For a typical light atom, z_l is 7. Thus a single heavy atom such as mercury with $z_h = 80$ could be found in a structure as large as 130 light atoms. It could not be found in a Patterson map of a typical protein with 1,000 to 10,000 atoms.

Testing agreement between calculated structure and observed data

How can one tell if a calculated structure is in good agreement with the measured x-ray diffraction? The most common measure of the agreement is the residual index R:

$$R = \sum||F_0| - |F_{\text{calc}}||/\sum|F_0| \tag{13-101}$$

where $|F_{\text{calc}}|$ represents structure factors computed from the model of the total structure by Equation 13-70. Thus the factor R essentially measures how the observed experimental data $|F_0|$ compare with the data that would be expected for the calculated structure, $|F_{\text{calc}}|$.

If the structure is very approximate, it may be simply a random arrangement of atoms of the correct numbers, types, and symmetry within the unit cell. In this case, it has been shown that R will be 0.59 for a space group without a center of symmetry, and 0.83 if a center is present. A very rough rule suggests that $R \cong 0.45$ means that the trial structure is not completely useless; $R \cong 0.35$ means definite convergence on the right track; and $R \cong 0.25$ means that most atoms are correctly placed to within the order of 0.1 Å. Small organic structures often can be refined to $R < 0.05$. Protein R values usually are large at early stages in the structure determination. This is because solvent and thermal motion effects usually are not taken into account until later stages.

Note that an R value of 0.25 actually implies a degree of disagreement between observed and calculated amplitudes that would be considered intolerable in most techniques. It probably is fair to say that the quantity of x-ray data makes up for the lack of quality of individual data points.

13-5 DETERMINING THE STRUCTURE OF A MACROMOLECULE

The method of multiple isomorphous replacements

The methods described earlier for small molecules are not successful with proteins or nucleic acids. These large molecules generally do not contain conveniently placed heavy atoms. Even where such atoms do occur naturally, the complexity of the structure demands different approaches. The steps in a typical macromolecular crystallographic study are the following.

1. One attempts to prepare suitable crystals of the native macromolecule. (This is the most difficult and time-consuming stage in protein or nucleic acid crystallography. In many cases, macromolecular crystals form that appear beautiful morphologically, but they have so much disorder that high-resolution diffraction data are unobtainable.) Using the crystals, one determines the space group and unit-cell dimensions, and then collects a set of scattering amplitude data.

2. One attempts to prepare several different heavy-atom isomorphous derivatives. These are crystals with the same unit cell, space group, and macromolecular structure as the parent crystal, except that one or more heavy atoms have been introduced at specific loci. For each derivative, one collects a new set of scattering amplitude data.
3. One attempts to find the locations of the heavy atoms in the crystal. A popular way to do this is the difference isomorphous Patterson synthesis we shall discuss.
4. One attempts to refine the positions assigned to heavy atoms, either by use of difference Fourier refinement techniques somewhat like those described for small molecules, or by more elaborate methods.
5. By comparing the structure factor data of the parent crystal with the corresponding data of one or more heavy-atom isomorphous derivatives, it is possible to estimate the phases of each $F(h, k, l)$ of the *parent* crystal. In general, the more heavy-atom derivatives available, the more accurate the phase estimates will be.
6. By using the phases estimated for the parent crystal, it often is possible to refine the positions of the heavy atoms further using least-square or difference Fourier techniques. This procedure in turn leads to better estimates for the phases of the parent crystal.
7. Using the estimated phases and observed amplitudes of each $F(h, k, l)$, one calculates an electron density map using Equation 13-93.
8. A model is built of the electron density map. Usually at this state, only low-resolution data (typically 5.5 to 7 Å) have been used, and so the map does not show well-resolved structural details. Then steps 4 through 8 are repeated with data at higher resolution (2.5 to 3 Å) until it is possible to construct a *molecular* model.
9. Sometimes, one attempts to refine the structure. For example, one can calculate phases from the atom positions in the molecular model by Equation 13-70 and use them instead of the phases determined in stages 5 and 6. The refinement can involve Fourier or least-squares techniques, and can treat just the x-ray data or can also include information about known energetics of protein conformation.

Most of these stages are discussed in more detail in the following subsections.

Preparation and properties of macromolecular crystals

For crystallographic studies on a macromolecule of 50,000 mol wt, one needs crystals about 0.3 mm in each dimension. To form these, one generally must prepare a supersaturated solution of the macromolecule and control the rate at which crystal nucleation and growth occur. Solubility can be altered by varying the pH, salt concentration, types of salts present, and temperature of the solution—or by adding organic solvents.

One convenient way to control the rate of change of many of these parameters is dialysis. Another is vapor diffusion. Here a droplet of macromolecular solution is allowed to equilibrate through the vapor phase with a reservoir of solution. If, for example, the reservoir has a higher salt concentration than the sample, the result will be a gradual removal of solvent from the sample. Further details about these and other techniques are given by T. L. Blundell and L. N. Johnson (1976).

Protein and nucleic acid crystals differ from small-molecule crystals in one important aspect. They contain a considerably quantity (typically 50%) of liquid solvent. Cases are known in which protein or nucleic acid crystals are more than two-thirds solvent by weight. A typical crystal has much less solvent than this, but it usually still resembles a two-phase system. A solid phase is composed of individual macromolecules that usually are touching each other in only a few places. In between is a series of open channels filled with solvent. Figure 13-32 shows an example.

The larger amount of solvent in crystals offers several advantages. It permits small molecules to be diffused into the crystals. As we shall see, this facilitates the incorporation of heavy atoms. It allows substrates or ligands to be introduced into a preformed crystal and thus makes possible the study of the structure of macromolecule–ligand complexes. Indeed, some enzymes actually are quite active in the crystalline state. Finally, the large amount of solvent present makes it likely that the structure determined for the crystalline molecule will closely resemble its structure when free in solution.

One disadvantage arises from the large solvent content. Some of the solvent quite close to the macromolecule is well-ordered. It contributes to the observed x-ray scattering, and it must be taken into account in solving the structure. On the other hand, once this is accomplished, it provides important clues to how macromolecules interact with solvent.

Preparation of isomorphous heavy-atom derivatives

The multiple isomorphous replacement technique has been used to solve almost all protein and nucleic acid structures known to date. It requires a set of three or more crystals of the sample: the parent crystal, and at least two other crystals identical in space group and molecular structure except for the presence of one or more heavy atoms. In general, the heavy atoms either can replace atoms normally present in the structure, or can be additions to the structure. We shall restrict our attention to the latter case because it is somewhat easier to treat mathematically, but ultimately both cases are fairly equivalent.

One approach to preparing an isomorphous derivative would be to attach covalently a heavy metal to the macromolecule in solution, and then to subject it to crystallization conditions. In practice, this approach is not necessarily effective. The factors that promote formation of good crystals are so fickle that frequently even a

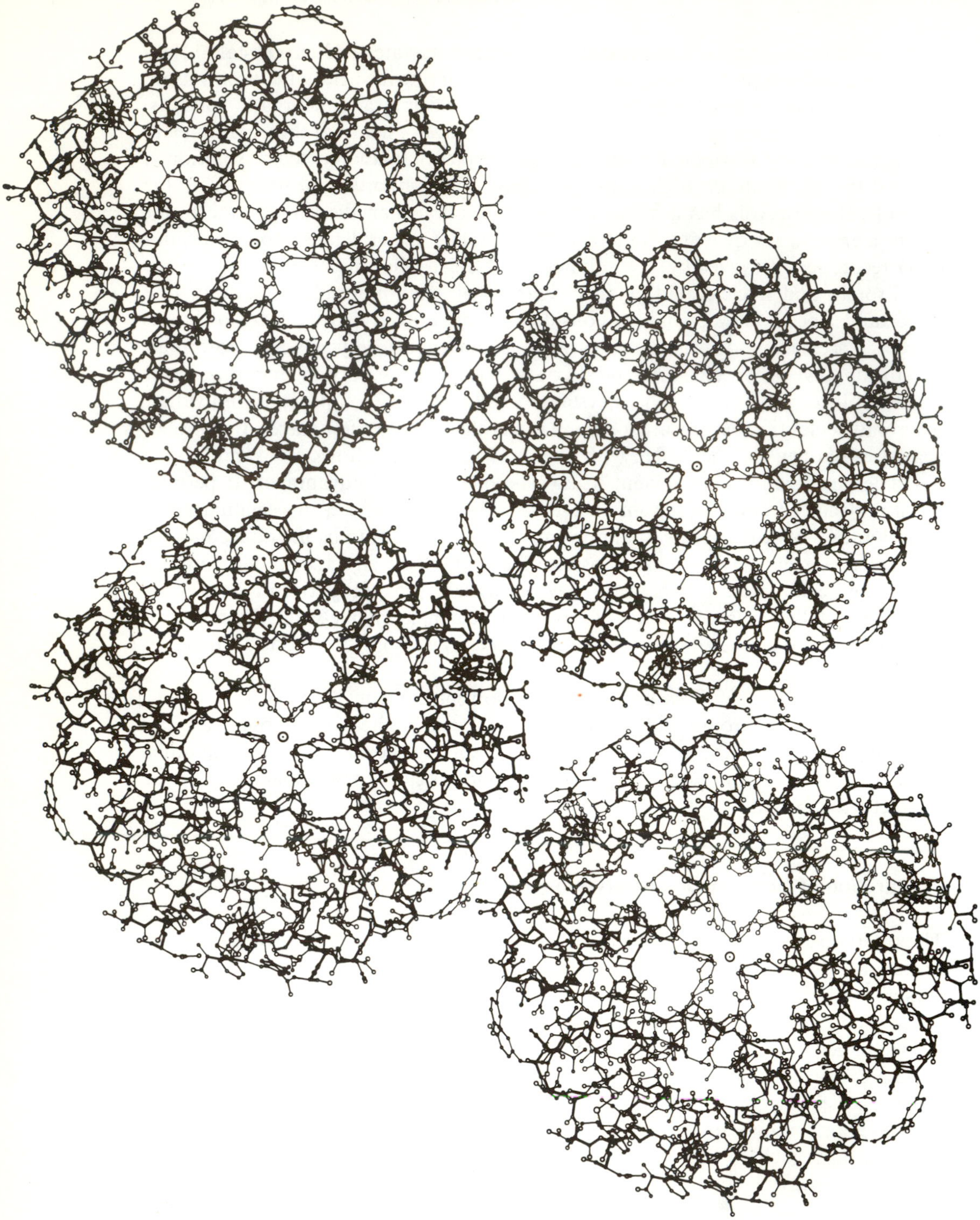

small chemical alteration of the structure will either block crystallization or lead to a crystal that no longer is isomorphous. Therefore, almost always one starts with a preformed crystal of nonmodified macromolecule. Reagents containing heavy atoms then are allowed to diffuse in the crystal. This technique provides fair assurance that crystal packing and molecular structure remain largely unaltered. Table 13-2 lists some of the types of reagents used, and Table 13-3 summarizes the results.

Table 13-2
Representative heavy-atom labeling reagents

Reagent	Binding sites
$AgNO_3$	SH groups
Xe	Noncovalent
$KI + I_2$	Tyrosines
PCMB: Cl—Hg—⟨◯⟩—COO⁻	SH groups
Na_2PtCl_4	Methionines, histidines, and others
Cl—Hg—⟨◯⟩—SO_2F	Active-site serines
$Hg(Ac)_2$	SH groups, histidines
$UO_2(NO_3)_2$ or $UO_2(Ac)_2$	Carboxyls
Mersalyl: HO—Hg—CH_2CH—CH_2—NH—CO	Histidines, SH groups

(Mersalyl structure: CH_3O substituent; lower portion shows phenyl ring with O—CH_2—COONa group)

SOURCE: After D. Eisenberg, in *The Enzymes*, 3d ed., vol. 1, ed. P. D. Boyer (New York: Academic Press, 1970).

Figure 13-32

A section through a crystal of insulin. Each wedge-shaped unit is one monomer. These monomers associate into dimers, which in turn aggregate into hexamers. The hexamers pack into the crystal. Note the large solvent channels and the relatively few direct contacts between hexamers. All atoms except hydrogen are shown. [From T. L. Blundell, D. C. Hodgkin, G. G. Dodson, and D. A. Mercola, *Adv. Protein Chem.* 26:279 (1972).]

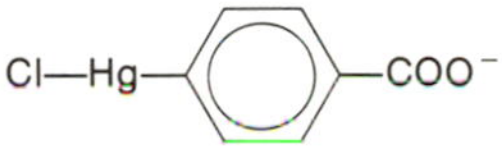

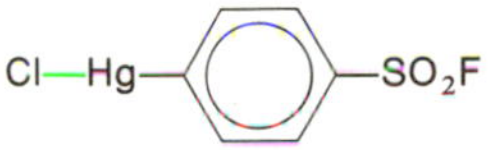

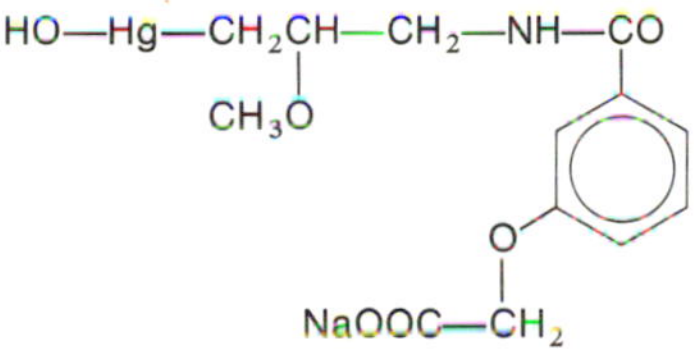

Table 13-3
Representative protein crystal structures determined with heavy-atom isomorphous derivatives

Protein	Molecular weight	Number of subunits	Space group	Molecules per asymmetric unit	Number of heavy atoms used	Resolution
Metmyoglobin, sperm whale	17,800	1	$P2_1$	1	8	1.4 Å
Oxyhemoglobin, horse	64,500	4	$C2$	1/2	7	2.8 Å
Ferricytochrome *c*, horse heart	12,400	1	$P4_3$	1	2	2.8 Å
Carboxypeptidase A, beef	34,600	1	$P2_1$	1	5	2.0 Å
α-Chymotrypsin, beef	25,000	1	$P2_1$	2	6	2.0 Å
Papain	23,000	1	$P2_12_12_1$	1	7	2.8 Å
Nuclease, *S. aureus*	16,800	1	$P4_1$	1	3	2.8 Å
Lactate dehydrogenase, dogfish	135,000	4	$I422$	1/4	5	2.0 Å
Lysozyme, hen egg	14,600	1	$P4_32_12$	1	8	2.0 Å

SOURCE: After D. Eisenberg, in *The Enzymes*, 3d ed., vol. 1, ed. P. D. Boyer (New York: Academic Press, 1970).

Once an isomorphous derivative is available, diffraction data are collected from it and compared with those from the unmodified crystal. The reciprocal lattice dimensions and symmetry should be unaltered, but the observed intensities of some of the reflections can change markedly (Fig. 13-33). These differences can make it possible to estimate the phases of the observed structure factors. However, it first is necessary to locate the heavy atoms; the next few subsections describe this process.

Structure factors for heavy-atom isomorphous derivatives

The electron density distribution of the heavy-atom isomorphous derivative is just the sum of the electron densities of the parent crystal and of the heavy-atom substitutions. Thus the structure factor F_{PH} of the heavy-atom isomorphous derivative must be related to the structure factor F_P of the parent crystal and the structure factor F_H of the heavy atoms along simply by

$$F_{PH}(h, k, l) = F_P(h, k, l) + F_H(h, k, l) \qquad (13\text{-}102)$$

because the additional scattering in the derivative is due simply to the presence of the heavy atoms.

Note, however, that all three quantities in Equation 13-102 are complex numbers. The significance of this equation can best be seen by expressing each number as a vector in the complex plane as described in Figure 13-4. To add two complex numbers,

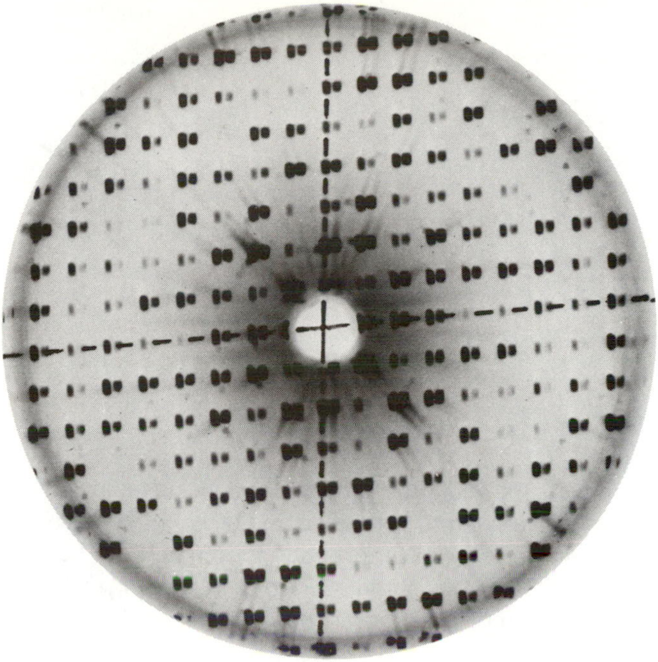

Figure 13-33

Isomorphous replacement. Two precession photographs of triclinic lysozyme crystals are superimposed, slightly out of horizontal register. The left spot of each pair is from a native lysozyme crystal; the right spot is from a crystal after diffusion of $HgBr_2$. This is a photograph of the $(0, k, l)$ plane of the reciprocal lattice. Note the differences in intensities. [From R. Dickerson, in *The Proteins*, 2d ed., vol. 2, ed. H. Neurath (New York: Academic Press, 1964).]

one simply adds the real and imaginary components separately. Therefore, represented as vectors, two complex numbers combine just as vectors do, component by component. The result is still a vector in the complex plane, as illustrated in Figure 13-34. One such vector equation holds for each value h, k, l of the structure factors.

If any two of the vectors shown in Figure 13-34 are known, the third can be calculated unambiguously. The principle obstacle in macromolecular crystallography is that only the *lengths* of *two* of the vectors, $|F_P|$ and $|F_{PH}|$, are directly measureable experimentally.

Location of heavy atoms by a difference Patterson map

An ordinary Patterson map (Eqn. 13-95) cannot be used to locate the heavy atoms in a macromolecular crystal. We showed earlier that the contrast between heavy

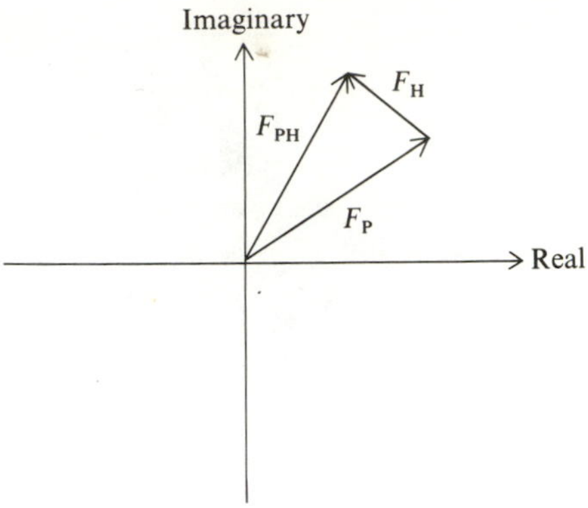

Figure 13-34
Structure factors plotted in the complex plane, for a parent-crystal diffraction spot, a heavy-atom isomorphous addition, and the expected derivative diffraction spot.

atom–heavy atom vectors and other vectors will be insufficient. However, when both an isomorphous heavy-atom derivative and a parent crystal are available, it is possible to calculate a difference isomorphous Patterson map between them, using the measured structure factor amplitudes $|F_{PH}(h, k, l)|$ and $|F_P(h, k, l)|$.

A true Patterson map of just heavy-atom vectors would be given by Equation 13-95 as

$$P_H = (1/V) \sum_{h=-\infty}^{\infty} \sum_{k=-\infty}^{\infty} \sum_{l=-\infty}^{\infty} |F_H(h, k, l)|^2 e^{-2\pi i(hx+ky+lz)} \tag{13-103}$$

we cannot calculate this map directly because $|F_H(h, k, l)|$ is not experimentally measurable. However, it turns out that $|F_H|$ often can be approximated fairly well by

$$|F_H| = ||F_{PH}| - |F_P|| \tag{13-104}$$

Thus we can calculate an estimate of $|F_H|$ from the measured amplitudes of the crystal and a heavy-atom isomorphous derivative. Then an isomorphous difference Patterson function ΔP is calculated:

$$\Delta P = (1/V) \sum_{h=-\infty}^{\infty} \sum_{k=-\infty}^{\infty} \sum_{l=-\infty}^{\infty} ||F_{PH}| - |F_P||^2 e^{-2\pi i(hx+ky+lz)} \tag{13-105}$$

In an ideal case, it can be shown that this function will display the heavy atom–heavy atom vector at one-half the expected intensity plus some contaminating noise due to light atom–light atom vectors (see Blundell and Johnson, 1976).

The accuracy and usefulness of Equation 13-105 depend on the validity of Equation 13-104. This in turn depends on relative phases and amplitudes of the three structure factors involved. We discuss a particular simplified case in the following subsection. Here, note the following observation. When the two vectors F_{PH} and F_P

in Equation 13-102 are parallel, then the phases cancel, and Equation 13-104 is exact. As long as the three vectors are near-parallel, Equation 13-104 should be an excellent approximation. The largest values of $||F_{PH}| - |F_P||$ will tend to be those that arise when F_{PH} and F_P are parallel. Thus, conveniently, the largest terms that enter Equation 13-104 will be those most likely to contain good estimates of $|F_H|$. By selectively including only the large terms in Equation 13-105, we often can produce an improved heavy-atom difference Patterson map. From the heavy atom–heavy atom vectors, we can attempt to find the actual heavy-atom locations by the methods described in Box 13-6.

If Equation 13-104 is to be useful, the presence of the heavy atom in the isomorphous derivative must cause a change in scattering intensity sufficient to yield measurable differences between $|F_{PH}|$ and $|F_P|$. For example, a single mercury atom with 80 electrons will produce an average change of 30% between $|F_P|$ and $|F_{PH}|$ in a 40,000 d protein. Thus, it is more than sufficient for the calculation of a difference Patterson map.

Using centrosymmetric projections to locate heavy atoms

A crystal of biological material cannot contain a center of symmetry because the molecules contain asymmetric carbon atoms. However, it frequently is possible to calculate a centrosymmetric projection (Box 13-5). For example, if a structure has a twofold screw or rotation axis, projection onto a plane perpendicular to this axis will result in a center of symmetry. In this two-dimensional projection, the phases of the structure factor must be either 0 or π, so that all structure factors are either parallel or antiparallel vectors. This greatly simplifies the use of Equation 13-102.

In a centrosymmetric projection, the structure factors for the parent crystal, heavy-atom isomorphous derivatives, and heavy atoms must be related by one of the arrangements shown in Figure 13-35. So long as F_{PH} and F_P point in the same direction, it is apparent that $|F_H| = ||F_{PH}| - |F_P||$, which allows $|F_H|$ to be calculated directly from experimental data. Only in those cases where $|F_H|$ is much larger than $|F_P|$ does Equation 13-104 become incorrect. These cases, which are called *crossovers*, are very rare and do not seriously compromise most heavy-atom isomorphous difference Patterson projections.

Figure 13-36 shows an example of three difference Patterson projections obtained for heavy-atom derivatives of cytochrome *c*. These were obtained by using Equation 13-105. The two relatively simple maps (Fig. 13-36a,b) result from a Pt and a Hg derivative. The more complex map (Fig. 13-36c) was obtained from a crystal into which both metals had been substituted. This map shows Hg–Pt vectors as well as Hg–Hg and Pt–Pt vectors. From these maps, estimates of both the Pt and Hg coordinates can be obtained (see Box 13-6).

In some cases, the heavy-atom positions found from a difference Patterson map are used directly to determine preliminary protein phases, as shown in the next subsection. In most cases, they must be refined first. (Refinement techniques are discussed in subsequent subsections.)

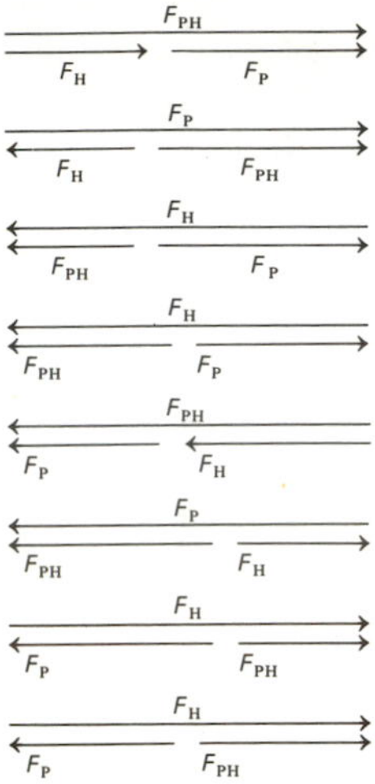

Structure factors	Signs F_P	Signs F_H	Measured amplitude change $\Delta F = \lvert F_{PH}\rvert - \lvert F_P\rvert$
	+	+	+
	+	−	−
	+	−	−
	+	−	+
	−	−	+
	−	+	−
	−	+	−
	−	+	+

Figure 13-35

Structure factors in a centrosymmetric projection. Shown as vectors are all of the possible arrangements of parent (P), heavy atom alone (H), and isomorphous derivative (PH) structure factors. A vector pointing from left to right is assigned a positive sign (zero phase angle).

Using heavy-atom positions to estimate phases of the structure factor

From the coordinates and identity of each known heavy atom, we can compute both the phase and amplitude of its contribution to the structure factor, using Equation 13-70. This computation yields F_H. The structure factor of a heavy-atom isomorphous derivative, F_{PH}, must be related to that of the parent crystal, F_P, and to F_H simply by Equation 13-102.

Once the heavy atom is found, F_H is known completely. However, only the amplitudes $|F_{PH}|$ and $|F_P|$ can be measured. Using all three quantities, it is possible to restrict the phase of F_P to only two possibilities (Fig. 13-37a). The possible values of F_P lie on a circle of radius $|F_P|$ centered at the origin. Possible values for F_{PH} will lie on a circle of radius $|F_{PH}|$ but, to satisfy Equation 13-102, the center of this circle must be displaced from the origin by the known vector F_H. Then the two circles intersect at two points. At each intersection, corresponding to phases of ϕ_a and ϕ_b, the conditions prescribed by Equation 13-102 are met.

The most common resolution of the remaining uncertainty in phase is to use a second isomorphous heavy-atom derivative. One estimates the position of the heavy

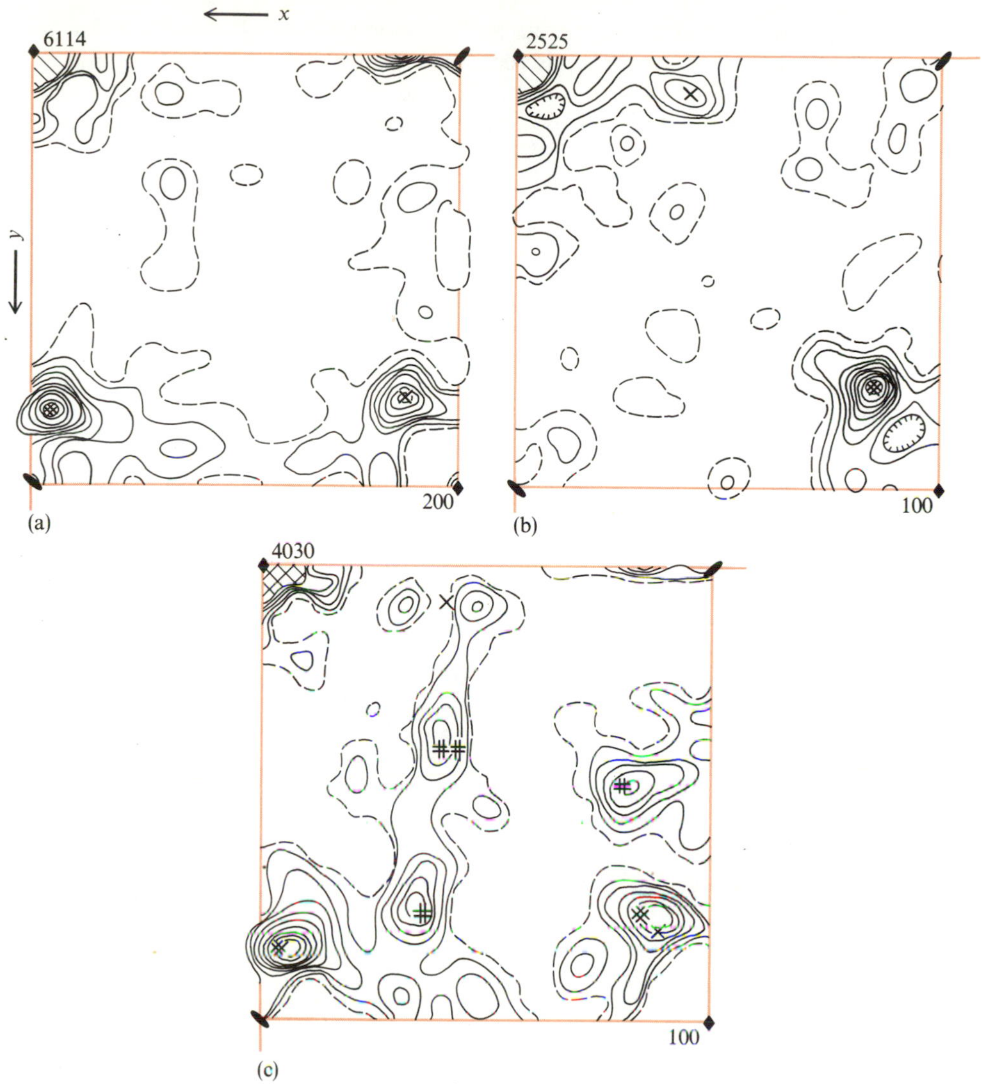

Figure 13-36

Difference Patterson maps, calculated for projections of crystals of cytochrome *c* into the x–y plane. The origin is at the upper left of each map. All maps are drawn to the same scale; contour intervals are marked at the lower right-hand corners; the height of the peak at the origin is indicated at the upper left-hand corners. The zero contours are dashed. The x and y coordinates are indicated; they run only from the origin to one-half the unit-cell dimensions. Single-weight Patterson peaks are shown by X, double-weight Patterson peaks by ⋇. **(a)** A platinum derivative. **(b)** A mercury derivative. **(c)** A derivative containing both heavy metals; platinum–mercury cross-vectors are shown by #. [From R. E. Dickerson et al., *J. Mol. Biol.* 29:77 (1967).]

atoms, computes F_H, then uses the analog of Equation 13-102: $F_{PH'} = F_P + F_{H'}$. The process of selecting a phase for F_P is repeated again by comparing $|F_{PH'}|$ and $|F_P|$, using the known $F_{H'}$ (Fig. 13-37b). In an ideal case, one of the two circle intersections will correspond either to ϕ_a or to ϕ_b, and the other will be some different value ϕ_c.

Box 13-6 AN EXAMPLE OF THE INTERPRETATION OF A DIFFERENCE PATTERSON PROJECTION

The difference Patterson map shown in Figure 13-36a was calculated for a projection into a plane perpendicular to the **c** axis of a tetragonal crystal of cytochrome *c* ($a = b = 58.45$ Å; $c = 42.34$ Å):

$$\Delta P(x, y) = (1/A) \sum_{h=-\infty}^{\infty} \sum_{k=-\infty}^{\infty} [|F_{PH}(h, k, 0)| - |F_P(h, k, 0)|]^2 e^{-2\pi i(hx+ky)}$$

where $|F_{PH}|$ and $|F_P|$ are the square roots of the measured intensities of the heavy-atom isomorphous Pt derivative and the parent crystal, respectively. Thus, $\Delta P(x, y)$ can be calculated from data collected for a single layer of the reciprocal lattice.

The space group of these crystals is $P4_1$. The asymmetric unit is one molecule of cytochrome *c*. There are four molecules per unit cell, and these are related by a fourfold screw axis. The projection of the structure perpendicular to the **c** axis places all four molecules in the **a**–**b** plane, where they are now related by a fourfold rotation axis. Because of this axis, the two-dimensional structure also has a center of symmetry.

We can use the symmetry to predict what the difference Patterson map should look like for a single heavy atom located at identical positions on each of the four molecules. For

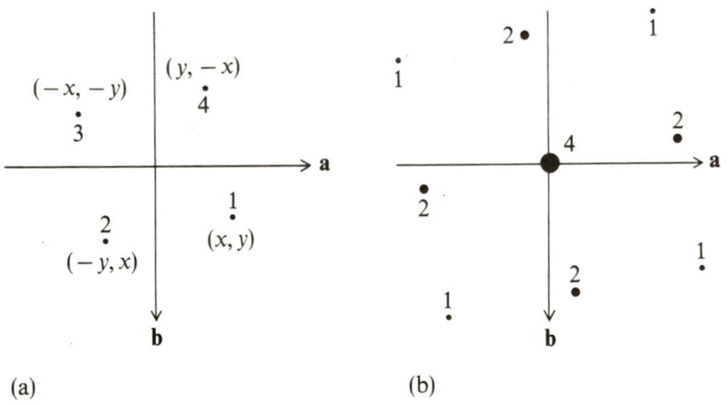

Thus, because F_P can have only one phase, it must uniquely be the angle derived in common from the two isomorphous derivatives.

In real life, the experimental data are not perfect; nor can the heavy atoms be located precisely. Therefore, the points of intersection of circles drawn from two

convenience, we choose the origin of the coordinate system right at the fourfold axis. Then, if the position of one heavy atom is $x\mathbf{a} + y\mathbf{b}$, the others must be located as shown in part a of the figure. The corresponding heavy atom–heavy atom vectors will be

$$\begin{array}{ll}
\mathbf{r}_{12} = (x + y, y - x) & \mathbf{r}_{21} = (-x - y, x - y) \\
\mathbf{r}_{13} = (2x, 2y) & \mathbf{r}_{31} = (-2x, -2y) \\
\mathbf{r}_{14} = (x - y, x + y) & \mathbf{r}_{41} = (y - x, -x - y) \\
\mathbf{r}_{23} = (x - y, x + y) & \mathbf{r}_{32} = (y - x, -x - y) \\
\mathbf{r}_{24} = (-2y, 2x) & \mathbf{r}_{42} = (2y, -2x) \\
\mathbf{r}_{34} = (-x - y, x - y) & \mathbf{r}_{43} = (x + y, y - x)
\end{array}$$

plus four heavy-atom self-vectors, which will lie at the origin.

The resulting difference Patterson map will be that shown in part b of the figure (for the relative values of x and y shown in part a of the figure), where the number adjacent to each peak gives its relative weight. Notice that the map has the same fourfold rotational symmetry as the structure that generated it. Concentrate just on the quadrant at lower right, and compare the result with Figure 13-36a. Notice the doubly-weighted peak near the vertical axis. This peak must correspond to the nearly vertical vector that forms two sides of the square of heavy atoms in the structure. The singly-weighted peak (near the lower right of Fig. 13-36a) is produced from a diagonal of the square. The hint of a peak near the horizontal axis arises from the nearby doubly-weighted vector in the upper right-hand quadrant of the map. Thus, a square structure of heavy atoms is fully consistent with the observed difference Patterson map. Once the vectors have been assigned, their locations yield the values of x and y, and thus the actual heavy-atom positions.

It would be a useful exercise for the reader to interpret the Hg difference map in Figure 13-36b and then, using the results of both maps, to attempt to explain the results shown in Figure 13-36c for the Hg, Pt double derivative.

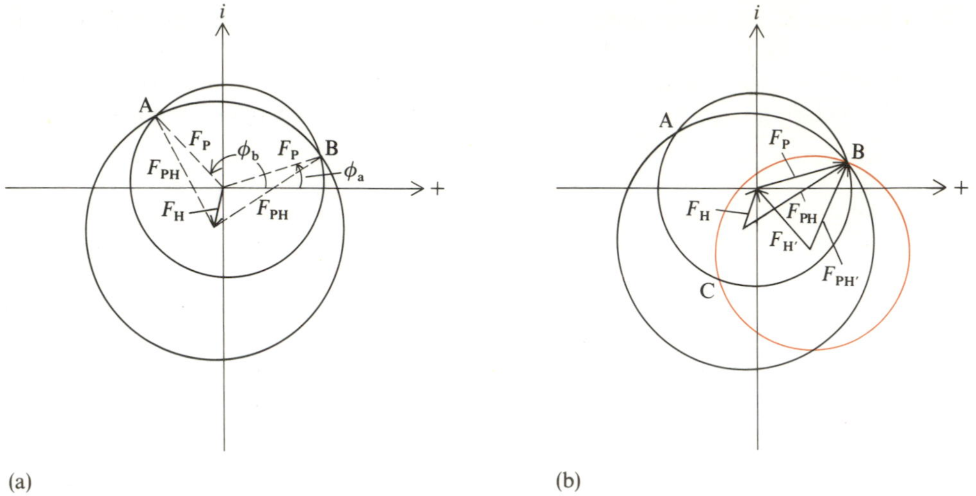

Figure 13-37

Phase determination by isomorphous replacement. Structure factors are plotted in the complex plane, as in Figure 13-34. **(a)** A single heavy-atom derivative. The circle with radius F_P represents the parent crystal, with measured intensity and unknown phase. The circle with radius F_{PH} represents the isomorphous heavy-atom-containing crystal, with measured intensity and unknown phase. The vector F_H is calculated from the heavy-atom position that had been determined from a difference Patterson synthesis. Because F_H is calculated, both its phase and its amplitude are known. Equation 13-102 will be satisfied by F_P, F_H, and F_{PH} when F_P lies at the origin, but F_{PH} is located at the end of F_H as shown. Therefore, the two circles are displaced, and they intersect at two positions: A and B. These positions define two possible values for the phase of F_P: ϕ_a and ϕ_b. **(b)** Inclusion of a second heavy-atom derivative. Its scattering amplitude yields a circle (*colored*) of radius $F_{PH'}$ centered at the end of the vector $F_{H'}$, which is calculated from the known position of the heavy atom. This circle also intersects F_P circle at two places: B and C. Because one intersection (B) is the same as an intersection found with the first heavy atom, the only phase choice for F_P consistent with both derivatives is ϕ_b. [After D. Eisenberg, in *The Enzymes*, 3rd ed., vol. 1, ed. P. D. Boyer (New York: Academic Press, 1970), p. 1.]

different isomorphous derivatives may not coincide exactly. Then, to resolve ambiguities, it usually is desirable to have additional derivatives to strengthen the accuracy of phase assignment and to guard against apparent agreement that is accidental. Naturally, the more derivatives available, the more accurately the phase angles are likely to be chosen. Statistical procedures for choosing the best phase estimates from multiple isomorphous derivatives are described by Blundell and Johnson (1976).

Once estimates of the phases of F_P are available, one can use Equation 13-93 to calculate an electron density map of the macromolecule by inserting the measured amplitudes $|F_P(h, k, l)|$ and calculated phases ϕ_{hkl}. However, in most cases, this map will not be very accurate unless the estimates of the heavy-atom positions are first refined.

Phase estimates with a center of symmetry

Suppose one can prepare only a single isomorphous derivative. The prognosis is not completely hopeless. In many cases, a projection of the crystal onto a plane will have a center of symmetry. The advantages of centrosymmetric projections in calculating the *amplitude* of F_H were described earlier. Here we show how such projections also assist the calculation of the *phase* of F_P. Note that only a single layer of reciprocal space is needed to compute the projections, so the structure problem now is purely two-dimensional.

Whenever a center of symmetry exists, the resulting phases can be only 0 or π, and so the only uncertainty remaining for F_P is the sign. The measured change in amplitude of one diffraction spot due to a heavy-atom isomorphous substitution is

$$\Delta F = |F_{PH}| - |F_P| \tag{13-106}$$

All the possible arrangements of F_{PH}, F_P, and F_H are shown in Figure 13-35. When the sign of ΔF (measured) is compared with the sign of F_H (calculated from the known heavy-atom positions), an interesting generalization emerges. Except for two of the rare crossover cases, whenever the sign of ΔF is the same as the sign of F_H, the sign of F_P must be positive ($\phi = 0$). Whenever ΔF and F_H have opposite signs, F_P is negative ($\phi = \pi$).

Thus, even with only a single isomorphous derivative, nearly all of the phases of a centrosymmetric projection of a structure can be computed correctly. Then the electron density of the projection, $\rho(x, y)$ can be calculated by a Fourier synthesis exactly analogous to Equation 13-93. X-ray crystallographers frequently use projections because they can be calculated at earlier stages in the analysis, and because less computer time is required to do two-dimensional sums than to do three-dimensional ones. However, bear in mind that a projection does not uniquely define the three-dimensional structure that produced it.

Narrowing heavy-atom positions with parent-crystal phase estimates

If we knew the phases of each of the diffraction spots of a parent crystal and of a heavy-atom isomorphous derivative, we could calculate an electron density map of each by using Equation 13-39. However, sometimes it is useful to display just the locations of the heavy atoms. This can be done using a difference Fourier synthesis:

$$\begin{aligned}\Delta\rho(x, y, z) &= \rho_{PH} - \rho_P \\ &= (1/V) \sum_{h=-\infty}^{\infty} \sum_{k=-\infty}^{\infty} \sum_{l=-\infty}^{\infty} \\ &\quad \times \left[|F_{PH}(h, k, l)|e^{i\phi_P(h,k,l)} - |F_P(h, k, l)|e^{i\phi_P(h,k,l)}\right] e^{-2\pi i(hx+ky+lz)}\end{aligned} \tag{13-107}$$

Here each structure factor has been explicitly shown as a phase plus an amplitude. The amplitudes, $|F_{PH}|$ and $|F_P|$, are measured. The phases of the parent crystal, ϕ_P, are estimated as shown in the preceding sections and are used for *both* amplitudes.

In principle, we could in similar fashion estimate the phases of the derivative, ϕ_{PH}, and use these in Equation 13-107; however, this would lead to problems. Fourier syntheses are dominated by phases, not by amplitudes (Fig. 13-28). Calculated phases of the derivative would contain a heavily weighted contribution from the heavy atoms. The resulting Fourier synthesis would simply give back the same heavy-atom positions one started with, and nothing would have been accomplished. Even the estimates of the parent-crystal phases, ϕ_P, are heavily contaminated with the heavy-atom phases.

In practice, when several heavy-atom derivatives are available, it is best to use parent phases estimated from one or more derivatives to compute the difference Fourier to find other derivatives; these are called cross-phase difference Fouriers. Figure 13-38 shows an example for the same cytochrome *c* derivatives discussed earlier. The Pt and Hg atoms show up clearly above a weak background. However, their apparent positions are not yet the true positions in the crystal.

A least-squares refinement of a structural model

There are adjustable parameters used in the calculation of an x-ray scattering pattern. In a least-squares refinement, one attempts to find the values for these parameters that minimize the difference between observed structure factor amplitudes and those calculated from any particular model or technique. We first illustrate this in the general case, and then show the specific application to isomorphous heavy atoms.

The experimental data are measured structure factor amplitudes $|F_0|$. The calculated $|F_c|$ values usually come from Equation 13-70. They are a function of all the structural parameters of the tentative model. These are the *x*, *y*, and *z* coordinates of each atom, and the atomic number of each atom.

In addition, for high-resolution structures, there is another effect we must worry about. Atoms are not fixed in space, even in a crystal. They are vibrating, and the amplitudes will vary for each atom. The x-ray scattering will be an average of the position of each atom. It can be shown that, for isotropic thermal motion, the atomic scattering factor will have the form $f = f_0 e^{-\beta|\mathbf{S}|^2/4}$, where **S** is the scattering vector, and β is related to the mean square amplitude of the atomic vibration, $\langle\mu\rangle$, by $\beta = 8\pi^2\langle\mu\rangle^2$. This relation introduces another parameter. The thermal factor β can be guessed from knowledge of the atom type, but in the most rigorous structure determination it too will be a variable. Furthermore, real vibrations are anisotropic and thus must be represented by a thermal ellipsoid defined by six parameters. So a total of anywhere from three to nine parameters, P_j, are needed for each atom; a very large number (n) of parameters are needed for the entire asymmetric unit.

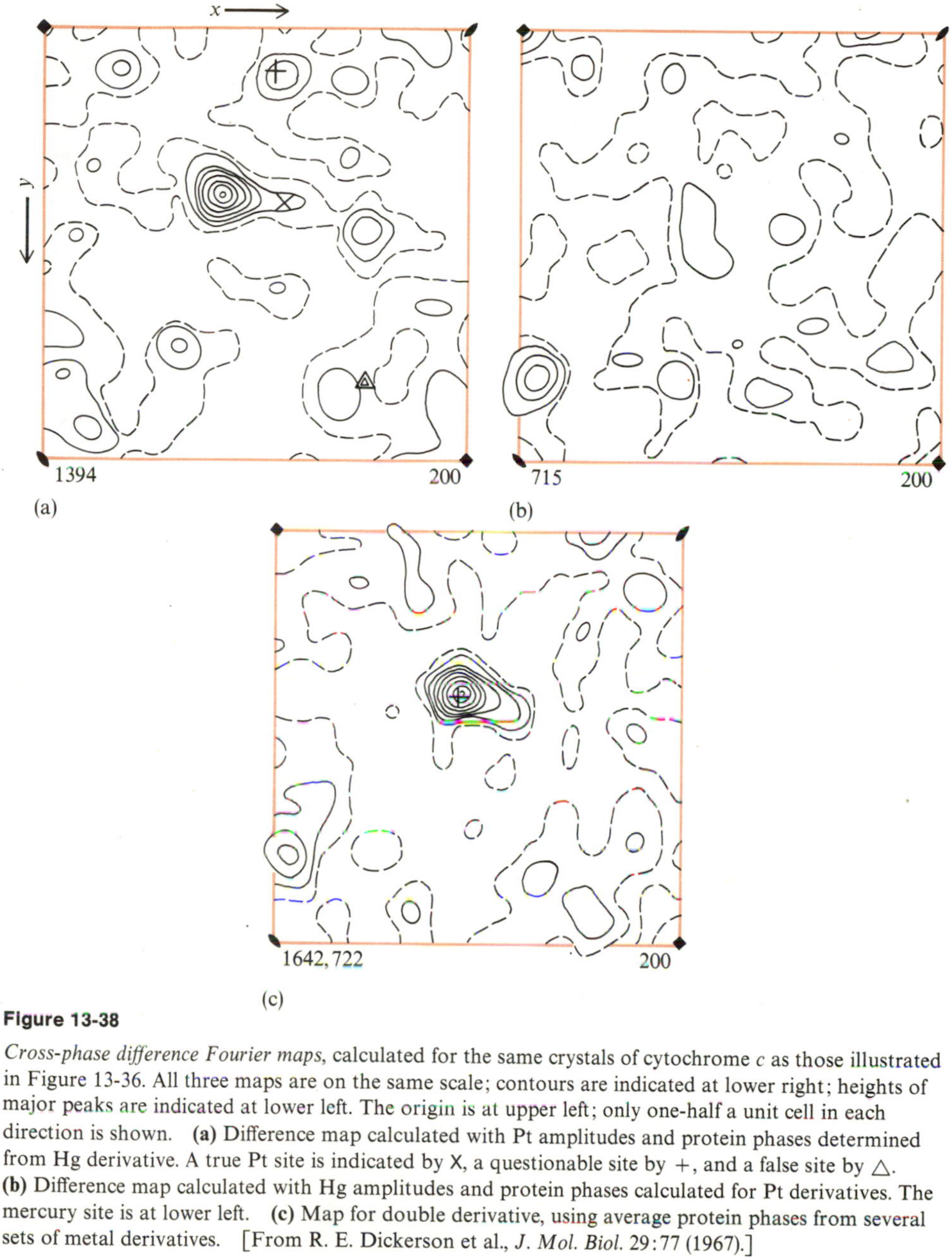

Figure 13-38

Cross-phase difference Fourier maps, calculated for the same crystals of cytochrome *c* as those illustrated in Figure 13-36. All three maps are on the same scale; contours are indicated at lower right; heights of major peaks are indicated at lower left. The origin is at upper left; only one-half a unit cell in each direction is shown. **(a)** Difference map calculated with Pt amplitudes and protein phases determined from Hg derivative. A true Pt site is indicated by X, a questionable site by +, and a false site by △. **(b)** Difference map calculated with Hg amplitudes and protein phases calculated for Pt derivatives. The mercury site is at lower left. **(c)** Map for double derivative, using average protein phases from several sets of metal derivatives. [From R. E. Dickerson et al., *J. Mol. Biol.* 29:77 (1967).]

In a least-squares refinement, one wants to adjust these parameters to minimize the difference between observed and calculated structure factor amplitudes. In practice, the actual quantity minimized is

$$D = \sum_{h=-\infty}^{\infty} \sum_{k=-\infty}^{\infty} \sum_{l=-\infty}^{\infty} W_{hkl}[|F_O(h, k, l)| - |\kappa F_c(h, k, l)|]^2 \qquad (13\text{-}108)$$

where W_{hkl} is a weighting factor measuring one's estimate of the reliability of a given experimental or calculated point, and κ is a scaling parameter. For each parameter P_j, one establishes the condition $\partial D/\partial P_j = 0$. This leads to n equations in the n unknown parameters. Solving these equations simultaneously produces the least-squares fit.

For an example of how such a calculation is set up in matrix form, see Section 8-1. The important thing is that an $n \times n$ matrix must be inverted. For linear equations, a single matrix inversion suffices. However, the equations that result from differentiating Equation 13-108 are not linear in the unknown parameters. Therefore, an iterative technique must be used. This involves inverting an $n \times n$ matrix, taking the resulting parameters, reinserting them into the equations, and repeating the matrix-inversion process. This routine is performed over and over again until the parameters converge on values that minimize D.

Least-squares refinement of heavy-atom positions

In the isomorphous replacement technique, $|F_P|$ and $|F_{PH}|$ have been measured, F_H has been calculated from an estimate of the heavy-atom positions, and ϕ_P has been calculated as described earlier. If all these results were correct, then F_{PH}, F_P, and F_H would form a triangle as shown in Figure 13-34. However, because of errors, the triangle usually is not closed.

We can calculate the structure factor expected for the heavy-atom derivatives as

$$F_{PH(calc)} = |F_H|e^{i\phi_H} + |F_P|e^{i\phi_P} \qquad (13\text{-}109)$$

To improve the location of the heavy atoms, one attempts to minimize the difference between the amplitude of this calculated structure factor and the observed amplitudes. The equation used, by analogy to Equation 13-108, is

$$D = \sum_{h=-\infty}^{\infty} \sum_{k=-\infty}^{\infty} \sum_{l=-\infty}^{\infty} W_{hkl}[|F_{PH}(h, k, l)| - |F_{PH(calc)}(h, k, l)|]^2 \qquad (13\text{-}110)$$

This minimization is done by allowing the heavy-atom positions and the thermal parameters to vary.

Such an approach is not necessarily the best one for every crystal, and alternative approaches are discussed by Blundell and Johnson (1976).

Once the heavy-atom positions have been refined, they are used to calculate a final set of phases for the parent crystal. Then, at last, the stage is set for a Fourier synthesis of the entire structure using Equation 13-93.

Anomalous dispersion of heavy atoms

One additional technique for exploiting the presence of heavy atoms has seen increasing use in protein and nucleic acid crystallography. This technique is anomalous dispersion. It occurs when the frequency of the x rays used falls near an absorption frequency of an atom. In practice, the technique is most useful for atoms heavier than sulfur.

Until now we have treated the atomic scattering factor f as a real number. In actuality, f is a complex number because the phase shift upon scattering is not necessarily an integral or half-integral number of oscillations:

$$f(S) = f_0(S) + if'_0(S) \tag{13-111}$$

The term f'_0 is significant only when the x-ray frequency is close to an atomic absorption frequency. It is related to the extinction coefficient of that particular atomic absorption.

When $f(S)$ was considered to be real, one of the implications was Friedel's law From Equation 13-18b, for parent-crystal atoms unaffected by anomalous dispersion, we can write

$$|F_P(h, k, l)| = |F_P(-h, -k, -l)| \tag{13-112}$$

However, for crystals containing heavy atoms, this relationship no longer holds. The breakdown of Friedel's law can be used in a number of different ways (see Blundell and Johnson, 1976). For example, suppose two different x-ray frequencies are used, one allowing anomalous dispersion and one not. The difference in scattered intensities should represent the anomalous scattering, and this is restricted to the heavy atom. Then an analog of the isomorphous methods described above can allow calculation of phases.

Interpretation of the electron density map

Here we describe some typical stages in the solution of the crystal structure of a protein (see Fig. 13-39). Several heavy-atom derivatives have been prepared and located. Isomorphous replacement has been used to estimate phases for all $F_P(h, k, l)$, and an electron density map has been calculated using these phases and all data to a certain resolution. The resolution chosen will be a function of the order of the actual crystal and of how isomorphous the derivatives are. Reliable phases are needed

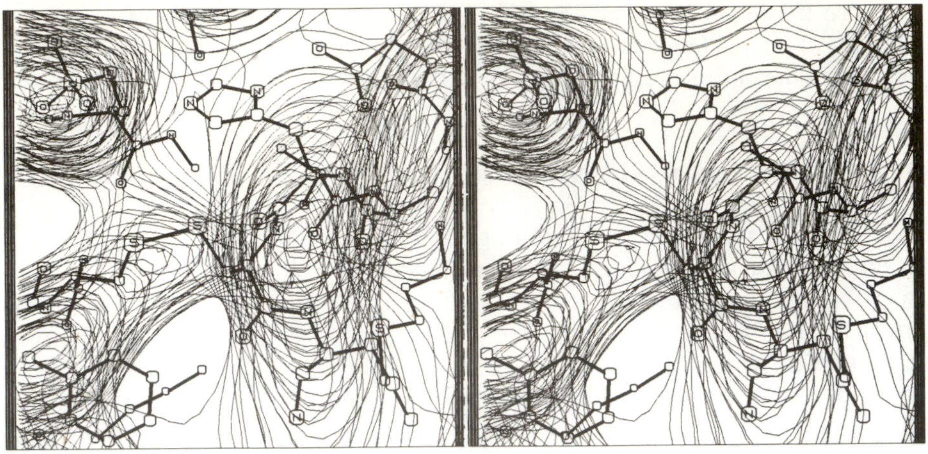

(a)

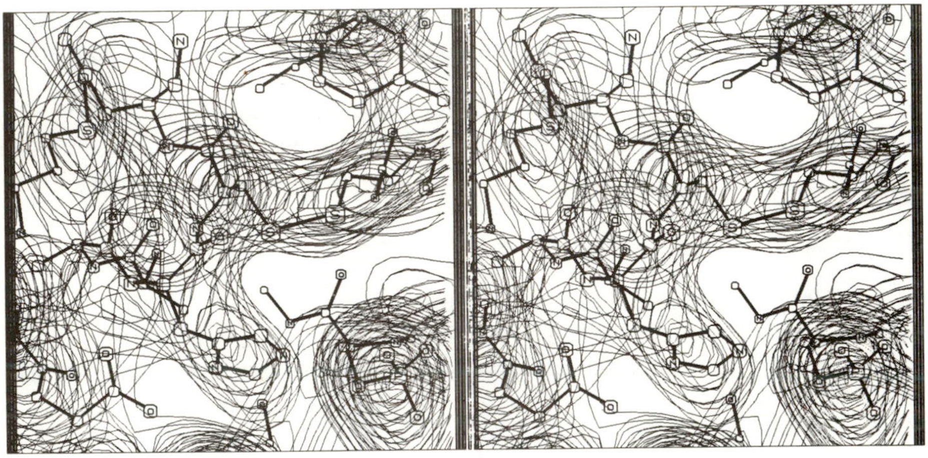

(b)

Figure 13-39

Protein electron density maps as a function of resolution. The maps are calculated from measured intensities and estimated phases. The protein is a diisopropyl fluorophosphate derivative of bovine trypsin. The view is down the *y* axis of the active site. A ball-and-stick model of the *final* best estimate of the structure is repeated in each map; note the phosphate at lower right and the active-site histidine at lower center; above these two features is a disulfide bond. **(a)** A map at 6.0 Å resolution, contoured

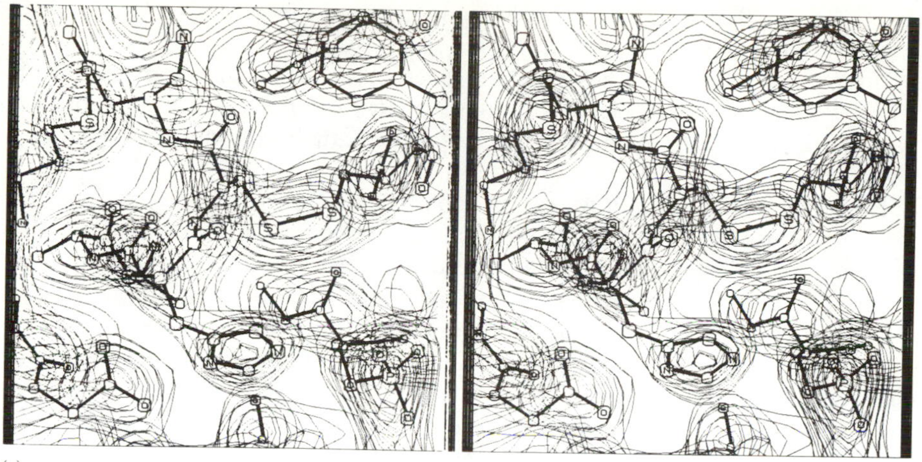

(c)

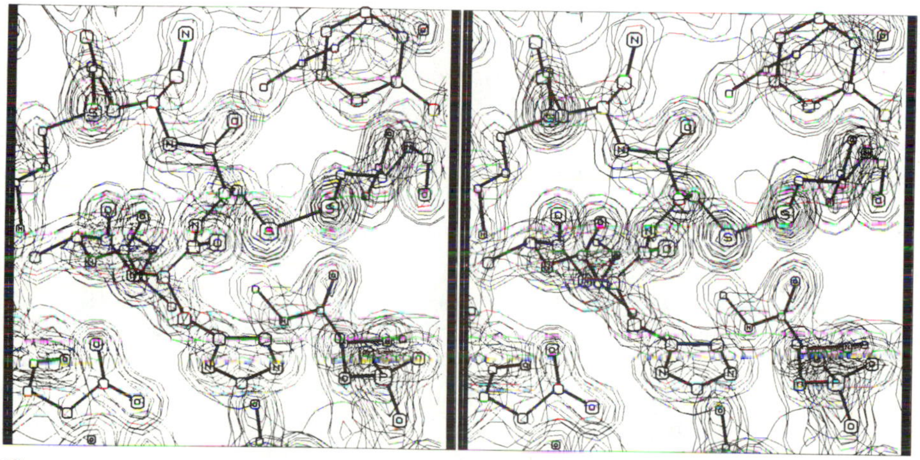

(d)

from 0.05 e Å^{-3} in steps of 0.05 e Å^{-3} **(b)** A map at 4.5 Å resolution, contoured from 0.10 e Å^{-3} in steps of 0.10 e Å^{-3}. **(c)** A map at 3.0 Å resolution, contoured from 0.35 e Å^{-3} in steps of 0.30 e Å^{-3}. **(d)** A map at 1.5 Å resolution, contoured from 0.50 e Å^{-3} in steps of 0.50 e Å^{-3}. All maps are shown as stereo pairs. [Courtesy of John L. Chambers. For further details, see his unpublished Ph.D. thesis, Calif. Institute of Technology, 1977.]

to justify the vastly increasing effort of using more and more scattering data in an attempt to obtain higher resolution.

If the result is around a 6 Å map, the macromolecule usually appears as a blob of electron density; Figure 13-39a shows a typical example. At this resolution, it usually is impossible to recognize the polymer chain backbone for a protein or nucleic acid.

Even at 6 Å resolution, however, considerable useful information emerges. One can learn a fairly detailed shape, and can spot crevices or subunits; α helices will appear as rods. If heavy-atom-labeled ligands or substrates are available, difference Fouriers can allow determination of the locations of their binding sites. If these results seem reasonable, it usually is worthwhile to attempt to proceed to higher-resolution analysis, providing that the quality of the data on available isomorphous derivatives justifies this.

At 3.0 Å resolution, it is possible to trace the path of the polymer chain backbone (Fig. 13-39c). In a protein, only the large amino acid side chains show up as discrete density peaks. It would be difficult to construct a meaningful molecular model at this stage. In nucleic acids, double helices will show up readily.

At 2.5 Å resolution, almost all protein side chains are visible. The carbonyl group of each peptide shows up as a protrusion from the main chain, and so it is possible to fix the orientation of each peptide plane.

If the amino acid sequence is known, one can begin to construct a model of the protein. Various techniques exist for doing this. The simplest is a Richards box, which uses a half-silvered mirror to superimpose a wire model of the structure onto a pile of lucite sections where the electron density map is plotted. Coordinates for atoms then are read off the model. Newer methods use computer searches to trace the most likely continuous paths of electron density and fit a peptide chain to these. Note that both approaches have the built-in assumption that the geometry of the peptide chain (except for dihedral angles) is known. This is a far cry from high-resolution small-molecule x-ray crystallography, where one determines bond lengths and angles *de novo*.

The original fit of the peptide chain to the electron density map at 2.5 to 3.0 Å will not be very precise. Many groups cannot be centered on the electron density peaks that presumably represent them. (See Fig. 13-39c for a typical example.) However, the preliminary model now can be used for one or more cycles of refinement. For example, in real-space refinement, one adjusts the model to try to minimize the difference between $\rho(x, y, z)$ calculated from the x-ray data and $\rho(x, y, z)$ calculated from the model. Alternatively, one can use Fourier refinement or least-squares techniques.

At higher resolution, individual atoms begin to be seen (see Fig. 13-39d for an example of a 1.5 Å map). Here it is actually possible to identify many amino acid side chains directly from the electron density map. In fact, crystal-structure work has revealed a number of serious errors in predetermined amino acid sequences. The more side chains one can see, the more accurate a model one can build and, in turn, the more likely it is that a further improvement in the electron density map can result from additional refinement.

Energetics of protein conformations in interpretation of the electron density map

A question of serious concern among crystallographers is how much knowledge about conformational analysis of proteins should be incorporated into the process of solving crystal structures. As shown in Chapter 5, we know with fair likelihood what ranges of dihedral angles are preferred by peptides. We know much about the forces that govern the interactions of nonbonded residues. Given a trial structure determined from a Fourier synthesis, this conformational knowledge could, by an energy minimization, be used to compute a structure more consistent with the body of acquired thermodynamic information.

M. Levitt and R. Diamond have shown that alternate cycles of Fourier and conformational energy refinement can be synergistic and can lead to convergence to a better structure. This is reasonable. The fact that a crystal forms implies that it must be in a crystal-wide free energy minimum. Alternatively, one can use conformational energies, not to try to obtain a minimal energy structure, but just as a guide on how to shift atoms slightly in the trial structure. A shift that simultaneously improves the fit of that atom to the calculated electron density map and also lowers the conformational energy is likely to be a step in the right direction.

Difference Fourier syntheses in studying ligand–macromolecule interactions

Difference Fouriers are extremely useful for comparing two structures. For example, the difference Fourier map of an isomorphous derivative and the parent should contain just the electron density of the added heavy atoms, as discussed earlier. Similarly, difference Fouriers have been used to locate substrate-binding or ligand-binding sites on proteins. Here the idea is to measure the diffraction intensities of the liganded protein. Then one calculates a difference Fourier using these measured intensities and calculated phases of the unliganded protein. The result can be just the electron density of the bound ligand, plus any difference in density due to changes in structure induced by the ligand binding. Naturally, the technique will work only so long as these differences are not too large.

The difference Fourier technique for locating a bound ligand works well because the presence of the ligand changes the phase of most structure factors relatively little. We can write (by analogy to Eqn. 13-102)

$$F_{PL} = F_P + F_L \tag{13-113}$$

where PL stands for the parent–ligand complex, and F_L is the structure factor of the ligand. Figure 13-40 shows this vector equation graphically. Here, F_P is known, and $|F_{PL}|$ is known. As long as $F_L < F_P$ and $F_L < F_{PL}$, the possible values of F_{PL} must lie within a small range of phase angles. Thus, as a first approximation, the parent phase ϕ_P is a good estimate of ϕ_{PL}.

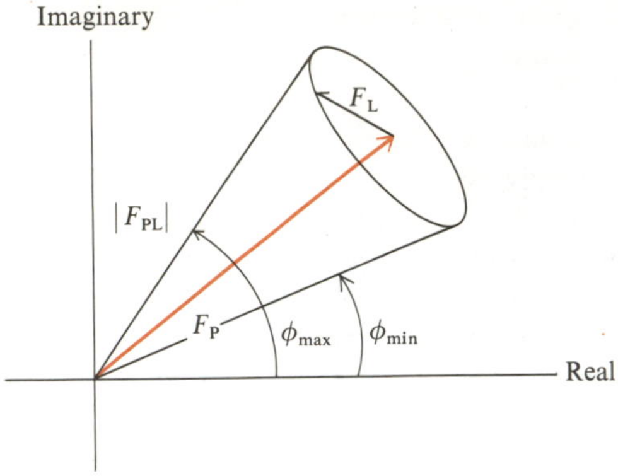

Figure 13-40

Effect on observed structure factors of an added ligand. The parent-crystal structure factor F_P is presumed to be known. Then, so long as the ligand structure factor $|F_L|$ is small, the phase of F_P is a good approximation of the phase of the ligand complex, F_{PL}.

A difference Fourier map thus can be calculated by analogy to Equation 13-107:

$$\Delta\rho(x, y, z) = (1/V) \sum_{h=-\infty}^{\infty} \sum_{k=-\infty}^{\infty} \sum_{l=-\infty}^{\infty} [|F_{PL}(h, k, l)| - |F_P(h, k, l)|] e^{i\phi_P(h,k,l)} e^{-2\pi i(hx+ky+lz)} \tag{13-114}$$

This map will show peaks that correspond to the position of the bound ligand. It also will show adjacent positive and negative regions of electron density that correspond to the movement of an atoms induced by ligand binding. Figure 13-41 shows this schematically in the one-dimensional difference Fourier.

An explicit justification for the validity of the difference Fourier map in representing the structure of the bound ligand can be seen by examining a centrosymmetric projection. The true structure of the bound ligand is

$$\rho_L(x, y, z) = (1/V) \sum_{h=-\infty}^{\infty} \sum_{k=-\infty}^{\infty} \sum_{l=-\infty}^{\infty} |F_L(h, k, l)| e^{i\phi_L(h,k,l)} e^{-2\pi i(hx+ky+lz)} \tag{13-115}$$

We need to know how well $|F_L|e^{i\phi_L}$ is approximated in Equation 13-114 by $||F_{PL}| - |F_P||e^{i\phi_L}$. In the centrosymmetric case as long as $|F_L|$ is small, Equation 13-114 is exact, as you can see by applying the same arguments used in Figure 13-35. For the general case, it is known that Equation 13-114 will correctly represent the electron density of the ligand, except that the peaks will be only half the correct height, and that there will be some noise in the data.

Much of our knowledge about the structure of enzyme active sites comes from difference Fourier calculations on crystals containing bound substrates or bound

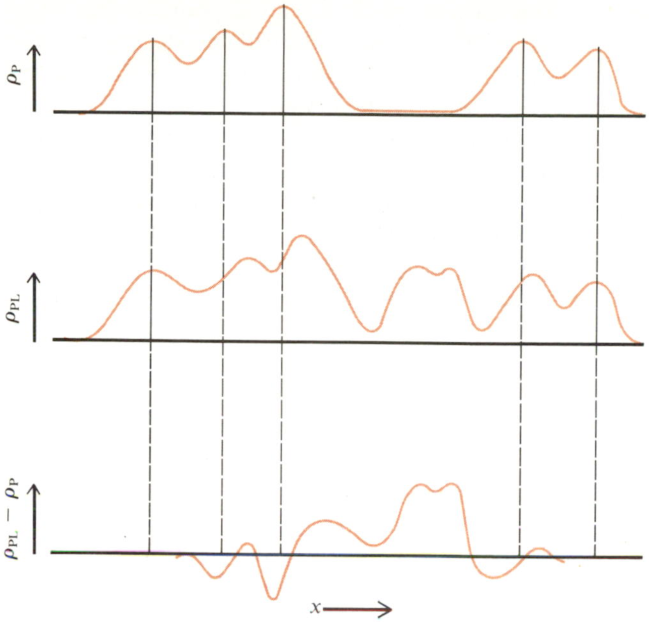

Figure 13-41

Using difference Fourier syntheses to study ligand binding. Such syntheses can be used to identify ligand binding sites and any conformational changes that accompany addition of the ligand. Shown are one-dimensional schematic drawings of the parent electron density map (ρ_P), the map that would be computed by solving the structure of the ligand complex (ρ_{PL}), and a difference Fourier ($\rho_{PL} - \rho_P$) that could be calculated in a relatively simple fashion (see text). Note that ligand atoms simply lead to increased density, whereas atom movements yield adjacent peaks and troughs in the difference Fourier. [After T. L. Blundell and L. N. Johnson, *Protein Crystallography* (London: Academic Press, 1976).]

inhibitors. The availability of this technique means that, in many cases, the determination of a macromolecular crystal structure is not so much the end of a massive effort as it is a starting point in the study of macromolecular function.

Summary

The x-ray scattering from an atom depends on its position in space and on the number of electrons it contains. The x-ray scattering from an array of atoms can be computed by summing the contributions of individual atoms. Periodic arrays of identical atoms restrict the observation of significant scattered intensity to only a discrete set of experimental geometries. Arrays of molecules can be treated in the same way as arrays of atoms. A crystal is a three-dimensional periodic array that consists of a unit cell replicated in space. The vertices of the cell define a crystal lattice. The periodicity of the array restricts observable scattering to a very limited set of geometries, which form the reciprocal lattice of the crystal.

The x-ray scattering can be described as the Fourier transform of the electron density of the object that generated it. Thus, if one could measure both the phase and the intensity of the scattered radiation, one could directly perform an inverse Fourier transfer and reproduce the structure. Unfortunately, all one can measure is the intensity. In principle, the pattern of scattered intensities still contains sufficient information to reconstruct the array that generated it. However, this information is not as easy to use or interpret. The inverse Fourier transform of the scattered intensity is called the Patterson function. It is a map of all interatomic vectors. Thus, if the structure contains n atoms per unit cell, there will be n^2 vectors per unit cell of the Patterson function.

The structure of macromolecular crystals usually is solved by the technique of multiple isomorphous replacement. Heavy-metal derivatives of a parent crystal are prepared, and the scattered intensities of the parent and the derivatives are compared. A difference Patterson map calculated directly from the intensities mostly shows just heavy atom–heavy atom vectors. This allows a preliminary estimate of the heavy-atom positions. Using these positions and the differences between scattered intensities of the parent crystal and the derivatives, it is possible to estimate the phase of the scattered radiation. Once this estimate is available, the estimates of the heavy-atom positions can be made more precise. The procedure is repeated, or other refinement techniques are used. Finally, phases are available accurate enough to use in conjunction with the measured intensities to compute an image of the structure.

Problems

13-1. Calculate the x-ray structure factor of the array of identical atoms shown in Figure 13-42, and compare it with the calculations in the text for similar arrays. Extend the result to the infinite array. Atoms are shown as circles, lattice points as dots.

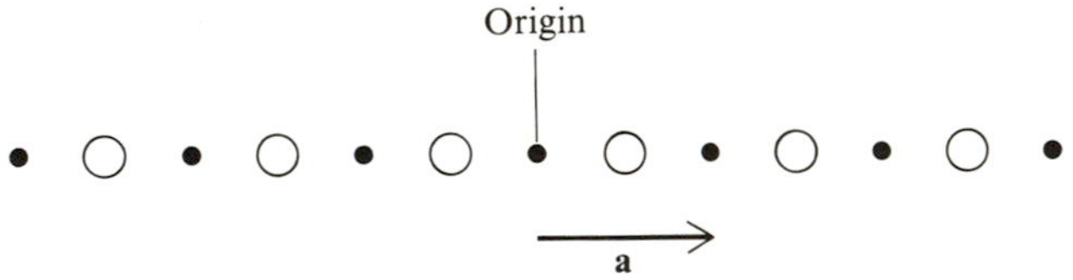

Figure 13-42

Array of atoms for Problem 13-1.

13-2. Calculate the x-ray scattering intensity expected from the infinite two-dimensional crystal shown schematically in Figure 13-43, where $|\mathbf{a}| = R$, and $|\mathbf{b}| = 4R$. You may

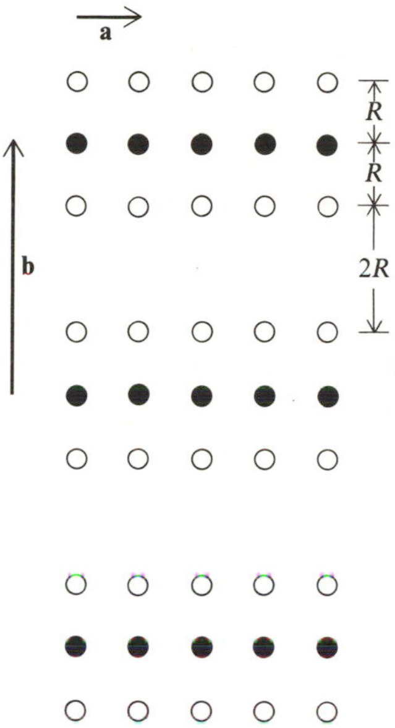

Figure 13-43
Crystal array for Problem 13-2.

assume that the atomic scattering factor of the central atom of each molecule is twice that of the other two atoms. Ignore the dependence of f on S. You should be able to denonstrate that, if the intensity is plotted on the reciprocal lattice, it is constant for all values of h but varies periodically with k, such that the strongest intensity is seen for $k = 0, 4, 8, \ldots$, and the weakest for $k = 2, 6, 10, \ldots$ HINT: Calculate the reciprocal lattice; calculate the scattering expected for one molecule placed at the origin; then use the principles of convolutions to compute the structure factor of the crystal; finally, square the amplitude to calculate the intensity. [This problem was adapted from one suggested by Bruno Zimm.]

13-3. Starting from the vector diagram in Figure 13-34, derive the following expression for the difference between the amplitude of the parent crystal (P) and that of a heavy-atom isomorphous derivative (PH):

$$|F_{\mathrm{PH}}| - |F_{\mathrm{P}}| = |F_{\mathrm{H}}| \cos(\phi_{\mathrm{PH}} - \phi_{\mathrm{H}}) - 2|F_{\mathrm{P}}| \sin^2[(\phi_{\mathrm{P}} - \phi_{\mathrm{PH}})/2]$$

Under what conditions can a comparison of the differences in observed intensities be used to obtain a good estimate of the heavy-atom structure-factor amplitude $|F_{\mathrm{H}}|$?

13-4. Draw the molecule that would produce the Patterson map shown in Figure 13-44. (Assume that all atoms are equal.) If you can't see how to do this, first choose a few arbitrary

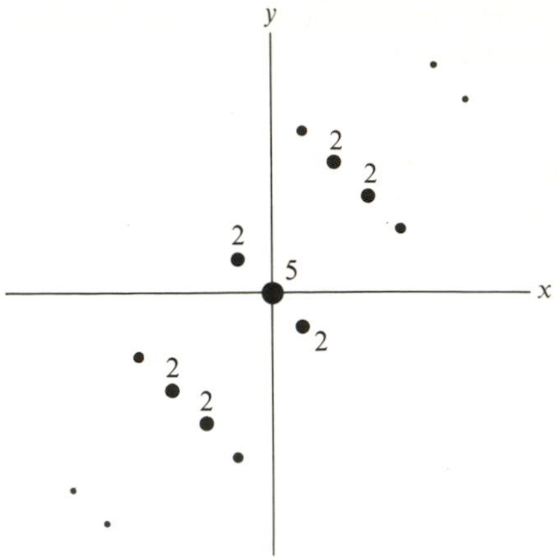

Figure 13-44

Patterson map for Problem 13-4.

small molecules and construct their Patterson maps. Multiple-weighted peaks are indicated in the figure; others have a weight of one.

13-5. A one-dimensional crystal has a unit cell 4 Å long. Measured x-ray scattering intensities—ignoring the dependence of $f(S)$ on S—are shown in Table 13-4. An isomorphous "heavy"-atom derivative can be prepared, in which one atom with an atomic scattering factor of 5 is added per unit cell; its scattering intensities also are shown in the table.

Table 13-4

Measured x-ray scattering intensities for problem 13-5

h	$\|F_P(h)\|^2$	$\|F_{PH}(h)\|^2$
0	49	144
1	5	20
2	25	0
3	5	20
4	49	144
5	5	20
6	25	0
7	5	20
⋮	⋮	⋮

a. Try to find the location of the heavy atom by using $|F_H| \cong ||F_{PH}| - |F_P||$ in a difference Fourier synthesis. Convince yourself that only the diffraction spots at $h = 0, 1, 2, 3$ need be considered. Assume for the moment that all phase terms are $+1$, and perform the calculation only for $x = 0, R/4, R/2$, and $3R/4$, where R is the length of the unit cell.

b. Because the result in part a gives two intensity maxima as possible positions for the heavy atom, calculate the contribution each makes to the structure factor. Use the *phases* calculated from each position, along with the *intensities* estimated as in part a, to repeat the difference Fourier. Is there any improvement?

c. Because the information that $|F_H| = 5$ has been given, use the isomorphous replacement replacement technique outlined in the text to estimate the phases of $|F_P(h)|$ for the two possible positions of the heavy atom. Now, using the criterion that $\rho(x)$ must be real for all x, select among these phases for four acceptable choices, and perform a Fourier synthesis of the $|F_P(h)|$ data to yield the structure. Note that each synthesis produces the same structure, except for changes in the origin of the unit cell and the direction of positive x.

References

GENERAL

Blundell, T. L., and L. N. Johnson. 1976. *Protein Crystallography*. London: Academic Press. [An excellent up-to-date monograph.]

Dickerson, R. 1964. X-ray analysis and protein structure. In *The Proteins*, 2d ed., vol. 2, ed. H. Neurath (New York: Academic Press), p. 603.

Eisenberg, D. 1970. X-ray crystallography and enzyme structure. In *The Enzymes*, 3d ed., vol. 1, ed. P. D. Boyer (New York: Academic Press), p. 1.

Glusker, J. P., and K. N. Trueblood. 1972. *Crystal Structure Analysis: A Primer*. London: Oxford Univ. Press. [A clear, elementary treatment.]

SPECIFIC

Fraser, R. D. B., and T. P. MacRae. 1969. X-ray methods. In *Physical Principles and Techniques of Protein Chemistry*, part A, ed. S. J. Leach (New York: Academic Press).

Guinier, A. 1963. *X-Ray Diffraction in Crystals, Imperfect Crystals, and Amorphous Bodies*. San Francisco: W. H. Freeman and Company. [An advanced mathematical treatise.]

Lipson, H., and C. A. Taylor. 1958. *Fourier Transforms and X-Ray Diffraction*. London: G. Bell & Sons.

Matthews, B. W. 1974. Determination of molecular weight from protein crystals. *J. Mol. Biol.* 82:513.

Stout, G. H., and L. M. Jensen. 1968. *X-Ray Structure Determination: A Practical Guide*. New York: Macmillan.

14
Other scattering and diffraction techniques

14-1 X-RAY FIBER DIFFRACTION

It would be unrealistic to hope that all macromolecules of biological interest could be induced to form crystals suitable for x-ray structural analysis. Long rod-shaped molecules, such as individual α helices or DNA, are reluctant to crystallize. However, concentrated gels formed by precipitates of such samples can be pulled mechanically into partially-ordered fibers. Figure 14-1 shows a schematic of several types of fibers.

X-ray scattering expected from fibers

Crystalline fibers actually are mosaics of microcrystals. These microcrystals have one axial direction in common, but the individual crystals can be oriented at random angles about this axis (Fig. 14-1a). Thus, an x-ray scattering measurement on such a fiber will sample the crystals at all orientations about the fixed axis. This is equivalent to a rotation photograph of a single crystal as shown in Figure 13-29a.

The x-ray scattering from one microcrystal is described by Equation 13-65, which we can write in abbreviated form as

$$F_{\mathrm{Tot}} = F_{\mathrm{m}}(\mathbf{S})F_{\mathrm{L}}(\mathbf{S}) \qquad (14\text{-}1)$$

Here $F_{\mathrm{m}}(\mathbf{S})$ is the molecular structure factor. It is the Fourier transform of the contents of the unit cell. $F_{\mathrm{L}}(\mathbf{S})$ is the sampling function generated by the lattice (see Eqn. 13-60).

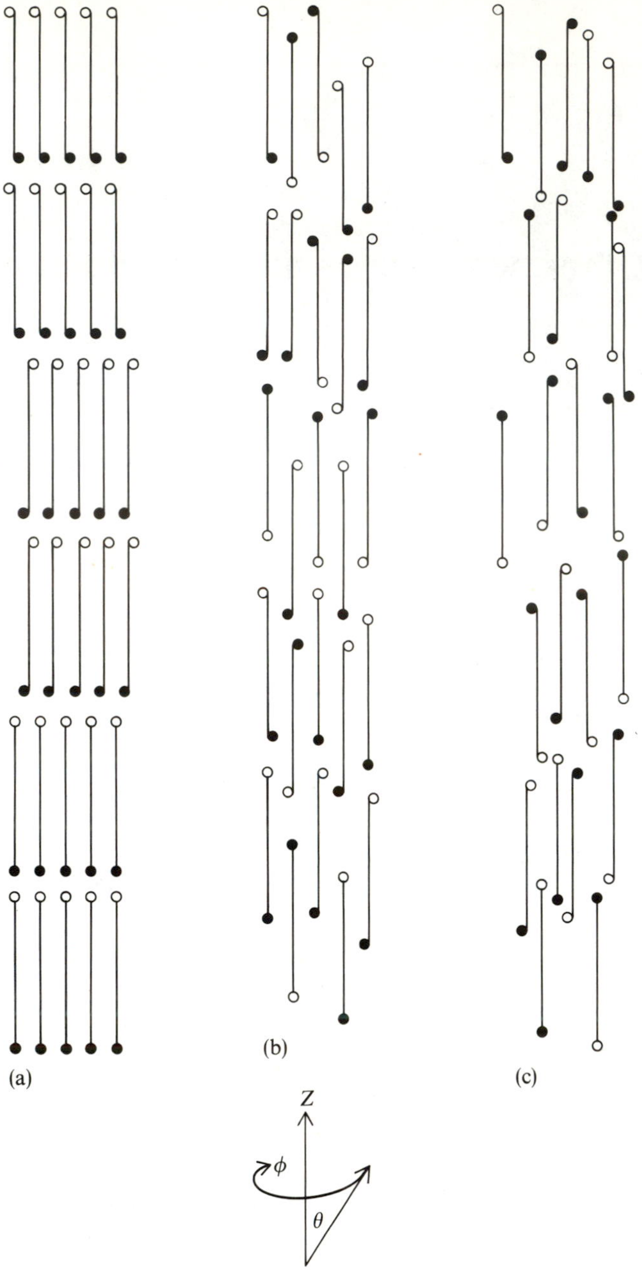

Figure 14-1

Schematic illustration of some of the types of fibers formed by rod-shaped molecules. **(a)** Crystalline fiber. Each molecule has the same orientation angle θ. The fiber consists of individual microcrystals, also ordered along Z and ϕ; however, these are packed in a random way about ϕ (and sometimes also about Z). **(b)** Semicrystalline fiber. Each molecule has the same orientation angle θ, but there are no microcrystals. There is still a fairly regular lattice perpendicular to Z, but individual molecules are disordered in the Z direction as well as the ϕ direction. **(c)** Noncrystalline fiber. The only order in this fiber is the constant value of θ for each molecule; there is no regular lattice.

It is the Fourier transform of the lattice. The observed intensity pattern from the entire fiber is $\langle|F_{\text{Tot}}|^2\rangle$, where the angle brackets indicate a rotational average around one axis of the lattice. In principle, sufficient x-ray scattering data could be collected on such a sample to solve the structure by the same methods described in Chapter 13 for crystals. However, in practice, this would be quite difficult, because at large scattering angles the diffraction spots overlap badly.

In semicrystalline fibers, the long axis of each rodlike molecule is well oriented. The molecules still pack together to form a fairly regular lattice when viewed perpendicular to the long axis. However, the molecules have random rotational orientations about their long axis and random translations along this axis (see Fig. 14-1b). The x-ray scattering from such a sample can be very tricky to analyze. Formally, it is described by

$$\langle I(\mathbf{S})\rangle = \langle|F_{\text{m}}(\mathbf{S})F_{\text{L}}(\mathbf{S})|^2\rangle \text{ lattice and molecular orientations} \tag{14-2}$$

where the angle brackets indicate that the intensity predicted for an ordered array must be averaged over the various types of disorder within the fiber.

At low resolution, the random orientation of molecules is not visible. The sample resembles an ordered lattice of rods. Thus the averaging denoted by Equation 14-2 will leave intact the lattice sampling of intensity at small $|\mathbf{S}|$. However, at high resolution, the random orientation will have a profound effect. At large $|\mathbf{S}|$, the lattice sampling will be eliminated, and all that will be left is the rotationally averaged molecular structure factor.

In a noncrystalline fiber, the only order is that long axes of the molecules share a common direction. There is random rotational orientation around this axis and, furthermore, there is no regular pattern of packing (Fig. 14-1c). Thus, the molecules are not necessarily equidistant. In effect, there is no lattice. The observed x-ray scattering intensity is simply related to the rotationally averaged molecular structure factor:

$$\langle I(\mathbf{S})\rangle = \langle|F_{\text{m}}(\mathbf{S})|^2\rangle \text{ molecular orientations} \tag{14-3}$$

Calculation of scattering from a helix

The experimental data provided by Equation 14-3 are not sufficient to compute a structure. Therefore, what is done is to use the observed data to evaluate models of the structure. For helical molecules, certain general features can be derived directly by inspection of the scattering data. In the next few subsections we concentrate on predicting the scattering from helices.

The basic structural parameters of most protein and nucleic acid secondary structures first were discovered from fiber studies. In addition, the same theory can be used to analyze the structure of helical arrays of protein subunits (such as those found in microtubules, actin, bacteriophages, and viruses) and of sheets of subunits (such as those found in muscle, bacteriophage heads, and membranes).

Consider a single-helical molecule consisting one type of atom repeated infinitely by a screw axis of symmetry. Such a molecule is essentially a miniature crystal. The lattice is a series of points arranged in a helix. The x-ray structure factor for this molecule will be given by the analog of Equation 13-60:

$$F_{\text{Tot}}(\mathbf{S}) = f(S)F_{\ell}(\mathbf{S}) \tag{14-4}$$

Here $f(S)$ is the atomic scattering factor of the atom; $F_{\ell}(\mathbf{S})$ is the sampling function due to the helical lattice—$F_{\ell}(\mathbf{S})$ is the Fourier transform of the helical lattice. A convenient way to compute this is to use the convolution theorem described in Equation 13-52.

The helical lattice can be generated mathematically by forming the product (intersection) of a helical line with a set of planes, each representing the position of single residue units (Fig. 14-2). The Fourier transform of a product is equal to the

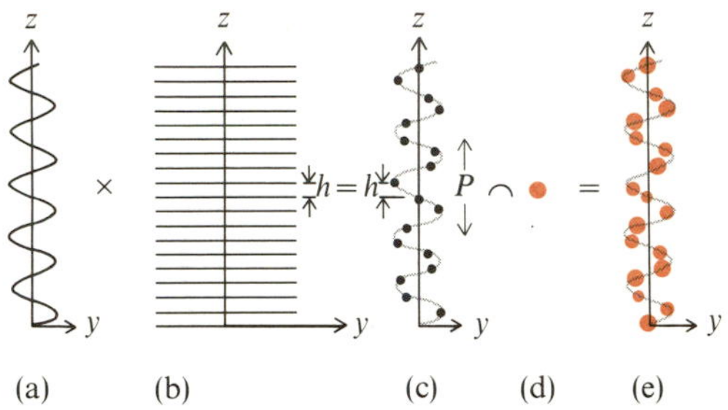

Figure 14-2
Steps in describing a helical polymer. **(a)** A helical line. **(b)** A set of parallel planes, perpendicular to the helix axis (z). **(c)** A helical point lattice (the intersection, or product, of parts a and b). **(d)** One residue (atom or molecule). **(e)** The helical polymer (the convolution of parts c and d).

convolution of the Fourier transforms of the two components (Eqn. 13-52). Therefore, the sampling function of the helical point lattice is simply given by

$$F_{\ell} = \widehat{F_c F_u} \tag{14-5}$$

where F_c is the Fourier transform of the helical line, and F_u is the Fourier transform of the set of planes. (Note from Eqn. 13-7 that F_c and F_u are also the structure factors expected from a helical line and from a set of planes, respectively, and that F_{ℓ} is the structure factor expected from a helical array of points.) The easiest approach to describing the sampling function of an oriented helical lattice is to derive a general expression for F_c and then convolute this with F_u.

Note that, in this simple model, we are considering each residue of the helix as a point. All residues are located a distance r from the helix axis, where r is the radius. To compute the x-ray scattering of a more realistic helix, we must place a residue consisting of a set of atoms at each point along the helix. The structure of such a molecular helix can be described as a convolution of the point lattice with the structure of one residue.§ Thus, if F_R is the structure factor of one residue, the total structure factor of the helix will be (by the convolution theorem in Eqn. 13-54) the product

$$F_{Tot} = F_R F_\ell \tag{14-6}$$

The structure factor of a helical line in cylindrical coordinates

It is easiest to calculate the scattering from a continuous helical line by using the cylindrical coordinate system shown in Figure 14-3. We set the helix axis parallel to

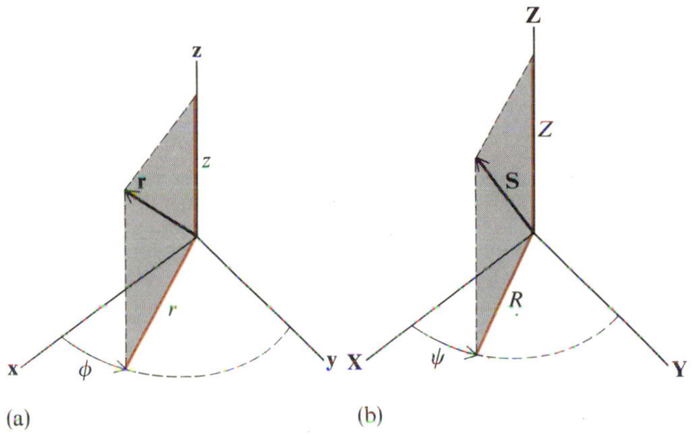

Figure 14-3
Cylindrical coordinate system used to describe diffraction from helices. **(a)** Real space. **(b)** Reciprocal space.

z in real space, and construct a Cartesian reciprocal-space coordinate system with axes X, Y, Z parallel to the real axes x, y, z. Then in cylindrical coordinates, there is a reciprocal-space variable R coresponding to the real-space variable r, which represents the radius. Similarly, the reciprocal-space angle ψ corresponds to the real-space angle ϕ.

The structure factor in Cartesian coordinates was given in Equation 13-7:

$$F(\mathbf{S}) = \int_{-\infty}^{\infty} d\mathbf{r}\rho(\mathbf{r})e^{2\pi i \mathbf{S}\cdot\mathbf{r}} \tag{14-7}$$

§ It may be more complex if the helix consists of alternating types or orientations of residues.

The $\mathbf{S} \cdot \mathbf{r}$ term is evaluated as the sum of components along x, y, and z axes:

$$\begin{aligned} \mathbf{S} \cdot \mathbf{r} &= (\hat{\mathbf{x}}S_x + \hat{\mathbf{y}}S_y + \hat{\mathbf{z}}S_z) \cdot \mathbf{r} = [\hat{\mathbf{x}}(\hat{\mathbf{X}} \cdot \mathbf{S}) + \hat{\mathbf{y}}(\hat{\mathbf{Y}} \cdot \mathbf{S}) + \hat{\mathbf{z}}(\hat{\mathbf{Z}} \cdot \mathbf{S})] \cdot \mathbf{r} \\ &= (\hat{\mathbf{x}} \cdot \mathbf{r})(\hat{\mathbf{X}} \cdot \mathbf{S}) + (\hat{\mathbf{y}} \cdot \mathbf{r})(\hat{\mathbf{Y}} \cdot \mathbf{S}) + (\hat{\mathbf{z}} \cdot \mathbf{r})(\hat{\mathbf{Z}} \cdot \mathbf{S}) \end{aligned} \qquad (14\text{-}8)$$

where $\hat{\mathbf{x}}, \hat{\mathbf{y}}, \hat{\mathbf{z}}$ and $\hat{\mathbf{X}}, \hat{\mathbf{Y}}, \hat{\mathbf{Z}}$ are unit vectors in real and reciprocal space, respectively. Then Equation 14-7 is transformed into cylindrical coordinates by recognizing that $\hat{\mathbf{z}} \cdot \mathbf{r} = z$; $\hat{\mathbf{Z}} \cdot \mathbf{S} = Z$; $\hat{\mathbf{x}} \cdot \mathbf{r} = r \cos \phi$; $\hat{\mathbf{X}} \cdot \mathbf{S} = R \cos \psi$; $\hat{\mathbf{y}} \cdot \mathbf{r} = r \sin \phi$; $\hat{\mathbf{Y}} \cdot \mathbf{S} = R \sin \psi$; and $d\mathbf{r} = r\, d\phi\, dr\, dz$. Using these results, we change Equation 14-7 to

$$F_c(R, \psi, Z) = \int_0^\infty r\, dr \int_0^{2\pi} d\phi \int_{-\infty}^{\infty} dz \rho(r, \phi, z) e^{2\pi i[rR(\cos\phi \cos\psi + \sin\phi \sin\psi) + zZ]} \qquad (14\text{-}9)$$

Equation 14-9 can be simplified in several steps. Note that the geometric terms in the exponential can be written as $\cos(\phi - \psi)$. For a helical line, $\rho(r, \phi, z)$ is zero except when $r = r_h$, the actual helix radius. Then it is constant independent of ϕ and z; that is, $\rho(r, \phi, z) = C\delta(r - r_h)$. Thus, the result of the integral over r is simply to produce a constant proportional to the electron density along the helical line. We shall omit this constant because it affects only the magnitude of F_c. With these simplifications, Equation 14-9 becomes

$$F_c(R, \psi, Z) = \int_0^{2\pi} d\phi \int_{-\infty}^{\infty} dz e^{2\pi i[r_h R \cos(\phi - \psi) + zZ]} \qquad (14\text{-}10)$$

Discontinuous structure factor of a helix

A helix is periodic in z. It repeats precisely after a distance P, the pitch. In other words, for any fixed values of r_h and ϕ, the electron density along z is just like that of a one-dimensional crystal. As shown in Chapter 13, any electron density distribution that repeats along z at an interval P can show x-ray scattering only in a set of planes perpendicular to z. Using the coordinate system in Figure 14-3, we can locate these planes in reciprocal space at $Z = n/P$, where n is an integer. Thus, the integral over z in Equation 14-10 will vanish except in these planes.

The helix rises by P for each 2π rotation about ϕ. Thus, the z coordinate is related to the pitch simply by $z = P\phi/2\pi$. Therefore, the term zZ in Equation 14-10 can be evaluated in terms of ϕ:

$$zZ = (P\phi/2\pi)(n/P) = n\phi/2\pi \qquad (14\text{-}11)$$

The only integral now remaining in Equation 14-10 is over ϕ:

$$F_c(R, \psi, n/P) = \int_0^{2\pi} e^{2\pi i r_h R \cos(\phi - \psi) + in\phi}\, d\phi \qquad (14\text{-}12)$$

This integral may look formidable at first, but it turns out that it can be related to a Bessel function. This is not too surprising; Bessel functions often appear when problems are solved in cylindrical coordinates. We can use the identity

$$J_n(x) = (1/2\pi i^n) \int_0^{2\pi} e^{ix \cos y + iny}\, dy \qquad (14\text{-}13)$$

where J_n is a Bessel function of the first kind. When we substitute $y = \phi - \psi$ and $x = 2\pi r_h R$ into Equation 14-12 (ignoring constant factors of 2π, which only change the magnitude of F_c, and using the fact that $i^n = e^{ni\pi/2}$), the resulting helical structure factor becomes

$$F_c(R, \psi, n/P) = J_n(2\pi r_h R) e^{in(\psi + \pi/2)} \qquad (14\text{-}14)$$

Thus, for a continuous helix of radius r_h and pitch P, the structure factor is a set of Bessel functions, each defined on a layer line $Z = n/P$ in reciprocal space. Note that the scattering amplitude, J_n, is radially symmetric. It does not depend on ϕ or ψ; only the phase does.

X-shaped pattern of scattering from a helical line

If a sample consisted of a single continuous helical line, the measured scattering intensity would be (from Eqn. 14-14)

$$I_c(R, n/P) = F_c^* F_c = J_n^2(2\pi r_h R) \qquad (14\text{-}15)$$

Figure 14-4 shows the appearance of this function for $n = 0$ to $n = 7$. It can be shown that $J_{-n}^2 = J_n^2$, and $J_n^2(-R) = J_n^2(R)$.

X-ray fiber data usually are collected with fiber oriented so that the Z axis is vertical and the R axis is horizontal. Then a set of equally spaced horizontal layer lines is seen. The spacing is $1/P$ for a helical line, so this allows P to be measured.

On each layer line, a series of intensity maxima is expected. The magnitude of each of the peaks falls off as R increases. Also, the position of the first and most intense maximum shifts to progressively larger values of R as n increases. Thus, the major feature of the diffraction pattern will resemble an X (Figure 14-5a). As n becomes very large, the arms of the X become straight lines that make an angle δ with the Z axis. The angle δ has the limiting value $\tan^{-1} P/2\pi r_h$ for $n \to \infty$. Thus, from the appearance of the diffraction pattern of a helical line, one immediately can learn the pitch and the radius.

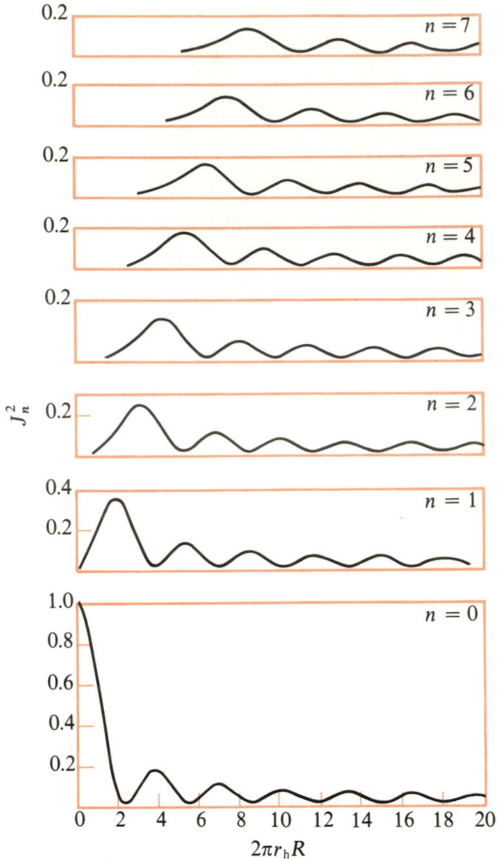

Figure 14-4

Bessel functions that can contribute to the x-ray scattering pattern of helices. Shown are $J_n^2(2\pi r_h R)$ for $n = 0$ to $n = 7$. Note the progressive shift and decrease in intensity of the first maximum as n increases. [From R. D. B. Fraser and T. P. MacRae, *Conformation in Fibrous Proteins and Related Synthetic Polypeptides* (New York: Academic Press, 1973).]

The structure factor of a discontinuous helix

The results thus far described are for a helix of continuous electron density. Equation 14-5 can be used to introduce the effect of residues spaced a distance h apart in the z direction.

We first need to calculate the structure factor, F_u, of a set of parallel planes perpendicular to z and spaced by h. The easy way to do this is to recall that the Fourier transform of a set of points spaced a distance **a** apart is a set of planes spaced at $1/a$ and perpendicular to **a**. From the symmetry of Fourier transforms and inverse transforms, it follows immediately that the structure factor arising from a set of planes spaced at h is a set of points along the Z axis and located at 0, $\pm 1/h$, $\pm 2/h$, and so on (Figure 14-5b,d). Thus, scattering amplitude can be observed from a set of points only at $Z = m/h$, where m is any integer:

$$F_u(Z) = \sum_{m=-\infty}^{\infty} \delta(Z - m/h) \tag{14-16}$$

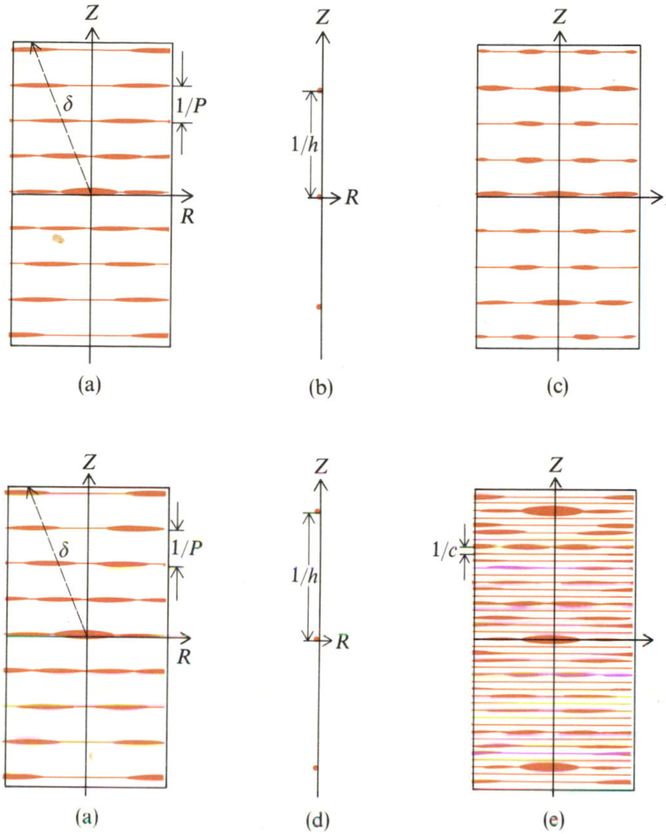

Figure 14-5

X-ray scattering from a helix oriented parallel to z in real space, but with orientation averaged over ϕ. **(a)** Intensity pattern that results from a continuous helix with pitch P. The angle δ is a function of P/r_h (see text). **(b)** The Fourier transform of a set of parallel planes oriented as shown in Figure 14-2b, and spaced at $h = P/3$. **(c)** The scattering intensity of a point helical lattice (such as that shown in Fig. 14-2c) with three residues per turn. It is computed as the convolution of the patterns shown in parts a and b. **(d)** The Fourier transform of a set of parallel planes spaced at $h = (5/18)P$, so that there are 18 lattice points (residues) in 5 turns. **(e)** The diffraction pattern of a point helix with 3.6 residues per turn. This is computed as the convolution of the patterns shown in parts a and d. The result is the x-ray scattering predicted for an α helix. In practice, disorder at short distances will smear out the diffraction pattern at large Z and large R. Only regions near the centers of the patterns in parts c and e are seen as well-defined intensity spots. [After R. D. B. Fraser and T. P. MacRae, *Conformation in Fibrous Proteins and Related Synthetic Polypeptides* (New York: Academic Press, 1973).]

The overall diffraction intensity of a discontinuous helix will be the convolution of two diffraction patterns: $F_c(R, \psi, Z = n/P)$ from Equation 14-14, and F_u from Equation 14-16. This means that one lays down an image of $F_c(R, \psi, n/P)$ at each value of $Z = m/h$. This is illustrated (actually for the corresponding rotationally averaged F_c) in Figure 14-5c,e.

Mathematically, we can express the convolution $\widehat{F_c F_u}$ as

$$F_\ell(R, \psi, Z) = \widehat{F_c F_u} = \sum_{n=-\infty}^{\infty} J_n(2\pi r_h R) e^{in(\psi + \pi/2)} \tag{14-17}$$

For each value of Z, the sum is carried out only over those values of n that satisfy the constraints on Z discussed above. A scattering layer line will be seen wherever

$$Z = m/h + n/P \tag{14-18}$$

This is because, for each value of m/h, the convolution $\widehat{F_c F_u}$ allows all integral values of n.

The structure factor of a helix with an integral number of residues per turn

Suppose a helix has an integral number k of residues per turn: $k = P/h$. Then the x-ray scattering will be a relatively simple pattern. Figure 14-5b,c shows a calculation of the structure factor for a helix with three residues per turn (actually, the rotationally averaged intensities are shown). In this case, the convolution $\widehat{F_c F_u}$ simply lays down an image of F_c, displaces the image three layer lines, lays down the image again, and so on. Thus, the pattern of scattering will repeat every fourth layer line. For an integral helix, the spacing between layer lines yields the pitch of the helix, and the repeat identifies the number of residues per turn.

To demonstrate these results mathematically, substitute k/P for $1/h$ in Equation 14-18:

$$Z = (1/P)(mk + n) = l/P \tag{14-19}$$

Because m, k, and n are integers, l must also be an integer.

To calculate the structure factor on a given layer line, we must add all terms in Equation 14-17 that satisfy the relationship $l = mk + n$. Suppose that k is three residues per turn. Then $F(R, \psi, Z = 0)$ will contain contributions from all J_n whose index n satisfies the equation $l = 3m + n = 0$, where m and n are integers. It is obvious that $n = -3m$, and therefore the structure factor on the zero layer line is

$$F(R, \psi, 0) = \sum_{m=-\infty}^{\infty} J_{3m}(2\pi r_h R) e^{[i3m(\psi + \pi/2)]} \tag{14-20}$$

It is easy to show that the same equation holds for any $F(R, \psi, l)$ where l is an integral multiple of three. If we write $l = 3$; then $n = 3(1 - m)$, which means that n can still take on the values 0, ± 3, ± 6, and so on. Therefore, the structure factor is a pattern that repeats every third layer line. The reader should attempt to work out the equivalent equation for $F(R, \psi, l, = 1,4,7, \ldots)$ and $F(R, \psi, l = 2,5,8, \ldots)$.

The structure factor of a helix with a nonintegral number of residues per turn

Many helices do not have an integral number of residues per turn. However, in most cases, it is possible to describe the structure as a helix with k residues in V turns. Then the overall structure repeats exactly after a distance $d = VP$. The spacing between adjacent residues is $h = VP/k$. Figure 14-5d,e is a schematic illustration of a calculation of the diffraction pattern of such a helix. In this example $k = 18$, and $V = 5$. The convolution $\widehat{F_c F_u}$ lays down an image of F_c spaced every $1/h = (18/5)(1/P)$. This leads to a much more complex diffraction pattern than that of an integral helix.

Note that the layer lines are much more finely spaced for a nonintegral helix than they are for an integral helix of equal pitch. Instead of layer lines every $1/P$, they will now appear every $1/VP$. The diffraction pattern repeats every kth layer line. To demonstrate these results mathematically, substitute the condition $h = VP/k$ into Equation 14-18. This leads to the prediction of layer lines spaced at

$$Z = l/VP \qquad \text{where } l \text{ is given by } mk + Vn \tag{14-21}$$

Note that this reduces to Equation 14-19 when $V = 1$.

Because m, k, V, and n are all integers, l must be an integer also. You should convince yourself by trial and error that for any nonintegral choice of k/V, values of m and n can be found that will enable l to take on any integral values 0, ± 1, ± 2, ... Thus the layer lines must be spaced every $Z = 1/VP$.

To compute the terms in Equation 14-17 that contribute to the structure factor at the lth layer line, we can rewrite the definition of l in Equation 14-21 as

$$n = (l - mk)/V \tag{14-22}$$

Only values of l and m that makes n an integer need be considered. For example, suppose k is 18, V is 5, and l is 3. Then Equation 14-22 is

$$n = (3 - 18m)/5 \tag{14-23}$$

The possible integral values of n are found by trying all possible integral values of m. (This is easiest by plotting Eqn. 14-23 graphically.) The result is

m	-9	-4	$+1$	$+6$	$\cdots$
n	$+33$	$+15$	-3	-21	$\cdots$

So J_{-3}, J_{+15}, J_{-21}, and J_{+33} all contribute to the third layer line.

The repeating nature of the structure factor can be seen if we write $l = kj$, where j is an integer. Then values of n that satisfy Equation 14-22 are $n = k(j - m)/V$, where $j - m$ must be chosen to make n integral. The allowable values of $j - m$ are the same, regardless of j. Thus the diffraction pattern must repeat every kth layer line.

X-ray scattering intensity from a rotationally averaged helix

Until now we have dealt with structure factors, whereas one can measure only the scattering intensities. For a single helix at an orientation angle ϕ in real space,

$$I(R, \psi, Z = l/VP) = F_\ell^* F_\ell = \sum_n J_n(2\pi r_h R) e^{in(\psi + \pi/2)} \sum_{n'} J_{n'}(2\pi r_h R) e^{-in'(\psi + \pi/2)} \quad (14\text{-}24)$$

where each sum is carried out over the values of n or n' allowed by Equation 14-22 for a particular layer line l. When the two sums are combined,

$$I(R, \psi, Z = l/VP) = \sum_n \sum_{n'} J_n(2\pi r_h R) J_{n'}(2\pi r_h R) e^{i(n-n')\pi/2} e^{i(n-n')\psi} \quad (14\text{-}25)$$

In practice, in a noncrystalline or semicrystalline fiber, all orientations of angle ϕ appear at random. Therefore, to compute the observed diffraction one must average Equation 14-25 over the corresponding angle ψ in reciprocal space. Integrating over ψ, we find

$$\langle I(R, Z = l/VP) \rangle = \sum_n \sum_{n'} J_n(2\pi r_h R) J_{n'}(2\pi r_h R) e^{i(n-n')\pi/2} \int_0^{2\pi} d\psi \, e^{i(n-n')\psi} \quad (14\text{-}26)$$

However, the integral over ψ is zero unless $n = n'$. Thus,

$$\langle I(R, Z = l/VP) \rangle = \sum_n J_n^2(2\pi r_h R) \quad (14\text{-}27)$$

This equation, together with the restrictions on n implied by Equation 14-22, is used directly to compute the experimentally measured diffraction pattern. The results predicted by this equation are what were actually plotted in Figure 14-5c,e.

A model for the α helix

We shall show how Equation 14-27 can be used to analyze the x-ray diffraction from an oriented fiber of α helices. The α helix has 18 residues in 5 turns. Layer lines will appear at $Z = l/5P$, where $l = 18m + 5n$. Intensities at these lines will be proportional to J_n^2 for all integers n satisfying this condition on l. In practice, only one value of n makes a significant contribution to the intensity at each l.

Figure 14-6 is a photograph of the diffraction pattern from an α-helical fiber.

(a)

(b)

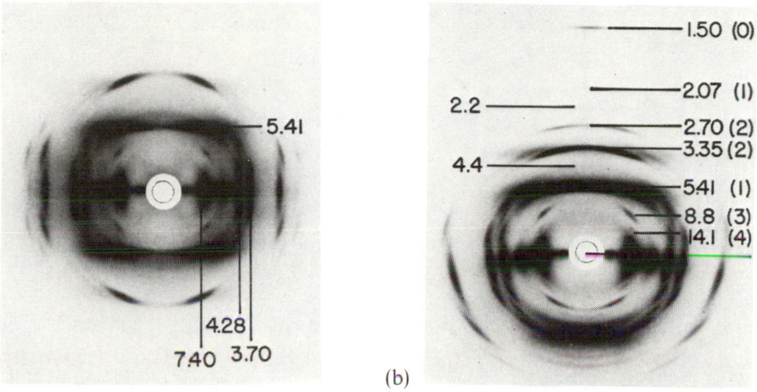

Figure 14-6

Measured x-ray scattering from fibers of α-helical poly-L-alanine. **(a)** Pattern seen with fiber perpendicular to direction of incident x rays. **(b)** Pattern seen with fiber tilted 31° away from the direction of incident x rays. Numbers give spacings in $Å^{-1}$; numbers in parentheses given the order of the Bessel functions contributing to the particular layer line. [Courtesy of L. Brown.]

Table 14-1 compares the predicted intensity on each layer line with that observed experimentally. Qualitatively the agreement is excellent. The diffraction is relatively strongest for those layer lines containing contributions from J_0^2, $J_{\pm 1}^2$, or $J_{\pm 2}^2$. This is just what is expected from the decrease in magnitude of J_n^2 as n increases (Fig. 14-4). For example, J_0^2 will appear only every 18th layer line, because when $n = 0$, one has $l = 18m$. $J_{\pm 1}^2$ occurs at $l = 18m \pm 5$. This contributes at $l = \pm 5$ ($m = 0$) and at $l = \pm 13$ ($m = \pm 1$).

Figure 14-7a shows the geometry of a typical x-ray fiber diffraction experiment. The photographic film has been unwrapped from its usual cylindrical shape for clarity. The fiber is parallel to the long axis of the cylinder. The vertical axis on the film (corresponding to the Z axis in reciprocal space) is called the meridian. The horizontal axis on the film (corresponding to the R axis in reciprocal space) is called the equator. Scattering intensity will be measured wherever the sphere of reflection intersects the reciprocal lattice (F_ℓ) of the fiber.

In practice, one usually cannot measure the scattering along the meridian where it crosses the equator (at $l = 0$) because this is directly in the x-ray beam, and any

Table 14-1

Observed intensities (I_{obs}) for layer lines of poly-γ-methyl-L-glutamate, and the Bessel functions (J_n^2) expected to contribute for an α helix

Layer line (l)	I_{obs}	n	Layer line (l)	I_{obs}	n	Layer line (l)	I_{obs}	n
0	vvs	0	10	w	2	20	—	4
1	—	7	11	—	5	21	—	3
2	vw	4	12	—	6	22	—	8
3	vvw	3	13	vvw	1	23	vvw	1
4	—	8	14	—	8	24	—	6
5	m	1	15	—	3	25	—	5
6	—	6	16	—	4	26	vvw	2
7	—	5	17	—	7	27	—	9
8	vvw	2	18	vw	0	28	vvw	2
9	—	9	19	—	7			

NOTE: Observed intensities are expressed as follows: vvs = extremely strong; m = moderate; w = weak; vw = very weak; vvw = extremely weak.

SOURCE: After A. G. Walton and J. Blackwell, *Biopolymers* (New York: Academic Press, 1973).

scattering is obscured by the transmitted beam intensity. From the form of the Bessel functions (Fig. 14-4), only J_0 has finite intensity on the Z axis (the meridian). Therefore, the first meridional reflection expected should occur at $l = 18$. The first really strong off-meridional reflections after the $l = 0$ layer will be at $l = \pm 5$.

Thus, just by examining these two aspects of the diffraction pattern, we can find the number of residues per turn of the helix. From this result and the actual value of Z for one of the layer lines, the pitch can be calculated as $P = l/5Z$. However, there can be serious pitfalls. Many layer lines have such weak intensity, and the diffraction spots from fibrous samples are so broad, that there can be considerable difficulty in indexing the layer lines seen.

X-ray scattering from a real α helix

Quantitatively, the agreement between the observed and calculated α-helix diffraction shown in Table 14-1 is not so excellent. This is made more evident by the typical diffraction pattern for the α-helical sample (Fig. 14-6a). Compare this with the theoretically expected pattern (Fig. 14-5e). As one moves vertically away from the center of the diffraction image, the observed pattern becomes weaker. Furthermore, the 13th layer line is much weaker than the 5th, although simple theory would predict them to be equal. The 18th layer-line meridional spot is not visible at all unless the fiber axis is tilted, and then it appears very weakly. Finally, additional diffraction spots appear along the layer lines at positions that do not correspond to maxima in J_n^2.

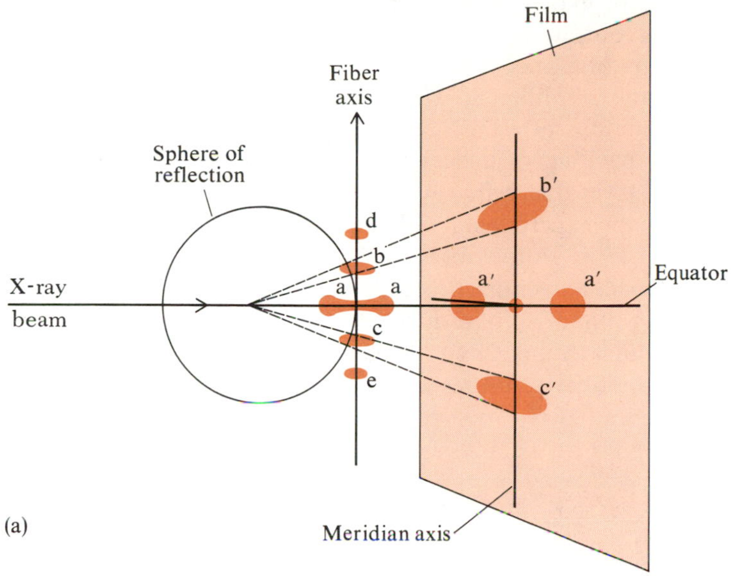

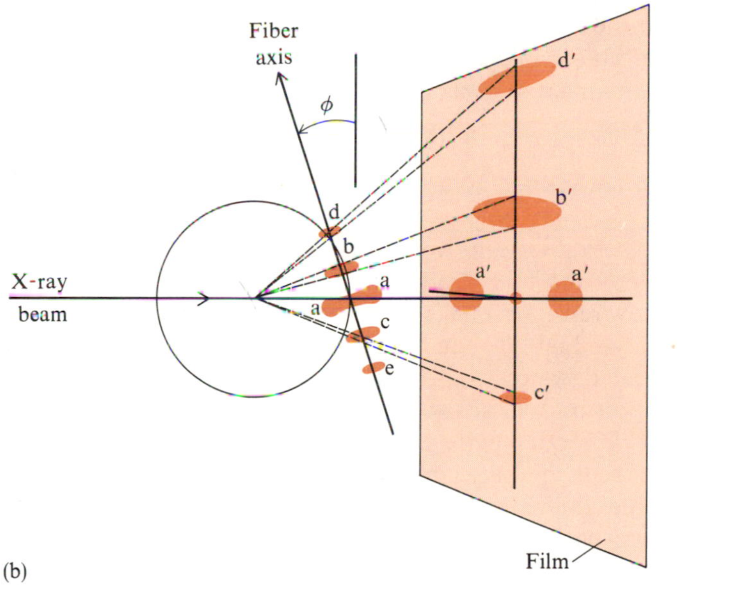

Figure 14-7

Schematic diagrams of x-ray fiber diffraction measurement. Shown along the fiber axis is the helix reciprocal lattice. Diffraction occurs only where this lattice intersects the sphere of reflection. **(a)** Fiber axis perpendicular to incident beam. **(b)** Fiber axis tilted to allow detection of the meridional spot at d′. [After R. E. Dickerson, in *The Proteins*, 2d ed., vol. 2, ed. H. Neurath (New York: Academic Press).]

All of this can be explained if we take into account how a real α-helical sample differs from the discontinuous-line helix we have used to model it. Each residue of a real helix is a group of atoms. Each corresponding atom lies on a helix of the same pitch P, but the radii are different. The scattering pattern will contain contributions from individual atomic scattering factors and from fringes resulting from interatom spacings. Finally, packing of helices in the fiber leads to additional interference fringes along each layer line, reflecting the spacing of the packed helices.

Let's examine the experimental diffraction pattern and see what conclusions can be drawn from it alone. In the absence of any prior knowledge about the α helix, we would index the first layer line ($l = 1$) as the intense diffraction at $Z = 1/5.41$ Å^{-1}. Assuming an integral number of residues per turn, this yields a helix pitch of 5.41 Å. With the fiber parallel to the Z axis, there are no diffraction peaks on the meridian. However, tilting the fiber produces a meridional reflection at $1/1.5$ Å^{-1}. The ratio of 5.41 and 1.5 gives 3.6 residues per turn. Because this ratio is not integral, the $1/5.41$ Å^{-1} line cannot be the first layer line. Recognizing that $3.6 = 18/5$ soon gives the correct indexing.

Why is it necessary to tilt to see the $1/1.5$ Å^{-1} reflection? All of the calculated diffraction we have shown is plotted in the reciprocal lattice. However, just as with crystals, one can observe only those reciprocal lattice points that intersect the sphere of reflection. As shown schematically in Figure 14-7b, the $1/5.41$ Å^{-1} reciprocal spot is close enough to the sphere of reflection, even with the fiber normal to the x-ray beam, to yield detectable intensity. However, the $1/1.5$ Å^{-1} spot is too far away, and the fiber must be tilted to bring it into coincidence with the sphere of reflection.

Effect of intermolecular packing on the α-helix diffraction pattern

At the beginning of this section, we argued that, in a semicrystalline fiber, intermolecular order may produce a sampling of the diffraction pattern at small $|\mathbf{S}|$, corresponding to low resolution. This effect appears in the α-helix data of Figure 14-6.

On the zero layer line (the equator), a set of discrete spots is seen at distances too long (reciprocal distances too short) to correspond to successive maxima of J_0^2. Hence, these spots must arise from helix packing. The zero layer line is the most likely region of the diffraction pattern to show the effects of lattice sampling, because here the data represent the projection of the structure into a plane perpendicular to the fiber axis. This eliminates any irregularities due to packing along the fiber axis.

The most usual packing for helices, or any other pseudocylindrical objects, is hexagonal. Figure 14-8 shows that the two principle lattice planes expected for closest-packed cylinders of radius a are $\sqrt{3}a$ and a. The periodicity in this packing will produce diffraction spots at $1/\sqrt{3}a$, $2/\sqrt{3}a$, and so on, and at $1/a$, $2/a$, and so on. In reciprocal space for the α-helical sample shown in Figure 14-6, zero layer line spots appear at $1/7.40$ Å^{-1}, $1/4.28$ Å^{-1}, and $1/3.70$ Å^{-1}. By trial and error, we find that the first two spots appear at positions with a ratio of $(1/\sqrt{3}):1$; the first and third clearly have a ratio of $(1/2):1$. Thus, if a is 4.28 Å, the three spots correspond to

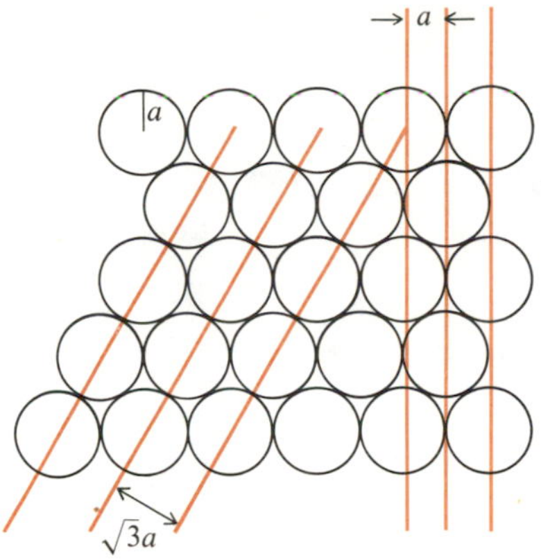

Figure 14-8

Hexagonal lattice resulting from the close packing of cylinders of radius a. This packing leads to periodicities of $\sqrt{3}\,a$ and a, as shown. These periodicities result in the extra diffraction spots on the equator of Figure 14-6a (see text).

$1/\sqrt{3}a$, $1/a$, and $2/\sqrt{3}a$, respectively. This confirms that they originate from hexagonal packing, and allows an assignment of the helix radius a.

X-ray scattering from nucleic acid fibers

The same principles that have been used earlier to analyze polypeptide helix x-ray scattering also apply to single-stranded nucleic acids. An example is the x-ray fiber photograph of a highly crystalline sample of poly C shown in Figure 14-9a. There is a strong meridional spot on the 7th layer line. This suggests a helix with six residues per turn. Then, the spacing between the layer lines indicates that in the helix there is an axial rise of 3.11 Å per residue. Further analysis provides some additional structural details and reveals that there are three poly-C helices per unit cell of the fiber. Figure 22-5 shows a structural model consistent with the observed x-ray data.

It is possible to extend this treatment to cover multiple-stranded helices and superhelices such as coiled coils. We do not do this here, but instead refer the interested reader to more advanced texts. The overall diffraction patterns still look qualitatively similar. Figure 14-9b shows, as one example, the diffraction pattern given by the B-form double helix of DNA. It is a classic X shape, actually much closer to that predicted for a continuous helix than was the α-helix diffraction. The ratio of the reciprocal-space distances from the origin to the first layer line and to the intense meridinal spot at 3.4 $Å^{-1}$ is 1:10. This reveals that the helix has 10 residues per turn.

Figure 14-9b also shows the x-ray scattering observed from a fiber of DNA in the A helical form. The pattern is much more complex than the B form. One feature

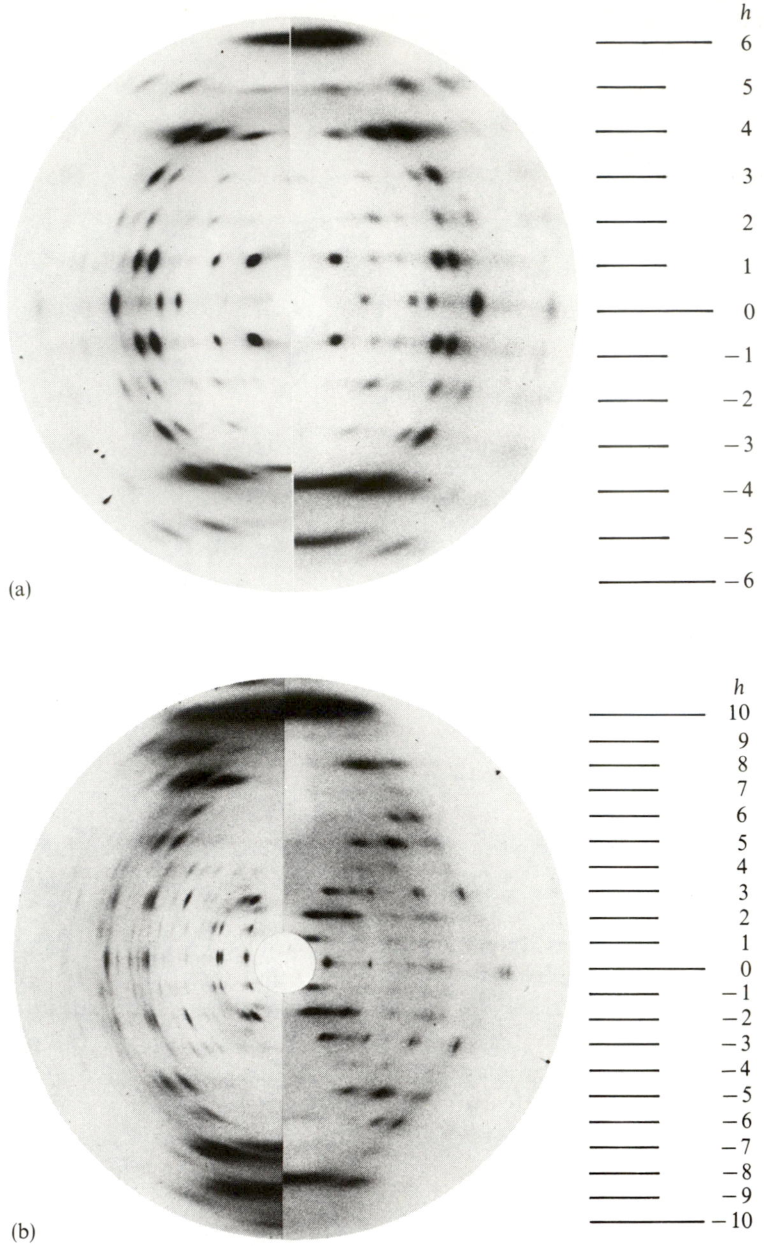

Figure 14-9

X-ray diffraction from nucleic acid fibers. Layer lines are indexed. **(a)** Sodium salt (*left*) and rubidium salt (*right*) of a single helical poly C. [After S. Arnott, R. Chandrasekaran, and A. G. W. Leslie, *J. Mol. Biol.* 106:735 (1976).] **(b)** *Clostridium perfringens* DNA. The A-form helix (*left*) and B-form helix (*right*) are shown. [After E. Selsing and S. Arnott, *Nucleic Acids Research* 3:2443 (1976).]

that is immediately apparent, however, is that the layer lines are spaced farther apart in DNA A than in DNA B. Because these spacings correspond to reciprocal distances, this is strong indication that the vertical rise per residue must be less in DNA A than in DNA B.

14-2 SOLUTION X-RAY SCATTERING

If one cannot obtain a suitable crystal or fiber, it is still possible to acquire useful information about macromolecular structure by measuring the x-ray scattering from a dilute solution. X rays are allowed to impinge on a sample described by the same geometry ($\hat{\mathbf{s}}_0$ and $\hat{\mathbf{s}}$) used in Chapter 13. Now, however, the molecules of interest are only a minority component. What one actually must measure is the additional scattering from macromolecule-plus-solvent over that of pure solvent alone. This leads to complications that will be treated later when contrast effects in neutron scattering are discussed.

Computing solution scattering by averaging over all molecular orientations

Let us assume it is possible to work at such high dilution that intermolecular interference effects are negligible. Then, for one molecule, the scattering (if we could observe it instantaneously) is just the molecular structure factor given by Equation 13-27:

$$F_{\mathrm{m}}(\mathbf{S}) = \sum_{n=1}^{N} e^{2\pi i \mathbf{S}\cdot\mathbf{r}_n} f_n(S) \tag{14-28}$$

where the sum is taken over all N atoms in the molecule.

We are not going to get much information about atoms, so it is more convenient to reexpress the structure factor as a sum over all the electrons in the molecule. Then $f_n(S)$ can be replaced by the scattering factor for a single *electron*, $f_0(S)$, and removed from the sum. Experimentally, what is measured is the intensity, or the square of the amplitude:

$$I(\mathbf{S}) = F_{\mathrm{m}}F_{\mathrm{m}}^* = f_0^2(S)\sum_{n=1}^{n_\varepsilon} e^{2\pi i\mathbf{S}\cdot\mathbf{r}_n}\sum_{m=1}^{n_\varepsilon} e^{-2\pi i\mathbf{S}\cdot\mathbf{r}_m} = f_0^2(S)\sum_{n=1}^{n_\varepsilon}\sum_{m=1}^{n_\varepsilon} e^{2\pi i\mathbf{S}\cdot\mathbf{r}_{nm}} \tag{14-29}$$

where n_ε is the number of electrons in the molecule, and we have defined $\mathbf{r}_{nm} \equiv \mathbf{r}_n - \mathbf{r}_m$ as the vector between electrons n and m.

During the time it takes to make an x-ray measurement in solution, each molecule will assume all possible rotational orientations. In addition, of course, one is observing a collection of macromolecules, and these will (at any one time) represent all possible orientations. Therefore, what is actually observable is

$$\langle I(S)\rangle = f_0^2(S) \sum_{n=1}^{n_\varepsilon} \sum_{m=1}^{n_\varepsilon} \int d\Omega e^{2\pi i \mathbf{S} \cdot \mathbf{r}_{nm}} \tag{14-30}$$

The integral is evaluated over all possible relative orientations of each vector $\mathbf{r}_{nm}$ and the scattering vector **S**.

We evaluated this integral when we computed the scattering from a single spherical atom in Chapter 13. Because all angles are integrated over, the integral has the same value for any choice of n and m. Using the result shown by Equations 13-20 and 13-21, we find what is called the Debye formula:

$$\langle I(S)\rangle = 4\pi f_0^2(S) \sum_{n=1}^{n_\varepsilon} \sum_{m=1}^{n_\varepsilon} [(\sin 2\pi S r_{nm})/2\pi S r_{nm}] \tag{14-31a}$$

Note that S is now a scalar because all intramolecular angular dependence has been integrated away. This means that the observed scattering intensity will be a function only of the scattering angle 2θ between incident and observed radiation. It will not depend on their orientation in space. This is perfectly reasonable because the sample itself has no preferred orientation.

Although we have expressed Equation 14-31a in terms of individual electrons, for some applications it more convenient to describe the scattering in terms of atoms. If $f_n(S)$ and $f_m(S)$ are the atomic scattering factors of atoms n and m, and if there are N atoms in all, then Equation 14-31a can be rewritten in the equivalent form

$$\langle I(S)\rangle = 4\pi \sum_{n=1}^{N} \sum_{m=1}^{N} f_n(S) f_m(S) (\sin 2\pi S r_{nm})/2\pi S r_{nm} \tag{14-31b}$$

Determining molecular weight and radius of gyration

The double sum in Equation 14-31 is formidable. However, it is possible to provide a simple interpretation of it in the limiting case of $S \to 0$. (In fact, the case one must actually consider is lim $S r_{nm} \to 0$.) Because S is $2 \sin \theta/\lambda$, this case corresponds to scattering in the laboratory frame in the limit of $\theta \to 0$. Technically, such experiments are quite difficult, because measurements at $\theta = 0$ are precluded by the presence of the transmitted beam. Therefore, special x-ray cameras must be used that work close to the incident beam without allowing any of it to enter.[§]

[§] All the equations in this section assume a point source of collimated x rays. In practice, to produce sufficient scattered intensity, a larger source must be used. This seriously complicates the analysis of the scattering data. For further details, see Pessen et al. (1973).

In the limit $\theta \to 0$, each term in the sum of Equation 14-31a can be expanded using $\sin x = x - x^3/3! + x^5/5! - \cdots$. Keeping only the first two terms, we find

$$\langle I(S) \rangle = 4\pi f_0^2(0) \sum_{n=1}^{n_\varepsilon} \sum_{m=1}^{n_\varepsilon} [1 - (2\pi S r_{nm})^2/6] \tag{14-32}$$

where we have replaced $f_0^2(S)$ by the value at $S = 0$. Note that r_{nm}^2 is just the distance r_{nm} squared. The first term in the sum is simply $\sum_n \sum_m 1 = n_\varepsilon^2$. The second term can be evaluated using some of the results presented in Chapter 18. There we indicate that the radius of gyration of a polymer consisting of $n + 1$ identical residues is $R_G^2 = [1/(n+1)^2] \sum_{i<j} r_{ij}^2$, where r_{ij} is the distance between residues i and j (see Eqn. 18-8). Viewing a macromolecule as a polymer of n_ε identical electrons, we can write the radius of gyration of electrons as

$$R_G^2 = (1/n_\varepsilon^2) \sum_{n<m} r_{nm}^2 = (1/2n_\varepsilon^2) \sum_n \sum_m r_{nm}^2 \tag{14-33}$$

This expression lets us evaluate the second term of the sum in Equation 14-32 as $(2/3)\pi^2 S^2 2n_\varepsilon^2 R_G^2$. For a flexible molecule, R_G will be the average radius of gyration.

The low-angle x-ray scattering expected for a macromolecule, then, is

$$\langle I(S) \rangle = 4\pi f_0^2(0) n_\varepsilon^2 (1 - 4\pi^2 S^2 R_G^2/3) \tag{14-34}$$

Scattering intensity per macromolecule usually is measured relative to that for a single electron at the origin. Averaged over all orientations, the zero-angle single-electron scattering is $4\pi f_0^2(0)$. When we substitute $2 \sin\theta/\lambda$ for S in Equation 14-34, the relative scattering intensity per macromolecule is

$$\langle I(\theta) \rangle = n_\varepsilon^2 (1 - 16\pi^2 R_G^2 \sin^2\theta/3\lambda^2) \tag{14-35}$$

This expression shows that the zero-angle scattering is proportional to the square of the number of electrons or, approximately, to the square of the molecular weight. Furthermore, a plot of intensity against $\sin^2\theta$ yields the radius of gyration.

It is popular among practitioners of small-angle scattering to rearrange Equation 14-35 somewhat. Note that a Gaussian for small argument can be expanded as $e^{-x^2/a} = 1 - x^2/a$. Then the small-angle scattering can be plotted as

$$\ln[I(\theta)/I(0)] = -(4\pi R_G \sin\theta/\lambda)^2/3 \tag{14-36}$$

This representation is called a Guinier plot. It yields the radius of gyration even in the absence of information about the molecular weight.[§]

[§] The Guinier plot actually is a somewhat more accurate approximation to Equation 14-31 than is Equation 14-35, because the expansion of a Gaussian $e^{-x^2/a} = 1 - x^2/a + x^4/2a^2$ includes most of the third term in the expansion $(\sin y)/y = 1 - y^2/6 + y^4/120$, as you can see by letting $a = 3$ and $y = \sqrt{2}x$, to make Equations 14-32 and 14-36 consistent.

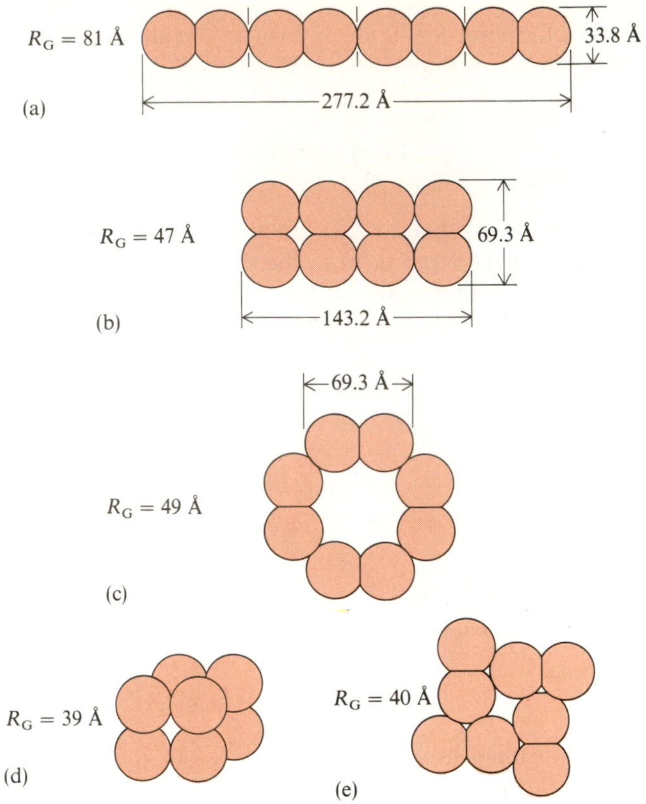

Figure 14-10

Several possible models for the α-lactoglobulin tetramer. The predicted radius of gyration is shown for each model. The experimental value of R_G obtained from low-angle x-ray scattering is 34 Å, indicating that structure d probably is the correct model. [After J. Witz, S. N. Timasheff, and V. Luzzati, *J. Amer. Chem. Soc.* 86:168 (1964).]

Figure 14-10 shows one simple application of radius-of-gyration data. Prior experiments had suggested that, at pH 4.5, α-lactalbumin is a tetramer of four dimeric subunits. The measured radius of gyration from small-angle x-ray scattering is 34.4 ± 0.4 Å. This measurement allows exclusion of many plausible models for this tetramer.

Using wider-angle scattering to evaluate models of molecular structures

By looking only at low angle, we lose much of the information potentially available from solution x-ray scattering. An alternative approach is to measure $\langle I(S) \rangle$ over a broad range of S and then to use Equation 14-31 (or equivalent forms) directly. Because of the orientational averaging, there clearly is not enough information in $\langle I(S) \rangle$ to determine a structure. Instead, one can assume various models, such as solid uniform ellipsoids of revolution, and compute the expected scattering. This computed result is then compared with the measured results.

Consider the simplest possible case. Suppose the sample can be modeled as a uniform sphere of radius R and constant electron density ρ_0. Then we can calculate the structure factor directly from Equation 13-21:

$$F(S) = 4\pi\rho_0 \int_0^R [(\sin 2\pi Sr)/2\pi Sr]r^2\,dr$$

$$= 4\pi\rho_0 R^3\left(\frac{\sin(2\pi SR) - 2\pi SR\cos(2\pi SR)}{(2\pi SR)^3}\right) \qquad (14\text{-}37)$$

This is simply a set of peaks that alternate in sign, and that progressively decrease in magnitude as S (or 2θ, the scattering angle) increases (see Guinier, 1963). The observed intensities, $|F(S)|^2$, show periodic maxima of decreasing intensity. From the location of the maxima, R can be determined.

Figure 14-11 shows a typical example of solution x-ray scattering. The scattering from R17 bacteriophage is fit fairly well, at low scattering angles, to a uniform sphere with $R = 133$ Å. However, the fit is not perfect, and more accurate models must be constructed.

In the general case, the electron density is not constant. To calculate $\rho(r)$ from $|F(S)|^2$, one usually proceeds by assuming, in the first instance, that the sign of $F(S)$ alternates at each successive peak. Then the data can be used directly for a Fourier

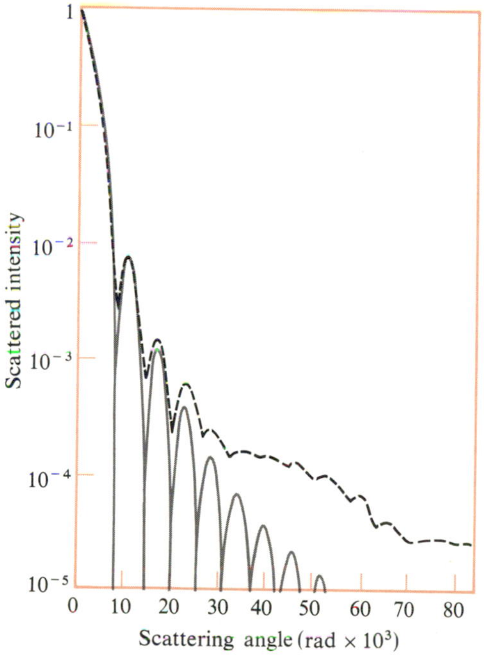

Figure 14-11

Solution x-ray scattering of R17 bacteriophage (*solid curve*), compared with the scattering calculated for a uniform sphere of radius 133 Å (*dashed curve*). [From I. Pilz, in *Physical Principles and Techniques of Protein Chemistry*, ed. S. Leach (New York: Academic Press, 1973).]

synthesis. We need to find the inverse of Equation 13-21. Note that this equation can be written as the Fourier sine transform of $r\rho(r)$ (see Box 14-1):

$$SF(S) = 2\int_0^\infty dr\,\sin(2\pi Sr)r\rho(r) \tag{14-38}$$

Box 14-1 FOURIER SINE TRANSFORMS

The Fourier transform of $f(r)$ is

$$g(S) = \int_{-\infty}^{\infty} f(r)e^{2\pi iSr}\,dr$$

Suppose $f(r)$ is an odd function: $f(r) = -f(-r)$. Then the Fourier transform can be rewritten as

$$g(S) = \int_0^\infty f(r)e^{2\pi iSr}\,dr - \int_0^\infty f(r)e^{-2\pi iSr}\,dr$$

This becomes

$$g(S) = 2i\int_0^\infty f(r)(\sin 2\pi Sr)\,dr$$

Note that $g(S)$ is an odd function of S.

Now start with the inverse Fourier transform

$$f(r) = (1/2\pi)\int_{-\infty}^{\infty} g(S)e^{-2\pi iSr}\,dS$$

Breaking up the integral as before, we obtain

$$f(r) = (-i/\pi)\int_0^\infty g(S)(\sin 2\pi\,Sr)\,dS$$

A more convenient form for these equations can be obtained if we define $g'(S) = -ig(S)$. Then

$$g'(S) = 2\int_0^\infty f(r)(\sin 2\pi Sr)\,dr$$

$$f(r) = (1/\pi)\int_0^\infty g'(S)(\sin 2\pi Sr)\,dS$$

These results are called the Fourier sine transform and the inverse Fourier sine transform, respectively.

The inverse Fourier sine transform of $SF(S)$ yields

$$r\rho(r) = (1/\pi) \int_0^\infty dS \sin(2\pi Sr)SF(S) \tag{14-39}$$

Assuming that each peak alternates in sign means that $F(S)$ in Equation 14-39 is replaced by $|F(S)|$ sign (S), where sign (S) alternates for every peak. In many cases, such as spherical shells or solid ellipsoids, the scattering still can qualitatively resemble Equation 14-37. However, the signs of successive peaks in $F(S)$ may no longer alternate.

Note that the probability of finding electron density a distance r from the origin is proportional to

$$D(r) = r^2\rho(r) = (r/\pi) \int_0^\infty dS \sin(2\pi Sr)SF(S) \tag{14-40}$$

$D(r)$ is called a radial distribution function.

Computing a radial Patterson function from solution scattering data

The difficulty with computing the electron density distribution $\rho(r)$ from Equation 14-39 is the unknown sign of $F(S)$. This is another example of the phase problem in x-ray scattering. If diffraction intensities measured on a crystal are subjected directly to an inverse Fourier transform, the result is the Patterson function (Chapter 13). It follows that, if intensity data of a spherically averaged sample are subjected to an inverse Fourier transform, the result will be a spherically averaged Patterson function $\langle P(r) \rangle$.

By analogy to Equation 14-39, we can write

$$r\langle P(r) \rangle = (1/\pi) \int_0^\infty dS \sin(2\pi Sr)SI(S) \tag{14-41}$$

Because the Patterson function is a map of all interatom vectors, the spherically averaged Patterson function will resemble this map rotated through all angles θ and ϕ in spherical coordinates. Figure 14-12 shows an example for a simple molecule.

It is hard to develop much of an intuitive feeling for $\langle P(r) \rangle$. A much more useful function is the radial Patterson function, $U(r)$. By analogy to Equation 14-40, this is

$$U(r) = r^2\langle P(r) \rangle = (r/\pi) \int_0^\infty dS \sin(2\pi Sr)SI(S) \tag{14-42}$$

Physically, the radial Patterson function $U(r)$ is the relative probability of finding electron density at a distance r. Thus it is similar in spirit to the Patterson

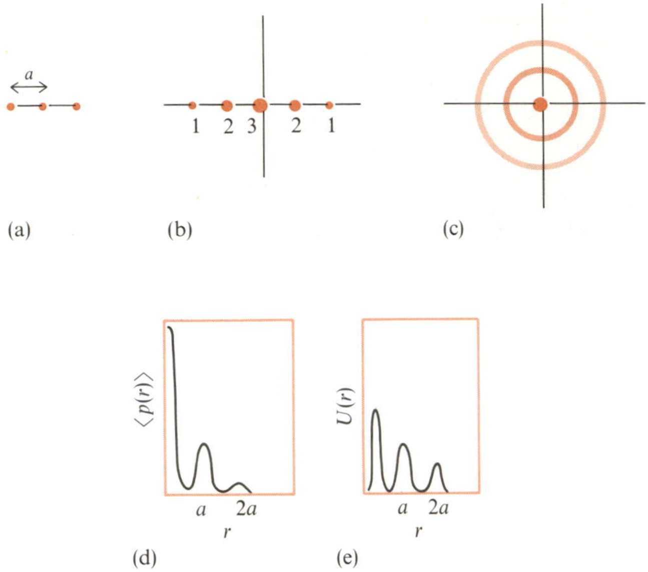

Figure 14-12

Patterson functions that can be calculated from solution scattering data. **(a)** A linear triatomic molecule. **(b)** Patterson function of structure shown in part a. **(c)** A spherical average of the Patterson function shown in part b. Notice how the density dimishes sharply with increasing radius. **(d)** A radial profile of the spherically averaged Patterson function for the molecule shown in part a. **(e)** A radial Patterson function for the molecule shown in part a. This function is computed by multiplying the data shown in part d by r^2. Note that the relative heights of the peaks at a and $2a$ are now proportional to the relative probabilities of finding interatomic distances a and $2a$ within the molecule.

function, whose peaks give the pairs of electron density spaced a vector **r** apart. In some cases, the radial Patterson is easy to interpret (as in Fig. 14-12e).

Extended x-ray absorption fine structure (EXAFS)

The availability of easily tunable high-flux x-ray sources from synchrotron radiation facilitates a variety of new types of x-ray spectroscopy. One technique that offers great promise for the analysis of biological samples is extended x-ray absorption fine structure (EXAFS). Here the absorption of x rays by an amorphous solid or solution sample is measured as a function of wavelength at energies just above the absorption transition of a particular atom. Elements that have relatively rare occurrence and atomic numbers greater than around 20 are particularly useful.

At energies just above the sharp absorption threshold, a series of rapid oscillations in absorbance is observed. It turns out that this pattern represents an interference effect from the neighboring atoms of the absorber. A Fourier transform of the oscillations can be analyzed in favorable cases to yield information on the number, types, and distances of the neighboring atoms. Usually this requires parallel studies on model compounds of known structure. Extremely accurate structural information can result, and the unique feature is that no ordered sample is required. However, the information is limited to fairly short distances.

EXAFS has been used successfully to analyze the structure around metal atoms in a series of metalloproteins including irons in rubredoxin and hemoglobin, and molybdenum in nitrogenase. For an introduction to this promising new technique, see Shulman et al. (1978).

14-3 SCATTERING OF OTHER TYPES OF RADIATION

The reader may be surprised that we have had to say very little about the physics of x rays or their interaction with matter in order to describe all the various kinds of scattering and diffraction discussed thus far. The reason is that almost everything that has been said is quite general and applies to any electromagnetic radiation. Fundamentally, all the same equations apply if one uses light, or microwaves, or gamma rays. However, because the ratio of molecular size to wavelength will be quite different in these cases, such radiation may not always be useful.

We have been able to treat the atomic interactions as spherically symmetrical for x rays. This will not be strictly true for light. Furthermore, we have largely ignored the necessity that the molecule have contrast with its surroundings. For light scattering, this consideration becomes important. The most essential results, however—the expression of scattering as a Fourier transform, and the interference effects of pairs of atoms—are retained precisely, regardless of the radiation used. In fact, it does not have to be electromagnetic radiation. Particles such as electrons and neutrons have the properties of transverse waves with a wavelength dependent on their energies. Thus, with only a few changes, electron and neutron scattering and diffraction also can be described by the equations we have derived. For electrons and neutrons, the atomic scattering factors must be replaced by other descriptions of the interaction of radiation with matter.

The useful wavelength range of various types of radiation

Why can one not solve a crystal structure using light scattering? In principle, there would be an enormous experimental advantage, because coherent light sources (lasers) now are available, and therefore avenues exist to measure phases as well as intensities. Consider a typical macromolecule crystal with a unit cell consisting of a 40 Å cube. The reciprocal lattice is a cubic array of points with a spacing of 1/40 $Å^{-1}$ in each direction. Suppose a scattering experiment is performed with 2,000 Å light. The sphere of reflection has a radius of $1/\lambda = 1/2{,}000\ Å^{-1}$. Thus, it contains only a single reciprocal lattice point, the origin. No diffraction pattern can be observed. If the crystal had a large enough unit cell (small enough reciprocal lattice), it would be possible to do a structural analysis, but the resolution would be poor, because the reciprocal lattice could be sampled only up to distances equal to the wavelength of the light.

On the other hand, suppose one needed to measure the radius of gyration of a virus. Here R_G might easily be 1,000 Å or more. With small-angle x-ray scattering, to derive the Guinier equation (Eqn. 14-36), we made the implicit assumption that $SR_G < 1$. To meet this condition means that $\sin\theta < \lambda/R_G$. With 1 Å x rays, this would require $\sin\theta \cong \theta \cong (1/1{,}000)$ radian $= 0.06°$, which is quite a small angle. With 6,000 Å light, there is no problem. So, depending on what one is seeking, different methods have different relative advantages. In the following sections, we explore briefly some unique capabilities of a few techniques that are fundamentally similar to x-ray scattering.

14-4 ELECTRON MICROSCOPY

Measuring electron diffraction of a solid with the electron microscope

Here we consider the determination of the structure of thin samples in the electron microscope by electron diffraction. Figure 14-13 is a schematic of the experiment.

The wavelength of the electron (or any other particle) is given by the de Broglie relationship:

$$\lambda = h/m_e v \tag{14-43}$$

where m_e is the mass, v is the velocity, and h is Planck's constant. If electrons originally at rest are accelerated by a voltage difference Φ, they will acquire a kinetic energy $(1/2)m_e v^2 = e\Phi$. Combining this expression with Equation 14-43, and substituting values for the electron mass, electron charge, and Planck's constant, we find

$$\lambda = 12.3/\sqrt{\Phi}\ \text{Å} \tag{14-44}$$

where Φ is expressed in volts. Thus, the wavelength of electrons accelerated by a 100,000-volt potential will be about 0.04 Å, more than short enough for the determination of the structure of molecules. Indeed, in principle, it allows more accurate structures to be determined than does the use of x rays, because of the shorter wavelength. In the gas phase, electron diffraction can yield extremely precise structures of small molecules.

Electrons are easily stopped or absorbed by thin layers of all materials. Hence, it is necessary to work in a vacuum. Samples usually are placed on a thin gold support film. The damage caused by collisions of electrons with the sample can be considerable. This means that only low irradiation intensities can be employed. The resulting contrast between a sample and the support film is poor.

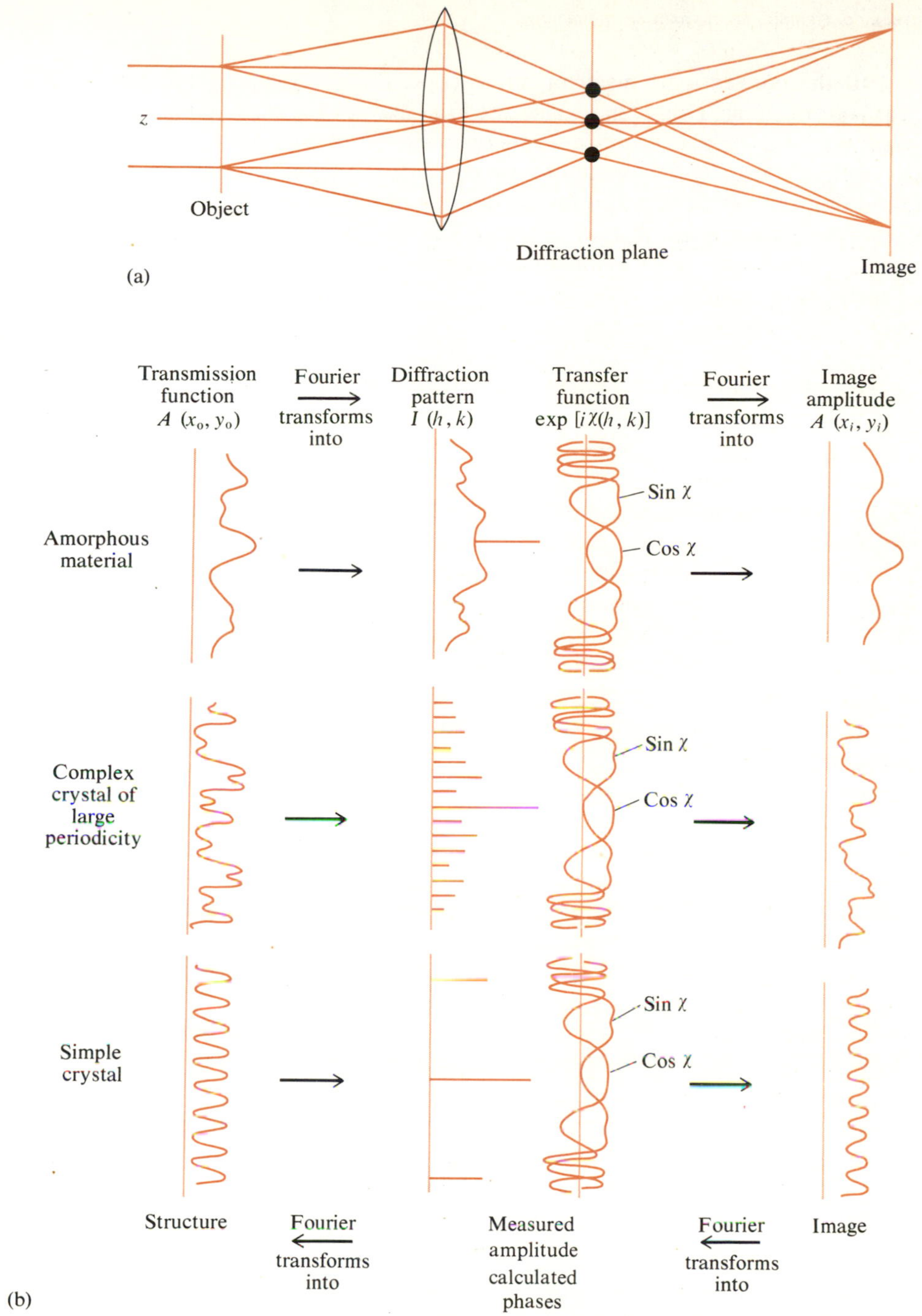

Figure 14-13

Electron diffraction and image formation in the electron microscope. **(a)** Geometrical optics. **(b)** Wave optics. [After J. M. Crowley and S. Iijima, *Physics Today* (March 1977), p. 32.]

In the conventional electron microscope, the contrast is improved by using heavy-atom stains. However, as discussed in Chapter 10, these stains lead to loss of resolution and can also produce severe distortion of the sample. For the applications that follow, it usually is necessary to use unstained samples. In addition, care must be taken that the drying required for vacuum operation does not distort the structure. One way to accomplish this for a thin crystal slice is to replace the aqueous solvent contained in the crystal by a nonvolatile liquid. The following discussion applies only to such a thin sample.

Determining molecular structure in the electron microscope

The thin sample is a two-dimensional array oriented in a plane perpendicular to the direction of incidence of the electrons. All of the diffraction pattern will appear in a plane parallel to the sample plane[§] (Fig. 14-13). However, the pattern is distorted by imperfections in the optics used to focus the electron beam. This distortion is represented by the transfer function shown in Figure 14-13. The image obtained by focusing the diffraction pattern will represent the structure. However, the image is distorted and, because the sample is unstained, the contrast is poor. Even with the Fourier averaging techniques described in Chapter 10, it is difficult to obtain high-resolution structural information.

Compared with the poor quality of the image, the data at the diffraction plane are excellent because they represent an average over many unit cells. Sharp spots are seen (Fig. 14-14a). The intensities of the diffraction pattern can be determined accurately. However, to reconstruct the structure accurately from this pattern, one would also need the phases and correction factors for the transfer function. D. DeRosier and A. Klug showed that it is possible to determine those phases by Fourier-transforming the image to compute a diffraction pattern. Then the structure is calculated by a Fourier synthesis using these phase estimates and the measured diffraction intensities, exactly as was done in x-ray diffraction.

N. Unwin and R. Henderson were the first to perform a high-resolution electron microscopic structure determination on unstained biological samples. Figure 14-14 illustrates three stages in the process. The sample is the purple membrane of a halophilic bacterium. This membrane consists of lipid and (essentially) one pure protein. The experimental diffraction pattern with the plane of the membrane perpendicular to the electron beam (Fig. 14-14a) is shown along with a portion of the diffraction computed from the electron microscopic image (Fig. 14-14b,c). Also shown is the final contour map of the projection of the structure in the plane of the membrane

[§] Although the actual diffraction pattern is still three-dimensional, all distances of interest are much longer than the wavelength of the electrons. Thus, for all practical purposes, the curvature of the sphere of reflection can be ignored.

(Fig. 14-14d). The current resolution is 7 Å, although there seems to be no serious obstacle to improving this. Numerous high density peaks 10 Å apart are seen. These are likely to be protein α helices viewed end-on. Thus the helices must be oriented roughly perpendicular to the membrane surface.

To obtain three-dimensional structure information, the whole process just outlined is repeated with tilted samples. Then the structure is reconstructed from a set of two-dimensional projections (Henderson and Unwin, 1975).

A mathematical treatment of electron diffraction

Let the sample be in the x–y plane with electrons incident along the z axis. Refer to Figure 14-15 for a summary of all the steps outlined below. The essential points are that the diffraction pattern is a Fourier transform of the sample and that the image is a Fourier transform of the diffraction pattern.

Just as the electrons reach the sample, we can describe them by a wave with unit amplitude and zero phase. Electrons passing through the sample will be subjected to the local potential $\phi(x, y, z)$ generated by electrons and nuclei. This will cause a phase shift in the original wave. Because the sample is very thin, we need consider only the projection of the potential into the x–y plane: $\phi(x, y) = \int_{-\infty}^{\infty} dz\phi(x, y, z)$.

The phase shift caused by this potential is proportional to its strength relative to the original accelerating voltage Φ_0. It can be shown to be $\delta = \phi(x, y)/2\lambda\Phi_0$. Thus, the wave immediately after it leaves the object is described by Equation 14-45 (a transmission function):

$$A(x_o, y_o) = e^{2\pi i\delta} = e^{i\pi\phi(x_o,y_o)/\lambda\Phi_0} = 1 + i\pi\phi(x_o, y_o)/\lambda\Phi_0 \qquad (14\text{-}45)$$

Here we have used the subscript o to indicate that the coordinates refer to the object, and we have expanded the exponential because the effect of the potential field of the sample is small.

The wavefunction $F(h, k)$ of the electrons at the diffraction plane (equivalent to the structure factor) results from integrating the effect of each scattered wave:

$$F(h, k) = e^{i\chi} \int_{-\infty}^{\infty} dx_0 \int_{-\infty}^{\infty} dy_0 e^{2\pi i(x_0 h + y_0 k)} A(x_o, y_o) \qquad (14\text{-}46)$$

where h and k represent reciprocal-lattice positions just as they did for x rays. The additional term $e^{i\chi}$ (a transfer function) must be introduced to account for defocusing of the beam by the sample and for spherical aberration (see Fig. 14-13).

The real part of $e^{i\chi}$ changes the amplitude of the radiation at the diffraction plane. It is called an amplitude contrast term, $\cos\chi$. The imaginary part, $i \sin\chi$, changes only the phase of the radiation and is called a phase contrast term. The

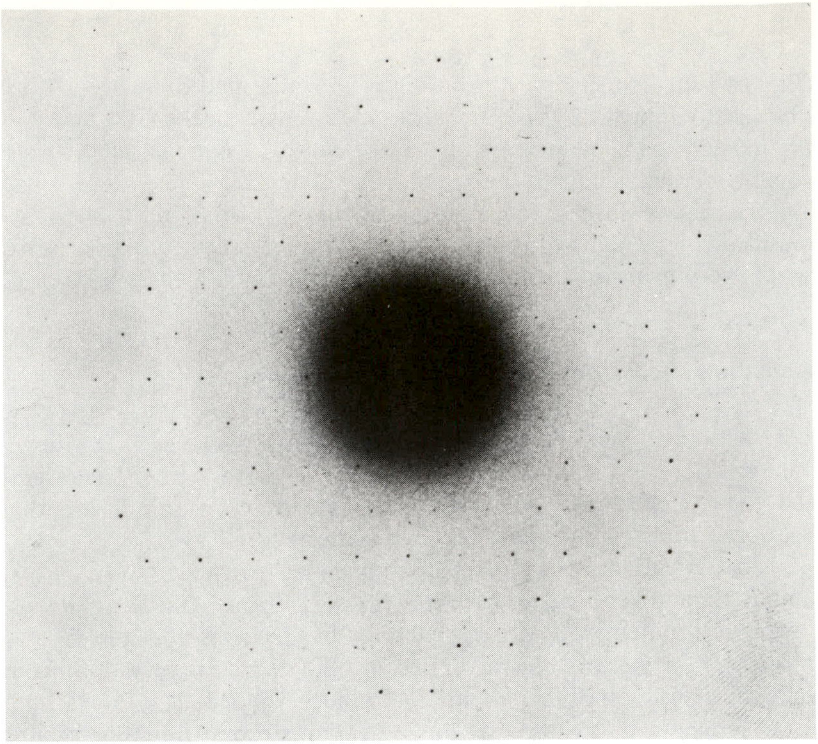

(a)

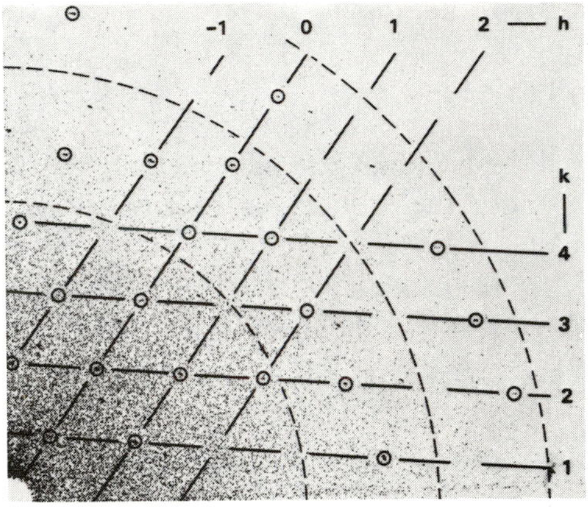

(b)

Figure 14-14

Stages in analysis of structure of purple membrane by electron diffraction in the electron microscope. **(a)** Experimentally observed diffraction pattern. **(b)** Part of the diffraction pattern calculated by Fourier-transforming the electron microscopic image. A low-dose micrograph; indexing of some of the spots in the reciprocal lattice is indicated. **(c)** A high-dose micrograph, dominated by the transfer function. **(d)** Electron density map at 7 Å resolution, calculated by a Fourier synthesis using the measured intensities in part a and the calculated phases from parts b and c. The unit-cell dimensions are 62 × 62 Å. [From P. N. T. Unwin and R. Henderson, *J. Mol. Biol.* 94:431 (1975).]

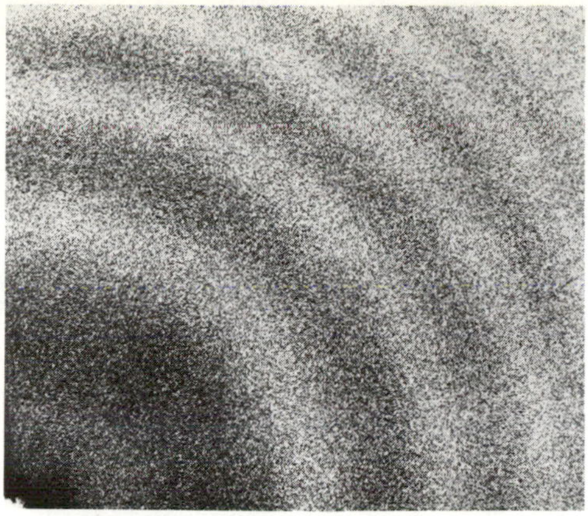
(c)

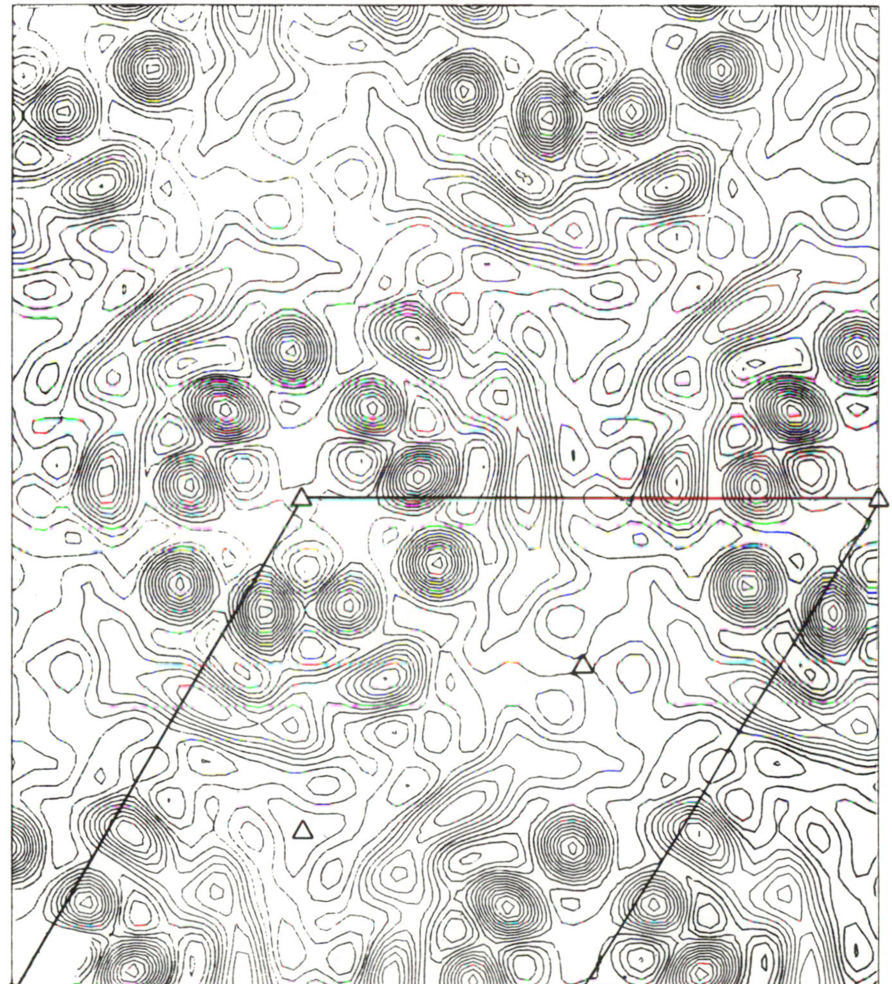
(d)

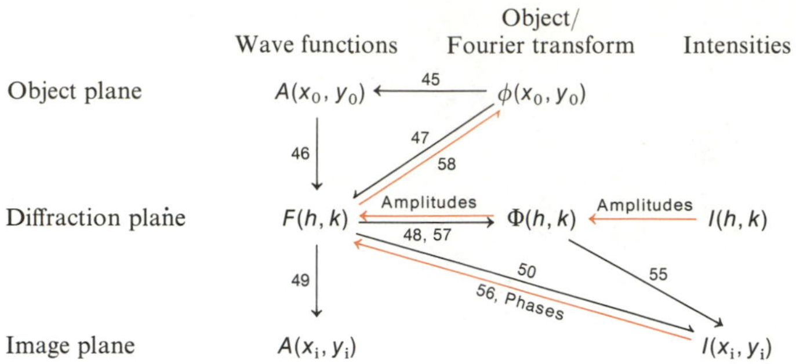

Figure 14-15

Structure determination in the electron microscope. The relationship among various observed or calculated quantities is shown. Numbers next to each arrow refer to equations in the text that allow computation in the directions shown. Colored arrows indicate paths used in actual structure solutions.

parameter χ is a function of the square of the scattering angle, and it is zero when h and k are zero. Otherwise, Equation 14-46 looks just like Equation 13-7 after the von Laue conditions are inserted. These conditions apply because the sample is periodic in x_o and y_o, and this periodicity leads to destructive interference of the scattering everywhere except at the reciprocal-lattice positions.

If we insert the approximate form given for $A(x_o, y_o)$ by Equation 14-45 into the expression for $F(h, k)$, the result is

$$F(h, k) = \delta(0, 0) + (i\pi/\lambda\Phi_0)(\cos\chi + i\sin\chi)\int_{-\infty}^{\infty} dx_o \int_{-\infty}^{\infty} dy_o \phi(x_o, y_o) e^{2\pi i(x_o h + y_o k)} \tag{14-47}$$

where the function $\delta(0, 0)$ is 1 if both h and k are zero, and is 0 otherwise. Note that $\delta(0, 0)$ arises from the Fourier transform of the constant term in Equation 14-45, and represents the unscattered electron beam passing through the origin. The integral remaining in Equation 14-47 is the Fourier transform of $\phi(x_o, y_o)$. We can define this to be $\Phi(h, k)$. Then

$$F(h, k) = \delta(0, 0) - (\pi/\lambda\Phi_0)\Phi(h, k)(\sin\chi - i\cos\chi) \tag{14-48}$$

What is measured experimentally at the diffraction plane is the intensity $I(h, k) = F^*(h, k)F(h, k)$.

Now, what happens when the diffraction pattern is focused on the image plane in an electron microscope? We must take the wavefunctions at the diffraction plane (Eqn. 14-48) and perform a Fourier synthesis to see how they combine to give the wavefunction at the image plane, $A(x_i, y_i)$:

$$A(x_{\text{i}}, y_{\text{i}}) = \sum_{h=-\infty}^{\infty} \sum_{k=-\infty}^{\infty} e^{-2\pi i(hx_i + ky_i)} F(h, k) \tag{14-49}$$

The subscript i means that the function is in the image plane. Once again, all that is measurable is the intensity, AA^*. This can be calculated as

$$I(x_{\text{i}}, y_{\text{i}}) = \sum_{h=-\infty}^{\infty} \sum_{k=-\infty}^{\infty} \sum_{h'=-\infty}^{\infty} \sum_{k'=-\infty}^{\infty} e^{-2\pi i(hx_i + ky_i - h'x_i - k'y_i)} F(h, k)F^*(h', k') \tag{14-50}$$

We will now demonstrate that Equation 14-50 can be simplified considerably. Substitute Equation 14-48 for $F(h, k)$ and $F^*(h', k')$, and consider one term of the resulting sum. Note that, because $\pi/\lambda\Phi_0$ is much less than $\delta(0, 0)$, we can neglect terms in $(\pi/\lambda\Phi_0)^2$.

$$\begin{aligned} F(h, k)F^*(h', k') \cong\ & \delta(0, 0)\,\delta(0', 0') - \delta(0', 0')(\pi/\lambda\Phi_0)\Phi(h, k)(\sin\chi - i\cos\chi) \\ & - \delta(0, 0)(\pi/\lambda\Phi_0)\Phi^*(h', k')(\sin\chi' + i\cos\chi') \end{aligned} \tag{14-51}$$

Inserting these terms into Equation 14-50, and evaluating the sums over the functions $\delta(0, 0)$ and $\delta(0', 0')$, we obtain

$$\begin{aligned} I(x_{\text{i}}, y_{\text{i}}) = 1 - (\pi/\lambda\Phi_0)\Bigg\{ & \sum_{h=-\infty}^{\infty} \sum_{k=-\infty}^{\infty} \Phi(h, k)(\sin\chi - i\cos\chi)e^{-2\pi i(hx_i + ky_i)} \\ & + \sum_{h'=-\infty}^{\infty} \sum_{k'=-\infty}^{\infty} \Phi^*(h', k')(\sin\chi' + i\cos\chi')e^{+2\pi i(h'x_i + k'y_i)} \Bigg\} \end{aligned} \tag{14-52}$$

Equation 14-52 can be considerably simplified. First, note that the indices h, k, h', and k' run over all integers, so h' and k' can be replaced by h and k, and χ' can be replaced by χ. Next, note that Φ is a Fourier transform of a real object: the sample. Therefore, it must obey Friedel's law (see Eqns. 13-12 through 13-17). Thus, describing Φ as an amplitude and a phase, we have $\Phi(h, k) = |\Phi(h, k)|e^{i\alpha_{h,k}}$, so we can write

$$|\Phi(h, k)| = |\Phi(-h, -k)| \tag{14-53a}$$

$$\alpha_{h,k} = -\alpha_{-h,-k} \tag{14-53b}$$

Because χ depends only on the square of the scattering angle,

$$\sin \chi(h, k) = \sin \chi(-h, -k) \quad \text{and} \quad \cos \chi(h, k) = \cos \chi(-h, -k) \qquad (14\text{-}54)$$

Using all of these results, we can rewrite Equation 14-52 by letting $h' \rightarrow h$ and $k' \rightarrow k$:

$$I(x_i, y_i) = 1 - (\pi/\lambda\Phi_0)\left\{ \sum_{h=-\infty}^{\infty} \sum_{k=-\infty}^{\infty} [|\Phi(h, k)|e^{i\alpha_{h,k}}(\sin \chi - i \cos \chi) + |\Phi(h, k)|e^{i\alpha_{h,k}}(\sin \chi + i \cos \chi)]e^{-2\pi i(hx_i + ky_i)} \right\} \qquad (14\text{-}55)$$

Thus, we can see that the term in cos χ drops out. The electron microscopic image will be proportional only to $2|\Phi(h, k)|e^{i\alpha_{h,k}} \sin \chi$. This result is true only if there are no heavy atoms. (See Frank, 1973, for further details.)

The term $I(x_i, y_i)$ represents the image actually measured in the electron microscope. Therefore, we know everything required to compute its Fourier transform back to the diffraction plane:

$$F_{\text{calc}}(h, k) = \int_{-\infty}^{\infty} dx_i \int_{-\infty}^{\infty} dy_i e^{+2\pi i(x_i h + y_i k)} I(x_i, y_i) \qquad (14\text{-}56)$$

However, $I(x_i, y_i)$ is simply a constant plus the Fourier synthesis of $|\Phi(h, k)|e^{i\alpha_{h,k}} \sin \chi$. The inverse Fourier transform in Equation 14-56 will simply give back the function:

$$F_{\text{calc}}(h, k) = \delta(0, 0) - (2\pi/\lambda\Phi_0)|\Phi(h, k)|e^{i\alpha_{h,k}} \sin \chi \qquad (14\text{-}57)$$

Note, however, the critical difference between this equation and Equation 14-48. Only the amplitude of $\Phi(h, k)$ can be measured from the diffraction pattern, but both the amplitude and the phase can be calculated from the image.

One can insert into Equation 14-48 the phases calculated from Equation 14-57 and the more accurate measured amplitudes. Finally, a Fourier synthesis of $F(h, k)$ is done, and this gives $\phi(x_o, y_o)$, the structure of the sample:

$$\phi(x_o, y_o) = \sum_{h=-\infty}^{\infty} \sum_{k=-\infty}^{\infty} F(h, k)e^{-2\pi i(x_o h + y_o k)} \qquad (14\text{-}58)$$

The results are only a projection of the electron density onto a plane perpendicular to the electron beam. But they are elegant and powerful because, in this case, the phase problem is solved.

14-5 NEUTRON SCATTERING

Neutron and x-ray scattering compared

Some nuclear reactors put out large fluxes of neutrons. The wavelength of the particles is given by Equation 14-43. In practice, it is possible to obtain nearly monochromatic beams with wavelengths on the order of 2 to 4 Å. Thus neutrons, viewed as waves, afford the possibility of structure determination at roughly the same level of resolution as typical x-ray sources.

The principal disadvantage of neutrons is that available fluxes require rather long exposure times to accumulate scattering intensities sufficient for analysis. Furthermore, at present, there are only several reactors in the world producing suitable fluxes at the desired energies. However, neutron scattering has some advantages over x rays. For a direct comparison of neutron and x-ray scattering, we need to introduce the idea of a scattering length. This is an absolute measure of the scattering power of a particle. Consider the x-ray scattering from a single electron. The intensity per unit solid angle, $I_e(\theta)$, will be $I_e(\theta) = 7.90 \times 10^{-26} I_0(1 + \cos^2 2\theta)/2$, where 2θ is the scattering angle, and I_0 is the incident beam energy flux per cm^2 (see Guinier, 1960). Thus the constant 7.90×10^{-26}, which determines the actual amount of scattering, has dimensions of cm^2. It is a scattering cross section quite analogous to the extinction coefficient (see Box 7-2).

If we write the corresponding equation for amplitudes of radiation, at $2\theta = 0$ the result is $|I_e(0)|^{1/2} = 2.8 \times 10^{-13} |I_0|^{1/2}$. The constant 2.8×10^{-13} has dimensions of cm and is called the scattering length of the electron. Recall that the atomic scattering factor was defined as the amplitude of scattering from an atom relative to that of a single electron. Thus the x-ray scattering length b of any atom can be computed just by $b = 2.8 \times 10^{-13} f(0)$, where $f(0)$ is the atomic scattering factor at zero angle. The neutron scattering length is computed in an equivalent way from absolute intensity measurements.

Table 14-2 compares the neutron scattering lengths b for several elements with corresponding x-ray scattering lengths. First, note that hydrogen or deuterium atoms have a neutron scattering length comparable in magnitude to other elements, even heavy ones. Thus, hydrogen will make a substantial contribution to the observed neutron scattering, although it contributes little to x-ray diffraction.

More striking is the fact that the scattering length of H is negative, whereas that of D is larger and positive. A negative scattering length means that the scattering from H is 180° out of phase with the scattering from virtually all other atoms. Measured intensities depend on b^2, so they are not affected directly; but interference terms due to hydrogen–other atom vectors are affected. Furthermore, when a mass density distribution is computed by Fourier syntheses, the negative b for hydrogen leads to negative density rather than the positive found for other elements. We explore here three ways in which the different neutron scattering of hydrogen and deuterium can be exploited.

Table 14-2
Neutron scattering lengths and atomic scattering lengths of various elements

Element	Neutrons $b \times 10^{13}$ (cm)	X rays $b \times 10^{13}$ (cm)
H	−3.74	3.8
D	6.67	2.8
C	6.65	16.9
N	9.40	19.7
O	5.80	22.5
P	5.10	42.3
S	2.85	45
Mn	−3.60	70
Fe	9.51	73
Pt	9.5	220

Locating hydrogens by neutron diffraction of crystals

It is possible to do neutron diffraction of crystals in a manner exactly analogous to x-ray diffraction. However, the small range of neutron scattering lengths makes the isomorphous replacement method impossible. In practice, it usually is necessary to work with a crystal where the molecular structure is already approximately known. Then neutron diffraction intensities are measured on the same crystal. A Fourier synthesis is computed by combining measured neutron intensities with phases calculated from the positions of all nonhydrogen atoms visible in the structural model. This Fourier map weights H or D much more heavily than does an electron density map, because the relative contribution of these atoms to the neutron scattering is so great. From the Fourier map, the locations of H (negative density) or D (positive density) can be determined. These can be inserted into the structural model and further refinement carried out if desired.

A variation on these studies allows a crystal originally formed in H_2O to soak in D_2O prior to measurement. Then neutron diffraction should reveal not only where hydrogens are, but which ones are capable of exchanging with deuterium. Because we know chemically that protons such as NH_2 or OH readily exchange whereas CH protons do not, such information can help to confirm structure assignments. Furthermore, if protons expected to be exchangeable show no exchange, we have some indication that they are located in a region impermeable to solvent.

A complete neutron diffraction analysis such as we have just described is not often done on macromolecules because of the large amounts of reactor time required.

Solvent contrast in neutron and x-ray scattering

Neutron scattering from solutions can be analyzed in ways analogous to small-angle x-ray scattering. A Guinier plot (Eqn. 14-36) of low-angle x-ray or neutron scattering will give the radius of gyration of scattering material as it contrasts with the scattering from the solvent.

It is convenient to introduce the notion of scattering density σ. For x rays, σ is proportional to electron density, which is in turn proportional to atomic scattering factors. For neutrons, σ is proportional to the average neutron scattering length of the atoms involved. With x rays, the solvent scattering density can be adjusted by salt or sucrose, but the accessible range is small. With neutrons, the contrast between solute and solvent is adjusted by changing H_2O to D_2O. Figure 14-16 shows the effect of solvent scattering density on the observed zero-angle scattering intensity of myoglobin. The wider range accessible with neutrons is quite clear.

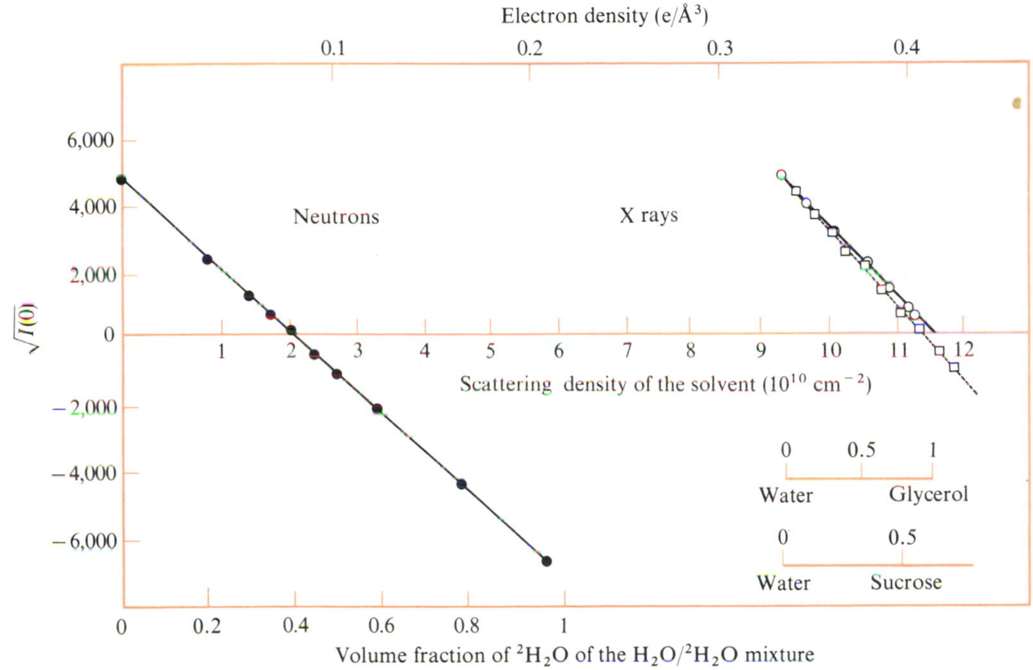

Figure 14-16

Scattering amplitude of myoglobin as a function of the contrast between the protein and solvent scattering densities. At left are neutron scattering data (●) in H_2O–D_2O mixtures. At right are x-ray scattering data in water–glycerol (○) and water–sucrose (□) solutions. What is actually plotted is the square root of the zero-angle scattering intensity. [From K. Ibel and H. B. Stuhrmann, *J. Mol. Biol.* 93:255 (1975).]

Consider a dissolved macromolecule consisting of a volume V_c impermeable to solvent. A uniform scattering density in the solution leads to no net scattering. Therefore, increasing the scattering density outside V_c is equivalent to decreasing the scattering density uniformly within V_c. The contrast $\bar{\rho}$ is the mean difference in scattering density between macromolecule and solvent:

$$\bar{\rho} = \langle \sigma \text{ macromolecule} \rangle - \langle \sigma \text{ solvent} \rangle \tag{14-59}$$

where the angle brackets indicate averages over the volume of the macromolecule. The significance of such contrast is illustrated pictorially in Box 14-2.

Using the preceding argument, we can write the scattering density $\sigma(\mathbf{r})$ of a dilute macromolecular solution as a sum of two terms. One describes the effect of excluding solvent from the region occupied by the macromolecule; this term is directly affected by the contrast between macromolecule and solvent. The other term covers any variations in scattering density within the macromolecule.

$$\sigma(\mathbf{r}) = \bar{\rho}\sigma_c(\mathbf{r}) + \sigma_s(\mathbf{r}) \tag{14-60}$$

The function $\sigma_c(\mathbf{r})$ simply describes the shape of the macromolecular surface. It is 1.0 inside the macromolecule, and 0.0 outside. Thus, $\int d\mathbf{r}\sigma_c(\mathbf{r}) = V_c$. The function $\sigma_s(\mathbf{z})$ arises from any internal structure within the macromolecule; it describes the variations in scattering density around $\langle \sigma$ macromolecule$\rangle$. The term $\sigma_s(\mathbf{r})$ vanishes at zero scattering angle but contributes at larger angle (corresponding to smaller distances).[§]

The structure factor is calculated from Equation 13-7, substituting the scattering density $\sigma(\mathbf{r})$ for the electron density $\rho(\mathbf{r})$. Using Equation 14-60, the result is

$$F(\mathbf{S}) = \bar{\rho}F_c(\mathbf{S}) + F_s(\mathbf{S}) \tag{14-61}$$

where F_c and F_s are the Fourier transforms of σ_c and σ_s, respectively. Thus the intensity of scattered radiation will be a quadratic function of $\bar{\rho}$:

$$I(S) = F(\mathbf{S})F^*(\mathbf{S}) = \bar{\rho}^2|F_c(\mathbf{S})|^2 + \bar{\rho}|F_c(\mathbf{S})F_s(\mathbf{S})| + |F_s(\mathbf{S})|^2 \tag{14-62}$$

However, because $F_s(\mathbf{S})$ vanishes at $S = 0$, the square root of scattering intensity at zero angle is linear in $\bar{\rho}$: thus $[I(0)]^{1/2} = \bar{\rho}|F_c(0)|$ (see Fig. 14-16).

Solvent contrast effect on the apparent radius of gyration

To examine the effect of contrast on the observed radius of gyration, it is convenient to describe the sample as a continuous scattering density distribution. Thus, instead

[§] If hydrogen-deuterium exchange occurs, either σ_c or σ_s or both can be affected. See Stuhrmann (1975) for further details.

of $R_G^2 = (1/2N^2)\sum_i \sum_j r_{ij}^2$ for N discrete scattering elements (Eqn. 14-33), we have

$$R_G^2 = \int d\mathbf{r} \int d\mathbf{r}' \, \sigma(\mathbf{r})\sigma(\mathbf{r}')(\mathbf{r} - \mathbf{r}')^2 \Big/ 2 \int d\mathbf{r} \int d\mathbf{r}' \, \sigma(\mathbf{r})\sigma(\mathbf{r}') \tag{14-63}$$

Insertion of Equation 14-60 into this equation yields, after some manipulation, a quadratic relationship between the contrast and the apparent R_G^2 that would be derived from a scattering measurement:

$$R_G^2 = R_c^2 + \alpha/\bar{\rho} - \beta/\bar{\rho}^2 \tag{14-64}$$

The constants in Equation 14-64 are evaluated as follows:

$$R_c^2 = (1/V_c) \int \sigma_c(\mathbf{r})\mathbf{r}^2 \, d\mathbf{r} \tag{14-65a}$$

$$\alpha = (1/V_c) \int \sigma_s(\mathbf{r})\mathbf{r}^2 \, d\mathbf{r} \tag{14-65b}$$

$$\beta = (1/V_c)^2 \iint \sigma_s(\mathbf{r})\sigma_s(\mathbf{r}')\mathbf{r} \cdot \mathbf{r}' \, d\mathbf{r} \, d\mathbf{r}' \tag{14-65c}$$

R_c is the radius of gyration that would be observed for the macromolecule at infinite contrast; it measures the shape (see Box 14-2). The term α is a measure of the internal structure. In particular, if the macromolecule can be approximated as a core surrounded by a spherical shell, then α is positive if the shell is more dense than the core and is negative if the core has the higher density. The parameter β is always positive; it describes the displacement of the apparent center of mass as a function of the contrast. For the simple shell–core case, β increases as the distance between the centers of mass of the core and shell increases. From Equation 14-64, you can see that R_c^2 is measured at $1/\bar{\rho} = 0$; the parameter α is obtained as the tangent of a plot of R_G^2 versus $1/\bar{\rho}$ at $1/\bar{\rho} = 0$.

Figure 14-17 shows an example of the application of Equation 14-64 to myoglobin. Note that R_c^2 obtained from neutron scattering is slightly smaller than the corresponding value from x rays. This is because sucrose and glycerol—used to vary x-ray contrast—cannot penetrate the protein, whereas some hydrogen–deuterium exchange, and thus some water penetration, occurs when the H_2O/D_2O ratio is varied for neutron scattering. The parameter α is positive for both methods. This is because hydrophilic groups are located preferentially near the surface. These groups have a higher electron density (leading to increased x-ray scattering density) and also a higher neutron scattering density than the interior (hydrophobic) residues.

Covalent deuterium as a neutron label

A special feature of neutron scattering is that, in addition to solvent variation, one can use covalent deuterium substitution as a label substantially to magnify or reduce the contributions of different components of the structure. Typical average scattering lengths for various deuterated or protonated samples are shown in Table 14-3. Suppose a sample is composed of two different components, 1 and 2, which are preferentially labeled, one with H and one with D. By adjusting the H_2O–D_2O composition of the solvent, one should be able to find conditions where $\langle \sigma_1 \rangle \cong \langle \sigma_{\text{solvent}} \rangle$ or $\langle \sigma_2 \rangle \cong \langle \sigma_{\text{solvent}} \rangle$. Then, the scattering of one component is blocked out, and all one sees is the other.

Box 14-2 EFFECT OF CONTRAST ON THE APPEARANCE OF AN OBJECT

All scattering depends on the spatial heterogeneities in a sample. Thus the contrast $\bar{\rho}$ in density or scattering power between an object and the surrounding medium can have a profound effect on the observed scattering. The text describes this in detail for neutron scattering, but it also is an important consideration for x-ray and light scattering.

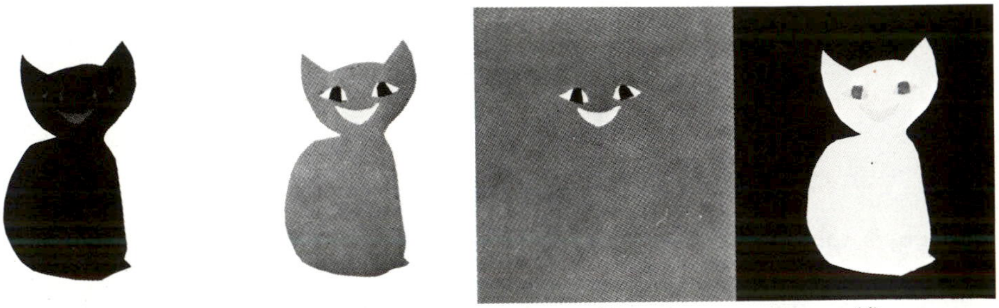

(a) (b) (c) (d)

The figure illustrates how variations in contrast between a simple object and its background can alter the apparent shape of the object. **(a)** When $\bar{\rho}$ is large and positive, the overall shape of the object is visible, but most internal detail is lost. **(b)** The appearance of the object in the absence of any background at all. **(c)** When $\bar{\rho}$ is zero, the overall shape of the object disappears, and only internal heterogeneities are visible. **(d)** When $\bar{\rho}$ is large and negative, the overall shape of the object is clear, but few internal features are visible. [Figure from R. Parfait, M. H. J. Koch, H. B. Stuhrmann, and R. R. Crichton, *Trends in Biochemical Sciences* (Feb. 1978), p. 4.]

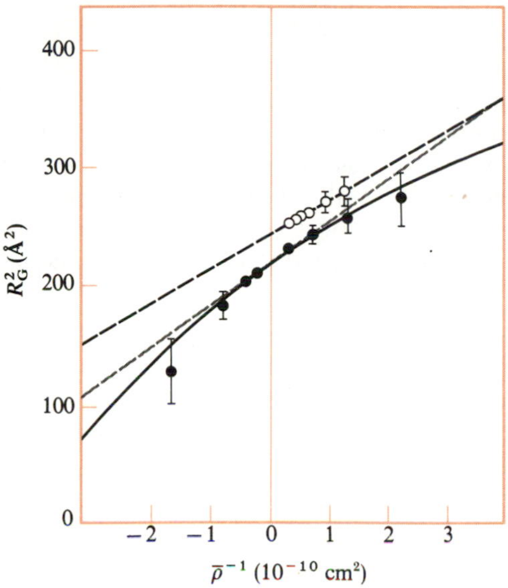

Figure 14-17

Apparent radius of gyration of myoglobin, plotted as a function of the increase in contrast between protein and solvent. Both neutron scattering (●) and x-ray scattering (○) data are shown. Dashed lines are straight lines with the slope $R_G^2/\bar{\rho}^{-1}$ at $\bar{\rho}^{-1} = 0$. The solid line is the expected scattering calculated from the known tertiary structure of myoglobin. [From K. Ibel and H. B. Stuhrmann, *J. Mol. Biol.* 93:255 (1975).]

Table 14-3

Average neutron and x-ray scattering lengths per unit volume

Substance	Neutrons $b \times 10^{14}$ (cm Å^{-3})	X rays $b \times 10^{14}$ (cm Å^{-3})
Fully protonated:		
H_2O	−0.6	9
Protein	3.1	12
Nucleic acid	4.4	16
Fatty acid	−0.0	8
Carbohydrate	4.3	14
Fully deuterated:		
H_2O	6.4	9
Protein	8.5	12
Nucleic acid	7.4	16
Fatty acid	6.9	8
Carbohydrate	8.1	14

SOURCE: After D. M. Engelman and P. B. Moore, *Ann. Rev. Biophys. Bioeng.* 4:219 (1975).

In real life, measurements are complicated by the fact that some protons exchange with solvent and others do not. This effect must be corrected for. In addition, one must know enough about the chemical composition of the sample to be able to convert measured scattering densities into physically tangible information about the structure, such as mass radial distribution functions.

Consider a sample in which two preferentially deuterated components are placed in an otherwise undeuterated structure. A specific case would be two deuterated ribosomal proteins reassembled into a 30S particle, with all other proteins and the 16S rRNA undeuterated. Compare the scattering intensity expected for this particle, $I_{12}(S)$, with that of singly-substituted equivalent samples $I_1(S)$ and $I_2(S)$, and with an unsubstituted 30S particle $I(S)$.

The total intensities will contain contributions from the nondeuterated particle, $I(S)$; from extra effects due to each deuterated component, $i_1(S)$ and $i_2(S)$; and from interference either between each deuterated component and the particle as a whole, $C_1(S)$ and $C_2(S)$, or between the two deuterated components, $C_{12}(S)$. Thus, we can write

$$I_{12}(S) = I(S) + i_1(S) + i_2(S) + C_1(S) + C_2(S) + C_{12}(S) \tag{14-66a}$$

$$I_1(S) = I(S) + i_1(S) + C_1(S) \tag{14-66b}$$

$$I_2(S) = I(S) + i_2(S) + C_2(S) \tag{14-66c}$$

Solving these three equations for C_{12}, we obtain

$$C_{12}(S) = I_{12}(S) + I(S) - I_1(S) - I_2(S) \tag{14-67}$$

Because all of the quantities on the right-hand side of Equation 14-67 are measurable separately, C_{12} (the protein–protein interference term) can be determined.

In the simple case of two spherical proteins, the scattering interference between them will have the form

$$C_{12}(S) = 2(\sigma_{\text{protein}} - \sigma_{\text{ribosome}})^2 F_1(S)F_2(S)\sin(2\pi d_{12}S)/2\pi d_{12}S \tag{14-68a}$$

where F_1 and F_2 are the spherically averaged structure factors of the two separate spheres (the neutron analogy of Eqn. 14-37), and d_{12} is the distance between their centers. The factors σ_{protein} and σ_{ribosome} are the average neutron scattering densities of the two proteins and of the remainder of the ribosome, respectively. We shall not derive this equation (see Problem 14-1) but, if you compare it with Equation 14-31, you should at least be comfortable with the term $\sin(2\pi d_{12}S)/2\pi d_{12}S$. This shows that, as S increases, the interference term will have a series of maxima and minima. The first zero value will appear at $S = d_{12}/2$, which allows easy estimation of d_{12}.

Figure 14-18a shows several experimental values of $C_{12}(S)$ for pairs of 30 S ribosomal proteins. The behavior of $C_{12}(S)$ is just as predicted from Equation 14-68a.

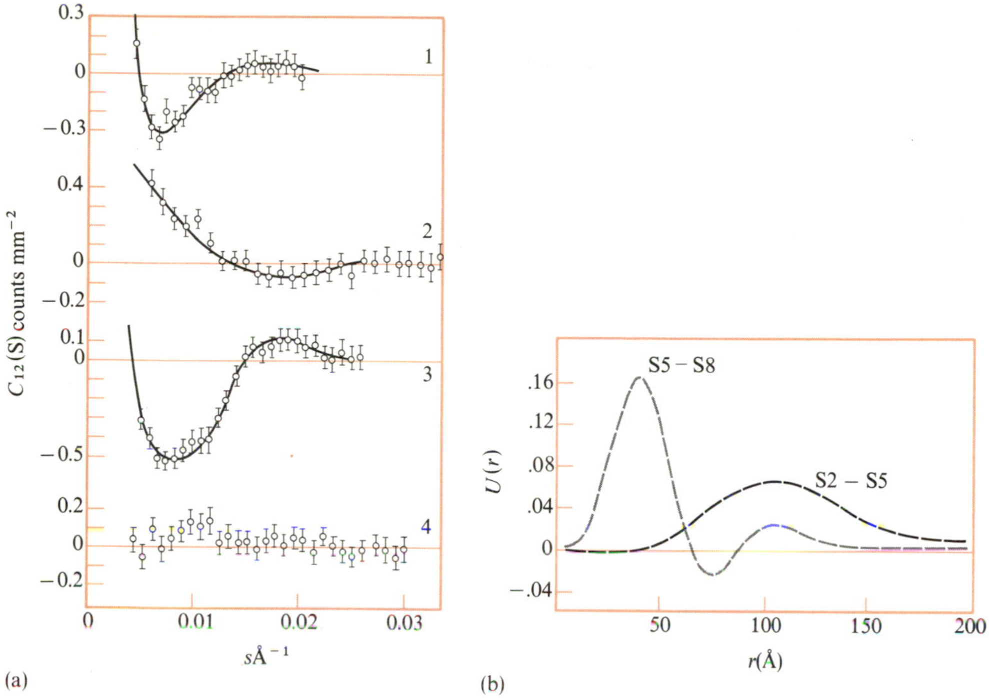

Figure 14-18

Neutron scattering of 30S *E. coli* ribosomes in which two deuterated proteins are substituted for normal proteins, but all of the rest of the ribosome constituents contain normal hydrogen. **(a)** The interference term $C_{12}(S)$ for samples with deuterated S2 and S5 (curve 1), S5 and S8 (curve 2), S3 and S7 (curve 3), and with undeuterated proteins only (curve 4). The term $C_{12}(S)$ is defined in Equations 14-67 and 14-68. **(b)** Radial Patterson function derived from the data in curves 1 and 2 of part a. The protein pair S5 and S8 shows a narrow distribution of distances, indicating that each might be a compact structure and that they are located close to each other. In contrast, the protein pair S2 and S5 shows a broad distribution of distances, indicating that at least one of these two proteins must be fairly elongated. [After D. Engelman, P. Moore, and B. Schoenborn, *Proc. Natl. Acad. Sci. USA* 72:3887 (1975).]

For example, the interference term between proteins S5 and S8 shows a first zero at $S = 70$ Å^{-1}, corresponding to $d_{12} = 35$ Å. This is quite consistent with what one would expect for two proteins the size of S5 and S8 in close contact.

Note that $C_{12}(S)$ can be inverted by a Fourier sine transform. By analogy to what we have shown earlier in this chapter, this inversion yields the radial Patterson function:

$$U(r) = (r/\pi) \int_0^\infty dS\, C_{12}(S)S \sin 2\pi Sr \tag{14-68b}$$

The function $U(r)$ gives the probability of finding interprotein vectors of length r. From $U(r)$, one can infer some details about the shape and orientation of the proteins. Figure 14-18b shows two examples.

14-6 LIGHT SCATTERING

Because of space limitations, only classical elastic light scattering is treated here in any detail. This technique is closely analogous to solution x-ray and neutron scattering. A sample is illuminated with a collimated beam of light at wavelength λ, and scattered radiation at the same wavelength is measured as a function of 2θ, the angle between the incident beam and the detector.

Single molecules much smaller than the wavelength

Classical light scattering is called Rayleigh scattering. In the limit of infinite dilution and small scattering angle, it will be given by an equation with the same form as Equation 14-34. Fundamentally, light scattering and x-ray scattering are the same phenomenon. However, it is useful to replace the single-electron scattering factor in Equation 14-34 by some measurable quantity. It also is necessary to consider explicitly the contrast between solute and solvent, just as we did for neutron scattering.

Light incident on a molecule much smaller than the wavelength induces an oscillating electric dipole $\boldsymbol{\mu} = \alpha\mathbf{E}$, where $\mathbf{E}$ is the electric field of the light, and α is the polarizability (see Chapter 7). The light radiated by this oscillating dipole is the origin of the scattered radiation. It can be shown that the intensity I of the scattered radiation detected through a 1 cm^2 slit is

$$I = I_0(8\pi^2\alpha^2/\lambda^4 r^2)(1 + \cos^2 2\theta) \qquad (14\text{-}69)$$

where 2θ is the angle between incident and scattered rays, λ is the wavelength, r is the distance between the sample and the detector, and I_0 is the intensity of the incident light per cm^2.

The interested reader with a good background in electricity and magnetism can find detailed derivations of Equation 14-69 elsewhere (see Berne and Pecora, 1976). Here we just try to justify a few of the terms. As the detector is moved away from the sample, the solid angle of radiation sampled by a fixed slit decreases as the square of the distance. This explains the appearance of the quantity $1/r^2$. The amplitude of light emitted by a dipole depends on the acceleration of the charges involved. By dimensional analysis, it is reasonable to suppose that this will vary as the square of the frequency of the radiation. Thus, the intensity of the scattered radiation (square of the amplitude) will depend on ν^4 or on λ^{-4}. Similarly, the amplitude will depend on the induced dipole $\boldsymbol{\mu}$, so the intensity will depend on $(\boldsymbol{\mu})^2$ or on α^2.

Experimentally, the ability to demonstrate λ^{-4} dependence is a good indication that Rayleigh scattering is being observed rather than absorption, Raman scattering, or other effects. However, when the sample contains molecules larger than λ, the λ^{-4} dependence is not exact. For example, very long rods will show a wavelength dependence of scattering that is closer to λ^{-3}.

Just in passing, note that the λ^{-4} dependence of Rayleigh scattering is a constant feature of daily life. It explains why the sky is blue and why sunsets are orange.

Effect of polarization on angular distribution of scattered light

The term $1 + \cos^2 2\theta$ in Equation 14-69 arises from the restrictions between light polarization and propagation directions. Consider unpolarized light incident along the z axis (Fig. 14-19). As described in Chapter 8, this light can be treated as a superposition of components polarized along x and y.

Incident light polarized along the x axis will induce a dipole that oscillates in the x direction. This dipole will radiate light polarized along the x axis. The scattered light can propagate along the z axis or the y axis. However, x-polarized scattered light cannot propagate along the x axis. Because light is a transverse wave, it can never have a field amplitude oriented along its propagation direction. Using a similar argument, we can conclude that y-polarized exciting light can induce y-polarized scattered components propagating along the x and z axes, but no component along the y axis.

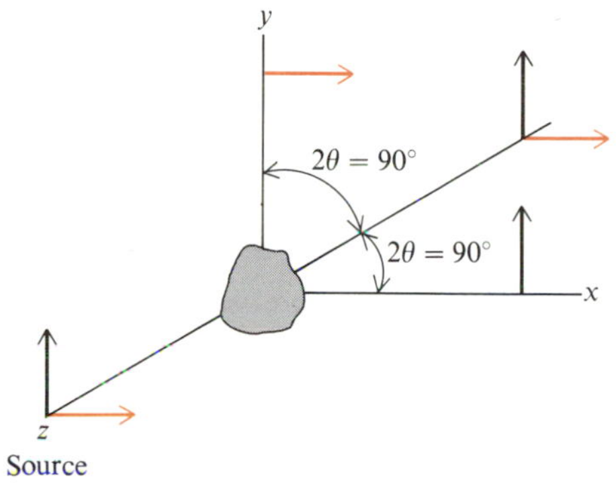

Figure 14-19
Light scattering from a particle, shown schematically for two possible polarization directions of the incident radiation. Note that, for unpolarized light, 90° scattering can have contributions from only one of the two initial polarized excitation components. Therefore, it can be at most only one-half as intense as 0° scattering.

These polarization effects mean that the scattering intensity observed along x and y will be half of that observed along z. The $1 + \cos^2 2\theta$ term accounts for this effect. With light incident along z, the angle 2θ is 0° for scattering along z, and $\cos^2 2\theta$ is 1.0. For scattering along the x or y axis, the angle 2θ is 90°, and $\cos^2 2\theta = 0$. Incidentally, exactly the same effect occurs with x-ray scattering, and observed intensities must be corrected for this effect before the equations developed in Chapters 13 and 14 can be used.

Populations of molecules much smaller than the wavelength

All the preceding discussion was for a single scattering particle. If there are N identical particles in the volume of solution under observation,

$$I = I_0 N(8\pi^2\alpha^2/\lambda^4 r^2)(1 + \cos^2 2\theta) \tag{14-70}$$

It is convenient to segregate all of the measurable quantities in Equation 14-70.

The Rayleigh ratio R_θ is the relative intensity of scattered light. It is defined as

$$R_\theta = \frac{r^2}{V(1 + \cos^2 2\theta)}\left(\frac{I}{I_0}\right) \tag{14-71}$$

where V is the volume of solution under observation. Thus, using Equations 14-70 and 14-71, the results of a light-scattering experiment can be described as

$$R_\theta = (8\pi^4\alpha^2/\lambda^4)N' \tag{14-72}$$

where N' is N/V, the number density of particles.

The polarizability α usually is not measured directly. Also, we are interested only in the excess polarizability of the solution over pure solvent, because what usually is measured is the excess scattered intensity relative to pure solvent. The excess polarizability is related to the difference between the dielectric constant of the solution, ε, and that of the pure solvent, ε_0:

$$\varepsilon - \varepsilon_0 = 4\pi N'\alpha \tag{14-73}$$

For dilute solutions, the left-hand side of Equation 14-73 can be expanded in a power series in the weight concentration c. Keeping only the first term,

$$\varepsilon - \varepsilon_0 = c\,\partial\varepsilon/\partial c \tag{14-74}$$

The number density of solute, N', is related to c by $N' = cN_0/M$, where M is the molecular weight. Using this relationship, we can write the polarizability (from Eqns. 14-73 and 14-74) as

$$\alpha = (1/4\pi)(M/N_0)\,\partial\varepsilon/\partial c \tag{14-75}$$

Because the radiation is at optical frequencies, we can replace the dielectric constant ε by n^2, where n is the refractive index. Then, in dilute solution,

$$\partial\varepsilon/\partial c = \partial(n^2)/\partial c = 2n(\partial n/\partial c) \cong 2n_0(\partial n/\partial c) \tag{14-76}$$

where n_0 is the refractive index of pure solvent, and $\partial n/\partial c$ is called the refractive increment. Both of these quantities are measurable and, using them, we can finally rewrite Equation 14-72 as

$$R_\theta = (2\pi^2 n_0^2/N_0\lambda^4)(\partial n/\partial c)^2\,Mc \equiv KMc \tag{14-77}$$

All of the quantities contained within R_θ and within the constant K can be measured separately. Thus, the quantity Mc can be determined and, if c is known, one can calculate the molecular weight.

If the sample is a heterogeneous mixture of species at concentrations c_i g cm^{-3}, with molecular weights M_i, the experimentally measured light scattering will yield $\sum_i M_i c_i$. Because the total weight concentration ($c_{\text{Tot}} = \sum_i c_i$) usually is known, a weight-average molecular weight $\bar{M}_w = \sum_i M_i c_i / \sum_i c_i$ can be computed.

Molecules comparable in size to the wavelength

All of the preceding discussion holds only for a macromolecule very much smaller than the wavelength of light. As molecular dimensions become significant compared with the wavelength (λ), intramolecular interference effects become important in light scattering just as they do in small-angle x-ray scattering. We can adopt Equation 14-35 directly to describe these (the physics is the same). Instead of $\langle I(\theta) \rangle$, we use R_θ, which corrects for polarization and experimental geometry. Instead of n_ε^2, which appears in Equation 14-35, we use KMc, which already appears in Equation 14-77. Thus, the light scattering predicted for large particles becomes

$$R_\theta = KMc(1 - 16\pi^2 R_G^2 \sin^2 \theta / 3\lambda^2) \tag{14-78}$$

It is customary to rewrite this result in a form convenient for extrapolation to infinite dilution:

$$\lim_{c \to 0} \frac{Kc}{R_\theta} = \frac{1}{M}\left(1 + \frac{16\pi^2 R_G^2 \sin^2 \theta}{3\lambda^2}\right) \tag{14-79}$$

The change in sign in the intramolecular interference term results from the use of the approximation $1/(1 - x) \cong 1 + x$ for small x.

Equation 14-79 is the basis of the Zimm plot, used to analyze most macromolecular light-scattering data. It is valid only for data extrapolated to zero angle (to justify the approximation used in treating intramolecular interference) and to zero concentration (to justify the neglect of intermolecular interference). What is done is to measure Kc/R_θ at various concentrations as a function of scattering angle 2θ. A grid of points is plotted (Fig. 14-20). Frequently an arbitrary constant $K'c$ is added to $\sin^2 \theta$ to make a convenient spread in the data points (and to account for intermolecular effects). Then the data are extrapolated separately to $c = 0$ and to $\theta = 0$.

The intersection of the two extrapolated curves yields the molecular weight M. The initial slope of the curve extrapolated to $c = 0$ can be related to R_G^2. Thus, Rayleigh scattering is potentially a very powerful method, because it can yield both the molecular weight and the radius of gyration. The major difficulty with light scattering is its inherent great sensitivity to any impurities of large molecular weight. Thus, great care must be taken to exclude aggregates or dust particles.

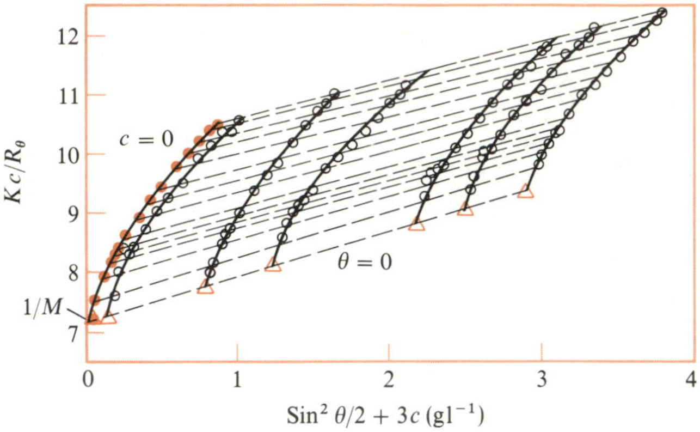

Figure 14-20

Zimm plot for light scattering measured on solutions of *Ascites* tumor-cell rRNA. Plotted points include experimental data (○), extrapolations to zero concentration (●), and extrapolations to $\theta = 0$ (△). [After M. J. Kronman et al., *Biochim. Biophys. Acta* 40:410 (1960).]

Other types of light scattering

Two other types of light scattering that can provide much useful information are Raman and dynamic light scattering. These are both examples of inelastic scattering, in which the detected radiation is at different wavelengths from those of the incident beam.

In Raman scattering, radiation is observed at energies equal to that of the incident beam plus or minus vibrational quanta. From the wavelength distribution of the scattered radiation, information is available about the vibrational energy levels of dissolved macromolecules (see Chapter 8).

Dynamic light scattering measures scattered radiation at wavelengths only very slightly different from that of the incident beam. This scattering can arise in several ways. For example, molecules in motion will alter the wavelength of the radiation due to Doppler shifts. Even if there is no net motion, these fluctuations in the number of scattering molecules will result in a distribution of wavelengths of scattered light. This distribution is quite narrow, and an extremely monochromatic light source—that is, a laser—is required to study it.

Dynamic light scattering can provide a direct measure of the diffusion constant of a macromolecule solute. The spectrum of scattered light produced by random diffusion will be a Lorentzian band:

$$I(S, v) = \frac{\langle N \rangle}{\pi}\left(\frac{S^2 D}{4\pi^2 v^2 + (S^2 D)^2}\right) \tag{14-80}$$

where v is the frequency of the light, $\langle N \rangle$ is the average number of solute molecules in the volume being sampled, S is $2 \sin \theta/\lambda$, and D is the diffusion constant. The half-width of the frequency distribution of scattered light is $S^2 D$. Thus, measuring the half-width as a function of $\sin^2 \theta$ yields the diffusion constant.

Summary

Various scattering techniques are available for samples that do not have three-dimensional periodic structures. Rod-shaped or helical molecules form ordered fibers. X-ray diffraction from such fibers usually can be analyzed to yield the pitch of the helix and the number of residues per turn. The diffraction pattern does not contain sufficient information to compute a detailed structure directly. However, it can be used as a test of various plausible structural models. Some information about how the molecules are packed in the fiber also is available.

Two-dimensional ordered arrays, such as thin crystals or crystalline membranes, can be examined in the electron microscope. An electron diffraction pattern and a direct image can be obtained on the same sample. The image can be Fourier-transformed to provide estimates of the phases of the structure factor. These can be combined with amplitudes measured from the diffraction pattern. A Fourier synthesis of these quantities can reconstruct an accurate picture of the array at high resolution.

In solution, all scattering techniques measure the spherically averaged structure. X-ray, neutron, and light scattering all can be treated by the same fundamental equations. Molecular weight estimates can be derived from the scattering intensity at zero angle. The angular dependence of the scattering intensity near zero angle yields the radius of gyration of the sample. If data are available over a wide range of angles, they can be used to calculate the radial Patterson function of the sample. This function represents the probability of finding two points separated by a distance r within the particle. Alternative models of the structure can be tested by calculating the expected scattering and comparing it with experiment.

Neutron scattering has the particular advantage that hydrogen atoms are weighted roughly the same as all other atoms, and deuterium scattering is fairly different from that of hydrogen. Thus, isotopic substitution can be used as a scattering label to highlight individual components. The contrast between scattering molecule and solvent is a critical feature in neutron scattering and light scattering, and it can also be important in interpreting x-ray scattering.

Problems

14-1. Show that the x-ray solution scattering of a sample consisting of two spherical proteins separated by a distance d_{12} is

$$I(S) = |F_1(S)|^2 + |F_2(S)|^2 + 2F_1(S)F_2(S)\sin(2\pi d_{12}S)/2\pi d_{12}S$$

where $F_1(S)$ and $F_2(S)$ are the structure factors that would be measured for each of the two proteins separately. HINT: adapt Equation 13-27 for this purpose, and place one protein center at the origin.

14-2. Figure 14-21 represents hypothetical radial Patterson functions calculated from neutron scattering data on pairs of deuterated ribosomal proteins. Using these data, and assuming

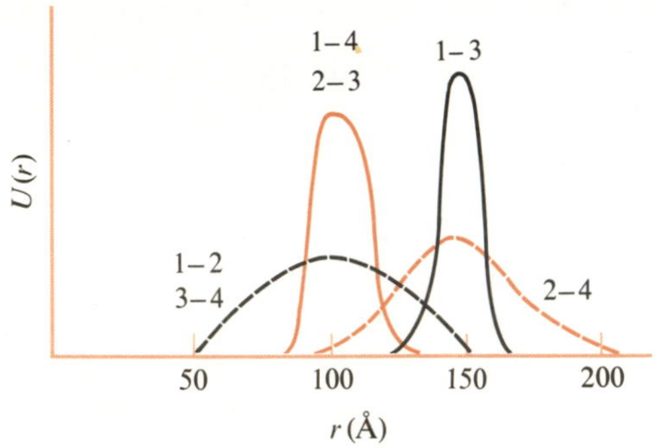

Figure 14-21
Radial Patterson functions for Problem 14-2.

that the proteins are either prolate ellipsoids or spheres, construct a detailed picture of their shapes and organization within the particle. Four proteins have been examined. The numbers beside each curve give the pair of proteins represented. [If you have difficulty, see P. B. Moore and D. M. Engelman, *J. Mol. Biol.* 112:228 (1977).]

14-3. Using the methods described in this chapter, compute the small-angle X-ray scattering expected from a structure that resembles a sphere of radius R sliced into infinitely thin sections spaced a distance h apart, with all the matter between the sections removed. (See Fig. 13-8c for an illustration of such a structure.) HINT: Use the convolution theorem.

14-4. Try to interpret the schematic fiber diagram for silk fibroin shown in Figure 14-22. You

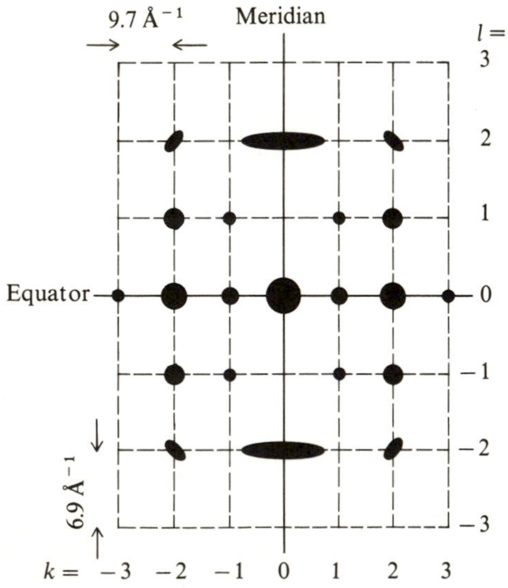

Figure 14-22
Schematic fiber diagram for Problem 14-4.

may assume that the sample is semicrystalline. (If you have trouble look at Fig. 13-27a,b.)

a. Why is there no meridional spot on the first layer line? (Remember that the meridian is diffraction from the projection of the structure onto the fiber axis.)
b. Why is the diffraction spot on the equator at $k = 2$ so strong? (Remember that the equator is the diffraction from the projection of the structure onto a plane perpendicular to the fiber axis.)

14-5. Light scattering sometimes interferes seriously with absorption and fluorescence measurements on large protein or nucleic acid particles.

a. One solution is to add sucrose to the solution. Explain why this helps.
b. Another solution, for fluorescence, is to use unpolarized exciting light but set a horizontal polarizer in front of the emission detector. Explain why this helps, and show why it works even when the fluorophore is rigidly attached to the particle (with parallel excitation and emission transition dipoles).
c. In absorption, one measures the attenuation of the incident beam, as viewed directly along the beam, by $A = \log I_0/I$. The apparent absorbance caused by scattering is called the turbidity, τ. This usually is defined as $\tau = \ln I_0/I$. Show that τ is given by $\tau \cong (16\pi/3)R_\theta$. HINT: First integrate over all angles to compute total scattering.

References

GENERAL

Eisenberg, H. 1971. Light scattering and some aspects of small angle x-ray scattering. In *Procedures in Nucleic Acid Research*, vol. 2, ed. G. L. Cantoni and D. R. Davies (New York: Harper & Row), p. 137.

Fraser, R. D. B., and T. P. MacRae. 1973. *Conformation of Fibrous Proteins and Related Synthetic Polypeptides*. New York: Academic Press.

Guinier, A. 1963. *X-Ray Diffraction in Crystals, Imperfect Crystals, and Amorphous Bodies*. San Francisco: W. H. Freeman and Company.

Holmes, K. C., and D. Blow. 1960. *The Use of X-Ray Diffraction in the Study of Protein and Nucleic Acid Structure*. New York: Interscience.

Schoenborn, B. P. 1977. Neutron scattering and biological structures. *Chemical and Engineering News* (Jan. 24), p. 31.

Tanford, C. 1961. *Physical Chemistry of Macromolecules*. New York: Wiley. [Contains a very complete discussion of light scattering.]

SPECIFIC

Berne, B. J., and R. Pecora. 1976. *Dynamic Light Scattering*. New York: Wiley.

Dickerson, R. E. 1964. X-ray analysis and protein structure. In *The Proteins*, 2d ed., vol. 2, ed. H. Neurath (New York: Academic Press), p. 603.

Engleman, D. M., and P. B. Moore. 1975. Determination of quaternary structure by small angle neutron scattering. *Ann. Rev. Biophys. Bioeng.* 4:219.

Erickson, H. P., and A. Klug. 1971. Measurement and compensation of defocusing and aberration by Fourier processing of electron micrographs. *Phil. Trans. Roy. Soc. London Ser. B* 261:105.

Frank, J. 1973. Computer processing of electron micrographs. In *Advanced Techniques in Biological Electron Microscopy*, ed. J. D. Kuehler (Berlin: Springer Verlag), p. 215.

Henderson, R., and P. N. T. Unwin. 1975. Three-dimensional model of purple membrane obtained by electron microscopy. *Nature* 257:28.

Pessen, H., T. F. Kumonsinski, and S. N. Timasheff. 1973. Small angle x-ray scattering. In *Methods in Enzymology*, vol. 27, ed. C. H. W. Hirs and S. N. Timasheff (New York: Academic Press), p. 151.

Pittz, E. P., J. C. Lee, B. Bablouzian, R. Townend, and S. N. Timasheff. 1973. Light scattering and differential refractometry. In *Methods in Enzymology*, vol. 27, ed. C. H. W. Hirs and S. N. Timasheff (New York: Academic Press), p. 209.

Schoenborn, B. P., and A. C. Nunes, 1972. Neutron scattering. *Ann. Rev. Biophys. Bioeng.* 1:529.

Shulman, R. G., P. Eisenberger, and B. M. Kincaid. 1978. X-ray absorption spectroscopy of biological molecules. *Ann. Rev. Biophys. Bioeng.* 7:559.

Stuhrmann, H. B. 1975. Small angle scattering of proteins in solution. *Brookhaven Symp. Biol.* 27:IV-3.

Unwin, P. N. T., and R. Henderson. 1975. Molecular structure determination by electron microscopy of unstained crystalline specimens. *J. Mol. Biol.* 94:425.

Appendix A
Review of elementary matrix algebra

A matrix is an array (or table) whose elements are numbers or symbols; for example,

$$\begin{pmatrix} 8 & 7 \\ 23 & 28 \end{pmatrix} \qquad \begin{pmatrix} a_{11} & a_{12} \\ a_{21} & a_{22} \end{pmatrix}$$

These are 2×2 matrices (having two rows and two columns); the symbol a_{12} denotes the element in the first row and the second column. In general, a_{ij} represents the element in the ith row and the jth column. For the matrix on the left, $a_{22} = 28$, $a_{21} = 23$, and so on. In this text, a matrix is denoted by a bold-faced letter with a "tilde" underscore. For example, we could represent the matrix at the right above as $\underset{\sim}{\mathbf{A}}$, where

$$\underset{\sim}{\mathbf{A}} = \begin{pmatrix} a_{11} & a_{12} \\ a_{21} & a_{22} \end{pmatrix}$$

A *row* matrix (also called a row vector) contains only one row:

$$\underset{\sim}{\mathbf{A}}_{\mathrm{r}} = (a_{11}, a_{12})$$

This matrix $\underset{\sim}{\mathbf{A}}_{\mathrm{r}}$ is a 1×2 row matrix (with one row and two columns). A *column* matrix (or column vector) contains only one column:

$$\underset{\sim}{\mathbf{A}}_{\mathrm{c}} = \begin{pmatrix} a_{11} \\ a_{21} \end{pmatrix}$$

The matrix $\underset{\sim}{\mathbf{A}}_c$ is a 2 × 1 column matrix (with two rows and one column). In general, a matrix may be of any size $n \times m$ (with n rows and m columns).

MATRIX MULTIPLICATION

Two matrices may be multiplied together to give a third matrix:

$$\underset{\sim}{\mathbf{a}}\underset{\sim}{\mathbf{b}} = \underset{\sim}{\mathbf{c}}$$

The element c_{ij} in matrix $\underset{\sim}{\mathbf{c}}$ is defined as

$$c_{ij} = \sum_k a_{ik} b_{kj}$$

That is, the elements in the ith *row* of a are multiplied by counterparts in the jth *column* of b to produce c_{ij}. For example, if

$$\underset{\sim}{\mathbf{a}} = (a_{11}, a_{12})$$

and

$$\underset{\sim}{\mathbf{b}} = \begin{pmatrix} b_{11} & b_{12} \\ b_{21} & b_{22} \end{pmatrix}$$

then

$$\underset{\sim}{\mathbf{a}}\underset{\sim}{\mathbf{b}} = (a_{11}, a_{12}) \begin{pmatrix} b_{11} & b_{12} \\ b_{21} & b_{22} \end{pmatrix}$$

$$= (a_{11}b_{11} + a_{12}b_{21}, a_{11}b_{12} + a_{12}b_{22})$$

Therefore,

$$\underset{\sim}{\mathbf{c}} = (c_{11}, c_{12})$$

where

$$c_{11} = a_{11}b_{11} + a_{12}b_{21}$$

$$c_{12} = a_{11}b_{12} + a_{12}b_{22}$$

It is clear that multiplication is possible only when the number of columns of $\underset{\sim}{\mathbf{a}}$ equals the number of rows of $\underset{\sim}{\mathbf{b}}$. In general, an $n \times m$ matrix can multiply only an $m \times p$ matrix, where n and p are arbitrary; the resulting matrix is $n \times p$. Thus, in the example just given, $n = 1$ and $m = 2$ for $\underset{\sim}{\mathbf{a}}$, and $m = 2$ and $p = 2$ for $\underset{\sim}{\mathbf{b}}$; multiplication of $\underset{\sim}{\mathbf{a}}$ times $\underset{\sim}{\mathbf{b}}$ yields $\underset{\sim}{\mathbf{c}}$, which is $n \times p$ or 1 × 2 (a row matrix).

As another example, let

$$\underset{\sim}{\mathbf{a}} = \begin{pmatrix} a_{11} & a_{12} \\ a_{21} & a_{22} \end{pmatrix}$$

$$\underset{\sim}{\mathbf{b}} = \begin{pmatrix} b_{11} & b_{12} \\ b_{21} & b_{22} \end{pmatrix}$$

$$\underset{\sim}{\mathbf{a}} = \underset{\sim}{\mathbf{a}}\underset{\sim}{\mathbf{b}}$$

$$= \begin{pmatrix} a_{11}b_{11} + a_{12}b_{21} & a_{11}b_{12} + a_{12}b_{22} \\ a_{21}b_{11} + a_{22}b_{21} & a_{21}b_{12} + a_{22}b_{22} \end{pmatrix}$$

Thus, when a 2×2 matrix multiplies a 2×2 matrix, the product also is a 2×2 matrix.

When a 1×2 matrix multiplies a 2×1 matrix, the result is a 1×1 matrix that is just a number (a scalar) and is no longer considered a matrix. For example:

$$\underset{\sim}{\mathbf{a}} = (a_{11}, a_{12})$$

$$\underset{\sim}{\mathbf{b}} = \begin{pmatrix} b_{11} \\ b_{21} \end{pmatrix}$$

$$\underset{\sim}{\mathbf{a}}\underset{\sim}{\mathbf{b}} = a_{11}b_{11} + a_{12}b_{21}$$

If the matrix is square (number of rows equal to number of columns), then the matrix may be raised to any power whatever. For example, for a square matrix $\underset{\sim}{\mathbf{M}}$,

$$\underset{\sim}{\mathbf{M}}^3 = \underset{\sim}{\mathbf{M}}\underset{\sim}{\mathbf{M}}\underset{\sim}{\mathbf{M}}$$

$$\underset{\sim}{\mathbf{M}}^N = \prod_{i=1}^{N} \underset{\sim}{\mathbf{M}}$$

(which is just $\underset{\sim}{\mathbf{M}}$ multiplied by itself N times). This procedure is possible only with a square matrix, because only in that case are the rules of matrix multiplication fulfilled. Note also that, if $\underset{\sim}{\mathbf{M}}$ is $n \times n$, then $\underset{\sim}{\mathbf{M}}^N$ also must be $n \times n$.

MATRIX INVERSION

For a square matrix $\underset{\sim}{\mathbf{a}}$, you can find an inverse matrix $\underset{\sim}{\mathbf{a}}^{-1}$ such that

$$\underset{\sim}{\mathbf{a}}^{-1}\underset{\sim}{\mathbf{a}} = \underset{\sim}{\mathbf{I}}$$

where $I_{ij} = 0$ for $i \neq j$, and $I_{ij} = 1$ for $i = j$. The matrix $\underset{\sim}{\mathbf{I}}$ is called the unit matrix.

For example, if $\underset{\sim}{\mathbf{a}}$ is a 2×2 matrix, then

$$\underset{\sim}{\mathbf{I}} = \begin{pmatrix} 1 & 0 \\ 0 & 1 \end{pmatrix}$$

Note that $\underset{\sim}{\mathbf{I}}$ has the property that any matrix multiplied by it is unchanged by the multiplication. Thus $\underset{\sim}{\mathbf{I}}$ is analogous to the number 1 in scalar algebra, so that $\underset{\sim}{\mathbf{a}}\underset{\sim}{\mathbf{I}} = \underset{\sim}{\mathbf{a}}$. The calculation of $\underset{\sim}{\mathbf{a}}^{-1}$ is relatively straightforward. For example, let

$$\underset{\sim}{\mathbf{a}} = \begin{pmatrix} a_{11} & a_{12} \\ a_{21} & a_{22} \end{pmatrix}$$

Then the rule for calculating $\underset{\sim}{\mathbf{a}}^{-1}$ is

$$\underset{\sim}{\mathbf{a}}^{-1} = \begin{pmatrix} a_{22}/\alpha & -a_{12}/\alpha \\ -a_{21}/\alpha & a_{11}/\alpha \end{pmatrix}$$

where

$$\alpha = a_{11}a_{22} - a_{21}a_{12}$$

which is the value of the determinant of $\underset{\sim}{\mathbf{a}}$. (Clearly, the determinant must be non-zero in order for $\underset{\sim}{\mathbf{a}}^{-1}$ to exist.)

MATRIX DIAGONALIZATION

For a 2×2 matrix $\mathbf{a}$, there is a matrix $\underset{\sim}{\mathbf{T}}$ such that

$$\underset{\sim}{\mathbf{T}}^{-1}\underset{\sim}{\mathbf{a}}\underset{\sim}{\mathbf{T}} = \begin{pmatrix} \lambda_1 & 0 \\ 0 & \lambda_2 \end{pmatrix}$$

where λ_1 and λ_2 are the eigenvalues of $\underset{\sim}{\mathbf{a}}$, and $\underset{\sim}{\mathbf{T}}$ is called a transformation matrix. The eigenvalues λ_1 and λ_2 are easy to find. Let

$$\underset{\sim}{\mathbf{a}} = \begin{pmatrix} a_{11} & a_{12} \\ a_{21} & a_{22} \end{pmatrix}$$

Then the eigenvalues of $\underset{\sim}{\mathbf{a}}$ are obtained by solving the determinant

$$\begin{aligned} 0 &= \begin{vmatrix} a_{11} - \lambda & a_{12} \\ a_{21} & a_{22} - \lambda \end{vmatrix} \\ &= (a_{11} - \lambda)(a_{22} - \lambda) - a_{12}a_{21} \\ &= \lambda^2 - (a_{11} + a_{22})\lambda + a_{11}a_{22} - a_{12}a_{21} \end{aligned}$$

Using the quadratic formula to solve for λ, we obtain

$$\lambda = \{(a_{11} + a_{22}) \pm [(a_{11} + a_{22})^2 - 4(a_{11}a_{22} - a_{12}a_{21})]^{1/2}\}/2$$

There are two roots, λ_1 and λ_2; let λ_1 be associated with the + sign, and λ_2 with the − sign. Thus we take $\lambda_1 > \lambda_2$. The rule for calculating $\underset{\sim}{\mathbf{T}}$ is simple. If

$$\underset{\sim}{\mathbf{a}} = \begin{pmatrix} a_{11} & a_{12} \\ a_{21} & a_{22} \end{pmatrix}$$

then

$$\underset{\sim}{\mathbf{T}} = \begin{pmatrix} (\lambda_1 - a_{22})/a_{21} & (\lambda_2 - a_{22})/a_{21} \\ 1 & 1 \end{pmatrix}$$

$$\underset{\sim}{\mathbf{T}}^{-1} = \begin{pmatrix} a_{21}/(\lambda_1 - \lambda_2) & (a_{22} - \lambda_2)/(\lambda_1 - \lambda_2) \\ -a_{21}/(\lambda_1 - \lambda_2) & (\lambda_1 - a_{22})/(\lambda_1 - \lambda_2) \end{pmatrix}$$

where λ_1 and λ_2 are the eigenvalues of $\underset{\sim}{\mathbf{a}}$. Two other useful facts for a 2×2 matrix $\underset{\sim}{\mathbf{a}}$ are

$$\lambda_1 + \lambda_2 = \mathrm{a}_{11} + \mathrm{a}_{22}$$

$$\lambda_1\lambda_2 = a_{11}a_{22} - a_{12}a_{21}$$

These relationships, together with the expressions just given for $\underset{\sim}{\mathbf{T}}$ and $\underset{\sim}{\mathbf{T}}^{-1}$, suffice to derive the expression for the transformation matrix in Equation 20-53 and its inverse in Equation 20-54. It also is a simple matter to generalize these results to $n \times n$ matrices, where $n > 2$.

For a clear and useful introduction to matrix algebra, see Chapter 1 of F. B. Hildebrand, *Methods of Applied Mathematics* (Englewood Cliffs, N.J.: Prentice-Hall, 1965).

Appendix B
Answers to problems

Chapter 7

7-1. The basic strategy throughout this problem is to evaluate trimer expectation values or integrals in terms of monomer integrals like $\langle \phi_{1a} | \phi_{10} \rangle = 0$. For example, the integral $\langle \phi_{10}\phi_{2a}\phi_{30} | \underset{\sim}{\mathbf{V}}_{12} | \phi_{10}\phi_{20}\phi_{3a} \rangle$ is zero because of the orthogonality of ϕ_{30} and ϕ_{3a}. The integral $\langle \phi_{10}\phi_{20}\phi_{30} | \underset{\sim}{\boldsymbol{\mu}}_1 | \phi_{10}\phi_{20}\phi_{30} \rangle$ is zero because the operator is odd while the square of the wavefunction is even.

a. Evaluate explicitly integrals like

$$\langle \Psi_{A^0} | \Psi_{A^0} \rangle = 1$$

$$\langle \Psi_{A^0} | \Psi_{A^+} \rangle = 0$$

b. Evaluate explicitly integrals like

$$\langle \Psi_{A^-} | \underset{\sim}{\mathbf{H}} | \Psi_{A^-} \rangle$$

The results are

State	Energy
0	$3E_0$
A^0	$2E_0 + E_a$
A^+	$2E_0 + E_a + \sqrt{2}V$
A^-	$2E_0 + E_a - \sqrt{2}V$

where E_0 and E_a are the energies of the monomer 0 and a states, respectively.

c. This follows directly from part b.

d. Show that integrals like $\langle \Psi_{A^+}|\underset{\sim}{\mathbf{H}}|\Psi_{A^-}\rangle$ are zero.

e. Evaluate integrals like $|\langle \Psi_0|\underset{\sim}{\boldsymbol{\mu}}_1 + \underset{\sim}{\boldsymbol{\mu}}_2 + \underset{\sim}{\boldsymbol{\mu}}_3|\Psi_{A^+}\rangle|^2 = D_{0A^+}$.

f. $D_{0A^0} + D_{0A^+} + D_{0A^-} = 3D_{0a}$

g. Show that integrals like $\langle \Psi_{A^+}|\underset{\sim}{\mathbf{H}}|\Psi_{A^0}\rangle$ are not zero.

7-2. The dipole strength and thus the absorbance is given by $A \propto (\boldsymbol{\mu}_D \cdot \mathbf{E})^2$, where $\boldsymbol{\mu}_D$, the transition dipole of the dimer, is given by $\boldsymbol{\mu}_{D^\pm} = \langle 0|\underset{\sim}{\boldsymbol{\mu}}_1|a\rangle \pm \langle 0|\underset{\sim}{\boldsymbol{\mu}}_2|a\rangle$ and **E** is the electric field vector of the light. For parallel and perpendicular polarization, just evaluate each absorbance separately. Then evaluate the linear dichoism using Equation 7-42. The results for the $0A^+$ and $0A^-$ transitions are $+0.5$ and -0.5, respectively.

7-3. It is a reasonable approximation to consider the aggregate as completely opaque. Then all you need to calculate is the fraction of the cell's cross-sectional area it occupies. This is 3.3×10^{-4}. See Box 7-2 if you need help.

7-4. Although hypochromism is usually defined in terms of oscillator strength, we won't go wrong here if we define it in terms of dipole strengths instead. The theory in Box 7-3 gives the following relationship between the dimer and monomer dipole strengths *per residue*: $D_{0A} = D_{0a} - C/R^3$, where C is a constant. We can determine C/R^3 for the particular dimer in the problem by writing

$$h = (2D_{0a} - 2D_{0A})/2D_{0a} = 0.1$$

Now, to compute the hypochromism of the tetramer, note that here there are three nearest neighbor interactions at a distance R, two next nearest neighbors at a distance $2R$, and a third neighbor interaction at a distance $3R$. The hypochromism of the tetramer is then calculated as

$$h = \frac{4D_{0a} - 4D_{0a} + 3C/R^3 + 2C/(2R)^3 + C/(3R)^3}{4D_{0a}}$$

The result is $h = 16.4\%$.

7-5. Find the energy of the 280-nm band in kcal mol^{-1}. The rest is easy, and the answer is 294 nm.

Chapter 8

8-1. There are a number of different ways to do this problem. Perhaps the simplest way is to recognize that the data given allow the donor lifetime τ_D and the rate of transfer between the donor and one acceptor k_T to be evaluated in terms of the donor fluorescence rate k_F. Then, since the rate of transfer from the donor to each of the two acceptors is the same, it follows that with two acceptors,

$$\phi_D = k_F/(2k_T + 1/\tau_D) = 0.167$$

8-2. A thermally induced conformation change at about 300°K is probably responsible for the nonlinear Perrin plot at variable temperature. In the limit of high temperature the effective hydrated volume of the protein is larger (smaller slope), but the fluorescent dye may have more local rotational freedom (higher extrapolated value of 1/A).

8-3. For the answer, see M. M. Warshaw and C. R. Cantor, *Biopolymers* 9:1079 (1970).

8-4. One must consider both the electric-magnetic coupling term and the exciton term as well as the splitting between the bands. Only when one term and the splitting are both nonzero will a dimer optical activity distinct from the monomer value be observed.

Case	Electric-magnetic	Exciton	V_{12}	CD
a	0	0	nonzero	monomer
b	nonzero	0	nonzero	dimer
c	0	nonzero	0	monomer
d	0	nonzero	nonzero	dimer
e	0	0	nonzero	monomer
f	nonzero	nonzero	0	monomer
g	nonzero	nonzero	nonzero	dimer

8-5. Since surface residues are mostly hydrogen-bonded to water in aqueous solution, adding a non-hydrogen-bonding solvent should cause spectral perturbations. The danger, of course, is that it may also denature the protein. Isotopic solvent perturbation, that is, adding D_2O, is potentially safer. However, hydrogen exchange even with interior protons is sufficiently rapid to make this approach difficult in practice.

The higher-frequency bands indicated in the second part of the question could arise from buried non-hydrogen-bonded peptide residues, although this is energetically unlikely.

Chapter 9

9-1. The cross-product $\mathbf{M} \times \gamma\mathbf{H}$ is easily obtained by constructing the determinant

$$\mathbf{M} \times \gamma\mathbf{H} = \begin{vmatrix} \hat{\mathbf{i}} & \hat{\mathbf{j}} & \hat{\mathbf{k}} \\ M_x & M_y & M_z \\ \gamma H_x & \gamma H_y & \gamma H_z \end{vmatrix}$$

$$= \gamma(\hat{\mathbf{i}}(M_yH_z - H_yM_z) - \hat{\mathbf{j}}(M_xH_z - M_zH_x) + \hat{\mathbf{k}}(M_xH_y - M_yH_x)$$

Addition of the relaxation terms (second line of Eqn. 9-22) to this expression gives

$$dM_x/dt = \gamma M_yH_z - \gamma M_zH_y - (M_x/T_2)$$

$$dM_y/dt = -\gamma M_xH_z + \gamma M_zH_x - (M_y/T_2)$$

$$dM_z/dt = \gamma M_xH_y - \gamma M_yH_x - (M_z - \bar{M}_z)/T_1$$

Substituting $H_x = H_{xy} \cos \omega t$ and $H_y = -H_{xy} \sin \omega t$ gives Equation 9-25.

9-2. Set Equations 9-27a, b, and c each equal to zero. This gives three simultaneous linear equations in u, v, and M_z. From Equation 9-27a, we have

$$u = -T_2(\omega_0 - \omega)v$$

From Equation 9-27b and the above expression for u in terms of v, we have

$$v = -\gamma T_2 H_{xy} M_z/(1 + T_2^2[\omega_0 - \omega]^2)$$

This expression, together with Equation 9-27c, enables us to solve for M_z (Eqn. 9-29a). It is then easy to combine the two expressions above with Equation 9-29a to give Equations 9-29b and c.

9-3. The data suggest that, as X is added, the protein eventually goes into a disordered form where all the C-2 histidine protons are magnetically equivalent. However, before it completely unfolds, it must pass through at least one intermediate state that coexists with the native species. This new form must be an organized structure because all four of its C-2 protons are at unique positions in the spectrum, indicating that they have magnetically different environments. Moreover, when present simultaneously with the native form, both the intermediate and the native structure must be relatively long-lived; otherwise, exchange between the two species would give a single line for each histidine and not two lines. If the resonance frequency of a given C-2 proton in the native form is ω and that for the same proton in the intermediate is ω', then the lifetime τ in each state is $\tau \gg |\omega - \omega'|^{-1}$.

9-4. Aromatic-ring protons will normally be downfield much further than -80 to -160 Hz, where we expect methylene- and methyl-group protons to resonate (Section 9-6). Thus, the affinity-labeling experiment targets a residue (tyrosine) that, as far as the aromatic ring is concerned, is not likely to contribute to the spectrum in the region (-80 to -160 Hz) that is perturbed by ligand binding. (An exception would be the β-methylene protons on the tyrosine residue that could be in this region.) The worry is that L′ may not be at the site where L normally binds. A further experiment would be to run a spectrum on the affinity-labeled protein. If the region where protons are affected by ligand binding (-80 to -160 Hz) is not affected by affinity labeling, then it is quite possible that L′ has not hit the site where L normally binds. If, on the other hand, the resonances in the range of -80 to -160 Hz are affected in the labeled protein in a way similar to the way they are affected in the unmodified protein that has bound L, then it is likely that L′ has labeled the active site region and influences it much as bound L does, even though tyrosine per se may not have the protons that are affected.

9-5. For three equivalent protons the total nuclear spin quantum numbers are $-3/2$, $-1/2$, $1/2$, $3/2$. Therefore, there will be four ESR transitions (one for each value of the total nuclear spin) and four ESR lines. The lowest-energy electron spin transition will occur with total nuclear spin at $-3/2$ and the highest-energy transition with total nuclear spin at $3/2$ (see Fig. 9-27). For n equivalent nuclei of spin I, the value of the total nuclear spin varies from $-nI$ to nI. Including $-nI$ and nI, when we go from $-nI$ to nI in steps of one unit, there are $2nI + 1$ steps that correspond to the $2nI + 1$ allowed values of the total nuclear spin. This gives us $2nI + 1$ lines in an ESR spectrum for an electron interacting with n equivalent nuclei of spin I.

Chapter 10

10-1. With only two subunits, Equation 10-34 takes on the simple form: $f/f_m = 2(1 + 1/2\alpha_{12})^{-1}$. When the two subunits are in contact, α_{12} is 2; whereas when they are very far apart, α_{12} becomes infinite.

10-2. Calculate V_h using Equation 10-11 and then f_{rot} using Equation 10-18. Next use the values of F_a and F_b for an 8:1 prolate ellipsoid from Figure 10-11b to calculate f_a and f_b. Finally use these quantities in Equations 10-23a and 10-23b to find $\tau_a = 1{,}220$ nsec and $\tau_b = 161$ nsec.

10-3. In Table 10-3 note that the f/f_{min} of tropomyosin is 3.22. Use the partial specific volume given in the table and the hydration stated in the problem to evaluate the term in brackets in Equation 10-70. Then the Perrin factor F can be calculated to be 2.86. From Table 10-2 this corresponds to an oblate ellipsoid with an axial ratio of about 90. The hydrated volume of the protein can be calculated from Equation 10-11 using the molecular weight given in Table 10-3. This volume must be equal to $(4/3)\pi a^2 b$ for an oblate ellipsoid. The axial ratio means that $a = 90b$, and thus b can be evaluated. The result is a short semiaxis of 1.66 Å. This is too small to correspond to a real peptide chain, and thus tropomyosin must be prolate.

10-4. Use Equations 10-36 and 10-46. The result is 2.08×10^{-5} cm sec^{-1}.

10-5. The largest possible diffusion constant occurs for a spherical molecule with no hydration. Using Equations 10-66, 10-68, and 10-69, the result is 13.2×10^7 cm sec^{-1}. A hydration of 0.40 g/g will reduce this by a factor of 1.16, and the longest reasonable prolate axial ratio of 100 will reduce the diffusion constant by another factor of 4 (from Table 10-2), so the result is about 2.8×10^7 cm^2 sec^{-1}.

Chapter 11

11-1. The density of copper is 8.96 g cm^{-3}. Using this value, the radius of the copper sphere is 0.299 cm. Use Equation 11-28 to calculate the sedimentation constant of the copper sphere: 15.8 sec. The velocity in the centrifuge is found with Equation 11-3 after remembering that the angular velocity must be converted to radians/sec. The result is 1.58 cm sec^{-1}. The sedimentation constant is the steady-state velocity per unit field. It can be used to calculate the free-fall velocity of the copper sphere in the lake just by recognizing that the acceleration field there is 980 cm sec^{-2}—the acceleration due to gravity at the earth's surface. The result is 15,500 cm sec^{-1}. The simple moral is: Don't stand under falling copper spheres. The more complex moral is that turbulence would set in long before a sphere reached this velocity in water.

11-2. One must calculate the total mass, M_T, transported across the surface at x_p during the ultracentrifuge run. Since x_p is in the plateau, diffusion can be neglected and the concentration at any time is given by Equation 11-17. The mass transport is given by

$$M_T = \int_0^t a\phi x_p^2 \omega^2 s c(t)\, dt$$

When Equation 11-17 is substituted for $c(t)$, the result is

$$M_T = a\phi\omega^2 x_p^2 s c_0 (1 - e^{-2\omega^2 st})/2$$

A second way to calculate M_T is to compute the total mass above x_p before and after the experiment. The change in total mass is

$$M_T = (c_0 - c_u)\phi a(x_p^2 - x_m^2)/2$$

These two expressions for M_T must be equal. When they are equated, the sedimentation constant can be written as

$$s = (-\tfrac{1}{2}\omega^2 t)\ \ln[x_m^2/x_p^2 - (c_u/c_0)(1 - x_m^2/x_p^2)]$$

11-3. Let the subscript o stand for the native protein, and p stand for the PCMB-treated sample.

a. The ratio of molecular weights is given by

$$M_o/M_p = (s_o/s_p)(D_p/D_o) = 3.98$$

Thus there are four subunits.

b. Using Equation 11-38 including Perrin shape factors, the ratio of the sedimentation constants can be written as

$$s_o/s_p = (M_o/M_p)^{2/3}(F_p/F_o)$$

From the known axial ratio of 20, F_o can be evaluated as 2.0 by using the data in Table 10-2. Thus F_p is 1.25. Alternatively, one can reach this conclusion by using the diffusion data.

c. The simplest model is a 5:1 prolate ellipsoid for the PCMB-treated protein. Straightase is apparently aptly named if it consists of four of these subunits in a row.

11-4. The frictional properties of the sphere are unchanged by removing the center, but the anhydrous mass is reduced by one eighth. Thus the predicted sedimentation constant is 7 S.

11-5. Simply use Equation 11-48, recognizing that although the molar volume of the deuterated protein is unchanged, because of its greater mass, the partial specific volume must be reduced by the same factor of 1.0155 with which the molecular weight increased. The partial specific volume is

$$\bar{V}_2 = \frac{1.0155 - S_D/S_H}{\rho_D - \rho_H(S_D/S_H)}$$

where S_D and S_H are the slopes of ln c_2 versus x^2 plots constructed as shown by Equation 11-48 and ρ_D and ρ_H are the solution densities in D_2O and H_2O.

Chapter 12

12-1. Define V_s as the volume of the supernatant including the volume excluded by the beads; define V_i as the internal volume of the beads. The protein concentration in the supernatant will be N_P/V_s. The ligand concentration in the sample with no protein will be $N_L/(V_s + V_i)$. These two samples allow the determination of V_s and V_i. The ligand concentration in the supernatant of the sample with both ligand and protein will be $\nu N_P[1/V_s - 1/(V_s + V_i)] + N_L/(V_s + V_i)$, where ν is the average number of ligands bound per proteins. This sample allows the determination of ν.

12-2. The Simha factor for an 18:1 prolate ellipsoid is 33, as shown in Table 10-2. If the partial specific volume is estimated as 0.73 $cm^3\ g^{-1}$, Equation 12-22 allows the hydration to be calculated as 0.39 g/g. This is quite a reasonable value.

12-3. The essence of this problem is to derive the equivalent of the Scheraga–Mandelkern equation by combining viscosity and diffusion data. Divide Equation 12-53 by the cube root of Equation 12-52 and substitute kT/D for f. Then take the ratio of the resulting equation for two samples. If the subscripts 1 and 2 refer to monomer and dimer, the result is

$$\frac{D_1[\eta_1]^{1/3}}{D_2[\eta_2]^{1/3}} = \frac{v_1^{1/3}M_2^{1/3}F_2}{v_2^{1/3}M_1^{1/3}F_1} = \frac{M_2^{1/3}\beta_1}{M_1^{1/3}\beta_2}$$

where we have used Equation 12-29 to substitute the ratio of two Scheraga–Mandelkern parameters for the ratio of Simha and Perrin factors. Since the monomer is oblate, one can guess that it has a Scheraga–Mandelkern parameter of 2.12. From the data given, this allows β_2 to be evaluated as 2.54, which corresponds to a prolate ellipsoid with an axial ratio of 16. It is obvious that the data given in the problem are highly artificial.

12-4. One must evaluate Equation 12-44 using a rotational relaxation time calculated by Equation 12-43. Assuming a hydration of 0.3 g/g, the viscosity data in Table 12-1 lead to a prolate axial ratio of 55. Use the known molecular weight to calculate the molar volume. Since the volume of a prolate ellipsoid is $4\pi ab^2/3$, combining this with the known axial ratio allows the long semiaxis a to be evaluated. The final result for χ is 0.749 radians or 42.9°. (Note that if hydration is neglected, the axial ratio is 68 and the angle χ becomes 0.729 radians or 41.8°.)

12-5. A protein of 15,000 d contains about 150 amino acids. If this protein formed mostly α helix, which seems likely in SDS, it would have a length of 1.5 Å per residue or 225 Å. From the diameter of 18 Å given in the text for SDS-protein micelles, one can estimate an axial ratio of about 12. At such axial ratios or smaller, the approximation used in the text that the frictional coefficient is proportional to the length breaks down.

Chapter 13

13-1. The result for the six-atom array shown in the figure can be written compactly as

$$f(S)2[\cos(\pi\mathbf{a}\cdot\mathbf{S}) + \cos(3\pi\mathbf{a}\cdot\mathbf{S}) + \cos(5\pi\mathbf{a}\cdot\mathbf{S})]$$

where $f(S)$ is the atomic scattering factor of the atom.

The extension to the infinite lattice is easily calculated using convolutions. By analogy to Equation 13-60, the result is

$$F_{\text{Tot}}(\mathbf{S}) = f(S)e^{\pi i\mathbf{a}\cdot\mathbf{S}} \sum_{n=-\infty}^{\infty} e^{2\pi i n\mathbf{a}\cdot\mathbf{S}}$$

This is exactly equivalent to Equation 13-31 because the first term when evaluated using the constraints of the von Laue conditions is either +1 or −1. Thus it cannot contribute to the observed intensities. This problem illustrates the fact that the observed scattering is independent of the choice of the unit cell origin.

13-2. The reciprocal lattice is

$$
\begin{array}{cccccc}
 & 1/R & 1/R & 1/R & 1/R & \mathbf{a*} \longrightarrow \\
1/4R & \cdot & \cdot & \cdot & \cdot & \cdot \\
1/4R & \cdot & \cdot & \cdot & \cdot & \cdot \\
1/4R & \cdot & \cdot & \cdot & \cdot & \cdot \\
\end{array}
$$

$\mathbf{b*} \downarrow$

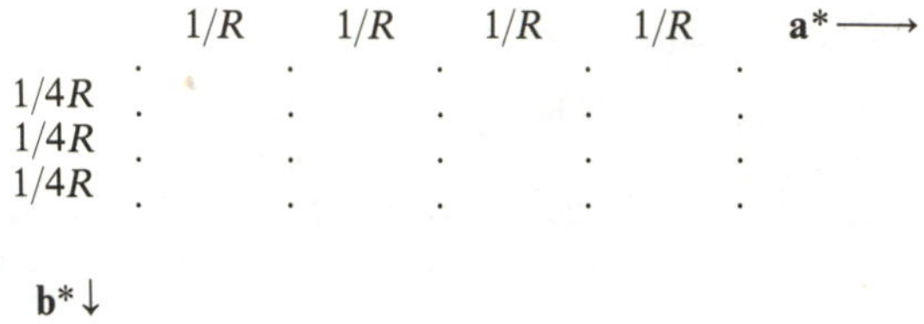

The molecular structure factor is

$$F_{\mathrm{m}}(\mathbf{S}) = f(S)[1 + 2e^{\pi ik/2} + e^{\pi ik}]e^{2\pi ih}$$

The distribution of scattering intensity is

$$I(\mathbf{S}) = f^2(S)[6 + 8\cos(\pi k/2) + 2\cos(\pi k)]$$

Clearly this does not depend on h. When it is evaluated for various k values, the result is

$$k = 0, 4, 8, \ldots \qquad I = 16f^2(S)$$

$$k = 1, 3, 5, 7, 9, \ldots \qquad I = 4f^2(S)$$

$$k = 2, 6, 10, \ldots \qquad I = 0$$

13-3. Drop a perpendicular line from the tip of F_{P} through F_{PH}. This divides F_{PH} into two (unequal) lengths that can be evaluated: one in terms of F_{P}, and the other in terms of F_{H}, by relatively straightforward geometric considerations. The sum of the two lengths is $|F_{\mathrm{PH}}|$.

$$|F_{\mathrm{PH}}| = |F_{\mathrm{P}}|\cos(\phi_{\mathrm{PH}} - \phi_{\mathrm{P}}) + |F_{\mathrm{H}}|\cos(\phi_{\mathrm{H}} - \phi_{\mathrm{PH}})$$

This can be transformed into the final result shown in the question by using the identity $\cos A = 1 - 2\sin^2(A/2)$. The best estimates of the heavy-atom amplitude will occur when the phases of the heavy atom and the heavy-atom isomorphous derivative are very similar.

13-4. The molecule has five equivalent atoms arranged as follows:

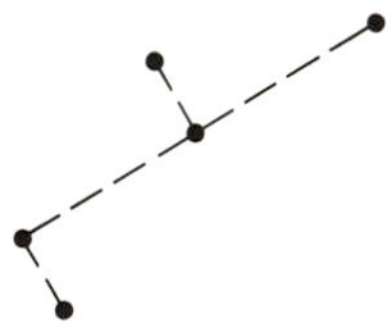

13-5. This problem is tedious but it will provide a good feeling for the manipulations involved in crystallography.

   a. Convert the intensities to amplitudes and then evaluate the heavy-atom electron density using the one-dimensional analog of Equation 13-93.

$$\rho_H(x) = \sum_{n=0}^{3} \big||F_{PH}(h)| - |F_P(h)|\big| e^{i\phi(h)} e^{2\pi i h x}$$

Assume that each phase term $e^{i\phi(h)}$ is 1. The result is electron-density maxima of $10 + 2\sqrt{5}$ at $x = 0$ and $10 - 2\sqrt{5}$ at $x = 1/2$.

b. With the heavy atom at $x = 0$, the phase terms are 1 for all h. Thus this is the case already evaluated in part a. With the heavy atom at $x = 1/2$, the phase terms are

$$+1, \ -1, \ +1, \ -1 \text{ for } h = 0, 1, 2, 3$$

This still leads to two electron-density maxima of $10 - 2\sqrt{5}$ at $x = 0$ and $10 + 2\sqrt{5}$ at $x = 1/2$. Thus no clear distinction is available.

c. Use the method shown in Figure 13-37a to compute possible phases for each $F_P(h)$ for the two possible heavy-atom positions computed in parts a and b. The results are

	h	0	1	2	3
$x = 0$		$+1$	$-\sqrt{5}/5 \pm 2\sqrt{5}/5(i)$	-1	$-\sqrt{5}/5 \pm 2\sqrt{5}/5(i)$
$x = 1/2$		$+1$	$+\sqrt{5}/5 \pm 2\sqrt{5}/5(i)$	-1	$+\sqrt{5}/5 \pm 2\sqrt{5}/5(i)$

where, in principle, each + and − represent independent possible choices of phase. In practice, if + is chosen for $h = 1$, one must choose − for $h = 3$, and vice versa in order that the calculated electron densities come out real. This leaves four possible sets of phases. Using each of these, compute the electron density using measured $|F_P|$ values and calculated phases in Equation 13-93, summing terms in $h = 0, 1, 2, 3$ for $x = 0, 1/4, 1/2$, and $3/4$ the unit cell. Divide all the results by 4 to correct the data for the difference in volumes of the cells used. The results are

Case	$x = 0$	1/4	1/2	3/4
Heavy atom at 0 $h = 1$, + choice	0	2	1	4
Heavy atom at 0 $h = 1$, − choice	0	4	1	2
Heavy atom at 1/2 $h = 1$, + choice	1	2	0	4
Heavy atom at 1/2 $h = 1$, − choice	1	4	0	2

These are all the same structure with just different choices of the origin and the positive direction. Note that in the parent crystal the position that will ultimately be occupied by the heavy-atom isomorphous derivative is empty.

Chapter 14

14-1. The structure factor is calculated by Equation 13-27, and then FF^* is used to compute the intensity of the molecule in one orientation. The result is

$$I(\mathbf{S}) = |F_1(\mathbf{S})|^2 + |F_2(\mathbf{S})|^2 + F_1^*(\mathbf{S})F_2(\mathbf{S})e^{2\pi i \mathbf{S}\cdot\mathbf{d}_{12}} + F_1(\mathbf{S})F_2^*(\mathbf{S})e^{-2\pi i \mathbf{S}\cdot\mathbf{d}_{12}}$$

Since the intensity is an observable, it must be real. Thus each of the terms that contribute to it must either cancel or be real. This allows us to substitute the complex conjugate of the fourth term, which shows it is identical to the third term. Next each remaining term must be spherically averaged, using Equations 13-20 and 13-21. Since the imaginary parts of complex numbers will cancel in such an average, we have the result given in the problem.

14-2. The four proteins are arranged in a 100 Å square with 1 and 3, and 2 and 4, respectively, at opposite diagonals. Proteins 1 and 3 are nearly spherical, whereas 2 and 4 are prolate ellipsoids about 100 Å long. The long axis of 2 is oriented along the 1–2 axis; the long axis of 4 is oriented along the 3–4 axis.

14-3. The sphere sliced into circles can be viewed as the intersection of a solid sphere and a set of parallel planes spaced by h. By the convolution theorem the structure factor of this object will be the convolution of the separate structure factors of the solid sphere and the set of planes.

$$F_{\text{Tot}}(\mathbf{S}) = \widehat{F_{\text{sph}}(\mathbf{S})F_{\text{pla}}}(\mathbf{S})$$

We have already evaluated $F_{\text{sph}}(\mathbf{S})$ in Equation 14-37. $F_{\text{pla}}(\mathbf{S})$ has also been calculated in Equation 14-16. To actually compute the scattering, you would have to explicitly perform the convolution, square to compute the intensity, and then spherically average.

14-4. The x-ray scattering pattern shown is that of a β sheet like that illustrated in Figure 13-27a. However, just by looking at the pattern, you should be able to draw the following inferences. The reciprocal lattice allows the construction of the real lattice of 9.7×6.9 Å. This is the true structure repeat. However, the absence of intensity at $l = 1$ on the meridian indicates that in projection onto the vertical axis the structure repeats every 3.45 Å. This implies that it is a twofold helix. Similarly, the absence of strong intensity at $k = 1$ on the equator suggests that in projection onto the horizontal axis the structure repeats every 4.85 Å. There are two strands of helix per unit cell.

14-5. This problem really explores all aspects of light scattering.

a. Addition of sucrose raises the index of refraction of the solvent and thus lowers the contrast between the macromolecule and the solvent. This substantially reduces scattering.

b. From the simple arguments given in the text, light scattered at 90° (the typical geometry used for fluorescence measurements) is expected to be all vertically polarized. In real life, scattering does cause some depolarization so there is a small horizontal component. The use of a horizontal polarizer will eliminate most of the scattered light. It will also eliminate all the vertically polarized fluorescence. Using the arguments developed in Chapter 8, you should be able to show that even for a rigid system one-third of the fluorescence will still be transmitted.

c. Start with Equations 14-70 and 14-72. It is easiest to call the sample volume 1 cm^3 since this is effectively the volume used in absorbance (with a 1-cm path length and intensities measured per cm^2). We want to compute the scattering averaged over all angles:

$$I_{av} = \int_0^\pi d(2\theta) I_0 R_\theta [1 + \cos^2(2\theta)] \sin(2\theta)/r^2 \Big/ \int_0^\pi d(2\theta) \sin(2\theta) = 4I_0 R_\theta / 3r^2$$

This is per cm^2 of detection. To consider all scattered radiation, take any spherical surface at a radius r larger than the dimensions of the sample. The area of the surface is $4\pi r^2$. Thus the total scattering is $I_s = 16\pi I_0 R_\theta/3$. The transmitted light is $I = I_0 - I_s$. When this is evaluated and logarithms are taken, using the expansion $ln[1/(1-x)] = x$ yields the final result.

Index to Parts I and II